U0321790

Architectural Material & Detail Structure

建筑材料与细部结构

（西）何塞普·费尔南多 编 常文心 译

Concrete 混凝土

辽宁科学技术出版社

Preface
前言
CONCRETE, Monolithic vs Fragmentary
Josep Ferrando

混凝土，整体与碎片
何塞普•费尔南多

Matter & Light
In the essence of the architectural project or the final built work resides an inherent duality: the place and the architect. The place always imposes itself, and architects adapt our method, gaze and, with it, a personal interpretation that will prevail during the process. This duality between the immovable state of reality – the context and the territory –, and our interpretation of it is combined in the project and, finally, is realised in the built work.

The architectural work, as a combination of matter and light. The volume (matter) in as far as the inanimate object: silent and still. And light (indefinite space), like a void defined by matter: alive, moving, dynamic.

In order to reach this symbiosis between matter and light, between the definition of space and the habitable space, the architect works in a place with a context, a place that imposes itself and that cannot be ignored and that must be addressed.

This architectural process is managed through many types of documents. A language is generated by the architect (personal and specific) from the place. It is the origin of a story that, in some cases, will arrive at its ultimate goal: to become spaces can be inhabited by people and can shelter new stories.

If the solid is not defined, there can be no habitable void. Without matter, there is no light, and therefore, no life. Nor architecture. Recognising a place and the process of adapting ourselves to it affords the architect the ability to provide it with life. It gives us the opportunity to bring matter to life through a correct interpretation of light.

Tectonic & Stereotomic
Construction materials owe their qualities to a specific technical procedure: tectonic (carpentry) or stereotomic (stonework, masonry). Therefore, the construction system that addresses the detail must be honest with the requirements of the material.

We can classify the materials into two groups according to the way it relate to each other. We work with aggregated materials (steel, wood, prefabricated ...) and mouldable materials (in situ concrete, cast steel, adobe ...). When it is about shaped or aggregated materials, is the joint's construction detail that defines the expression of the aggregation unit relative to the set. In the case of amorphous or malleable, detail expresses unwillingness to recognise the moment when two materials meet and that is why these are considered stereotomic or monolithic materials. We see in some examples the ability of architecture to generate ambiguity between the construction system used and its final appearance; always through the development of the constructive detail in the design process.

Concrete "in situ" & Precast Concrete
The difference between the reinforced concrete "in situ" and the use of precast concrete is that in the former case an amorphous material is set up in the construction site and in the second, one that is already shaped. One has the advantages of monolithic, while the other allows manipulation. The first takes advantage and accepts the imperfection of the construction site and the second benefits from the precision required by the design process.

In the case of precast concrete construction the weather conditions (temperature or humidity) do not affect its use in construction, but in the transport of its parts.

Precast concrete has many advantages for the architect and the project's execution:
Planning: it allows the architect to face up a

物质与光
建筑或建成设施的本质存在着固有的二元性：场所与建筑师。场所总是自己证明自己，而建筑师则不断适应我们的建筑方法，在建造过程中演绎出个人对建筑的解读。这种存在于不能移动的实体（环境和地域）与我们对其的解读之间的二元性在项目中相互融合，最终体现在建成的作品中。

建筑作品是物质与光的结合体。体块（物质）是静物，沉默静止；光（不确定的空间）则更像是物质所定义的空间，鲜活生动。

为了实现物质与光、空间与居住空间之间的共生，建筑师在置身于场景中的场所内工作，而场所则针对自身产生不可忽视的影响。

建造过程需要通过各种类型的文件进行管理，建筑师会针对场所的特点生成一套设计语言。有时候，故事的开始就决定了它的结局：经过处理的空间会成为人们的居住场所，在那里会衍生出全新的故事。

如果没有实体，就没有可居住的空间。如果没有物质，就没有光，也没有生命，更不用提建筑。认同一个场所并让自身适应它的过程就是建筑师赋予场所生命的过程。通过对光的正确解读，我们获得了把物质变成生命的机会。

构造与切割
建筑材料必须经过特殊的工艺处理才能体现出价值，即构造（主要针对木材）或切割（主要针对石材、砌体）。因此，注重细节的建造系统必须正确应对材料的处理要求。

我们可以根据建筑材料的相互关系将其分为两类：集料型材料（钢材、木材、预制材料）和可塑性材料（现浇混凝土、铸钢、砖坯）。在使用造型或集料型材料时，由节点构造细节来决定集合体的表现力。在使用无定形或可塑型材料时，细节会表现出两种材料的自然相斥，因此，它们又被称为切割材料或单体材料。在一些案例中，建筑的建造系统和最终外观能够通过设计过程中对建造细节的开发实现一种微妙的不确定感。

现浇混凝土与预制混凝土
现浇混凝土与预制混凝土的区别在于：前者是无定形材料，在施工现场进行塑形；后者是已经塑形完毕的预制材料。前者具有整体感，而后者有利于装配处理。前者能因地制宜，适应施工现场的缺陷；后者精密准确，可满足细节设计要求。

在预制混凝土施工中，天气条件（温度或适度）不会影响

project from its inception along with the craftsman-builder, optimising obstacles, reducing costs and avoiding contingencies.

Customisation: it allows customising the finished material textures and more complex mixtures. The use of architectural concrete provides versatility and flexibility in the design, also as regards sizes, shapes, colours, etc.

Time and Cost Control: it allows talking about prices and closed deadlines since the time of contract signing.

Optimising Performance: it decreases of the execution work's time, guarantees the safety during manufacture and installation, achieves reliability in all the phases of the project and minimises deviations from the project.

Absence of Residues: precast concrete work does not provide residue in the construction site nor in its manufacture, since it can be reused as raw material without affecting the characteristics of the finished product.

Continuity & Discontinuity

Monolithic concrete expression "in situ" gives the appearance of a moulded object to the construction, that is, the arising of a work through extracting materials.

A key feature of the precast concrete is its discontinuity. Opposite to reinforced concrete "in situ", a prefabricated building will never be a single piece, and the resolution of its joints will determine the system behaviour.

Mould & Material

Whereas the construction system is based on industrial production, which considerably increases the quality of their physical and chemical characteristics: strength, surface finish, adhesion, corrosion resistance, etc.

It is a material manufactured by moulding and capable, without forgetting the characteristics of the raw material, to adopt any shape and thus to conform the pattern that will be repeated.

The surface finishes that can be achieved with precast concrete can be summarised as follows:

1. Smooth: surface finish of the mould directly.
2. Texturised: finishing of the panels by employing the negative moulds of the textures to be obtained. Choosing reliefs or finishes that prevent the formation of deposits of dirt is recommended.
3. Sand blasting: with this treatment it can be obtained a façade's surface finish on which can be appreciated from a fine sand finish up to a coarse one, depending on the degree of intensity of mechanical treatment, consisting on the sandblasting pressure over the panel's face side.
4. Arid relief: exposed arid finish, ranging from a few tenths of a millimetre to several millimetres, by the employment of a paper or primer that retards the setting of concrete.
5. Acid washing: this is achieved by applying a dilute acid or other products.
6. Polished: with this mechanical treatment panels with a completely smooth appearance can be obtained.

Joint & Monolithic

In the section "Tectonic & Stereotomic" it points to the concept of ambiguity in architecture relative to that the eyes can see. Ambiguity cannot get confused with being dishonest or inconsistent with the construction systems and materials we use. The concept of ambiguity responds to the possibility for an architect to get a result – connected to the concept of the building and the expected atmosphere to be transmitted – through work and development of the constructive detail and the conformation of the materials going beyond its characteristics by definition.

The classical played with the effect of perspective and a single material. Today we have the opportunity, thanks to industrialisation, to refine the details of the union between different materials and pieces.

In the precast concrete exists the problem – or not – of the joints. Far from thinking that initially with it the sense of monolithic is rejected, it should unite their efforts, as we said, the study of the details of joints that generates the union of pieces or patterns. By studying the shadows produced by the textures on a surface we can hide and therefore perceive a single volume or, on the contrary, accentuate and even modify. The possibilities are endless. In this book a few examples of buildings with precast concrete façades that meet any of the abovementioned characteristics are collected.

施工效果，但是会影响混凝土零件的运输。

从建筑师和项目实施的角度来说，预制混凝土具有以下优势：
规划：它能让建筑师与施工人员共同直面设计规划，从而优化障碍、减少成本、避免意外事件。

定制：它能定制饰面材料的纹理和更复杂的混合质地。建筑混凝土的应用让设计更丰富、更灵活，使其在尺寸、造型和色彩上具有多样性。

时间与成本控制：它能控制成本和施工时间，保证合同的执行。

优化性能：它能缩短执行工作的时间，保证制造和安装的安全性，实现项目在各个阶段的可行性，并减少项目的偏差和失误。

无残留物：预制混凝土在施工现场不会留下任何残留物，因为它可以作为原材料被反复使用，不会影响最终成品的特性。

连续性与非连续性

单体混凝土通过"现场浇注"的方式赋予建筑物成型的外观，即通过提取材料来创造作品。

预制混凝土的主要特征在于它的非连续性。与钢筋混凝土的现场浇注相反，预制建筑绝不会是一个单一的体块，它的连接方式将决定整个建筑系统的性能与造型。

模具与材料

建筑系统的基础是工业生产，工业生产的进步大幅提升了材料的物理和化学特性，例如，强度、表面装饰、附着力、抗腐蚀性等。

在保持原材料的特性下，由模具制造出来的材料可以被塑造成各种形状，从而组成可以反复出现的图案。

预制混凝土材料可以实现以下几种表面装饰：
1. 光滑表面：直接通过模具获得。
2. 纹理表面：通过应用带有纹理的阴模获得板材饰面。建议选择不易积灰的浮雕或装饰。
3. 喷砂：喷砂处理能够让建筑立面获得细砂至粗砂的饰面效果，其效果取决于机械加工的强度，即喷砂压力作用于板材的强度。
4. 干纹浮雕：外露的干纹装饰，其宽度在零点几毫米至几毫米之间，通过在表面上使用纸或底漆来延缓混凝土的凝固获得。
5. 酸洗表面：通过应用稀酸或其他产品获得。
6. 抛光表面：通过极细处理板来实现彻底光滑的外观。

连接体与单体

在上文的"构造与切割"中，我指出建筑的不确定感与人们鉴赏它的方式有关。不确定感并不是不真实或建造系统与所选材料的不一致。不确定感所对应的是建筑师通过建造细节和材料构造所获得的结果的可能性，它与建筑的设计理念和建筑师所期望表现的氛围有关，超越了可定义的特征。

传统建筑通过透视和单一材料来实现不同的效果。今天，得益于工业化的进步，我们可以进一步改良不同材料与组件之间的组合效果。

预制混凝土存在着连接的问题。单独预制混凝土组件不具备整体感，但是对连接细节的研究能够组合成整体或图案。我们可以通过研究表面纹理的阴影来隐藏连接点，从而获得单一的整体；也可以反其道而为之，突出连接甚至对其进行修饰。设计有无限可能。本书中以混凝土作为主要材料的建筑案例将向读者展示上面提到的各种特征与应用。

Contents 目录

Contents 目录

Overview

Basic Information of Concrete

概述 混凝土简介

Concrete is a composite material composed mainly of water, aggregate and cement. Often, additives and reinforcements are included in the mixture to achieve the desired physical properties of the finished material. When these ingredients are mixed together, they form a fluid mass that is easily molded into shape. Over time, the cement forms a hard matrix which binds the rest of the ingredients together into a durable stone-like material with many uses.

混凝土是一种主要由水、骨料和水泥混合而成的复合材料。通常，混合物种会包含添加剂和强化剂，以实现饰面材料的理想物理性能。这些原料混合起来，形成容易塑形的流体团。随着时间的流逝，水泥形成坚硬的基质，将其他原料胶合起来，形成具有多种用途、经久耐用的石样材料。

1

1. History of Concrete

Concrete additives have been used since 6500BC by the Nabataea traders or Bedouins who occupied and controlled a series of oases and developed a small empire in the regions of southern Syria and northern Jordan. They later discovered the advantages of hydraulic lime – that is, cement that hardens underwater – and by 700 BC, they were building kilns to supply mortar for the construction of rubble-wall houses, concrete floors, and underground waterproof cisterns. The cisterns were kept secret and were one of the reasons the Nabataea were able to thrive in the desert. In both Roman and Egyptian times it was re-discovered that adding volcanic ash to the mix allowed it to set underwater. Similarly, the Romans knew that adding horse hair made concrete less liable to crack while it hardened, and adding blood made it more frost-resistant.

Concrete is one of the most durable building materials. It provides superior fire resistance compared with wooden construction and gains strength over time. Structures made of concrete can have a long service life. Concrete is used more than any other manmade material in the world.

Famous concrete structures include the Hoover Dam, the Panama Canal and the Roman Pantheon. The earliest large-scale users of concrete technology were the ancient Romans, and concrete was widely used in the Roman Empire. The Colosseum in Rome was built largely of concrete, and the concrete dome of the Pantheon is the world's largest unreinforced concrete dome.

After the Roman Empire collapsed, use of concrete became rare until the technology was re-pioneered in the mid-18th century. Today, concrete is the most widely used man-made material (measured by tonnage).

2. Applications in Architectural Field

(See Figure 1 to Figure 5)

1. 混凝土的历史

混凝土添加剂早在公元前6500年就已经被纳巴泰商人或贝都因人所使用，他们占领并控制了许多绿洲，在南叙利亚和北约旦地区建立了一个小王国。后来，他们发现了水硬石灰（即在水中变硬的水泥）的优点，公元前700年前后，他们建造了窑炉来制作灰浆，用于建造碎石墙住宅、混凝土地面以及地下防水蓄水池。这些秘密的蓄水池是纳巴泰人在沙漠中兴旺的原因之一。在罗马和埃及时代，人们发现在混凝土中添加火山灰能使其在水下凝固。同样的，罗马人知道添加马毛能让混凝土在变硬的过程中不易裂开，而添加血能让混凝土更抗霜。混凝土是最耐久的建筑材料之一。与木材结构相比，它具有卓越的防火性能，并且能随着时间而变得更加坚固。混凝土结构拥有很长的使用寿命，是应用最广泛的人造材料。

胡佛水坝、巴拿马运河以及罗马万神殿都是著名的混凝土建造物。最早大规模使用混凝土技术的是罗马人，混凝土在罗马帝国有着广泛的应用。罗马斗兽场主要由混凝土建造而成，而万神殿的混凝土圆顶则是世界上最大的无钢筋混凝土圆顶。

在罗马帝国陷落之后，混凝土的使用率变得很低。直至18世纪中期，新技术的发展实现了混凝土的复兴。现在，混凝土是应用最为广泛的人造材料（以吨数计量）。

2. 建筑领域中的应用（见图1~图5）

混凝土是最古老、应用最广的合成建筑材料，当前每年

Concrete is the oldest and the most widely used synthetic building material, currently produced at a rate of over five billion cubic yards per year and reportedly the second most consumed substance after water. It is easily taken for granted as the surface of everyday elements of infrastructure such as streets and sidewalks. It is also strongly associated with utilitarian structures such as parking garages and power plants, along with a variety of cheaply built and often poorly designed public and commercial buildings of the mid-twentieth century.

This common and apparently mundane material also, however, makes possible structures of extraordinary beauty and creativity. Concrete has been the indispensable medium for numerous architects and engineers attracted by its sculptural and expressive possibilities, and indeed, reinforced concrete is arguably the quintessential material of the Modern Movement in architecture. Its strength and versatility have allowed unprecedented experimentation with forms, surfaces, and structural frames, yielding numerous beloved landmarks ranging from Frank Lloyd Wright's Falling water, with its audacious cantilevered balconies, to Australia's highly evocative Sydney Opera House.

的生产量在38亿立方米以上，是仅次于水的第二大消耗物质。街道、人行道等大量日常基础设施的表面都是混凝土构成的。停车场、发电厂等市政设施以及20世纪中期所建造的大量廉价的公共建筑和商业建筑也大多采用混凝土作为主要材料。

然而，这种常见且看起来平淡无奇的材料能够打造出具有非凡的美丽和创意的结构。无数建筑师和工程师被混凝土的雕塑潜能和表现力所吸引，将其看成一种不可或缺的介质。事实上，钢筋混凝土可能是现代建筑运动中的典型材料。它的强度和广泛用途让建筑师在造型、表面和结构框架上进行了前所未有的实验。从弗兰克・劳埃德・赖特的流水别墅到悉尼歌剧院，混凝土成就了无数深受人们喜爱的地标性建筑。

2

4

3

5

Chapter 1
Cast-in-place Concrete
第一章 现浇混凝土

Cast-in-place concrete walls are made with ready-mix concrete placed into removable forms erected on site. Historically, this has been one of the most common forms of building basement walls. The same techniques used below grade can be repeated with above-grade walls to form the ground floor and upper levels of homes.

现浇混凝土墙由预拌混凝土浇注现场在可拆除模板中而成。这曾是建造建筑地下室墙壁最常见的形式之一。地下所用的技术同样可以应用在地上空间的墙壁，如一楼或住宅的上层墙壁等。

1.1 Components, Including Insulation

(See Figure 1.1 to Figure 1.4)

Cast-in-place (CIP) concrete systems are relatively straightforward. Steps required include the placement of temporary forms and placing fresh concrete and steel reinforcement. Although it is possible to batch concrete on site, ready mixed concrete is widely available and is usually delivered by a ready mix supplier.

Although uninsulated walls were common in the past, changing energy code requirements are more or less eliminating walls without insulation in most climates. This is the case with all types of systems, including concrete, wood, and steel. Energy is simply too important in terms of its cost and environmental impact. Concrete's thermal mass helps moderate temperature swings, but cannot provide the improved energy performance mandated by codes unless the wall system contains insulation. In the past, therefore, insulation may have been an optional component of a cast-in-place system, but it is increasingly included in contemporary construction.

The most common formwork materials for casting concrete in place are steel, aluminum, and wood. Many wood systems are custom manufactured and may be used only once or a few times. Steel and aluminum forming systems, on the other hand, are designed for multiple reuses, saving on costs. Metal panel forms are usually two to three feet wide and come in various heights to match the wall. Most common are eight and nine foot tall panels.

1.2 Installation, Connections, Finishes

Casting concrete in place involves a few distinct steps: placing formwork, placing reinforcement, and pouring concrete. Builders usually place forms at the corners first and then fill in between the corners. This helps with proper alignment of forms and, therefore, walls. Reinforcement bars ("rebar" for short) can be erected before either form face as a cage or after one side of the formwork is installed. Once both form faces are tied together and braced, concrete is placed in the forms via truck chute, bucket, or pump. Forms should always be filled at an appropriate rate based on formwork manufacturer recommendations to prevent problems. Although blowouts are uncommon with metal and wood forms, misalignment could potentially occur.

For single-family residential construction, wall thicknesses can range from four to 24 inches. Uninsulated walls are typically six or eight inches thick. Walls with insulation are generally thicker when they contain an internal layer of insulation: either the inner or outer wall layer has to serve a structural function. Cast-in-place walls are generally thicker than frame walls (wood or steel).

Reinforcement in both directions maintains the wall strength. Vertically, bars are usually placed at one to four feet on centre, and tied to dowels in the footing or basement slab for structural integrity. Horizontally, bars are typically placed at about four foot spacings in residential applications. Additional bars are placed at corners and around openings (doors,

1.1 构成组件，包括隔热层（见图1.1~图1.4）

现浇混凝土系统相对简单，制作步骤包括临时模板的摆放、新鲜混凝土的浇注以及钢筋加固。尽管可以在现场分批处理混凝土，大多数项目仍选用通过预拌供货商所提供的预拌混凝土。

尽管过去无隔热层的墙壁占主流，变化的能源标准正逐渐使没有隔热层的墙壁退出历史舞台。各种类型的系统都是一样的，如混凝土、木材、钢材等。从成本和环境影响来讲，能源至关重要。混凝土的热容量有助于缓和温度波动，但是不能提供良好的能源绩效，因此不许在墙面系统上安装隔热层。过去，隔热层是现浇混凝土系统的任选成分；现在，越来越多的建造结构都包含了隔热层。

最常见的现浇混凝土模板材料是钢、铝和木材。许多木材系统都是特别定制的，只能使用单次或数次。钢模板系统和铝模板系统则可以反复利用，节约了成本。金属板模板通常2~3英寸（约5.1~7.6厘米）宽，高度与墙壁的高度相匹配，大多数为8~9英尺（约20.3~22.9厘米）高。

1.2 隔热层、连接件、饰面

现场浇注混凝土包含一系列独立的步骤：摆放模板、放置钢筋、浇注混凝土。施工人员通常先将模板放在拐角，然后在拐角之间进行填充。这有助于保持模板的同轴度，从而建造更精准的墙壁。钢筋可以在模板面前方作为笼子造型架起，也可以在模板一侧安装之后进行安装。当模架的两面固定起来支撑牢固之后，就可以通过货车斜槽、桶或泵进行浇注。为了避免产生问题，一定要根据模板制造商的推荐规范进行匀速浇注。尽管金属模板和木模板很难爆裂，错位等情况也可能发生。

对于独户住宅的建造，墙壁的厚度可以在4~24英寸（约10~60厘米）之间。无隔热层的墙壁通常是15厘米或20厘米厚。带有隔热层的墙壁通常更厚，它们在内部加入了隔热层，由内层或外层墙壁担当结构功能。现浇混凝土墙通常比框架墙（木框或钢框）要厚。

双向钢筋加固能维持墙壁的强度。纵向钢筋通常放置在中心1~4英尺（约30~120厘米）处，系在底脚或底板的木钉上来保持结构完整性。在住宅建筑中，横向钢筋通常间隔4英尺（约120厘米）。额外的钢筋被配置在转角和门窗开口四周，有助于防止开裂，增加强度。

1.1

1.2

1.3

1.4

windows) to help control cracking and provide strength.

Openings for doors and windows require bucks to surround the opening, contain the fresh concrete during placement, and provide suitable material for fastening window or door frames.

Floors and roofs can be concrete or wood and light-gauge steel. Ledgers are anchored by bolts adhered into holes in the concrete. For heavy steel floors, weld plates are installed inside the formwork so they become embedded in the fresh concrete. This provides an attachment for steel joists, trusses, or angle irons.

Finishes on CIP systems are dependent on the presence of insulation and on the formed face. Finishes can alternately be attached with furring strips. Almost any type of finish can be used with removable form concrete wall systems. Wallboard remains the most common interior finish. Exteriors are much more varied and depend on customer preference. Form liners attached to the exterior form face can impart any type of texture; alternately, other traditional finishes such as masonry or siding can be attached to the wall following form removal.

门窗开口的四周需要添加加固，并且需要合适的材料来固定门框和窗框。

地面和屋顶可以由混凝土或木材及轻钢龙骨构成。横木通过螺栓固定在混凝土的打孔上。在重型钢地面建造过程中，模板内安装了焊接板，嵌入鲜混凝土内部。这有利于连接钢龙骨、桁架或角铁。

现浇混凝土系统的饰面取决于墙面是否有隔热层及其成形面。饰面可以交替地附在横筋上。几乎所有类型的饰面都可以与模板可拆除式混凝土墙系统一起使用。墙板是最常见的室内饰面。外墙饰面有更多的形式供消费者选择。与外表面相连的模板衬层可以被赋予各种质感。另外，在拆除模板之后，墙壁上也可以附着砖石或护墙板等其他传统饰面。

隔热层可以放置在墙壁的内外两侧或中间部分。在墙壁表面上安装隔热层时，塑料配件被嵌入泡沫板里，

1.5

1.6

Insulation can be placed on inside or outside faces or in the centre portion of the wall. To place the insulation on the face, plastic fittings are inserted into the foam board and become embedded in the concrete. These are flanged to hold the foam and the flanges provide an attachment for finishes and fixtures. Face insulation can also be applied after the formwork is stripped. If foam is embedded in the formwork prior to concrete placement, composite fittings are used to tie together the two concrete faces (through the foam insulation layer). The inner wall is usually the structural layer, so it's thicker and contains the rebar, whereas the outer concrete layer has the finish applied. Foam insulation is most often expanded polystyrene (EPS). It can be extruded polystyrene (XPS), which is stronger, but also more costly.

1.3 Sustainability and Energy

A major appeal of insulated cast-in-place walls is reduced energy to heat and cool the building. Insulation, thermal mass, and low air infiltration contribute to the energy saving. Typical R-value for EPS and XPS foams are, respectively, four and five per inch. Thermal mass acts like a storage battery to hold heat or cold, moderating temperature swings. Cast-in-place walls have 10 to 30 percent better air tightness than comparable framed walls – because the concrete envelope contains few joints. In addition to saving energy and money associated with heating and cooling, concrete walls also provide more consistent interior temperatures for occupants, increasing their comfort. Cast-in-place systems are also suited to the use of recycled materials. Concrete can be made using supplementary cementing materials like fly ash or slag to replace a portion of the cement. Aggregate can be recycled (crushed concrete) to reduce the need for virgin aggregate. Most steel for reinforcement is recycled. Some polystyrene is made with recycled material as well. Some of these techniques contribute toward achieving points in certain green rating systems such as LEED®.

1.4 Applications in Building Façade

1.4.1 Exposed Concrete (See Figure 1.5 to Figure 1.8)

Exposed concrete is an expression of Modernism, which is also called decorative concrete for its decorative effect. The basic idea is to apply nothing such as coating, enclosing or cladding after the concrete is placed. It shows concrete without any makeup.

Since exposed concrete doesn't have any finishes, its weather resistance must be consistent to regulations related to reinforced concrete cover. In construction process, it needs careful management, so contractors and constructors always keep away from it. However, exposed concrete's unique strength, cleanliness and sculptural feelings are favoured by designers with strong design senses. In addition, exposed concrete lacks the finishing works after form removals, so its quality is largely determined in forming process.

然后嵌入混凝土。混凝土被安装了法兰凸缘，用于固定泡沫，同时也可以固定饰面和其他装置。贴面隔热层也可以在模板拆除后安装。如果泡沫在浇注混凝土前就嵌在模板中，可以用负荷配件将混凝土的两面捆绑起来。内墙通常是结构层，因此更厚，且包含钢筋，而外层混凝土则配有饰面。最常用的泡沫隔热材料是泡沫聚苯乙烯（EPS）。挤塑聚苯乙烯（XPS）也是不错的选择，它更强韧，也更贵。

1.3 可持续性与能源绩效

带隔热层的现浇混凝土墙最主要的优点之一是能够缩减建筑供暖和制冷所需的能源。隔热层、热容量和低空气渗透都有助于节能。泡沫聚苯乙烯和挤塑聚苯乙烯的标准热阻值分别为4英寸和5英寸（约10.2厘米和12.7厘米）。热容量作为一个蓄热体，能够存储冷和热，调节温差。现浇混凝土墙的气密性比框架墙要好10%~30%，因为混凝土外壳的接缝更少。除了节约与供暖和制冷相关的能源和金钱之外，混凝土墙还能为使用者提供更稳定的室内温度，从而提升他们的舒适度。现浇混凝土系统还有利于回收材料的使用。混凝土可以用粉煤灰或矿渣等材料作为替代水泥的胶结材料。压碎混凝土中的骨料可以进行回收，以减少原始骨料的需求。大多数用于加固的钢筋都是可回收的。一些聚苯乙烯也是由回收材料制作而成。这些技术应用有助于建筑在一些绿色等级评级系统中获得更高的得分，如LEED®绿色建筑认证。

1.4 建筑表皮中应用

1.4.1清水混凝土（见图1.5~图1.8）

清水混凝土是建筑现代主义的一种表现手法，因其极具装饰效果也称装饰混凝土。基本的想法于混凝土浇置后，不再有任何涂装、贴瓷砖、贴石材等材料，表现混凝土的一种素颜的手法。

由于清水混凝土不会再有其他装饰用的材料，因此对于风雨的抵抗力要非常注意施工规范中关于钢筋混凝土保护层的规定，因此在施工阶段要非常细心的管理，发包者、施工者往往对此敬而远之。然而，清水混凝土构造独特的强力、清洁感、素材感等出色的美学表现，促使个性强烈的设计者会特别喜欢它。另外，由于清水混凝土缺乏拆除模板后的装修工程，因此在组模的阶段就大致决定其好坏。

1.4.2 装饰混凝土（见图1.9、图1.10）

装饰混凝土中的混凝土不仅是建造结构的实用媒介，

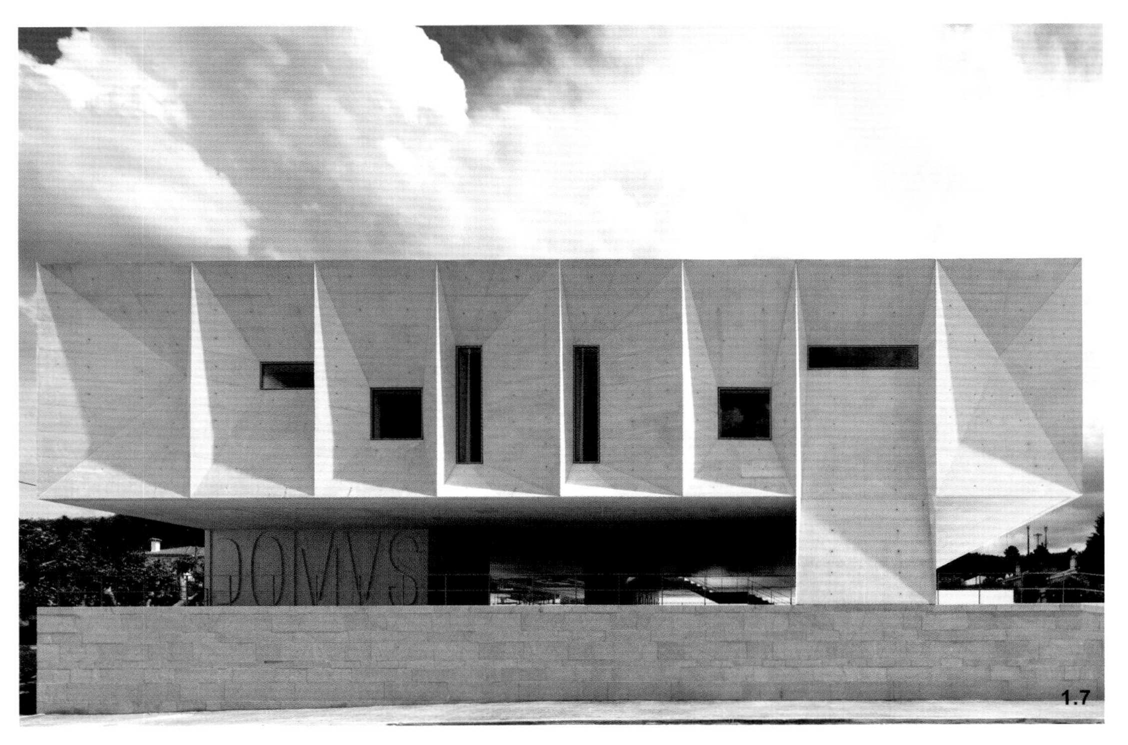

1.7

1.8

1.4.2 Decorative Concrete(See Figure 1.9, Figure 1.10)

Decorative concrete is the use of concrete as not simply a utilitarian medium for construction but as an aesthetic enhancement to a structure, while still serving its function as an integral part of the building itself such as floors, walls, driveways and patios.

The transformation of concrete into decorative concrete is achieved through the use of a variety of materials that may be applied during the pouring process or after the concrete is cured, these materials and/or systems include but are not limited to stamped concrete, acid staining, decorative overlays, polished concrete, concrete countertops, vertical overlays and more.

Stamped Concrete

Stamped concrete is the process of adding texture and colour to concrete to make it resemble stone, brick, slate, cobblestone and many other products found in nature including wood, fossils, shells and many more. This limitless array of possibilities combined with great durability and lower cost than natural products makes stamped concrete an easy choice for new construction and renovation projects.

The installation consists of pressing molds into the concrete while the concrete is still in its plastic state. Colour is achieved by using dry shakes or colour hardeners, powder or liquid releases, integral colours or acid stains. All these products may be combined to create even more intricate designs.

Stamped concrete may be used on driveways, patios, commercial roads and parking lots and even interior floors. Stamped concrete is a desirable finish to

还能够提升结构的美感，可以用于建筑中楼面、墙面、走道、露台等场所的建造。

从普通混凝土向装饰混凝土的转变需要在浇注过程中或固化结束后使用一系列不同的材料。这些材料或系统包含但不限于模压混凝土、酸性染色剂、装饰覆盖层、抛光混凝土、混凝土台面、垂直覆盖层等。

模压混凝土（见图1.11、图1.12）

模压混凝土能为混凝土添加纹理和色彩，使其看起来类似石材、砖、板岩、卵石或自然界中的其他产品，如木材、化石、贝壳等。模压混凝土将无限的可能与良好的持久性和低成本结合起来，是许多新建项目和翻修项目的必然之选。

模压混凝土要求在混凝土仍处于塑性状态时进行压模。可以用干撒面、彩色强化粉/液、整合色或酸性染色剂为其添加色彩，也可以将它们混合起来营造更复杂的设计。

模压混凝土可以使用在走道、露台、商业道路、停车场乃至室内地面。模压混凝土是混凝土材料极好的装饰，但是随着时间的流逝，染料会褪色，表面会变得斑驳不一。许多再染色方法都能彻底地还原和复原色彩，让褪色的模压混凝土重新焕发活力。

concrete areas, however with time and wear the colour dyes fade and the surface looks patchy and unpresentable. There are many re-colouring options which can completely restore and rejuvenate the colour and presentation of faded stamped concrete.

Concrete Dyes (See Figure 1.11, Figure 1.12)
Concrete "dyes" take many different forms and compositions and can be used on both residential and commercial concrete applications, including sound/retaining walls, bridges, countertops, floors, etc.

Early concrete dyes consisted of generic printing inks that were dissolved in mild solutions of alcohol and applied to concrete surfaces to add a wide array of colour to plain grey concrete. When alcohol-based dyes are exposed to sunlight, the colour either lightens or fades out completely. Therefore, alcohol-based dyes were more prevalent in interior applica-

1.9

1.10

tions where direct sunlight or other forms of ultraviolet (UV) lighting was not present.

Manufacturers later began dissolving the same printing inks in different carriers, such as acetone, lacquer thinner and other solvents, hoping to achieve increased penetration levels. In addition, UV inhibiting agents were added to new dyes to help with the UV instability issues. However, slight fading (5-8% per year) still occurs when the dye is exposed to direct sunlight.

Coloured concrete can be obtained from many ready mix concrete companies and many have colour charts available.

1.11

1.12

混凝土染料

混凝土染料的形式和组合方法数不胜数，可以应用在民用和商用混凝土结构中，包括隔音墙、挡土墙、桥梁、台面、地面等。

早期的混凝土染料由普通印刷用油墨溶解在乙醇中构成，能够为素面灰色混凝土带来缤纷的色彩。当醇基染料暴露在阳光下，色彩会变浅或完全消褪。因此，醇基染料更适用于没有直射阳光和其他紫外线照明的室内应用。

后来，制造商开始在不同的媒介中溶解印刷用油墨，如丙酮、油漆稀释剂等，希望获取更好的渗透性。此外，新染料中还添加了紫外线抑制剂来应对紫外线不稳定的问题。然而，当染料直接暴露在阳光下，还是会产生轻微的褪色（每年5%~8%）。

许多预拌混凝土公司都生产彩色混凝土，有多种不同的色彩选择。

酸性染色

酸性染色过程并不是印染或色素上色，而是一种化学反应。它在混凝土表面施加了一层水、矿物盐和少量王水的混合物。混合物与混凝土中的原有材料（主要

Acid Staining

Acid staining is not a dyeing or pigment-base colouring systems, but a chemical reaction. A mixture of water, mineral salts and a slight amount of muratic acid is applied to the concrete surface. This chemical reaction with the existing minerals (primarily lime) in the concrete over a period of one to four hours creates new earth tone colours on the concrete surface. The concrete surface is later scrubbed to remove excess stain and neutralised by a basic solution of ammonia and water or baking soda (less likely to cause whiting later) to help raise the PH level back to normal level. Due to inconsistencies in the surface level of concrete floor, acid staining creates a variegated or mottled appearance that is unique to each slab. The colour penetration ranges from 1/16 to 1/32 of an inch. Older exterior concrete surfaces may not colour as well as interior surfaces because the environment has leached or percolated out the mineral content. As well, any exposed aggregate (rocks) in worn concrete will not accept staining.

Chemicals commonly used in acid staining include hydrochloric acid, iron chloride and sodium bicarbonate.

Water Based Staining

Water based stains are similar to acid based stains in the sense that one can still achieve a translucent look like acid; some stains are able to achieve an opaque colour and/or a translucent effect. The main difference is that acid stains react to the concrete and change the physical make up of the concrete material, whereas water based stains are more of a "coating" that bonds with the concrete. There are many variations of water based stains that have come into the decorative concrete industry that perform in a number of different ways. Some are polymer based, acrylic and epoxy.

Polishing

Concrete can be polished with mechanical grinders and diamond pads of increasing grit sizes. Diamond pads come in many grit or mesh sizes. Common sizes start with 6 grit and can go up to 8500 grit although concrete can only maintain a shine of about 800 grit, it can be helped by adding a concrete hardener such as sodium silicate or lithium silicate which will allow concrete to hold an 1800 through 3000 grit shine. The work is accomplished in multiple stages by passing over the concrete with successive grit diamond pads until it has a hard-glassy finish. Both acid stains and concrete dyes can be used during the polishing process.

Engraving

Existing concrete can be remodeled by cutting lines and grooves into its surface. Geometric patterns, straight or curved lines and custom designs can be cut directly into the surface of the concrete creating the look of tile, flagstone, cobblestone and many other common surface patterns. This is usually achieved using tools like an angle grinder fitted with diamond blades, but is more effectively accomplished using specialty tools designed specifically for cutting designs into the concrete.

是石灰）在一至四小时内发生化学反应，在混凝土表面形成一种褐土色调。然后，对混凝土表面进行擦洗，清除多余的着色，用氨水溶液或小苏打水（不易泛白）使酸碱值（pH值）恢复到正常水平。由于混凝土表面并不均匀，酸性染色会形成斑驳的外观，让每块楼板都与众不同。色彩渗透的范围在每英尺1/16和1/32之间。较旧的室外混凝土表面的上色可能没有室内表面好，因为环境已经对矿物含量进行了渗透或过滤。同样，混凝土上的露出的骨料或碎石都不能染色。

酸性染色所使用的化学制品通常包括盐酸、氯化铁和碳酸氢钠。

水性染色

水性染色剂与酸性染色剂类似，都能实现一种半透明的效果。一些染色剂还能实现不透明色或半透明效果。它们的主要区别在于酸性染色剂与混凝土反应，改变混凝土材料的物理构成，而水性染色剂则更多地是给混凝土附着一层“包层”。在当前的装饰混凝土产业中，有许多中水性染色剂，它们通过不同的途径起效。一些甚至是高分子的丙烯酸材料和环氧基树脂材料。

抛光

混凝土可以用机械螺纹磨床和金刚石砂轮进行抛光。金刚石砂轮有许多不同的粗粒度和网眼大小进行选择。一般的尺寸从6粒开始，可上升至8500粒。混凝土所能承受的粗粒度约为800粒，可以在混凝土中添加硅酸钠或硅酸锂等硬化剂使其可承受的粗粒度达到1800~3000粒。抛光工作包含多个步骤，在混凝土表面连续地使用金刚石砂轮，直至形成具有玻璃光泽的饰面。抛光过程中可以使用酸性染色剂和混凝土染料。

雕刻

已有的混凝土可以通过在表面切割线条和凹槽进行改造。可以在混凝土表面直接雕刻出几何图样、直线、曲线以及定制设计，形成瓷砖、石板、卵石等许多种不同的表面效果。雕刻过程经常使用带有金刚石刀片的角磨机，但是专门用于切割混凝土的特殊工具会更加有效。

Gouveia Law Court

戈维亚法院

Location/地点: Gouveia, Portugal/葡萄牙，戈维亚
Architect/建筑师: Barbosa & Guimarães
Photos/摄影: Jose Campos
Site area/占地面积: 2,143.25 m²
Built area/建筑面积: 5,701.90m²
Completion date/竣工时间: 2011
Key materials: Façade – concrete
主要材料：立面——混凝土

Overview

Gouveia, the entrance to Serra da Estrela, will be serviced by new Law Court. The land set for its construction, is set between two public gardens, at the end of the Rampa do Monte do Calvário, substituting an existing building.

The project takes advantage of the demolished building that occupied the totality of the plot, to draw a new Square, with scale and dignity to receive the Law Court. In a dialogue with the granite walls that define its surroundings, the Square assumes itself as a stone base, upon which rests the Tribunal.

The building, set on four columns, guarantees transparency and connection between the two gardens that bind it at north and south. The dignity, and the symbolism that a building such as a Law Court should always have, is achieved at the cost of the monolithic and singular character, which the volume in white concrete acquires, above all in the expression of its solid elevations, with deeply "carved" openings as if suspended above the square.

A generously proportioned stair open to one courtyard, dignifies the access to the Tribunal level. The atrium/foyer, crosses the whole building lengthways, communicating directly with the garden at north, establishing a relation of proximity with the existing treetops through a horizontal opening.

From the volumetry, the hearing room stands out, with a set of vertical skylights that light up the whole space in a subtle manner. The registry services, that function autonomously, are installed in the base of the building, open to the interior courtyard, which communicates directly with the garden at north. The project also foresees a public car park, hidden below the square with access from the adjoining streets.

项目概况

作为埃什特雷拉山脉的门户，戈维亚建造了一座新法院。项目场地位于两个公园之间，卡尔瓦里山山脚，取代了一座原有的建筑。

项目充分利用了被拆除的建筑所占的地块，打造了一个新广场，提升了法院庄重感。为了与周围的花岗岩建筑墙面相对应，广场采用了石材底座，法庭就建在这个底座之上。

建筑采用四柱支撑，保证了南北两个公园之间的通透感和连接性。空间的整体感突出了建筑作为一座法院的庄重感和象征意义。深嵌的雕塑感开口让稳重的白色混凝土结构仿佛架空于广场上方一样。

宽阔的楼梯面向一边的庭院开放，增加了法庭通道的威严感。中庭兼门厅纵向贯穿了整座建筑，直接与北面的公园相连，并且通过水平开口拉近了与周围树木顶端的联系。

在空间上，听证厅向外凸出，一系列垂直天窗提升了整个空间的亮度。自助注册服务区被设置在建筑底座，与内庭相连，与北面的花园也有直接联系。项目还预留了一个公共停车场，隐藏在入口广场下方，与街道相连。

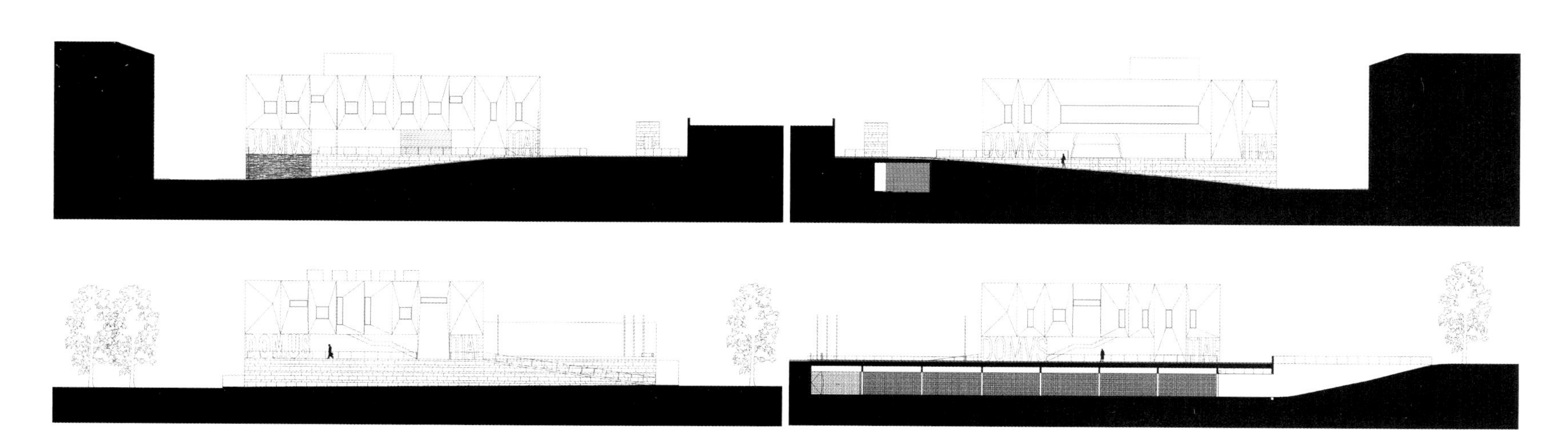

IVSTITIA
IVSTITI

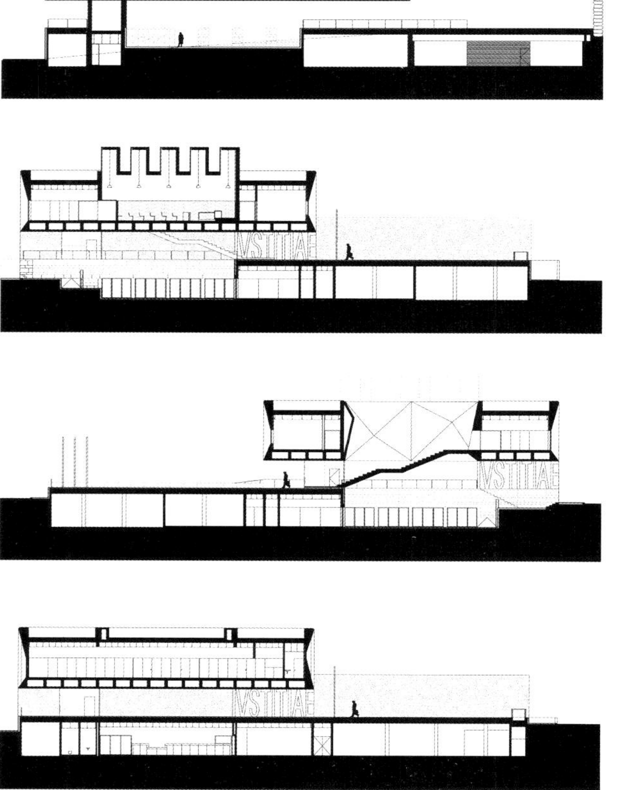

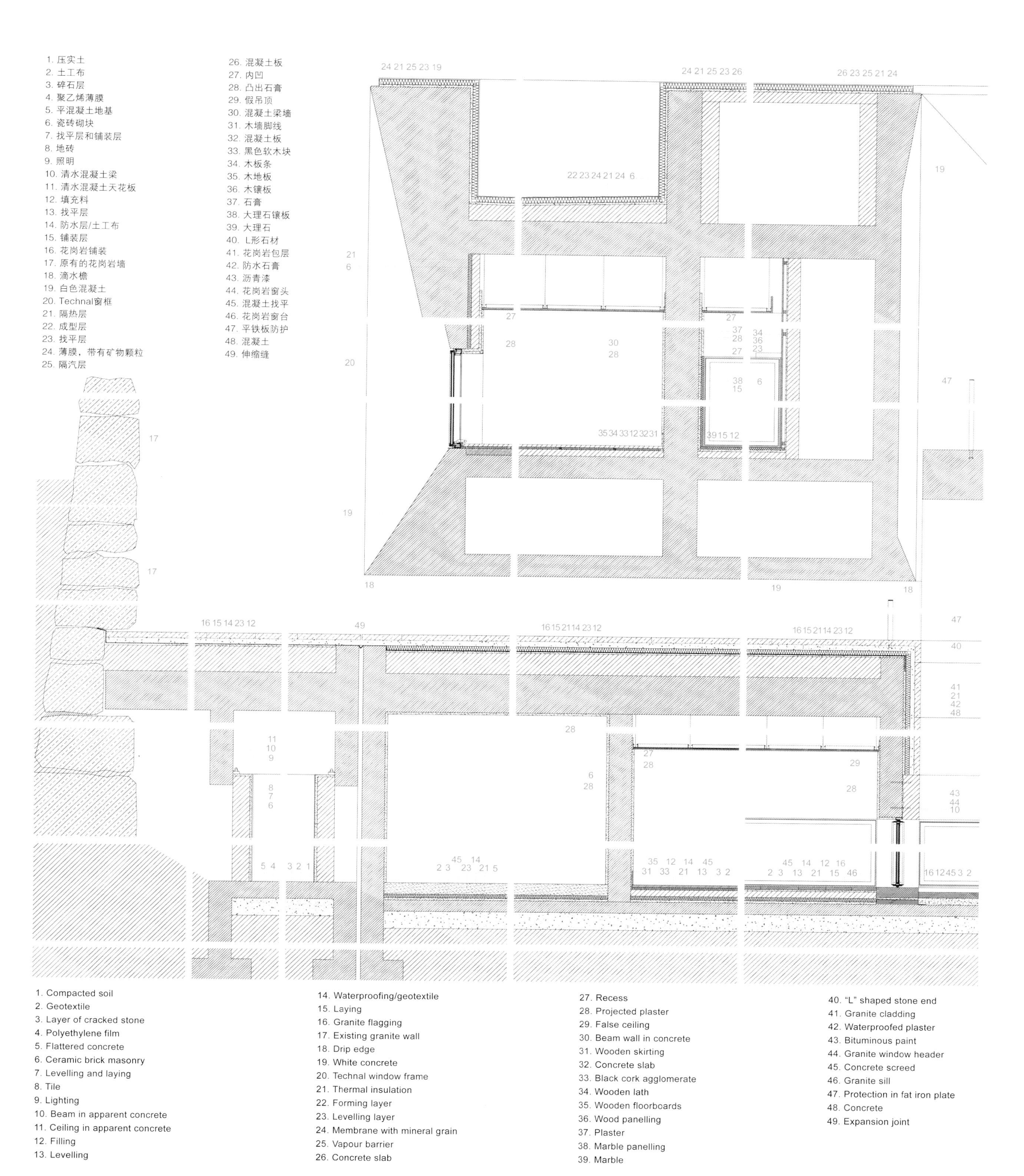
1. 压实土
2. 土工布
3. 碎石层
4. 聚乙烯薄膜
5. 平混凝土地基
6. 瓷砖砌块
7. 找平层和铺装层
8. 地砖
9. 照明
10. 清水混凝土梁
11. 清水混凝土天花板
12. 填充料
13. 找平层
14. 防水层/土工布
15. 铺装层
16. 花岗岩铺装
17. 原有的花岗岩墙
18. 滴水檐
19. 白色混凝土
20. Technal窗框
21. 隔热层
22. 成型层
23. 找平层
24. 薄膜，带有矿物颗粒
25. 隔汽层
26. 混凝土板
27. 内凹
28. 凸出石膏
29. 假吊顶
30. 混凝土梁墙
31. 木墙脚线
32. 混凝土板
33. 黑色软木块
34. 木板条
35. 木地板
36. 木镶板
37. 石膏
38. 大理石镶板
39. 大理石
40. L形石材
41. 花岗岩包层
42. 防水石膏
43. 沥青漆
44. 花岗岩窗头
45. 混凝土找平
46. 花岗岩窗台
47. 平铁板防护
48. 混凝土
49. 伸缩缝
1. Compacted soil
2. Geotextile
3. Layer of cracked stone
4. Polyethylene film
5. Flattered concrete
6. Ceramic brick masonry
7. Levelling and laying
8. Tile
9. Lighting
10. Beam in apparent concrete
11. Ceiling in apparent concrete
12. Filling
13. Levelling
14. Waterproofing/geotextile
15. Laying
16. Granite flagging
17. Existing granite wall
18. Drip edge
19. White concrete
20. Technal window frame
21. Thermal insulation
22. Forming layer
23. Levelling layer
24. Membrane with mineral grain
25. Vapour barrier
26. Concrete slab
27. Recess
28. Projected plaster
29. False ceiling
30. Beam wall in concrete
31. Wooden skirting
32. Concrete slab
33. Black cork agglomerate
34. Wooden lath
35. Wooden floorboards
36. Wood panelling
37. Plaster
38. Marble panelling
39. Marble
40. "L" shaped stone end
41. Granite cladding
42. Waterproofed plaster
43. Bituminous paint
44. Granite window header
45. Concrete screed
46. Granite sill
47. Protection in fat iron plate
48. Concrete
49. Expansion joint

Health Faculty of San Jorge University

圣乔治大学医学院

Location/地点: Zaragoza, Spain/西班牙，萨拉戈萨
Architect/建筑师: Taller Básico de Arquitectura
Photos/摄影: José Manuel Cutillas
Built area/建筑面积: 8,853 m^2
Completion date/竣工时间: 2012
Key materials: Façade – concrete and limestone
主要材料: 立面——混凝土与石灰石

Overview

The new Health Faculty of San Jorge University is located on a campus on the outskirts of Zaragoza city. Although it is a rural campus, the nature in it is scarce. The forest along the campus is the result of a man created operation. The surrounding buildings, the Rectory and Communications Faculty, respond to a contemporaneous architecture that lives besides that nature.

The Health Faculty joins the development of that little nature to reinforce the place where the existing buildings rest and where new buildings will do. The new faculty is not only another building; it becomes part of the new place. Architecture is thought as part of a new nature.

The building programme is organised in three concave lines. These white and scaled lines unfold on the campus as part of its landscape. Inside, on two floors, classrooms and laboratories are organised for teaching and research. Each scale catches the light needed for each room. The dimensions and shape of rooms allow a big variability of use. Consequently, it is possible an academic reorganisation in an easy way. Light coming through scales can be controlled, so digital technologies can be used inside rooms. The minor creases of each line contain the most public rooms of the new faculty: cafeteria, conference room and multipurpose rooms.

The three lines enclose a big room open to the sky. All the access corridors to laboratories and classrooms face this big room. The square gives access to the three lines. Lines look at each other through the square, which discovers the inside of this mineral complex. The inside and outside relation of the faculty gets inverted. The concave outside happens to be the most interior room, and the convex inside becomes

the most exterior place.
The mineral nature of this faculty in San Jorge University offers a new landscape of white scales breathing light on the outside, and it offers a big room opened to the sky on the inside.

Detail and Materials

The three buildings that form a complex show a scaled image that gives plenty of light to the educational spaces. This layer is made with "white capri" limestone 120x60cm and supported with mortar and staples. This stones show their trapezoidal shape through the cutting. The careful placement of the pieces, gives an organic texture image to the outside of the building.

The main entrance to the building is limited by a smooth concrete skin with large holes that provide views and light to the rest areas and the crossing points. Formwork panels were covered by phenolic panels to minimise the impact of the joints, looking for the maximum continuity in this concrete skin.

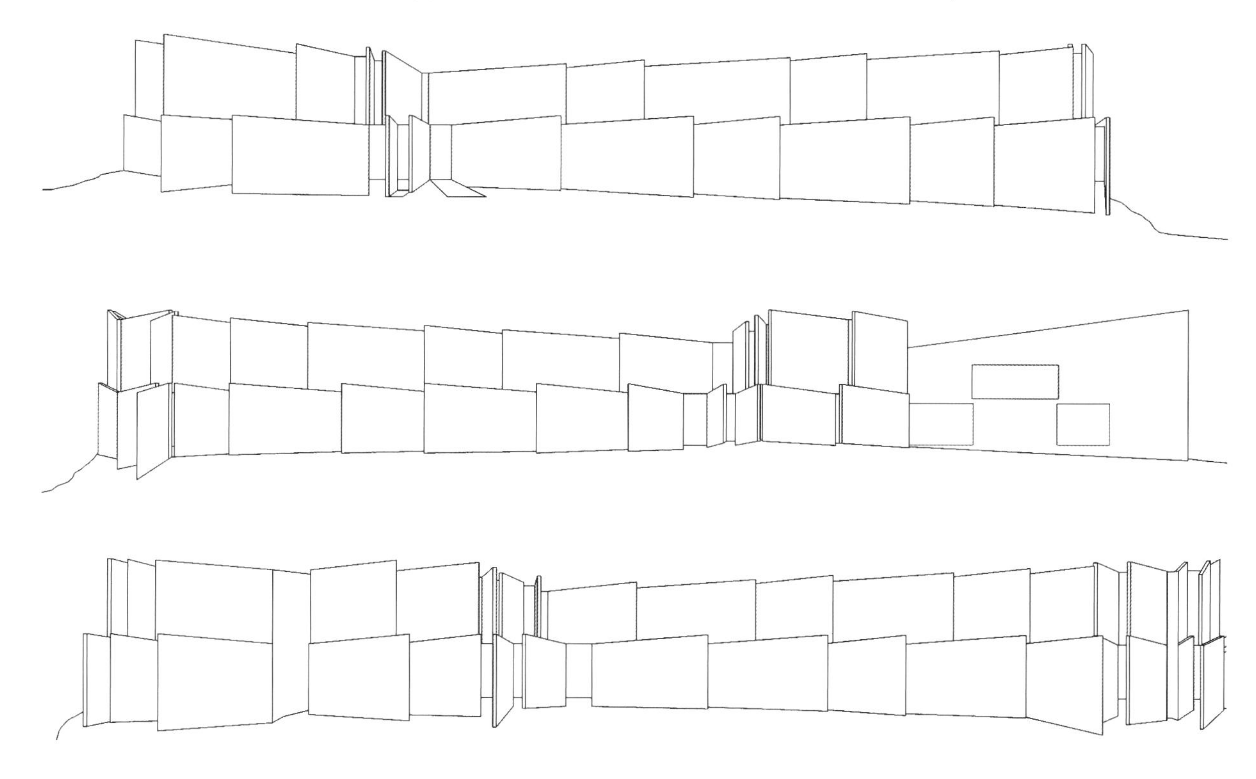

项目概况

圣乔治大学医学院位于西班牙萨拉戈萨城郊的校园内。虽然校园位于郊区，自然植被却十分稀缺。校园内的树林是人工种植的。院长行政楼和通讯学院等周边建筑都是同样的建筑形式，坐落在自然之中。

医学院融入了这个整体环境之中，与原有建筑和新建建筑共同组成了新的空间。新建的学院楼不仅是一座建筑，而是整个新环境的一部分。建筑已经融入了环境之中。

建筑规划通过三条折线展开。这些鳞次栉比的白色线条在校园中展开，形成了景观的一部分。建筑内部，教室和实验室在两层楼中展开，用于教学和科研。错落的墙面让阳光进入了各个房间。房间的尺寸和形状保证了空间的灵活运用。因此，教学性的空间充足将变得十分简单。可以从窗口控制进入的光线，因此可以在房间内使用数码技术设备。各个线条的褶皱内是学院的公共空间：餐厅、会议室和多功能室。

三条折线围起了一个巨大的中庭空间。所有通往实验室和教室的走廊都在此处交汇，经由广场可以进入三条折线内的空间。折线通过广场联系起来，可以相互看到各自的内部。学院楼的内部和外部相互交错起来。建筑外部的凹形空间形成了最里面的房间，而内部的凸出空间则是最外面的房间。

圣乔治大学学院楼的矿物特质为校园带来了清新的景观，错落的白色建筑结构将光线引入内部，而中央的广场则朝向外面的天空开放。

细部与材料

三座相互交错的建筑结构共同组成了一个建筑整体，将光线引入了教学区域。建筑外墙由120x60厘米尺寸的“白卡普里”石灰石混合砂浆构建而成。这种石材在切口处会呈现出不规则的四边形。石块的巧妙摆放为建筑外墙带来了天然纹理图案。

建筑的主入口采用了带有大孔的光滑混凝土表皮，这些圆孔为休息区和交叉路口提供了风景和光线。酚醛树脂板上覆盖着模块板，尽量消除了接缝处的影响，实现了混凝土表皮连续性的最大化。

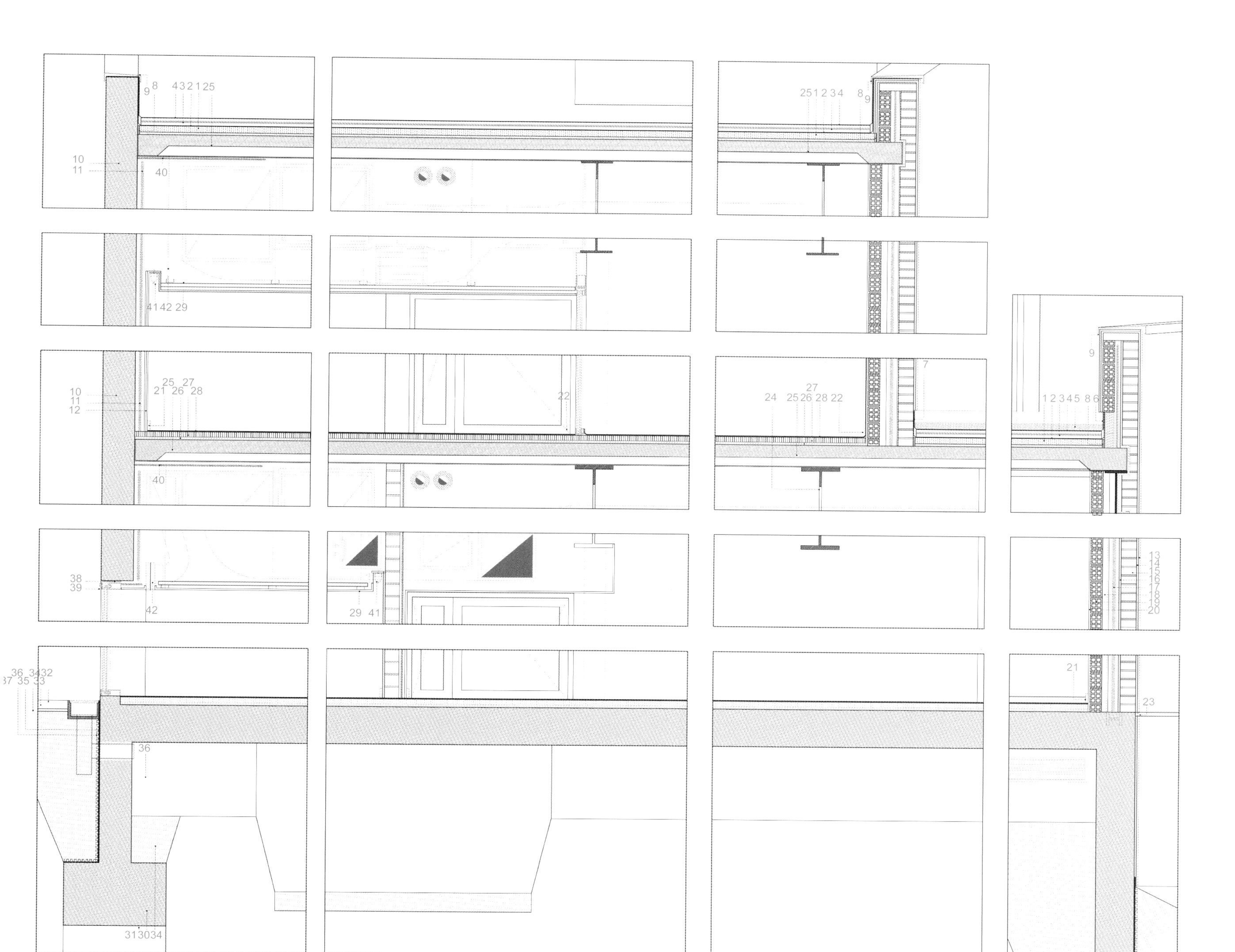

1. Lightweight concrete bed 5cm
2. Waterproofing layer
3. High Density XPS Insulation 6cm Thk
4. Waterproofing layer
5. White gravel
6. Galvanised steel L Profile
7. Galvanised steel folded sheet for aluminium gutter 2mm + High Density XPS Insulation 10mm
8. EPS join
9. Galvanised steel gutter
10. Concrete reinforced 25cm
11. Double plasterboard 13+13mm fixed to galvanised steel hidden structure
12. Insulation ISOVER ARENA 60 60mm thk
13. Limestone honed in white colour 60 x 120cm 2 thk
14. Fixing mortar
15. Reinforced ceramic brick wall 12cm
16. Plastering cement mortar
17. EPS thermal insulation 40+40mm Thickness
18. Cavity
19. Reinforced ceramic brick wall 90mm
20. Trim and plaster 15mm
21. MD skirting board 7 x 1.2cm
22. Half round resin skirting board
23. Join
24. Metal beam
25. Steel decking slab PL 79/H10
26. Mortar
27. Insulation ISOVER PST 22 mm
28. Epoxy resin grout 3-4mm Thk
29. Suspended ceiling
30. Concrete footing H-25/B/40 IIa
31. Pit foundation of cyclopean concrete
32. Concrete finished
33. Concrete reinforced bed 5cm
34. Gravel backfill
35. Geotextile layer for protection
36. Bituminous protection
37. Watherproofing polythene sheeting
38. Neoprene elastic join
39. Metal gutter
40. XPS Insulation
41. Lighting ODEL-LUX
42. Lineal aluminium grid ventilation

1. 轻质混凝土底座5cm
2. 防水层
3. 高密度XPS隔热层，6cm厚
4. 防水层
5. 白色碎石
6. L形镀锌钢
7. 镀锌折叠钢板，铝槽2mm+高密度XPS隔热层10mm
8. EPS连接处
9. 镀锌钢槽
10. 钢筋混凝土25cm
11. 双层石膏板13+13mm，固定在镀锌钢隐式结构上
12. 隔热层ISOVER ARENA 60，60mm厚
13. 白色亚光石灰石，60x120cm，2厚
14. 固定砂浆
15. 加固陶瓷砖墙12cm
16. 石膏水泥砂浆
17. EPS隔热层，40+40mm厚
18. 气腔
19. 加固陶瓷砖墙90mm
20. 装饰和石膏15mm
21. MD墙脚板7x1.2cm
22. 半圆树脂墙脚板
23. 连接处
24. 金属梁
25. 钢板层PL79/H10
26. 砂浆层
27. 隔热ISOVER PST 22 mm
28. 环氧树脂薄浆，3~4mm厚
29. 吊顶
30. 混凝土底脚H-25/B/40 IIa
31. 毛石混凝土坑基
32. 混凝土饰面
33. 钢筋混凝土层5cm
34. 碎石回填
35. 土工布保护层
36. 沥青保护层
37. 防水聚乙烯板
38. 氯丁橡胶弹性接合处
39. 金属槽
40. XPS隔热层
41. 照明ODEL-LUX
42. 线形铝格栅通风

Beit-Halochem

老兵之家

Location/地点: Be'er Sheva, Israel/以色列，贝尔谢巴
Architect/建筑师: Kimmel Eshkolot Architects
Photos/摄影: Amit Giron
Site area/占地面积: 18,000m²
Built area/建筑面积: 6,000m²
Completion date/竣工时间: 2011
Key materials: Façade – concrete and glass
主要材料：立面——混凝土、玻璃

Overview
On the outskirts of Be'er Sheva, where the city ends and the desert begins, is the site of a new building: Beit Halochem (Veterans' Home).

The scorching desert sun and the parched scenery served as inspiration. The structure was designed as an arrangement of "rock" like units grouped together. Between them a thin horizontal roof forms a courtyard – intimate, inviting and protected, to serve the functions of the building – a home for disabled veterans and their families.

While studying the various three-dimensional expanses, a unique relationship with the project emerged, based on relations between light and shadow, closed versus open, positive and negative. The bright sunlight makes it possible to achieve a three-dimensional richness by reflections from the rough frontal surfaces.

Detail and Materials
The building divides the site into new topographies. This allowed the design of two ground floors on two different levels, interlocking with each other, as an integral part of the building architecture. Thus achieving maximum accessibility as is appropriate for the special needs of users of the building.

The "rocks" enclose rooms for private and more intimate functions, while in-between spaces serve as public areas in the building. Light bridges spanned over those areas enable passage between public spaces, which reinforces the

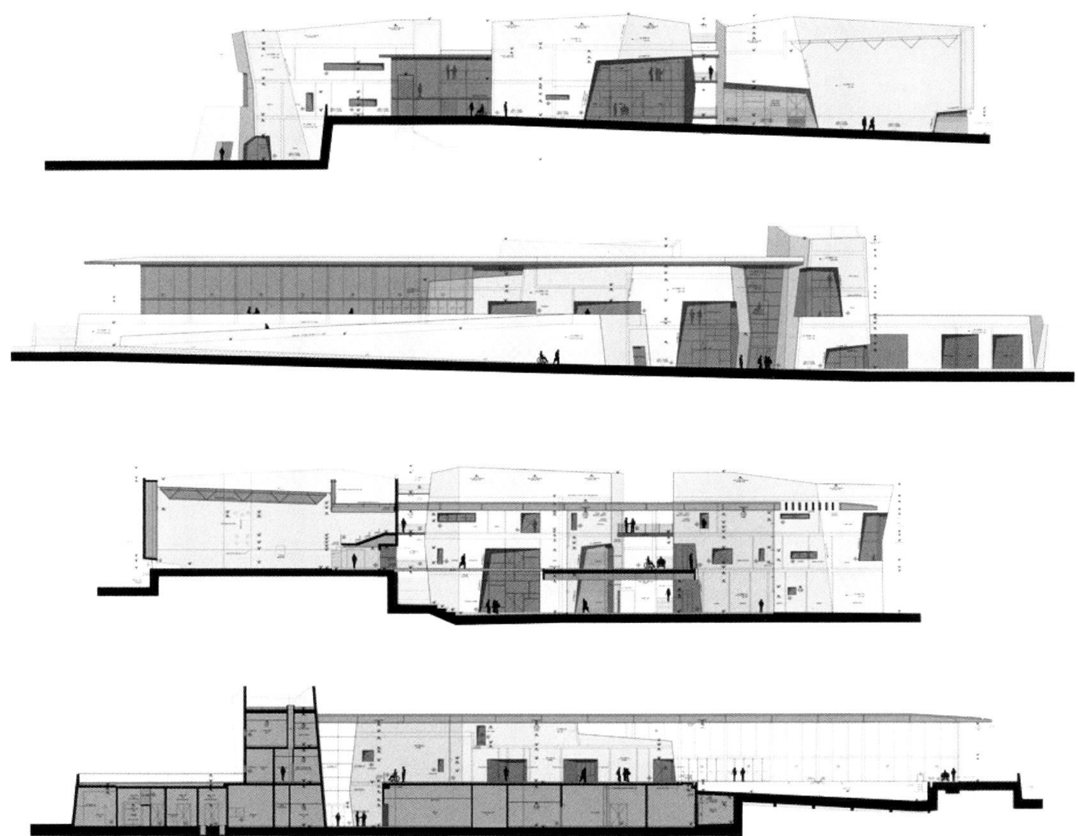

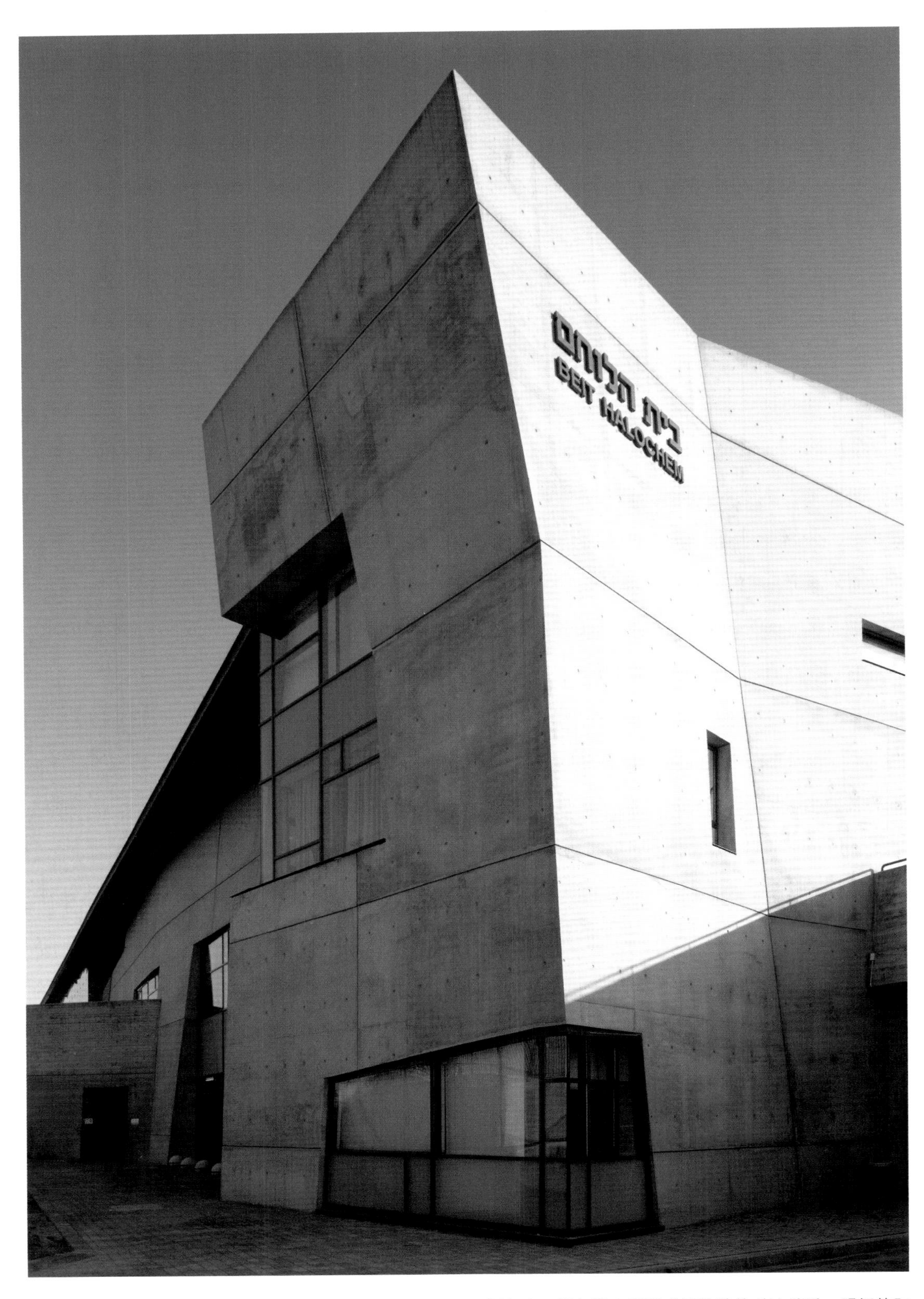

"experience" of the building for the users.

In the private areas, thick walls provide climate protection, which is so essential in the Negev desert. In contrast, in the public areas the light roof that caps the building provides shade and protection of the interior regions, and also creates a variety of external spaces where it is pleasant to relax.

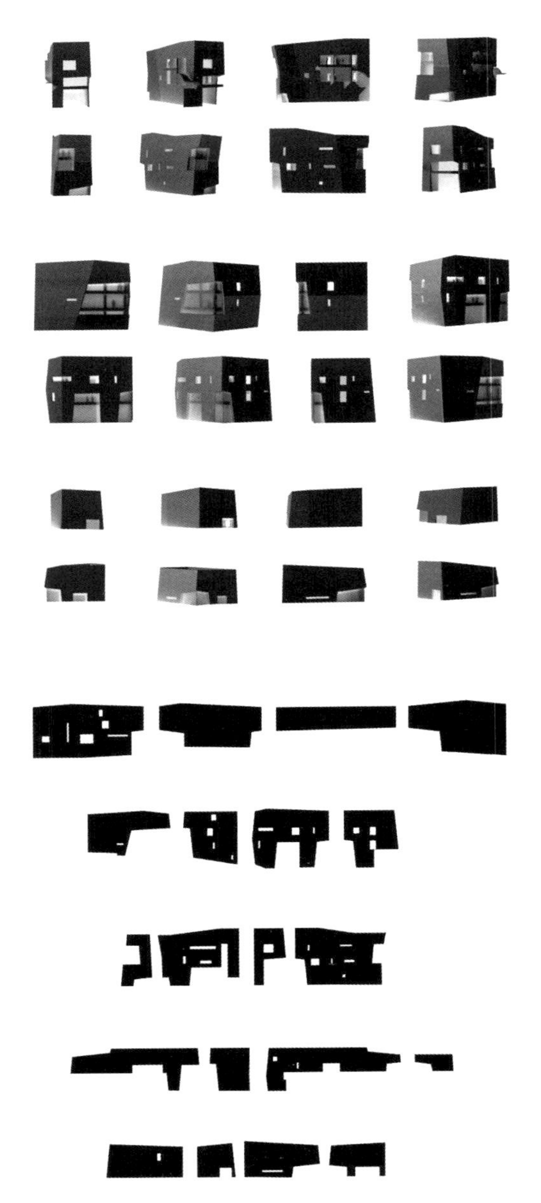

项目概况

这个老兵之家位于贝尔谢巴城郊，城市的尽头，沙漠的开端。

灼热的沙漠烈日和干燥的沙漠景象是设计的灵感来源。建筑结构被设计成排列成伍的岩石块。它们之间薄薄的水平屋顶形成了庭院，为伤残退伍军人及他们的家人提供了私密、友好而安全的生活空间。

在对各种三维维度进行研究的过程中，项目在光与影、开与合、阴与阳之间形成了独特的对比关系。明媚的阳光通过在粗糙的建筑表面上的反射而形成了丰富的三维效果。

细部与材料

建筑为项目场地塑造了全新的地形。这让建筑可以在不同的高度上进行建造，使不同的楼层相互交错，共同整合为建筑的一部分，从而实现全方位的无障碍通行，以便满足建筑使用者的特殊需求。

"石块"内部的房间有着良好的私密性，而石块之间的空间则被用作建筑的公共区域。这些区域之间由轻质天桥连接，保证了公共空间的连通性，也进一步提升了建筑使用的体验。

在私人区，厚厚的墙壁形成了一种气候保护，这在内盖夫沙漠是十分必要的。相反，公共区域的轻质屋顶既能为室内提供阴凉和保护，又打造了各种各样的室外空间供人休闲娱乐。

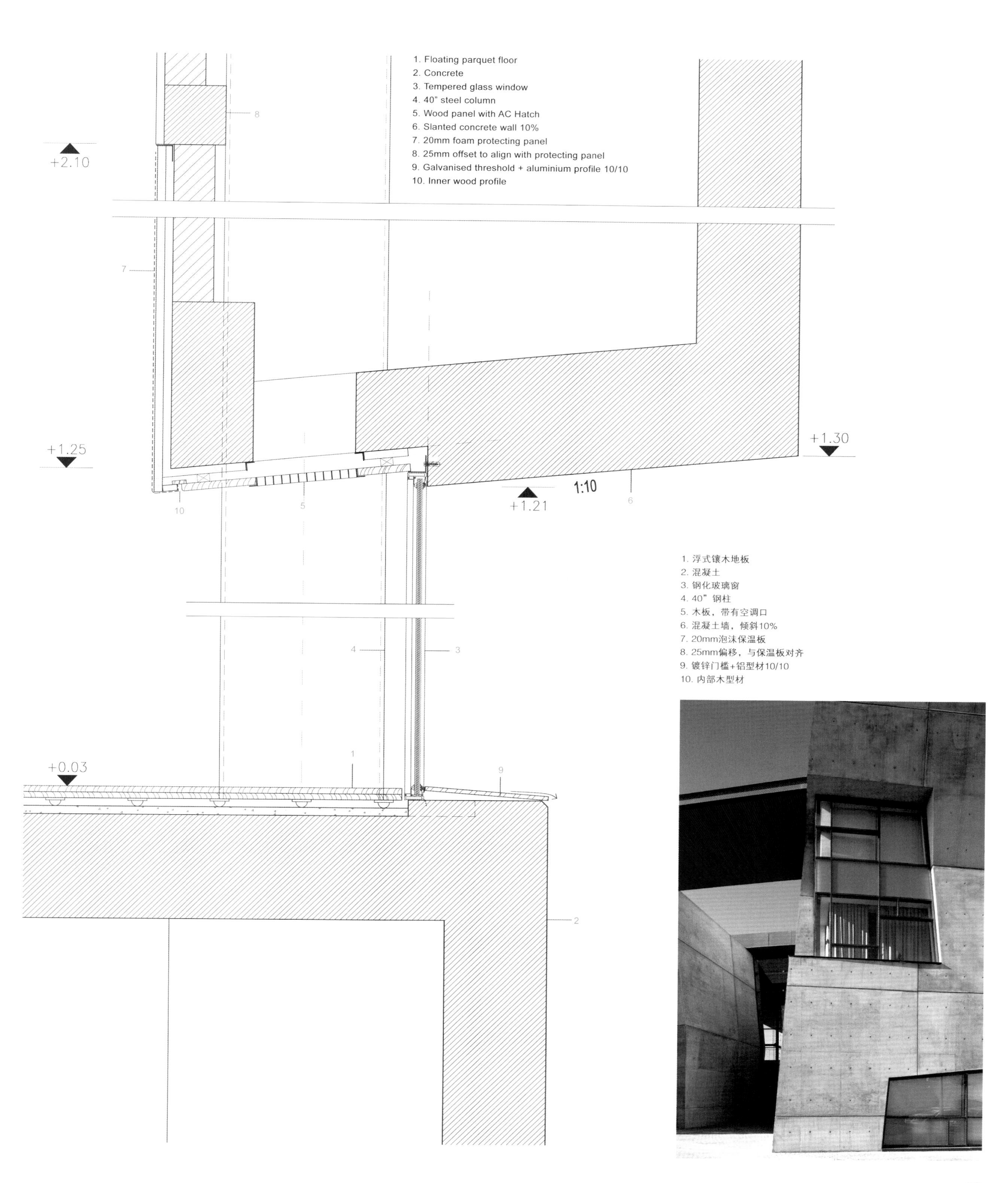
1. Floating parquet floor
2. Concrete
3. Tempered glass window
4. 40" steel column
5. Wood panel with AC Hatch
6. Slanted concrete wall 10%
7. 20mm foam protecting panel
8. 25mm offset to align with protecting panel
9. Galvanised threshold + aluminium profile 10/10
10. Inner wood profile
+2.10
+1.25
+1.30
1:10
+1.21
+0.03
1. 浮式镶木地板
2. 混凝土
3. 钢化玻璃窗
4. 40" 钢柱
5. 木板，带有空调口
6. 混凝土墙，倾斜10%
7. 20mm泡沫保温板
8. 25mm偏移，与保温板对齐
9. 镀锌门槛+铝型材10/10
10. 内部木型材

Music School and Areas for Culture

音乐学校及文化中心

Location/地点: Maizieres-les-Metz, France/法国，梅茨
Architect/建筑师: Dominique Coulon & Associés
Photos/摄影: Eugeni PONS, Guillaume WITTMANN
Built area/建筑面积: 3,400m^2
Key materials: Façade – in-situ concrete
主要材料: 立面——现浇筑混凝土

Overview

The music school is a monolithic block 100 metres long and 40 metres wide. It is sited perpendicular to the main road, projecting into the public area by 16 metres. The building is set against a forest of giant sequoias, also aligned perpendicular to the main road. The group forms a doorway marking the entrance to the town.

There is a broad forecourt area that disappears underneath the building. The public uses the monumental staircase leading to the inside courtyard and the main foyer. This is a wide area, open to the sky, treated with phosphorescent paint. In the evening it continues to glow with a strange light.

The building houses a mixed programme. It comprises premises for local teenagers, an extra-curricular centre for schoolchildren, a community hall, an auditorium, and a music school. These functions are brought together in a monolithic building. The programming complexity is managed on the inside in a single building. The juxtaposition of the combined programmes greatly enriches the building, with each entity standing out in contrast to the others.

The outside of the building reveals little of the programme on the inside – only the large bay windows allow a glimpse of the community hall. It is possible to catch sight of the ephemeral movements of the dancers. There is abundant natural light, with the highly coloured patios providing their own special light. This configuration of patios also protects the areas from disturbance from the nearby motorway.

The building is not designed merely as an elongated monolith, however. The outside curls round progressively, finally absorbing the two levels devoted to the music school. This curling adds dynamic impetus to the general outline, and the vanishing lines of the volumes seem strangely disturbed.

The materials used for the interior are precious. The main hall is in light-coloured wood, while the ceiling allows glimpses of wonderful gilded surfaces through the large cavities, which

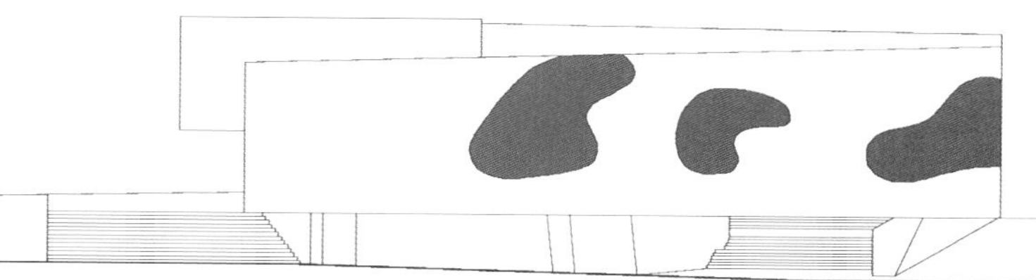

MUSIQUE
ENFANCE
SALLE
FESTIVE
LE TRAM

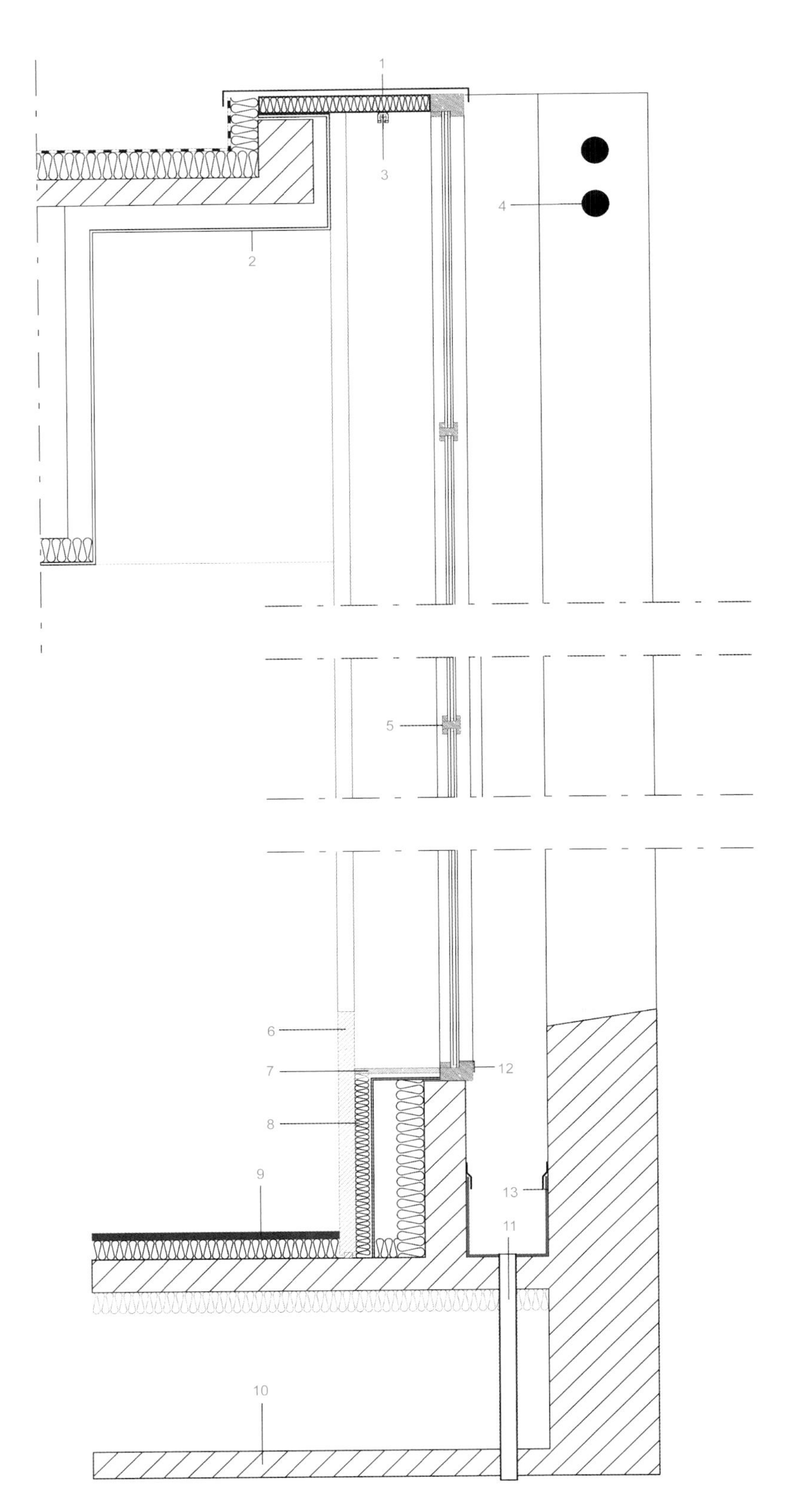

1. Steel tube for mounting on carpentry concrete parapet
2. Plasterboard
3. Curtain rail
4. Steel drawing
5. Exterior carpentry in lacquered wood
6. Laminated panel board of spruce
7. Lacquered medium shelf
8. Against-plaster wall with 50mm mineral wool
9. Hardwood timber batten
10. Architectural concrete soffit
11. Evacuation of rainwater
12. Lacquered aluminum angle to protect the low amount
13. Tightness of liquid system

1. 钢管，用于装配混凝土护墙
2. 石膏板
3. 窗帘轨道
4. 钢拉线
5. 涂漆木窗框
6. 云杉层压板
7. 涂漆中间架
8. 反石膏墙，配有50mm矿物棉
9. 硬木板条
10. 建筑混凝土底面
11. 排水
12. 涂漆铝角材
13. 液封层

gives the light a warm tinge. The auditorium is hung with tensed wires on its three sides. The walls move with the slightest breath of air, revealing their thickness. The adjustable acoustic (controlled shutters) disappears behind this elegant filter. The precious wood used for the flooring (wenge) reinforces the effect of a presentation box. The extracurricular centre for schoolchildren is monochrome; the orange colour saturates the space, and the shiny resin flooring reinforces its highly artificial aspect. The primary logic consists of implementing very marked contrasts among the different areas: contrasting materials, contrasting colours, contrasting light.

The interior and the exterior are totally dissociated, with the rustic look of the outside being the diametrical opposite of the precious interior.

Detail and Materials

The building is in reinforced concrete, cast on the spot, resting on piles. The outer casing has the rustic appearance of everyday concrete – concrete that assumes its defects.

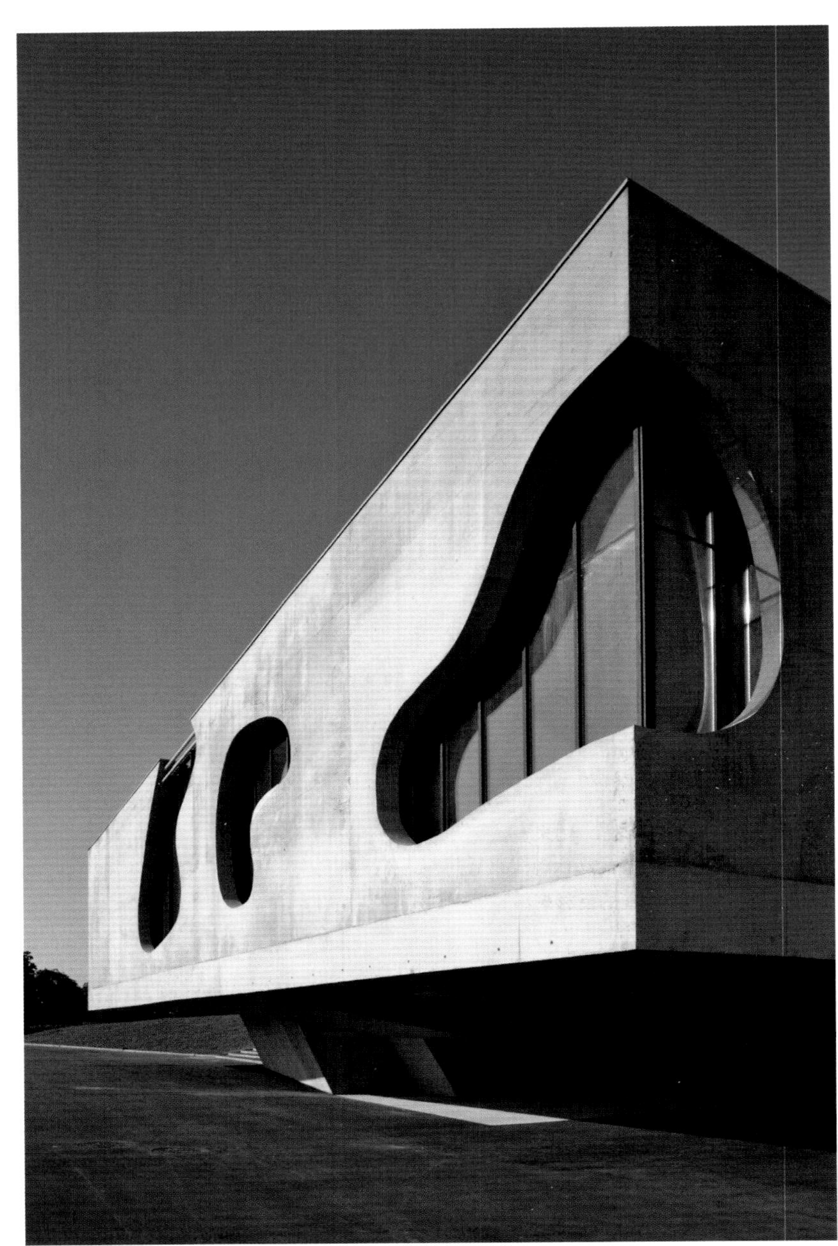

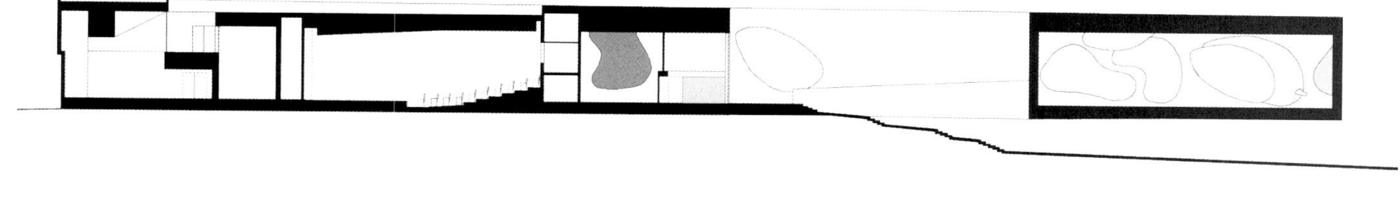

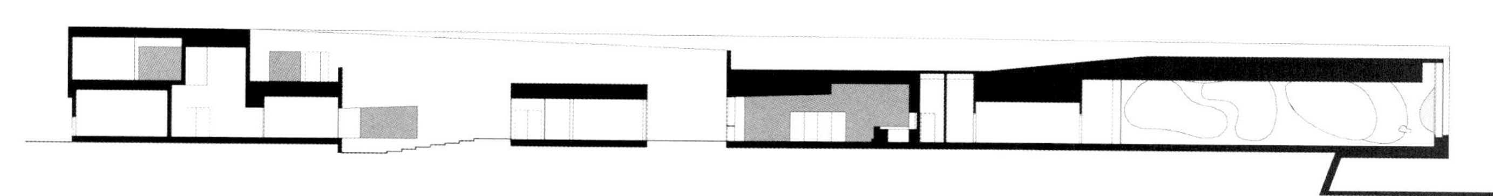

项目概况

音乐学校是一座100米长、40米宽的大型砌块建筑，它与主路垂直，有16米长的楼体融入了公共区域。建筑背靠一片同样与主路垂直的巨杉树林，二者共同标志出城镇的门户。

宽敞的前院一直插入建筑底部。公众可经由宽阔的楼梯进入内院和主门厅。这片宽敞的露天区域表面采用了磷光涂料进行处理。夜晚，它会散发出奇异的光芒。

建筑内设有多功能空间，包含本地青少年活动区、小学生课外活动中心、社区大厅、礼堂以及一所音乐学校。这些功能被整合在一座单一的建筑中，它们的交错并置极大地丰富了建筑，各个功能体之间相互形成了对比。

从建筑外观很难了解内部的功能设置，只有从宽大的飘窗才能瞥到社区大厅的一角，可以看到舞者们转瞬即逝的舞姿。室内阳光充足，而彩色天井也将室内映得五彩缤纷。天井的设置还能保护室内空间不受附近的车道影响。

建筑不仅是一座简单的长方体，建筑外墙向上卷起，形成了两层高的音乐学校。卷曲结构为建筑的整体轮廓增添了动感，将建筑的线条变得丰富多变。

室内所选用的装饰材料十分珍贵。主大厅采用浅色木材覆面，天花板上则装饰着的鎏金，流露出温馨淡雅的色调。礼堂的三面通过拉力电缆吊起，会随着微风轻轻摆动，十分有趣。可调节隔音板被隐藏在优雅的墙面背后。地板所选用的珍贵鸡翅木进一步突出了“礼盒”式的空间效果。小学生课外活动中心采用单一色调，橙色充满了整个空间，而闪亮的树脂地板则突出了人造效果。对比材料、对比色、对比光，这三者的合理运用实现了不同区域的区分。建筑的室内外空间是彻底分裂的，外观粗犷质朴，室内则细致精巧。

细部与材料

建筑采用现场浇筑的钢筋混凝土建成，底部由立柱支撑。外层的混凝土表面显得十分粗糙，呈现出各种常见的瑕疵。

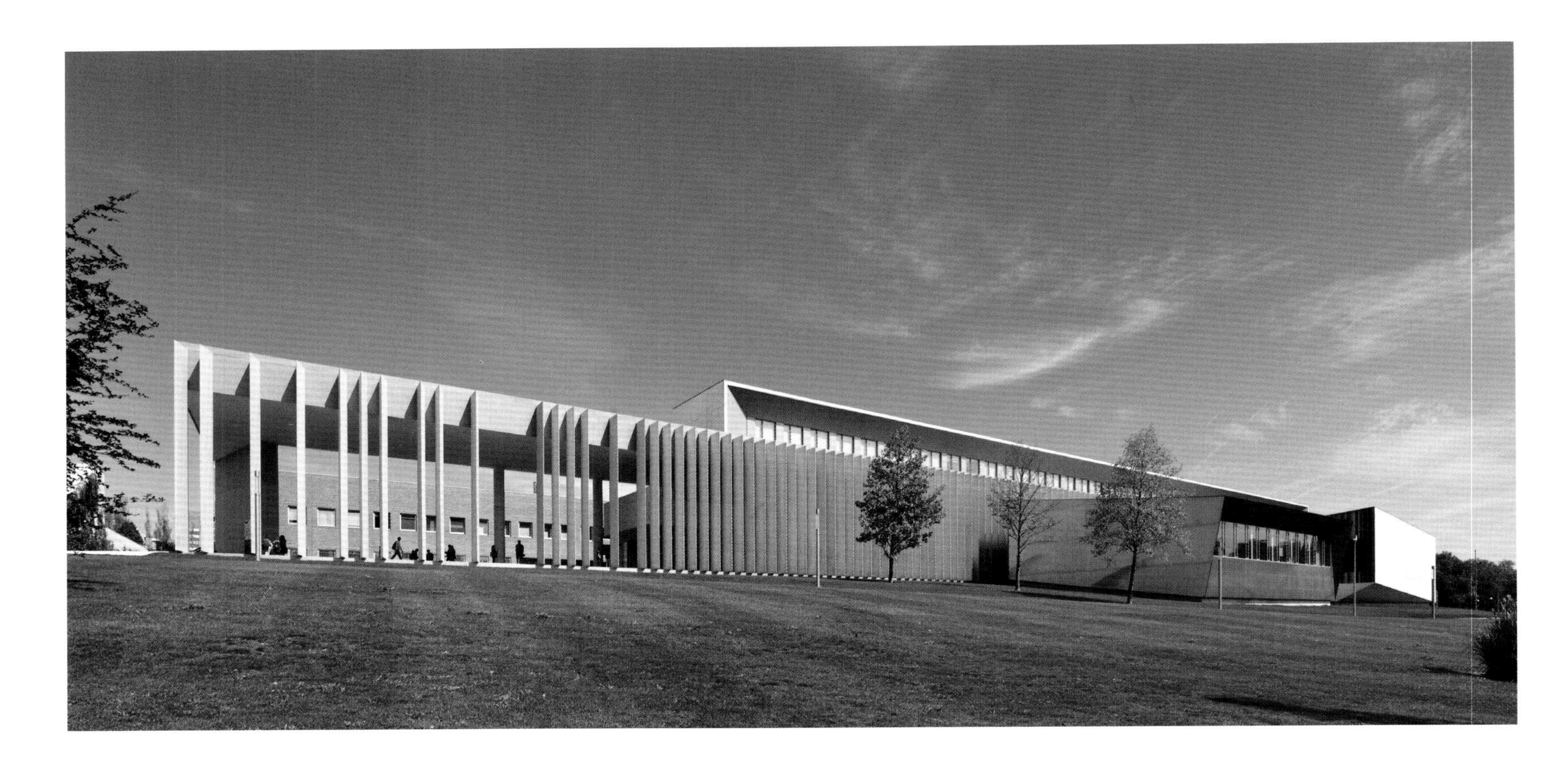

School of Economics and Business Administration

潘普洛纳经济与商务管理学院

Location/地点: Pamplona, Spain/ 西班牙，潘普洛纳
Architect/建筑师: Juan M. Otxotorena
Photos/摄影: Juan Rodríguez, Pedro Pegenaute, Rubén Pérez Bescós, José Manuel Cutillas
Built area/建筑面积: 15,529.60m^2
Completion date/竣工时间: 2012
Key materials: Façade – Concrete and Glass
主要材料: 立面——混凝土和玻璃

Overview

The new building is the definitive headquarters of the Faculty of Economy and Business Administration of the University of Navarra. It has a large number of classrooms available to increase the teaching possibilities of the University; and contains also several offices, study and seminar rooms.

The new built volume is located in the campus of the University, in Pamplona, next to the current Law building, with which it is connected by means of different accesses in the East and South façades. The close relationship between both buildings is strongly contemplated. Their academic offers are complementary and they will share students and spaces. In short, the maximum interconnection between both buildings is attempted, so they will have a single access from the outside.

The new construction appears also marked from the beginning by the new role acquired by the Postgraduate courses (Master's and Doctorate programs).

From the point of view of the shape of the building, it is necessary to highlight the desire to respect as much as possible the green meadow which goes down to the river, which is especially important in the visual and environmental heritage of the campus.

The construction is attached to the Law building, as mentioned, and it has a similar size, alignments and a new common façade. This is justified by the solution offered in access, and acquires outstanding importance in the result of its exceptional length, which is the one providing the background to the green campus.

The inside layout is based on elementary geometry that makes regular repetitions and alignments of rooms and constructive elements. The distribution of spaces turns around a large central covered patio.

The project is distributed in its largest part in the ground and first floor, which are extended in the surface occupied; these stories are spatially connected with the large central patio. The ground

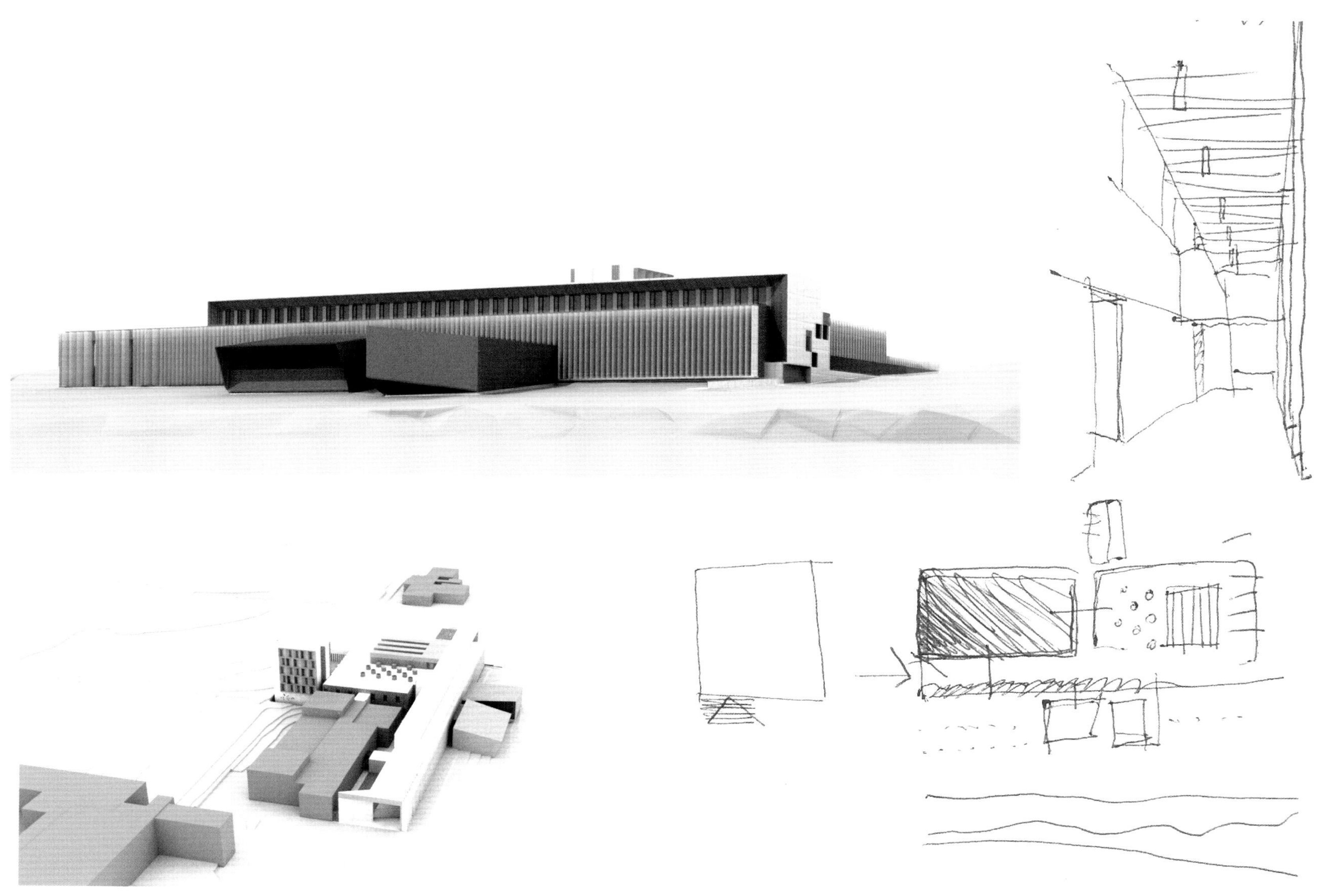

1. Waterproof mortar t = 1.5Cm
2. Brick wall t = 12cm
3. Rock wool insulation t = 6cm
4. Gypsum plasterboard 46+2x15mm
5. Profile "I" for window support
6. Galvanized steel folded sheet t = 2mm
7. Glass with air chamber 4+4/12/4+4
8. Aluminum window with thermal break
9. Geotextile sheet
10. Epdm waterproof sheet
11. Concrete foundation
12. Lean concrete t = 10cm
13. Extruded polystyrene insulation t = 4cm
14. Gravel layer t = 2cm
15. Mortar t = 3cm
16. Concrete slab
17. Galvanized steel rain guard t = 2mm
18. Clay board t = 4cm
19. Stainless steel profile "I" 100x15x2mm
20. Folded metal sheet t = 2mm
21. Alucobond substructure
22. Alucobond
23. Mortar layer t = 8cm
24. Hollowcore slab t = 25cm
25. Lintel steel t = 5mm
26. Silicone sealing joint
27. Gypsum plasterboard false ceiling t = 1.5Cm
28. Tubular profile #40.40.2Mm
29. Expanded polystyrene joint t = 2cm
30. Linear drain □16cm
31. Steel sheet t = 5mm for louvers support
32. Clay panel support
33. Clay panel t = 1.5Cm
34. Concrete wall t = 30cm
35. Gypsum plasterboard 70+2x15mm
36. Gravel layer t = 15cm
37. Brick wall t = 9cm
38. Profile heb200
39. Lacquered steel louvers
40. Concrete cantilever "in situ"
41. Insulation extruded polystyrene t = 8cm
42. Waterproof sheet
43. Geotextile sheet
44. Black acoustic insulation
45. Mortar layer t = 5cm
46. Hollowcore slab t = 50cm
47. Mortar layer t = 5cm
48. Double glass with middle layer for images
49. Marbel t = 3cm
50. Glue cement t = 1cm
51. Air chamber t = 50cm
52. Waterproof paint
53. Fiberglass mortar

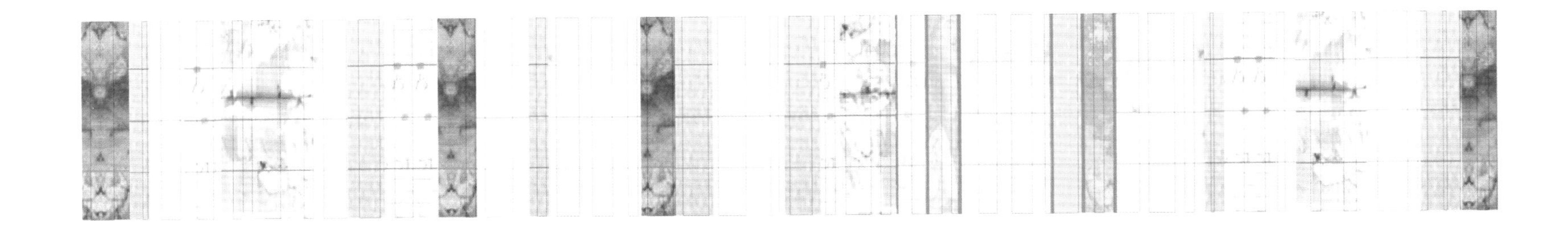

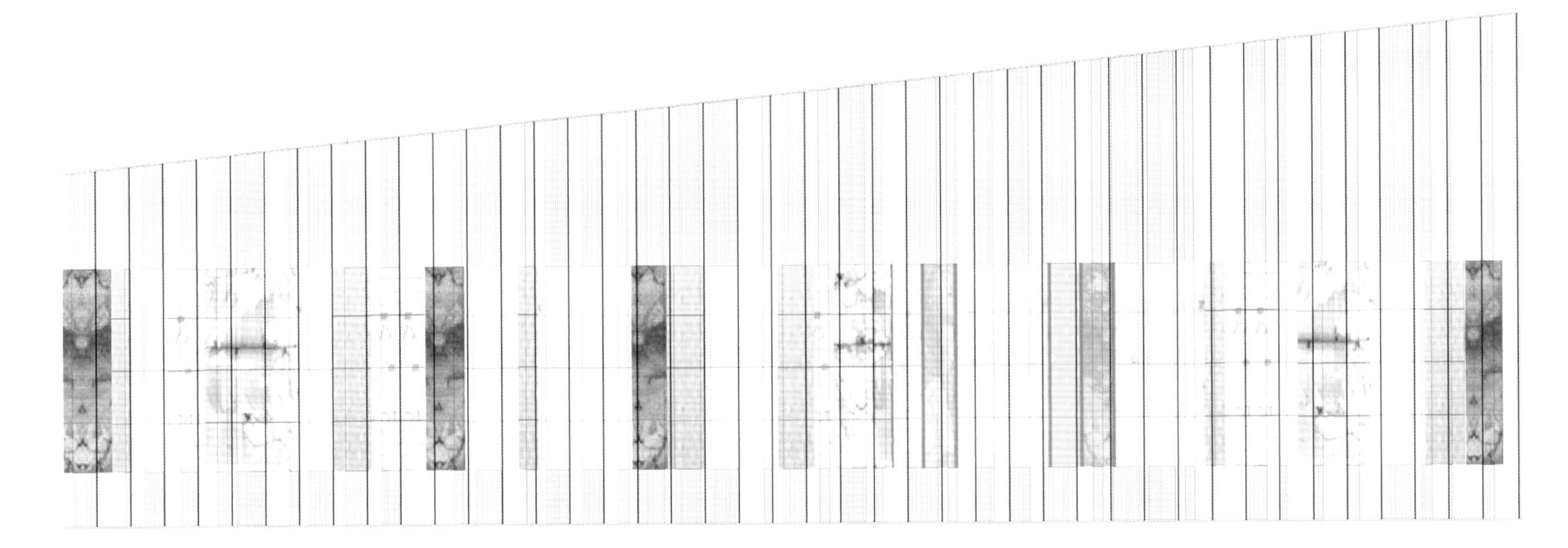

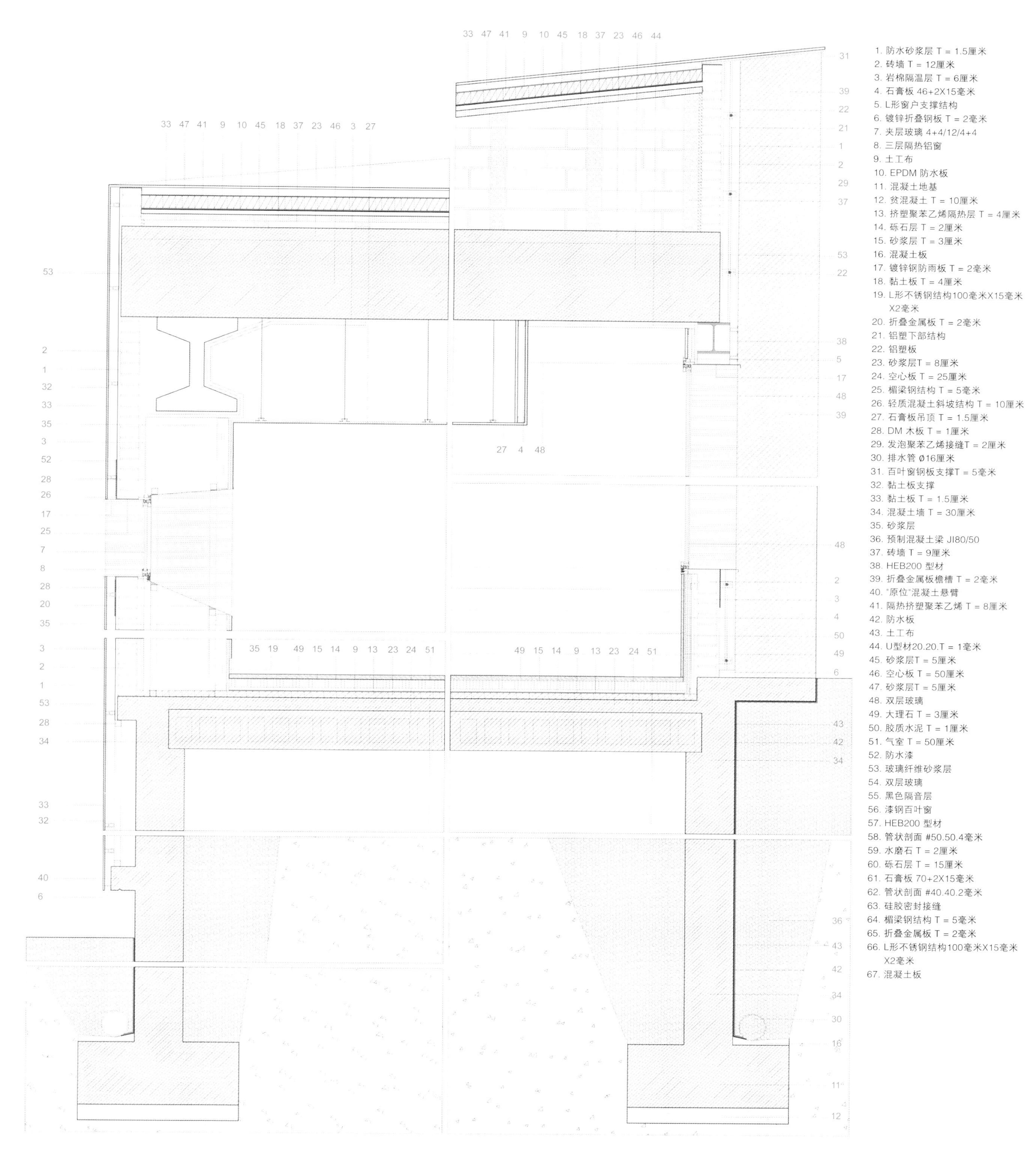
1. 防水砂浆层 T = 1.5厘米
2. 砖墙 T = 12厘米
3. 岩棉隔温层 T = 6厘米
4. 石膏板 46+2X15毫米
5. L形窗户支撑结构
6. 镀锌折叠钢板 T = 2毫米
7. 夹层玻璃 4+4/12/4+4
8. 三层隔热铝窗
9. 土工布
10. EPDM 防水板
11. 混凝土地基
12. 贫混凝土 T = 10厘米
13. 挤塑聚苯乙烯隔热层 T = 4厘米
14. 砾石层 T = 2厘米
15. 砂浆层 T = 3厘米
16. 混凝土板
17. 镀锌钢防雨板 T = 2毫米
18. 黏土板 T = 4厘米
19. L形不锈钢结构100毫米X15毫米X2毫米
20. 折叠金属板 T = 2毫米
21. 铝塑下部结构
22. 铝塑板
23. 砂浆层T = 8厘米
24. 空心板 T = 25厘米
25. 楣梁钢结构 T = 5毫米
26. 轻质混凝土斜坡结构 T = 10厘米
27. 石膏板吊顶 T = 1.5厘米
28. DM 木板 T = 1厘米
29. 发泡聚苯乙烯接缝T = 2厘米
30. 排水管 ø16厘米
31. 百叶窗钢板支撑T = 5毫米
32. 黏土板支撑
33. 黏土板 T = 1.5厘米
34. 混凝土墙 T = 30厘米
35. 砂浆层
36. 预制混凝土梁 JI80/50
37. 砖墙 T = 9厘米
38. HEB200 型材
39. 折叠金属板檐槽 T = 2毫米
40. "原位"混凝土悬臂
41. 隔热挤塑聚苯乙烯 T = 8厘米
42. 防水板
43. 土工布
44. U型材20.20.T = 1毫米
45. 砂浆层T = 5厘米
46. 空心板 T = 50厘米
47. 砂浆层T = 5厘米
48. 双层玻璃
49. 大理石 T = 3厘米
50. 胶质水泥 T = 1厘米
51. 气室 T = 50厘米
52. 防水漆
53. 玻璃纤维砂浆层
54. 双层玻璃
55. 黑色隔音层
56. 漆钢百叶窗
57. HEB200 型材
58. 管状剖面 #50.50.4毫米
59. 水磨石 T = 2厘米
60. 砾石层 T = 15厘米
61. 石膏板 70+2X15毫米
62. 管状剖面 #40.40.2毫米
63. 硅胶密封接缝
64. 楣梁钢结构 T = 5毫米
65. 折叠金属板 T = 2毫米
66. L形不锈钢结构100毫米X15毫米X2毫米
67. 混凝土板

floor, which is the one used for access is the highest in certain points. The main volume is crowned with a row of offices for the professors on the top floor. The volume is completed with a series of different bodies amongst which we could highlight those corresponding to the deanery, chapel and 'tower' of offices.

Detail and Materials

It is designed as a façade with a serial and rhythmic structure coherent with its scale. It is made up by a system of vertical concrete elements able to act as a filtering device of space and views, and as a parasol.

The other façades are open, with a discontinuous glass outside wall. This presents a paused visual expression by its vertical structural elements: it is projected to the exterior in order to ensure its protection against the sun and improve geometrical and clear spaces. The option for the starring role of the bare concrete, which is extended all over the building, is justified by obvious reasons of consistency, stability and solidity, and links with the neighboring constructions. The external image of the volume is supplemented by the use of metal, opaque or more or less permeable cladding, in whose design technical precision is sought, as well as expressive contrast and visual quality.

项目概况

本项目是潘普洛纳大学经济与商务管理学院的办公教学楼，容纳许多教室，方便教学活动的开展，同时还包含多个办公室、自习室和研讨室。

新工程位于潘普洛纳的校园内部，紧挨法学院大楼，二者由东墙和南墙上的不同通道连接。设计师着重考虑了两个建筑的密切关联，由于两个学院的教学、研究内容互补，可以实现学生和空间共享。简而言之，设计方案意图实现两个建筑之间最大程度的连通，共享一个出入口。

设计之初，新工程就从研究生（硕士和博士）课程中获得了灵感。建筑外形方面，设计师认为有必要突出绿油油的草坪元素。它无论在视觉还是环境层面都对校园有着特别的价值。

正如前面提到的，新建筑将与法学院相连，采用相近的建筑规格、排列以及共同的全新外墙设计。入口设计也与之配合，与超长的楼体结构对应。

建筑内部布局以简单的几何形状为基础，对各个房间和建筑元素进行规则的重复和排列。空间分布围绕一个中庭展开。

功能区主要分布在一楼和二楼，这些区域都与中庭相连。入口所在的一楼有几处的举架最高。建筑的主结构顶部是一排办公室，供教授们使用。建筑的几个补充结构分别作为院长室、礼堂和办公“塔楼”使用。

细部与材料

外墙结构由一系列有规律的结构单元连贯而成，其中垂直混凝土元素作为空间和视角的过滤设备，也可以起到遮阳的作用。

其他的几面外墙采用了开放式设计以及不连续的玻璃外墙。垂直结构在此呈现出一种暂停的视觉表现手法：影子投射在外墙上，遮挡阳光，并且提高空间的几何感。

建筑的各个部分都使用到了裸露混凝土，这是考虑到材料一致性、稳定性、坚固性的特点，与相邻建筑很好地融合。建筑外部还使用了金属质地的，基本不透明的覆盖材料。这一设计对技术精度有很高的要求，同时突出材料对比，强调视觉质感。

01. Brick wall t = 12cm
02. Rock wool insulation t = 6cm
03. Gravel layer
04. Geotextile layer
05. Epdm waterproof sheet
06. Extruded polystyrene insulation t = 4cm
07. Mortar t = 3cm
08. Terrazzo t = 2cm
09. Galvanized steel rain guard t = 2mm
10. Prefabricated concrete lintel
11. Clay board t = 4cm
12. Formation of ramps with light concrete t = 10cm
13. Skirting in stainless steel profile "l" 100x15x2mm
14. Tubular profile #200.700.15Mm
15. Aluminum window with thermal break and triple air chamber series altres ae0941 lacquered ral9006
16. Glass with air chamber 4+4/12/4+4
17. Expanded polystyrene joint t = 2cm
18. Rolling shutter support
19. Gypsum plasterboard 70+2x15mm
20. Aluminum rolling shutter
21. Clay panel t = 1.5Cm
22. Reinforced concrete slab 30cm
23. Cement mortar layer
24. Reinforced concrete wall t = 30cm
25. Chamfer
26. Waterproof paint
27. Mortar with concrete sand layer t = 2cm with electrowelded steel lath 20.30.4
28. Extruded polystyrene insulation t = 8cm

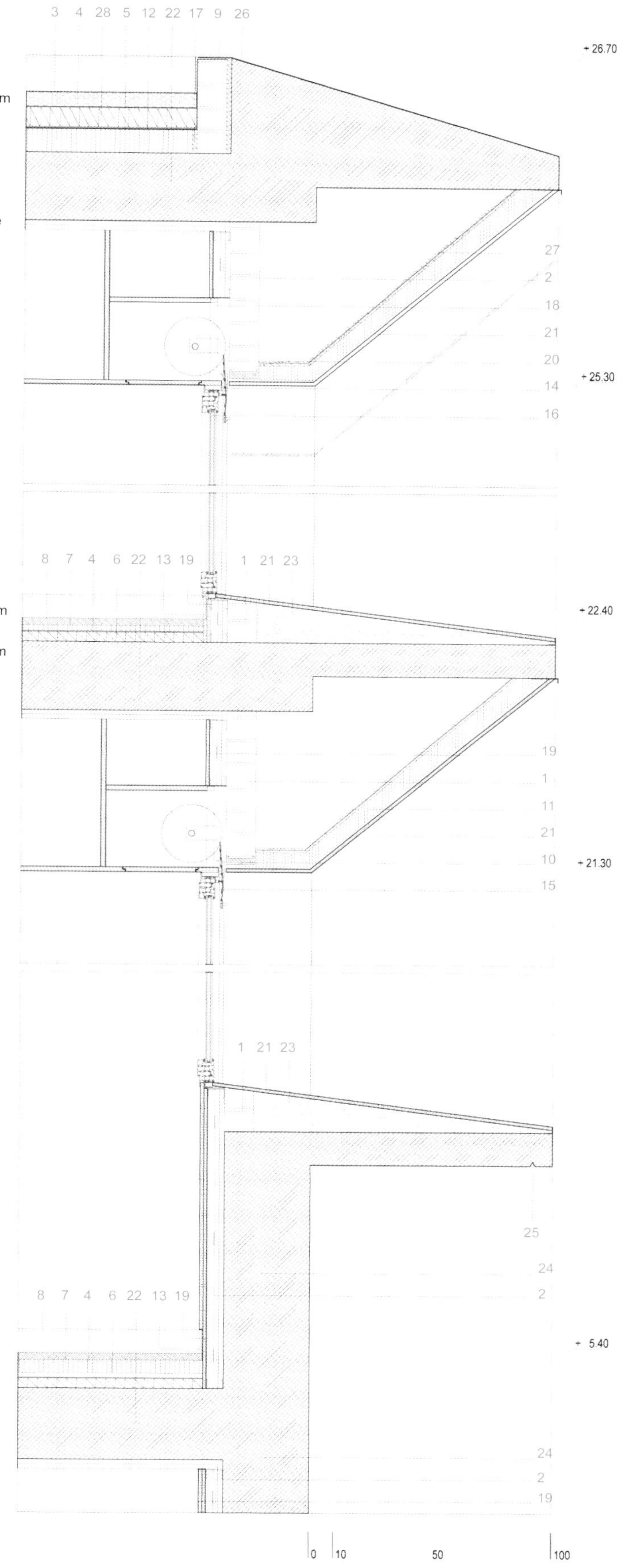

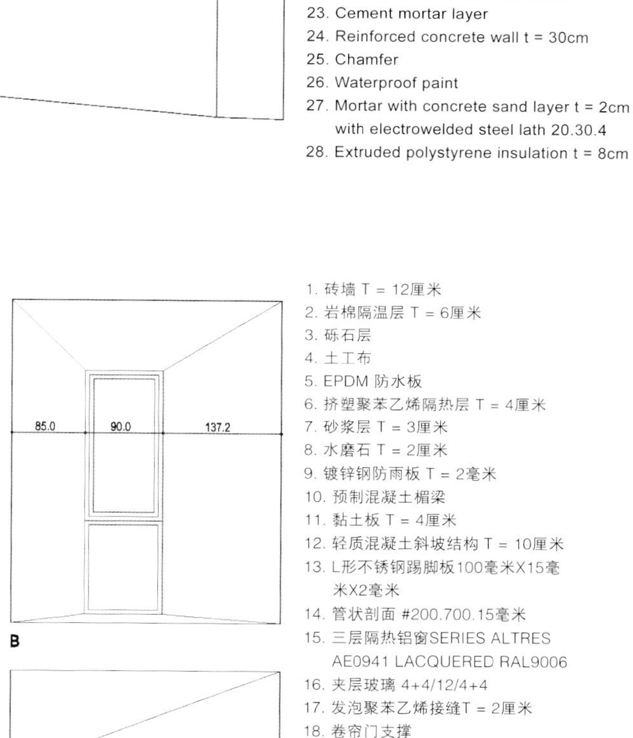

1. 砖墙 T = 12厘米
2. 岩棉隔温层 T = 6厘米
3. 砾石层
4. 土工布
5. EPDM 防水板
6. 挤塑聚苯乙烯隔热层 T = 4厘米
7. 砂浆层 T = 3厘米
8. 水磨石 T = 2厘米
9. 镀锌钢防雨板 T = 2毫米
10. 预制混凝土楣梁
11. 黏土板 T = 4厘米
12. 轻质混凝土斜坡结构 T = 10厘米
13. L形不锈钢踢脚板100毫米X15毫米X2毫米
14. 管状剖面 #200.700.15毫米
15. 三层隔热铝窗SERIES ALTRES AE0941 LACQUERED RAL9006
16. 夹层玻璃 4+4/12/4+4
17. 发泡聚苯乙烯接缝T = 2厘米
18. 卷帘门支撑
19. 石膏板 70+2X15毫米
20. 铝质卷帘门
21. 黏土板 T = 1.5厘米
22. 钢筋混凝土板 30厘米
23. 混凝土砂浆层
24. 钢筋混凝土墙 T = 30厘米
25. 倒角
26. 防水漆
27. 混凝土砂浆层 T = 2厘米，包含钢质板条 20.30.4
28. 挤塑聚苯乙烯隔热层 T = 8厘米

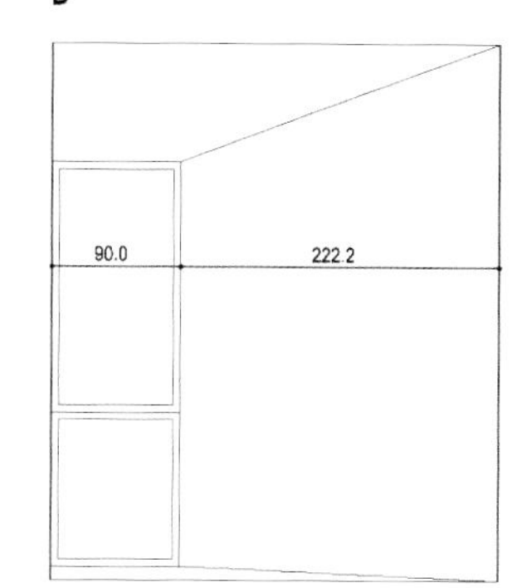

Caneças High School

加奈萨斯高中

Location/地点: Caneças, Portugal/葡萄牙，加奈萨斯
Architect/建筑师: ARX PORTUGAL, Arquitectos Lda. Nuno Mateus and José Mateus
Built area/建筑面积: 32,600m^2
Completion date/竣工时间: 2013
Key materials: Façade – concrete
主要材料： 立面——混凝土

Overview

The existent school, that is object of renovation and extension, is located in the outskirts of Caneças, Odivelas, in a territory of intense discontinuities. There the architects found a decadent cluster made up of six square pavilions, disconnected amongst themselves, that conveyed the idea of a fragmented school organism.

The proposal is structured by a double interpretation of the learning concept: formal learning and informal learning. Those two dimensions are translated in the building by two different architectural approaches, and yet maintaining a dialogue between each other. As a result, the stiffness of the existent blocks, where the class rooms are placed and arranged like efficient "learning machines", contrasts with the informal nature of the new built structures, enabling the "informal learning" in these collective spaces. This new intervention shapes itself in order to embrace the old volumes, which were previously segregated, and bring them all together through the new proposed spaces of learning.

Considering that in a school, every space is a teaching environment, each one with its own importance, the organisation and articulation between all spaces is meant to be fluid, with physical and visual permeability, allowing a more spontaneous and creative appropriation, leading to the will to learn through space. Human relations and activities are, in the end, in the base of all knowledge. From a tectonic point of view, the solutions adopted give the building an idea of matter unity and grant the space an elementary and abstract character.

Detail and Materials

Concrete is a common, inexpensive and lasting material. The architects intended to propose a protective perimeter for the school environment, where smaller volumes are finished in white painted plaster, a material somehow softer.

Like any exposed concrete wall, the rigour of the pattern of the formwork panels is very important in the final result. This is more so when the budget is very tight, as in this case.

escola

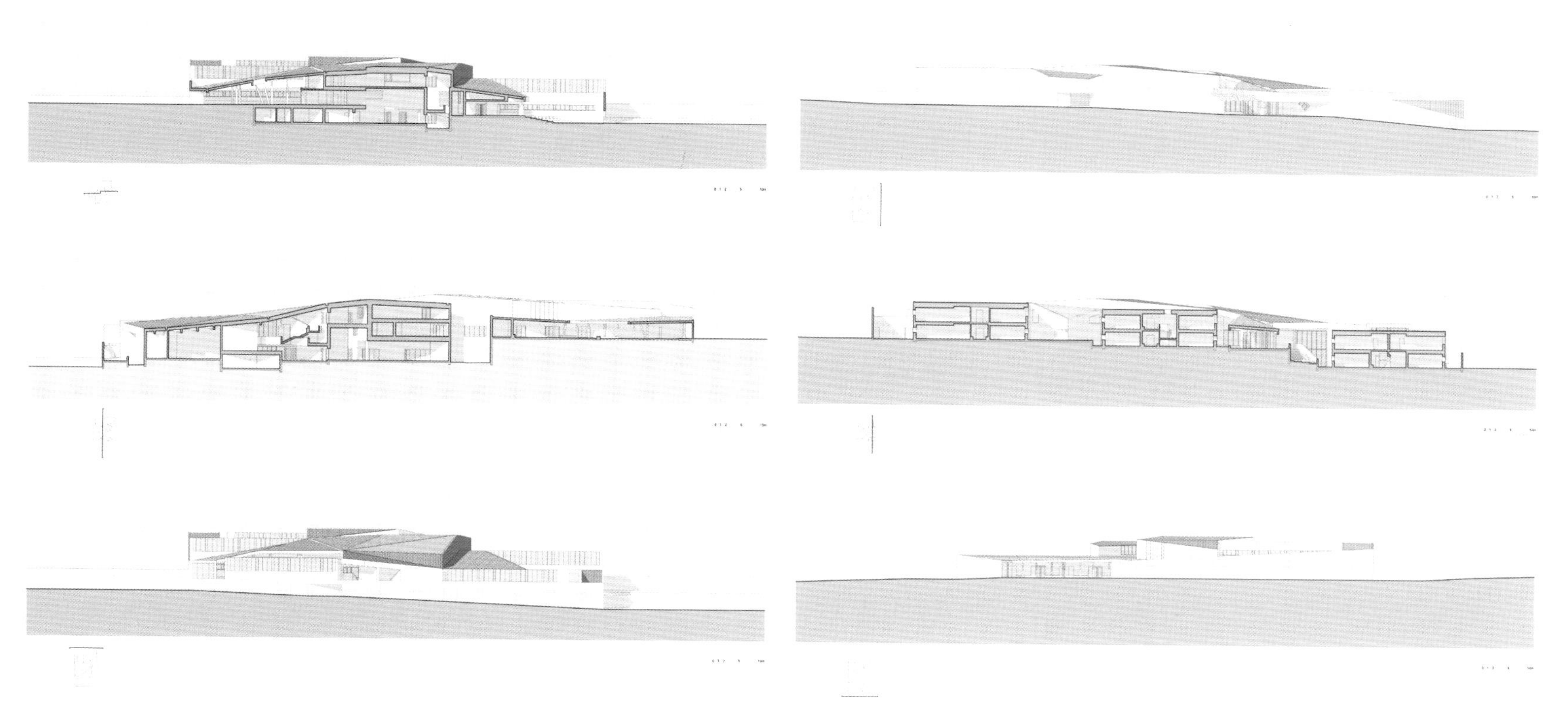

项目概况

进行翻修和扩建的学校位于加奈萨斯市的城郊，原有建筑的连续性较差。学校由六座方块建筑所组成的建筑群构成，它们的位置十分分散，整个学校空间呈现为碎片化。

项目设计以学习概念的两种不同诠释方法为基础：正式学习和非正式学习。这两种方法被转化为两种不同的建筑设计形式，但是二者之间又保持了良好的对话。在最终完成的设计中，原有建筑的僵硬感（教室像高效的"学习机器"一样布置）与新建结构非正式的自然特征（在集体空间实现了多样化的"非正式学习"）形成了有趣的对比。新建结构通过造型将旧结构包围起来，通过新式学习空间使它们从分离的状态转化为整体的状态。

在学校里，每个空间都是教学空间，都有其独特的重要价值。因此，无论是在物质上还是视觉上，空间之间的组织和连接必须要流畅，实现更自然而创新的空间设

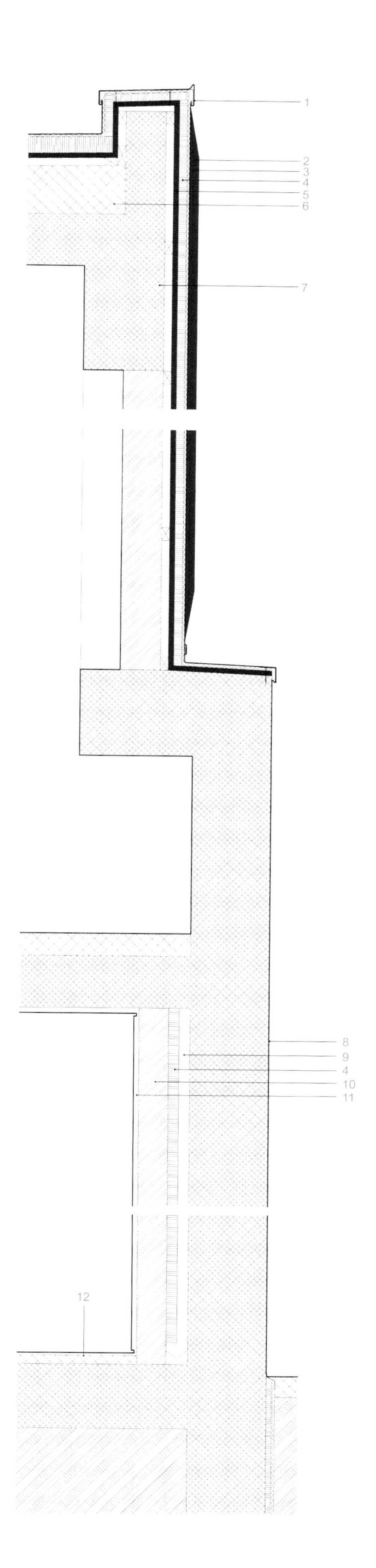

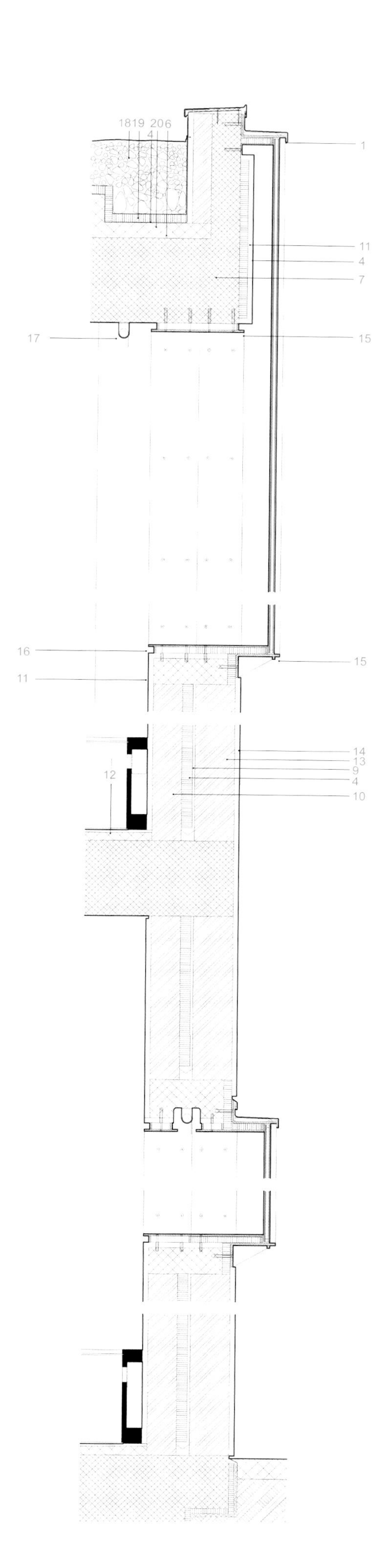

1. Capping natural zinc plate
2. Natural zinc plate
3. Breather membrane
4. Extruded polystyrene
5. Boarding structure
6. Gradient forming concrete
7. Reinforced concrete
8. Exposed concrete
9. Air cavity
10. Brick 11cm
11. Painted stucco
12. Hydraulic mosaic
13. Brick 15cm
14. Acrylic paint
15. Slide post and lintel 8mm thick steel plate, painted
16. Parapet in 8mm thick steel plate
17. Microperforated PVC canvas roll
18. Gravel
19. Geotextile
20. Waterproofing asphalt membrane

1. 天然锌板顶盖
2. 天然锌板
3. 透气膜
4. 挤塑聚苯乙烯
5. 木板结构
6. 渐变成型混凝土
7. 钢筋混凝土
8. 素面混凝土
9. 空气腔
10. 砖11cm
11. 粉饰灰泥
12. 液压马赛克
13. 砖15cm
14. 丙烯酸涂料
15. 滑动门套柱和门梁，8mm厚涂漆钢板
16. 8mm钢板护栏
17. 微孔PVC卷帘遮阳
18. 碎石层
19. 土工布
20. 防水沥青膜

计，让学生随时随地都能学习。毕竟，人际关系和活动是所有知识的基础。从建筑层面上来说，设计所采用的方案让建筑更加统一，赋予了空间简洁抽象的个性。

细部与材料

混凝土是一种常见且物美价廉的材料，耐久性很强。建筑师决定为学校环境设计一个具有保护性的外围结构，较小的空间全部采用白色石膏涂层装饰，显得更为柔和。

与其他素面混凝土墙一样，模架板图案的精准性对最终的建造成果来说十分重要，尤其是在项目的预算十分紧张的情况下。

Biokilab Laboratories
百奇拉伯实验基地

Location/地点: Miñano Mayor, Spain/西班牙，米尼亚诺
Architect/建筑师: Taller Básico de Arquitectura
Photos/摄影: José Manuel Cutillas
Construction area/建筑面积: 342.8m^2
Completion date/竣工时间: 2010
Key materials: Façade – concrete
主要材料: 立面——混凝土

Overview
The technologic Park of Vitoria colonises a little bit of nature. Two hollow boxes of concrete inhabit this new place on the structure. The whole complex in a permanent flight reveals a new gravity.

The metallic structure that raises the boxes in the air is a quadruped structure. Its two horizontal elements form a cross inscribed in the square floor of the boxes. The sides of these floors measure twelve and thirteen metres respectively. The horizontal beams where the boxes rest avoid any interlocking. Consequently, the structure is visible in its entirety. The ends of the beams join vertical elements, which become the legs of this quadruped anatomy. Legs are as wide as beams, managing a continuity that makes all the pieces be understood as a unique element. Different lengths of the legs let the slope remain unaltered.

The box is thought as a second structure that replaces walls with beams and roofs with double slabs. The vertical faces of the box are beams as high as the box. These wall-beams have only one hole, defined by the maximum dimensions that let the beams work properly. Outside, the concrete structure is visible on all faces of the box. Inside, plasterboards cover the structure. The window frame, drawn as a single line, stays hidden between both sheets. The gap between sheets, both in walls and slabs, contains all building systems, as plumbing, electricity, voice and data. This net of systems solves the flexibility needed by the laboratory for its continuous transformation.

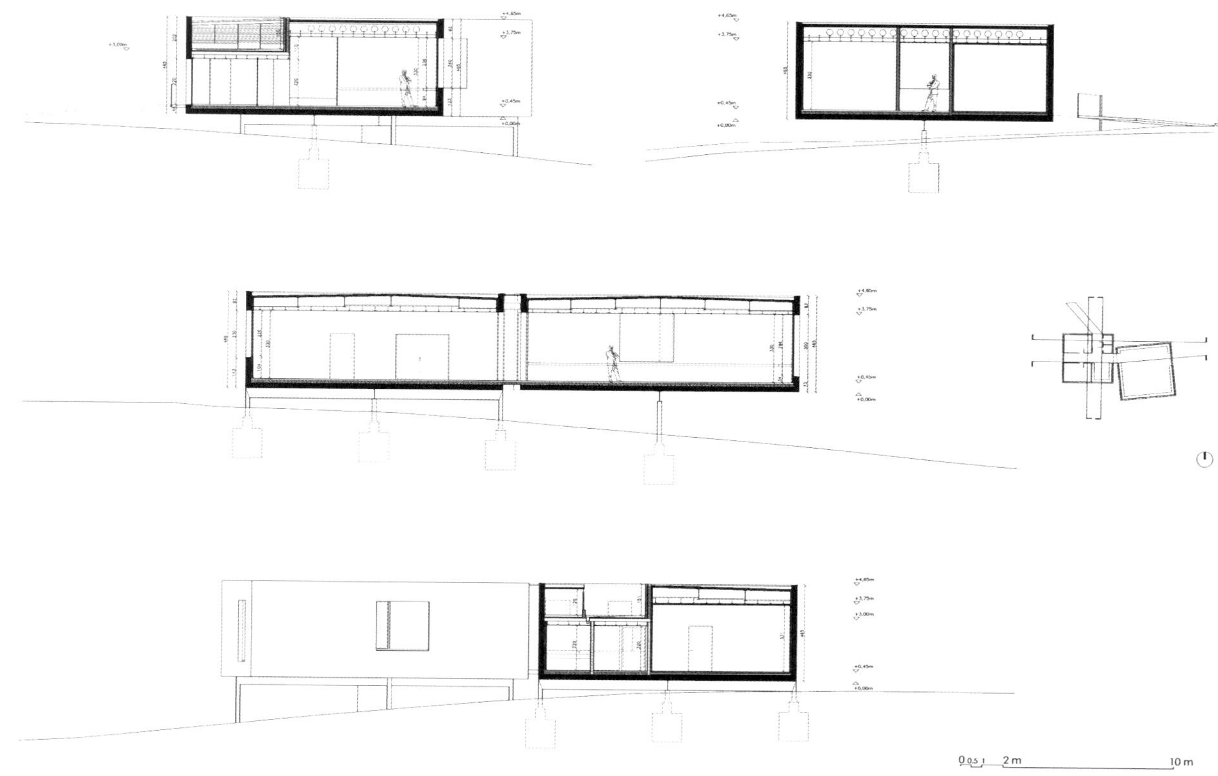

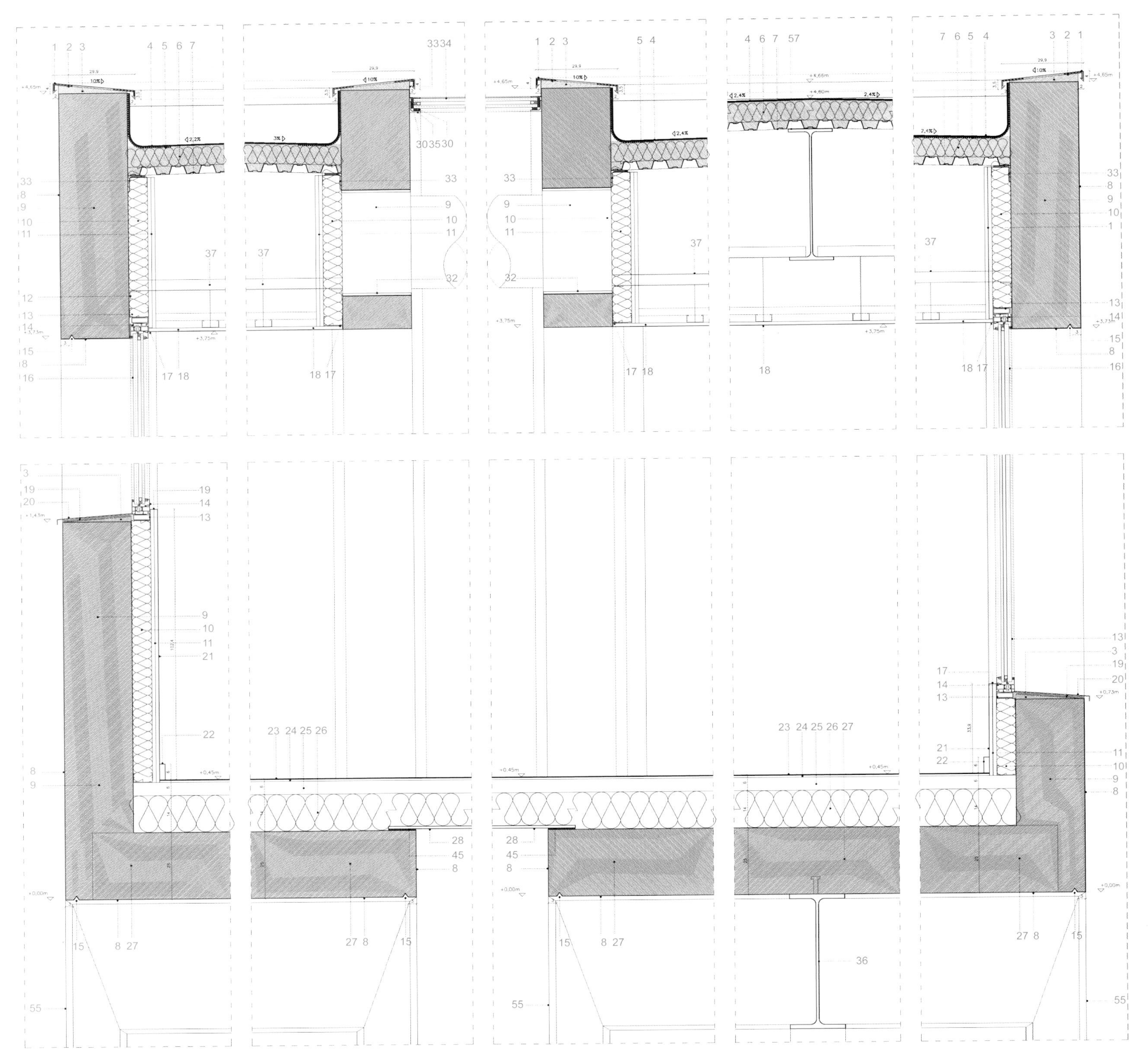

1. Coping in thermo lacquered galvanised steel folded sheet metal 2mm THK (Thickness). RAL 9006
2. Galvanised steel folded sheet metal 3mm THK fixed to wall
3. Floor mortar bed for slope
4. PVC waterproof membrane
5. PVC waterproof membrane reinforcement
6. High Density XPS Insulation 8cm THK (Thickness)
7. Steel decking slab 1mm THK PL 30/209
8. Anti-graffiti colourless acrylic finished
9. Exposed reinforced concrete wall 25cm THK
10. High Density XPS Insulation 7cm THK
11. Double plasterboard 13+13mm THK fixed to galvanised steel hidden structure
12. Steel profile as suspended celling steel frame
13. Galvanised steel frame
14. Aluminium windows frame with thermal break
15. Drip edge
16. Double glazing (8+8) /12/(5+5)
17. Thermo lacquered galvanised steel L Profile 2mm THK. RAL 9010
18. MF Suspended ceiling complete with 15mm painted plasterboard in with colour
19. SPF insulation reinforced 1cm THK
20. Aluminium gutter
21. Plastic paint satin finish RAL 9010
22. Pinewood skirting board 2x6cm painted in with colour
23. Vinyl flooring
24. Self-levelling mortar bed 1 THK
25. Concrete reinforced bed 5cm THK
26. Installation cavity
27. Exposed concrete reinforced slab 25cm THK
28. Thermo lacquered galvanised steel sheet metal 1mm THK (Thickness). RAL 9011
29. EPOXY paint
30. Neoprene elastic joint
31. INOX steel feedthroughs
32. Galvannized steel feedthroughs 10 mm THK
33. Steel profile L profile 60.5
34. Sanitation facility
35. Thermo lacquered galvanised steel profile L40.5. RAL 9006
36. Lacquered and galvanised steel profile beam IPE 500
37. Iron frame ACERALIA #40.4 for suspended ceiling substructure
38. Outdoor plastic paint satin finish RAL 9010
39. Plastering cement mortar screed
40. Ceramic brick wall 7cm
41. EPS thermal insulation 50mm

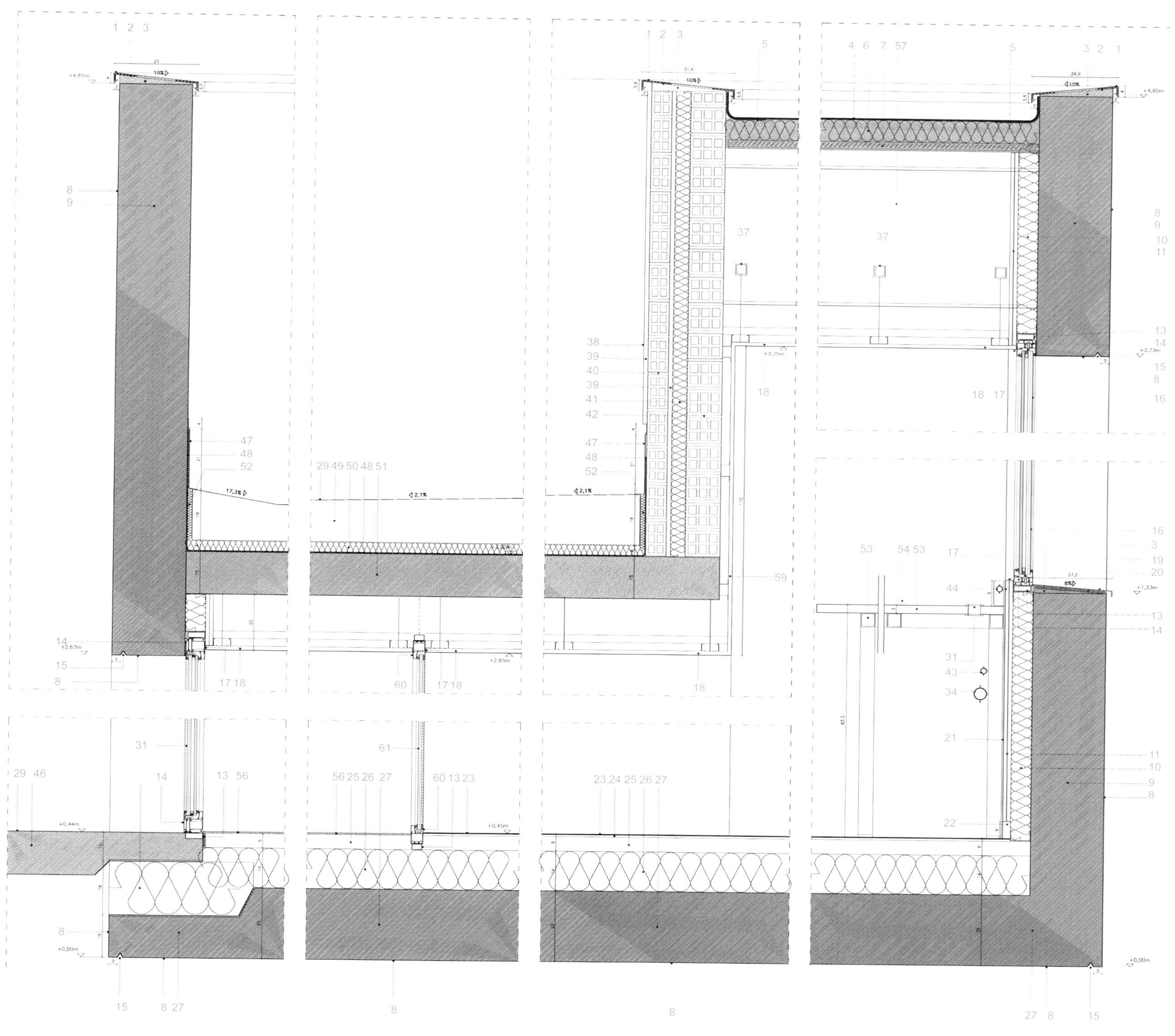

42. Ceramic brick wall 12cm THK
43. Plumbing installation
44. Electrical installation
45. Elastic seal
46. Concrete reinforced ramp 15cm THK, polished finished
47. Galvanised steel folded sheet metal 3mm
48. Rubber waterproofing layer
49. Reinforced concrete slab 10cm
50. Cork insulation 4 cm
51. Reinforced concrete slab 10cm
52. Cork seal
53. Structure for sustain worktop
54. (MDF) Medium density fibreboard worktop with finished in plastic laminated
55. Paint protection for metallic structure
56. Aluminium foil doormat
57. Steel profile beam IPE 330
58. Laminated glazing (5+5)
59. Plasterboard 13mm fixed to galvanised steel hidden structure
60. Aluminium windows frame

1. 涂漆镀锌折叠钢板顶盖，2mm厚，RAL 9006
2. 镀锌折叠钢板，3mm厚，固定在墙上
3. 坡面砂浆层
4. PVC防水膜
5. 加强PVC防水膜
6. 高密度XPS隔热层，8cm厚
7. 钢板层，1mm厚，PL 30/209
8. 防涂鸦无色丙烯酸饰面
9. 清水钢筋混凝土墙，25cm厚
10. 高密度XPS隔热层，25cm厚
11. 双层石膏板，13+13mm厚，固定在镀锌钢隐式结构上
12. 钢型材，作为吊顶钢架
13. 镀锌钢架
14. 断热铝窗框
15. 滴水檐
16. 双层玻璃（8+8）/12/（5+5）
17. 涂漆镀锌L形钢，2mm厚，RAL 9010
18. MF吊顶，配有15mm涂漆石膏板
19. SPF加强隔热层，1cm厚
20. 铝槽
21. 塑性涂料光面装饰，RAL 9010
22. 松木壁脚板2x6cm，涂有彩漆
23. 乙烯基地面
24. 自平砂浆层，1厚
25. 加固混凝土层，5cm厚
26. 隔热腔
27. 清水钢筋混凝土板，25cm厚
28. 涂漆镀锌钢板，1mm厚，RAL 9011
29. EPOXY涂料
30. 氯丁橡胶弹性垫圈
31. INOX钢穿通件
32. 镀锌钢穿通件，10mm厚
33. L形钢材60.5
34. 卫生设施
35. 涂漆镀锌L形钢材40.5，RAL 9006
36. 涂漆镀锌钢型材，重蚊木500
37. 铁架ACERALIA #40.4，作为吊顶下层结构
38. 室外塑性涂料光面装饰，RAL 9010
39. 石膏水泥砂浆层
40. 瓷砖墙7cm
41. EPS隔热层，50mm
42. 瓷砖墙，12cm厚
43. 管道装置
44. 电气装置
45. 弹性密封
46. 钢筋混凝土坡道，15cm厚，光面
47. 镀锌折叠钢板，3mm
48. 橡胶防水层
49. 钢筋混凝土板，10cm
50. 软木隔热层，4cm
51. 钢筋混凝土板，10cm
52. 软木密封
53. 工作台支撑结构
54. 中密度纤维板工作台，层压塑料饰面
55. 涂层保护金属结构
56. 铝箔门垫
57. 钢型材，重蚊木330
58. 夹层玻璃（5+5）
59. 石膏板13mm，固定在镀锌钢隐式结构上
60. 铝窗框

Detail and Materials

The façade is considered not only by their side walls, but the lower part of the two volumes that set in a wire structure. During the construction process, a special attention was considered to the continuity of the façade that involved the side walls and the lower part of the building. The formwork was built with metal panels in such a way that these elements could be used both for the façade and the slabs, keeping always the continuity in the cutting lines.

The two square boxes are composed of several layers that can be seen between them. The outer layer is used as structure. There are four deep main beams that support each square-box. The bay windows have been opened where the structural efforts allow them to be done. Behind this thick layer, there are some others of different qualities that give insulation and a comfortable finishing.

项目概况

这个实验基地项目坐落在维多利亚公园的山坡处，所以建造过程中如何将这两个悬空且相连接的白色盒状实验室与当地的地形地貌相结合是建造过程中的关键所在。

两个相连的白色盒状建筑为金属结构，通过四根柱子架空起来。两个水平结构相互交叉，形成了十字造型。楼板的侧面分别为12米和13米长。支撑盒状结构的水平梁没有任何交叉。因此，整个结构都裸露在外。横梁的两端连接垂直构件，即整个建筑结构的四根支柱。支柱与横梁同宽，实现了整个项目的统一感。不同长度的支柱让建筑结构在斜坡上也能保持水平。

作为次级结构，盒状结构用横梁和双层平板屋顶替代了墙壁。盒状结构的垂直面是与结构等高的横梁。这些墙面梁上只有一个孔，为保证横梁正常工作的最大尺寸。外部混凝土结构裸露在外。内部结构由石膏板覆盖。线条简洁的窗框隐藏在两层板材之间。板材（外墙和面板）之间的缝隙内是所有建造系统，包括水管、电线、隔音和数据设备。这个系统网络解决了实验室空间设计所需灵活性，便于未来的改造。

细部与材料

建筑立面不仅是侧墙，还包括两个建筑结构在网状结构里的底面。在施工过程中，项目十分注重侧墙和底面的统一性。建筑的模架有金属板制成，所有构件均可以被立面和底板所使用，保持了切割线的连续感。

两个正方形盒子由若干个层次组合而成。外层是结构框架。每个盒子由四根主梁支撑。在结构施工完成后，建筑师才在墙面上打开窗口。在这层厚层后面设置着一些其他的设施，为建筑提供了保温隔热和舒适的饰面。

Horse Museum

马博物馆

Location/地点: Jeju-do, Korea/韩国，济州岛
Architect/建筑师: Jegong Architects
Photos/摄影: Yoon Joon Hwan
Site area/占地面积: 5,050.00m^2
Built area/建筑面积: 569.88m^2
Completion date/竣工时间: 2012
Key materials: Façade – exposed concrete
主要材料： 立面——清水混凝土

Overview
This low-budget museum is located at small village called Gasiri in Jeju island, the volcanic island in Korea. The villagers have initiated various cultural enterprise to boost local economy of the region and the horse museum is one of their main works whose goal is to attracts tourists.

A doughnut shaped gallery space is elevated on the zigzagged walls of the ground floor in the vast prairie, which allows space underneath for rest, not only for the visitors but also for the workers of the farm. The doughnut shape was inspired by small volcanoes in the region, called Oreum, which have hollowed spaces on their summits. On the roof of the museum, visitors can enjoy a beautiful panoramic view of Jeju Island, just like they do on top of the volcanic mountains.

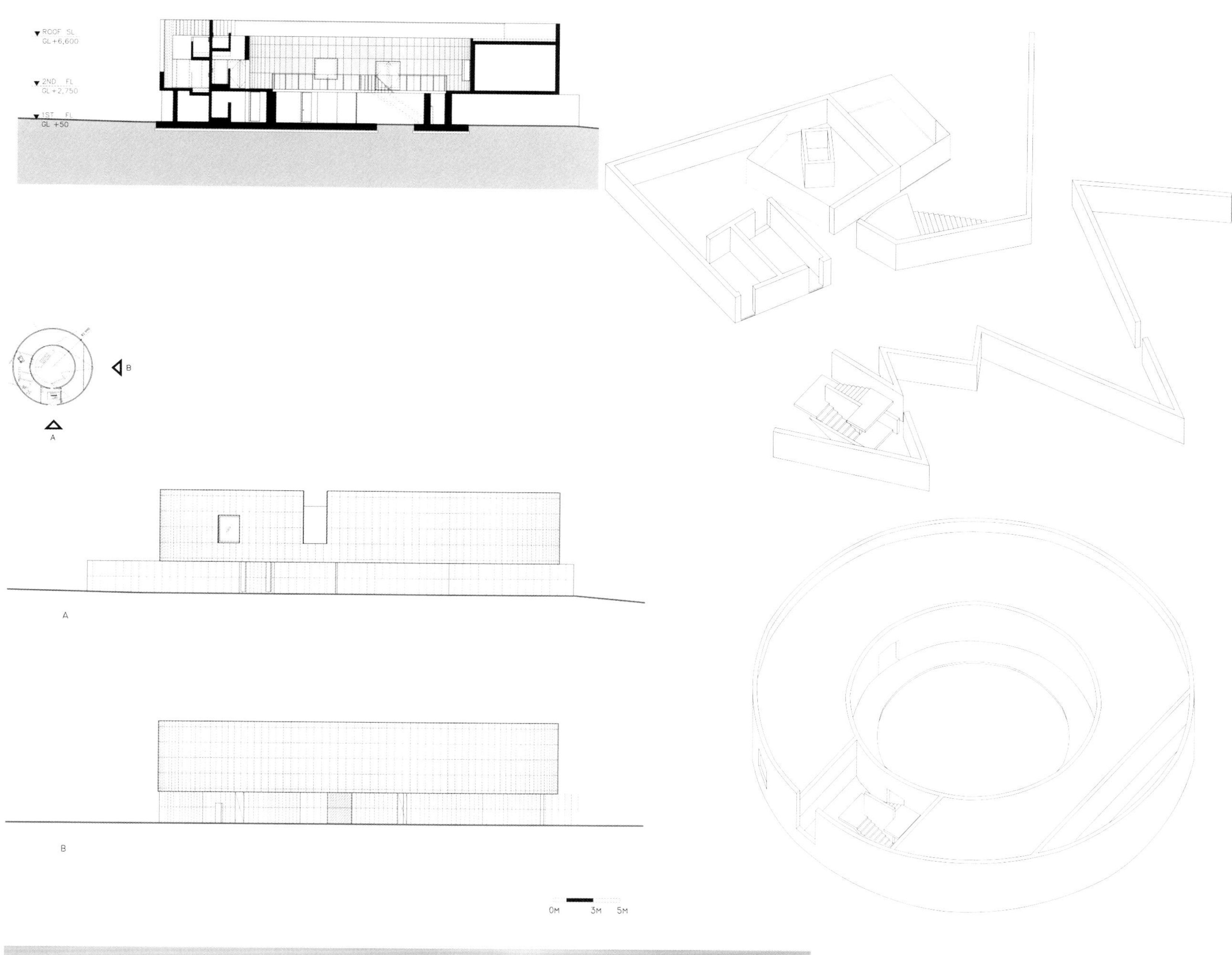
ROOF SL
GL +6,600
2ND FL
GL +2,750
1ST FL
GL +50
B
A
A
B
0M
3M
5M

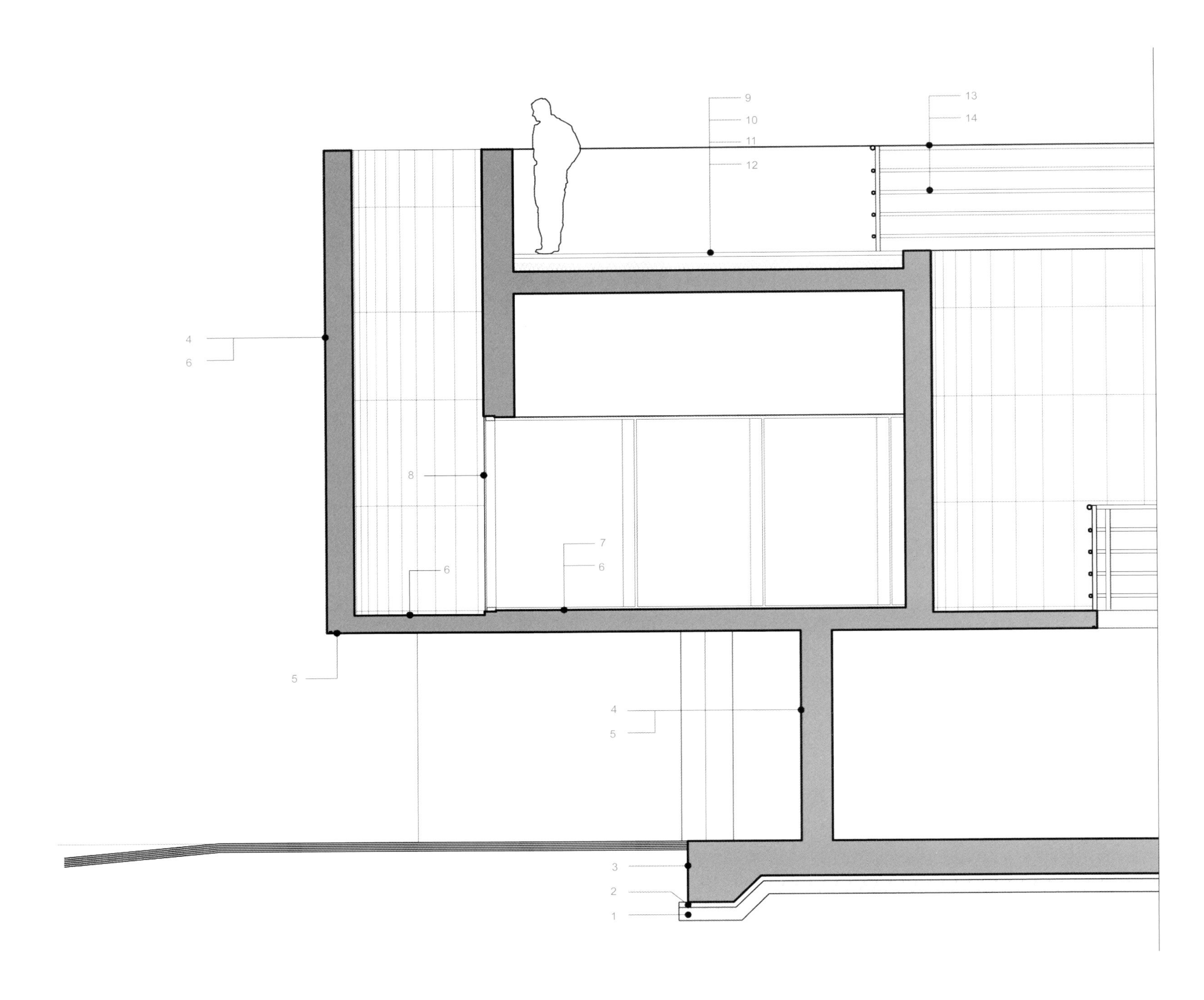

1. Bed of gravel 150 mm
2. Lean concrete 60 mm
3. Concrete footing
4. Euro-form exposed concrete
5. Water-repellent coating
6. Exposed concrete
7. White epoxy 3 mm
8. Double glazing in aluminum – profile frame
9. Plain concrete 80 ~ 100 mm
10. Polystyrene thermal insulation 150 mm
11. Bed of mortar 22 mm
12. Waterproof asphalt 3 mm
13. Ø 50 steel pipe railing
14. Ø 38 steel pipe railling

1. 碎石层150mm
2. 少灰混凝土60mm
3. 混凝土底脚
4. 欧式清水混凝土
5. 防水涂层
6. 清水混凝土
7. 白色环氧漆3mm
8. 双层玻璃，配铝框
9. 素混凝土80~100mm
10. 聚苯乙烯隔热层150mm
11. 砂浆层22mm
12. 防水沥青3mm
13. Ø50钢管栏杆
14. Ø38钢管栏杆

项目概况

这座低成本博物馆项目位于韩国济州岛的加西里村。村民们曾发起各种文化产业来宣传当地的经济，马博物馆正是他们的作品之一，其主要目标是吸引游客。

在辽阔的草场上，甜甜圈造型的展览空间被架高在一楼的Z字形墙壁上，这种设计在建筑下方留出了空间，给予了游客和农场工作人员更多便利。甜甜圈造型的设计灵感来自于当地的小型火山（当地称其为Oreum），它们的山顶上是中空的。在博物馆的屋顶上，游客可以一览济州岛的优美景色，就像站在火山顶上一样。

Expansion of Biomedical Research Centre in University of Granada

格拉纳达大学生物医学研究中心扩建项目

Location/地点: Granada, Spain/西班牙，格拉纳达
Architect/建筑师: Miguel Martínez Monedero, arquitecto
Photos/摄影: Javier Callejas Sevilla
Site area/占地面积: 3,730.38m^2
Built area/建筑面积: 1,243.73m^2
Completion date/竣工时间: 2012
Key materials: Façade – concrete, glass, steel
Structure – steel and vegetation
主要材料： 立面——混凝土、玻璃、钢材、植物
结构——钢材

Overview

The need to enlarge the existing Biomedical Research Centre (Health Technological Park of Granada, University of Granada) occurs in its own plot. This new building contains one of the most advanced basic research complex in Andalusia and Spain. The building programme includes a biological protection room type P3, something like a "biological bunker" where contamination can't enter or leave. Also contains a Magnetic resonance imaging (MRI) room, another "bunker", in this case radioactive, to research small rodents, and different basic research laboratories.

The new building has 5 levels, reserving the lowest (-1) for applications with more need for biological isolation. Laboratories, which rise in other plants, need a generous lighting, having a double orientation. We've all seen, on occasion, the images produced by the microscope in basic research. Increased thousands of times, abstracted from their original context, they become abstract coloured stains, without any comprehension for a layperson, forming as attractive as strange compositions. These images usually come from use in basic research, of a substance called Cyber-Green. This is the

idea that supports the image of the main façade, the most exposed and flashy.

The limited economical resources of the building were a decisive argument in search of good architecture with few materials. Formally, the building is composed of a continuous band of grey concrete, which contains a translucent glass box that floats above the ground. The use of glass and vegetation produces light filters that create a dynamic and changing space over time and seasons.

The construction, as an extension of a parent building, is located physically connected to it, but slightly separated by an outdoor courtyard, which allows illumination of both buildings, old and new. The new building technologically depends on old building, sharing their installations, reducing the need for equipment and costs.

Basic research laboratories are located throughout the south-east front of the building with direct views of the snowy mountains (Sierra Nevada) and the University Technology Park. Each laboratory has a small outside garden where researchers can make their own. This is a strategy that links research and leisure, work and rest.

The most visual space of the project is undoubtedly the diaphanous ground floor porch. Clearly the building links with relevant examples of modern architecture with the use of formwork concrete tables, the ground floor garden and the absence of interfering elements. Also, the diaphanous ground floor porch respects the visual continuity for the existing building, framing, with its pillared double height, the view through it.

Detail and Materials

This façade employs some bioclimatic control actions that should be reviewed. Direct insolation is protected by a "hanging" of autochthonous deciduous vegetation, irrigated by rainwater, planted on flowerpots supported by a maintenance walkway. Precisely the colour of the different pots achieves the chromatic game on the south façade.

项目概况

本项目是格拉纳达大学生物医学研究中心的扩建项目。这座新建筑内设置着安达卢西亚和西班牙最先进的基础研究设施。项目包含一个P3型生物防护室，这个类似“生物舱”的空间能隔绝污染物；一个核磁共振成像室，具有辐射屏蔽功能，用于研究小型啮齿动物；还有各种基础研究实验室。

建筑分为五层，地下室更多地应用于需要生物隔离的项目。实验室需要明亮的照明，基本采用双向朝向。我们都看过显微镜下的图像，经过数千倍放大之后，这些物体从它们的原始环境中分离出来，演变成外行人看不懂的抽象彩色斑点，形成了诱人而奇异的组合。这些图像通常来自于基础研究中一种称为“网际绿”的物质。这就是建筑最明显、最闪亮的主立面的基本设计概念。

由于建筑的建造资金有限，建筑师必须用最少的材料打造最好的建筑。在造型上，建筑由连续的灰色混凝土带构成，上方嵌有悬浮于地面之上的透明玻璃箱。玻璃和植物的运用过滤了光线，形成了一种随着时间和季节不断变化的动态空间效果。

项目作为主楼的扩建部分，与主楼相连，但是又通过一个露天庭院巧妙地隔开。这个庭院保证了新旧两座建筑的自然采光。新建筑在技术上依赖主楼，共享基础设施，减少了设备和成本的需求。

基础研究实验室沿着建筑的东南面展开，享有雪山和大学技术园的景色。每间实验室都有一个小型露天花园，研究者可自行布置。这种设计连接了研究与休闲、工作与休息。

项目最显眼的空间无疑是通透的一楼门廊。建筑通过模架混凝土板、一楼花园和无干扰元素体现了优秀的现代建筑特征。此外，通透的一楼门廊还尊重了原有建筑的视觉连续型，让后方的建筑透过双层高的门廊享有开阔的视野。

细部与材料

建筑立面采用了一些值得注意的生物气候控制措施。悬挂着的本土植物能遮挡部分直接日光。这些植物种植在检修走道上的花池里，通过雨水灌溉。不同色彩的花池实现了南侧立面斑斓的色彩变化。

1. Inverted deck:
   Aerated concrete H-200, thickness: 20cm, waterproofing membrane
   Thermal insulation XPS thickness: 8cm, 0.036 W/m°K
2. Reinforced concrete wall, thickness: 20cm
3. One-way floor of concrete joists with concrete pieces in between, thickness: 25+5cm
4. Luminary
5. Plasterboard ceiling with interior insulation by mineral rock wool thickness:50mm, 32kg/m³
6. External joinery formed by:
   double U-glass, width: 26cm, over aluminum frame
7. High density thermal insulation covered by anodized aluminum board
8. One-way floor of concrete joists with XPS pieces in between concrete ceiling
9. Metal railing
10. Skylight formed by concrete slab, thickness: 25cm + waterproofing membrane + concrete covering, thickness: 5cm
11. Anodised aluminum joinery. Glass: 6/16/4
12. Reinforced concrete wall H.A-40, thickness: 30cm
13. Joint between buildings. XPS thickness: 5cm
14. Existing wall
15. Epoxy resin floor, thickness: 3mm
16. Foundation slab, HA-25 concrete, thickness: 100cm
17. Waterproofing membrane
18. White rolled gravel. Diameter: 16 to 32mm; thickness: 5cm
19. Stainless steel grid: L 40.04
20. Grating floor
21. Water tank
22. Aerated concrete
23. Inverted concrete slab
24. Steel wire
25. Autochthonous decidious vegetation
26. Fixed panel of anodised aluminum + thermal insulation ISOCELL 50mm, 32kg/m³+ gypsum plasterboard
27. Planter: Steel board, thickness: 5 mm. Gravel φ10-60mm, thickness: 15cm
28. Structural anchorage: steel board, thickness: 1cm
29. Gangway for maintenance
   Grating floor on a steel structure IPE80
30. Drilled steel board φ5mm
31. Reinforced thermal insulation
32. Solar curtain
33. PVC pipe
34. Tensor for steel wire
35. Bottom ceiling of stainless steel grid
36. Fluted concrete pillar
37. Waterproofing membrane SBS 4kg/m²
38. Concrete slab H-20, thickness: 10cm
39. Mass concrete, thickness: 6cm
40. Graded aggregate, thickness: 20cm
41. Reinforced concrete wall HA-25 protected with waterproofing membrane SBS 4kg/m²
42. Laboratory panel, thickness: 60mm
43. Drain system: PVC pipe φ150mm + rolled gravel + geotextile sheet
44. Gravel. Diameter: 16 to 32mm; thickness: 5cm
45. Steel: L 40.04
46. Hole for water collection connected to downpipe

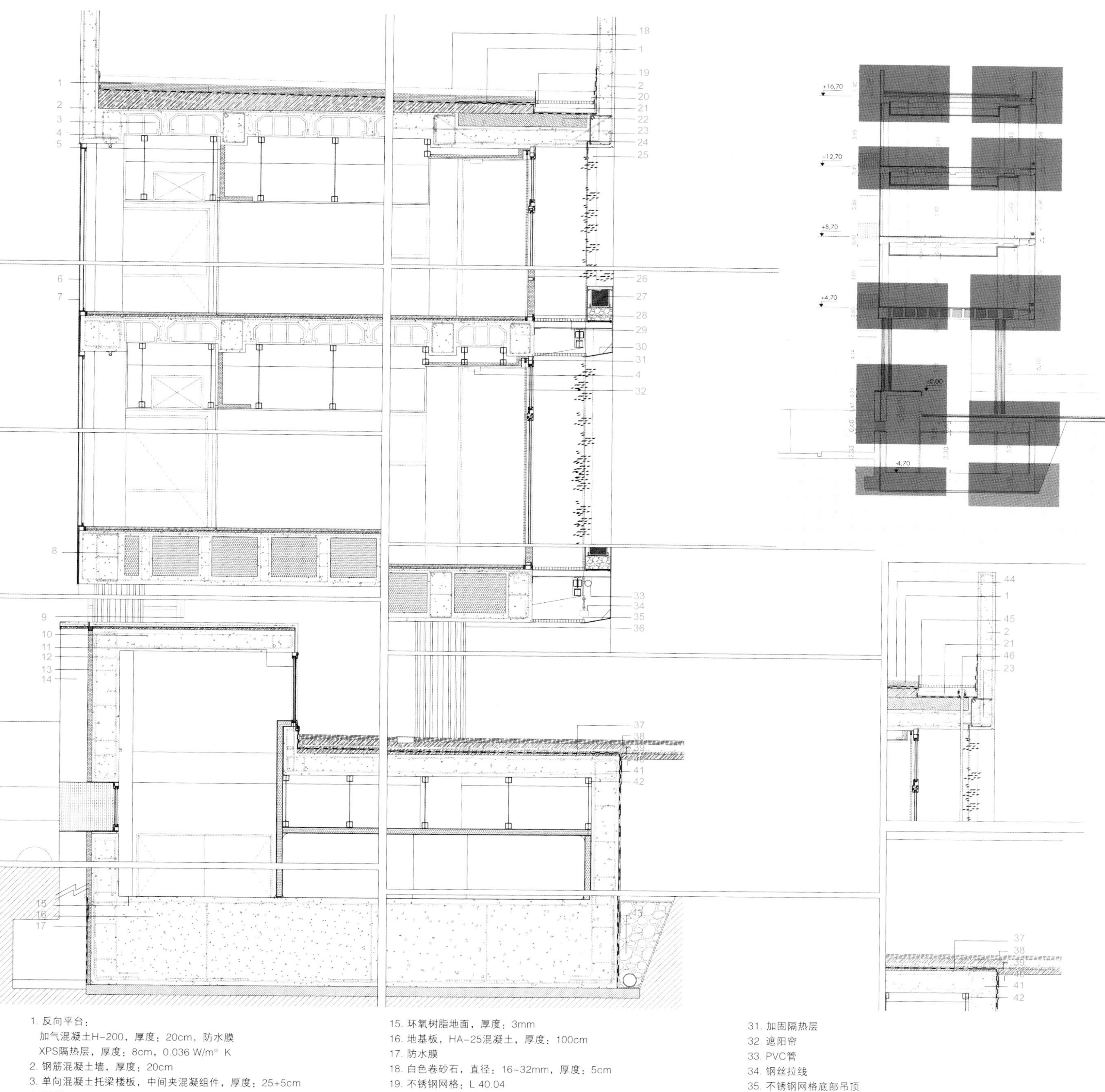

1. 反向平台：
   加气混凝土H-200，厚度：20cm，防水膜
   XPS隔热层，厚度：8cm，0.036 W/m° K
2. 钢筋混凝土墙，厚度：20cm
3. 单向混凝土托梁楼板，中间夹混凝组件，厚度：25+5cm
4. 照明
5. 石膏板吊顶，内部矿物棉隔热，厚度：50mm，32kg/m³
6. 外部窗：
   双层U形玻璃，宽度：26cm，铝框
7. 高密度隔热层，外覆阳极氧化铝板
8. 单向混凝土托梁楼板，混凝土吊顶中间夹XPS组件
9. 金属栏杆
10. 混凝土板形成的天窗，厚度：25cm+防水膜+混凝土保护层，厚度：5cm
11. 阳极氧化铝窗框，玻璃：6/16/4
12. 钢筋混凝土墙H.A-40，厚度：30cm
13. 建筑间接缝，XPS厚度：5cm
14. 原有墙壁
15. 环氧树脂地面，厚度：3mm
16. 地基板，HA-25混凝土，厚度：100cm
17. 防水膜
18. 白色卷砂石，直径：16~32mm，厚度：5cm
19. 不锈钢网格：L 40.04
20. 格栅地面
21. 水槽
22. 加气混凝土
23. 方向混凝土板
24. 钢丝
25. 本土落叶植物
26. 阳极氧化铝固定板+ISOCELL隔热层50mm，32kg/m³+石膏板
27. 花槽：
    钢板，厚度：5mm；碎石，直径：10~60mm，厚度：15cm
28. 结构锚点：钢板，厚度：1cm
29. 维护通道
    格栅地面，钢结构IPE80
30. 钻孔钢板 φ 5mm
31. 加固隔热层
32. 遮阳帘
33. PVC管
34. 钢丝拉线
35. 不锈钢网格底部吊顶
36. 凹槽混凝土柱
37. 防水膜SBS 4kg/m²
38. 混凝土板H-20，厚度：10cm
39. 大块混凝土，厚度：6cm
40. 分级粒料，厚度：20cm
41. 钢筋混凝土墙HA-25，防水膜SBS 4kg/m²保护
42. 实验板，厚度：60mm
43. 排水系统：PVC管，直径150mm +卷砂石+土工布
44. 碎石，直径16~32mm，厚度：5cm
45. 钢材：L 40.04
46. 集水孔，与落水管相连

Town hall in Échenoz-la-Méline

艾什诺兹拉梅利纳市政厅

Location/地点: Échenoz-la-Méline, France/法国，艾什诺兹拉梅利纳
Architect/建筑师: BQ+A – Quirot / Vichard / Lenoble / Patrono architects associated
Photos/摄影: Luc Boegly
Site area/占地面积: 2,920 m^2
Built area/建筑面积: 1,200 m^2
Completion date/竣工时间: 2013
Key materials: Façade – concrete and wood
主要材料：立面——混凝土和木材

Overview

Pursuant to a feasibility study leading to an impossible restructuring of existing facilities, The City of Echenoz has decided on a new construction project with ambitious environmental specifications and including the following program needs: City Hall, Extracurricular activities, Associations housing and Public library.

Due to a lack of an existing sufficiently dense and built up urban environment, the project is based upon topography. It is constituted by a building firmly laid out on the natural ground, with two separated levels forming a set of planted terraces following the ground natural slope.

On the east, at the upper side, the building is partially facing the nursery school through a courtyard, continuous with extracurricular activities spaces. On the west, at the lower side, the building is facing a large public space hanging over the Joseph Rougetstreet. These two levels are connected by an exterior public walkway, crossing the building centre while allowing views on inside activities.

The exterior areas are also planned with detailed care. The south square is paved and planted with a few high trees, while the western hillside partially houses an amphitheatre surrounded by a large lawn. The functional purpose of these areas is wide: City Hall square, exterior continuation for various festive events, games of bowls, meetings, shows, extracurricular schoolyard, etc...

These different choices concerning the architectural design as well as the building materials contribute to give the project a sobriety and strength well expressing the institutional nature of its program.

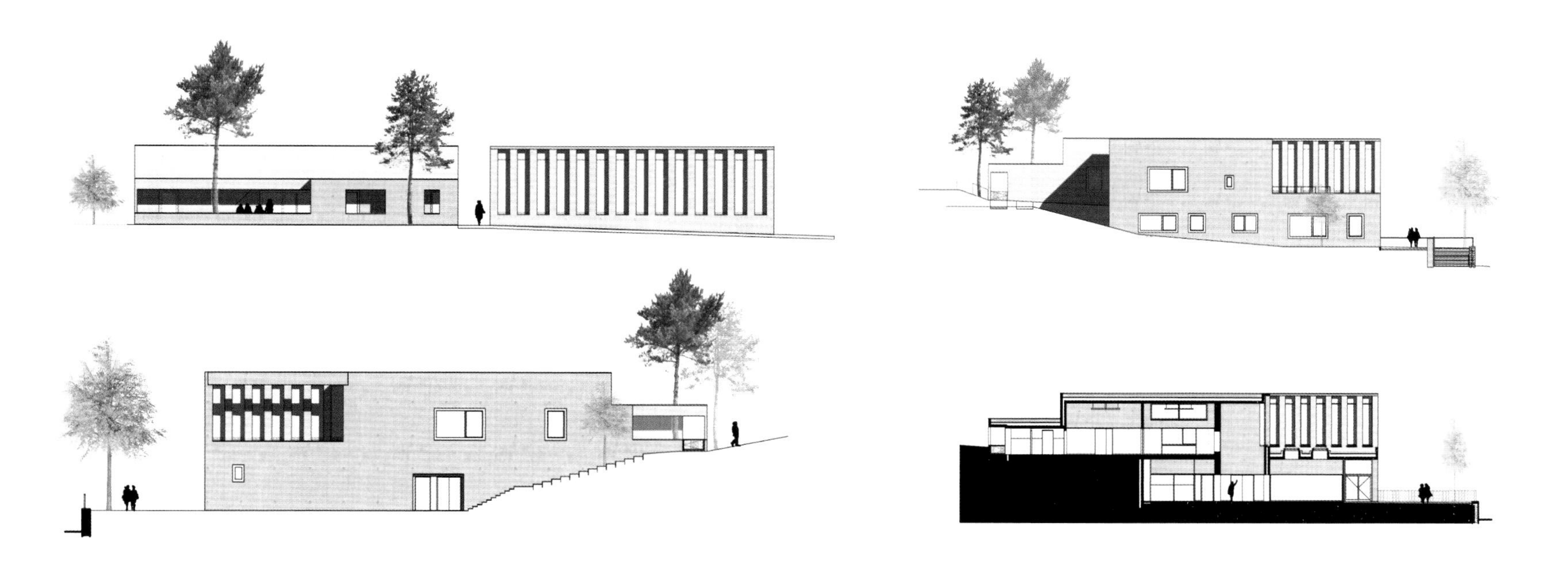

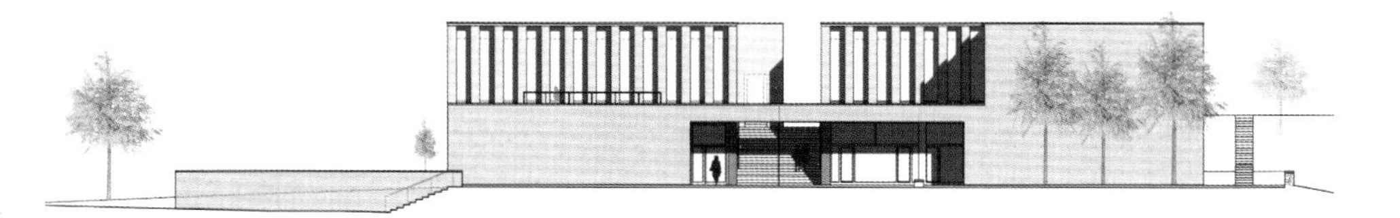

Detail and Materials

While the flat façade at square level is simply pierced by a large porch leading to the various programs as well as the public walkway, the upper floor façade alternates with windows and columns, thus creating a rhythmic pattern shaping a sort of large pediment.

The outside and inside of the building are constructed with the same material: a clear, well finished facing concrete, while the interior atmosphere is warmed through a large use of wood and stone.

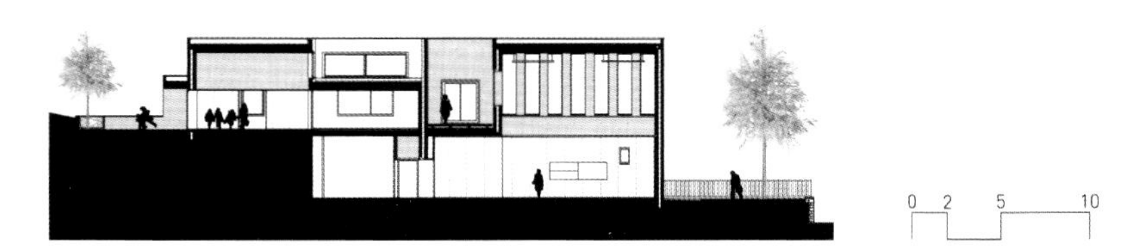

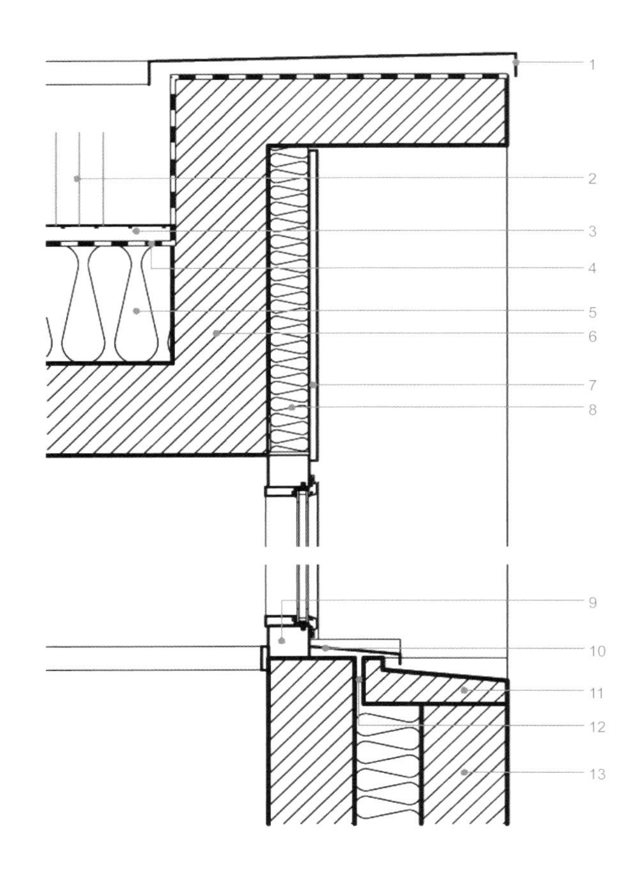

项目概况

在经过了可行性研究之后，艾什诺兹市决定开发一个雄心勃勃的新项目，包含市政厅、学生课外活动设施、相关住宅和公共图书馆。

由于现有的城市环境空间相对紧张，项目设计以地形为基础。一座建筑稳稳地坐落在自然地面上，分开的两层空间随着坡形地势形成了一系列的阶梯式种植平台。

在较高的东侧，建筑与幼儿园之间隔着一个庭院，延续了课外活动空间。在较低的西侧，建筑朝向约瑟夫罗吉特街的一个大型公共空间，二者由一条室外公共走道连接。这条走道穿过建筑中央，让人能看到建筑内部的活动。

室外空间的设计同样注入了巧思。南广场进行了地面铺装并装点着一些高大的树木；西山坡则顺地势设置了一个露天剧场，四周由大草坪环绕。这些区域的功能十分广泛：市政厅广场、庆典的户外活动空间、滚木球游戏场地、集会空间、表演场地、课外活动空间等。

建筑设计与建筑材料的选择共同赋予了项目庄重、大气的感觉，充分表现了项目作为公共机构的特质。

细部与材料

广场层的平板立面通过一个简单的大门廊与各项活动空间以及公共走道相连；上层的立面则交替布置着窗户和柱子，形成了一种富有节奏感的大山墙造型。

建筑内部全部采用了同样一种材料：简洁、光滑的混凝土。大量木材和石材的使用让室内氛围更显温馨。

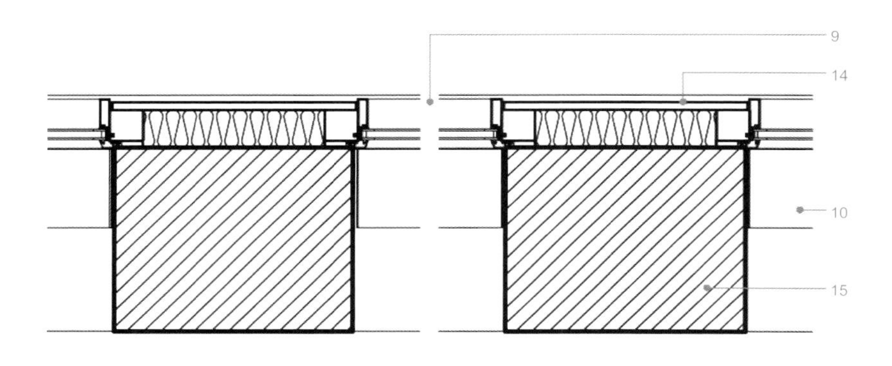

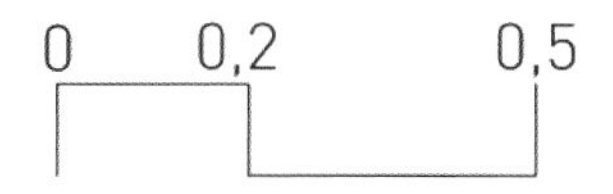

1.aluminium couvertine
2.vegetation
3.drainage layer
4.sealing
5.roof insulation 260mm
6.concrete parapet
7.aluminium cladding panel
8.insulation 80mm
9.wood carpentry
10.metal flap
11.concrete support
12.insulation 20 mm
13.raw concrete 180 mm
insulation 140 mm
raw concrete 180 mm
14.wood trim
15.concrete post

1. 铝顶盖
2. 植物
3. 排水层
4. 密封
5. 屋顶绝缘层
6. 混凝土护墙
7. 铝包层板
8. 隔热层80mm
9. 木框
10. 金属翻门
11. 混凝土支架
12. 隔热层20mm
13. 素混凝土180mm+隔热层140mm+素混凝土180mm
14. 木装饰
15. 混凝土柱

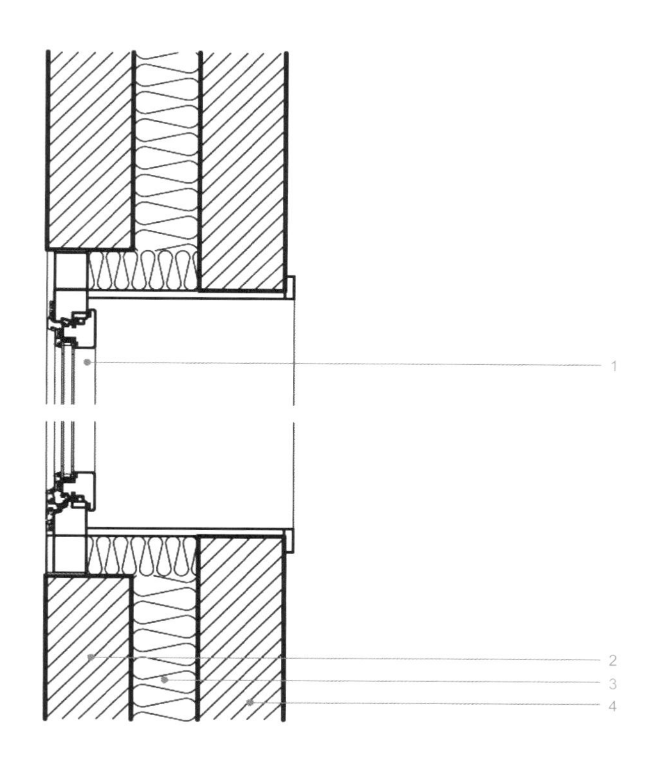

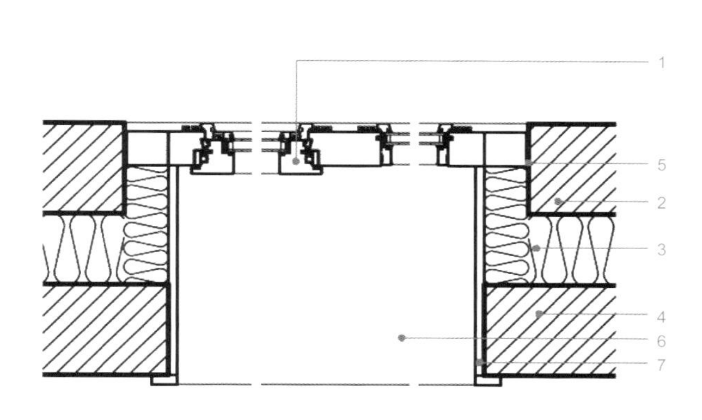

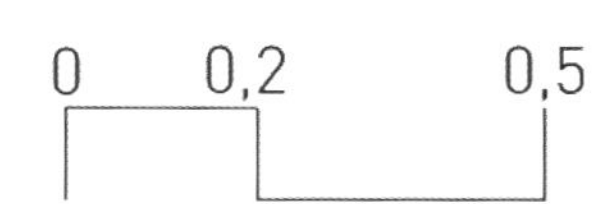

1. wood carpentry
2. concrete (exterior) 180mm
3. insulation 140mm
4. concrete (interior) 180mm
5. gasket join
6. wooden shelf
7. wooden frame

1. 木框
2. 混凝土（外层）180mm
3. 隔热层140mm
4. 混凝土（内层）180mm
5. 接口垫片
6. 木架
7. 木框

Expansion of Parish Church Carvico

卡尔维科教区教堂扩建

Location/地点: Bergamo, Italy/意大利，贝加莫
Architect/建筑师: Gianluca Gelmini
Photos/摄影: Andrea Martiradonna, Gianluca Gelmini
Built area/建筑面积: 1,300 m^2
Completion date/竣工时间: 2011
Key materials: Façade – concrete
主要材料： 立面——混凝土

Overview

The parish centre is located in the middle of Carvico, a village near the city of Bergamo. The actual parish centre is made by some small buildings from different ages. The monumental volume of the parish church stands out in the old central courtyard.

The project building is like a pavilion in the park, an independent entity in an irregular form garden in the north of the centre. The design of the new block answers to different and accurate considerations about the urban form, in particular to the relationship between the existent volumes of the centre. This building is connected with the old centre by a series of ways, they are not only paths but become a kind of space.

The addition is composed by three levels: the basement with the services, the ground floor with the entrance, the secretary and the café, in the first floor with three new classrooms and the direct connection with the others classrooms of centre. The system of the access and connections provides flexibility in the use of the entire structure as well as its individual parts.

The large windows on the ground floor, which open expansive views of the park and on the street, giving the new building an extrovert character open to the village and creating a place of transition between the liveliness

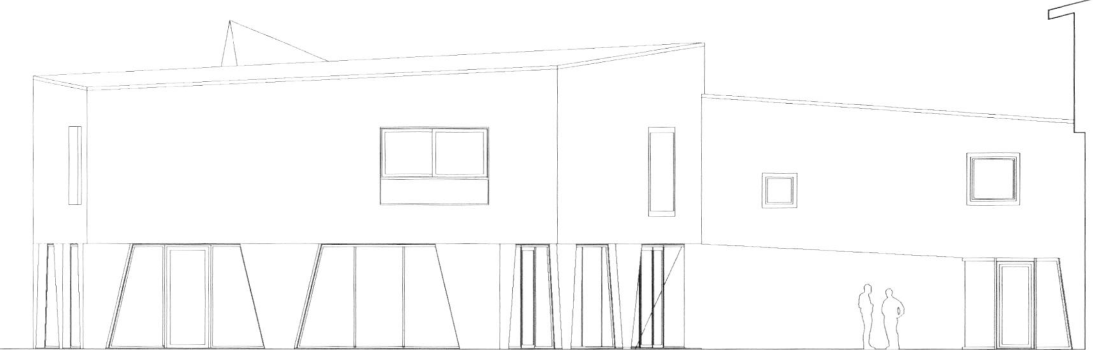

of the street and more protected space of the park. At the first level, the walls become more closed and compact, favouring a bigger introversion of the space allocated to the activities of catechesis.

Detail and Materials
The use of concrete is the natural consequence of the initial concept of a sculptural block in which the architecture coincides with the structure of this new element and the involucres denounces the internal spatial organisation.

The use of beam-wall molds with self-supporting and smooth formworks (PERI formwork Fin-Ply Maxi) has made it possible to match the structure in the shape of the building enhancing its plasticity.

项目概况

教区中心位于贝加莫城郊的小镇卡尔维科的中心。事实上，教区中心由若干个建于不同年代的小型建筑组成。教区教堂的宏伟结构在中央庭院中显得极为突出。

项目像是公园里的一座凉亭，以独立的不规则形态位于教区中心的北侧。建筑设计呼应了不同的城市规划形式，特别注意了与现有建筑结构之间的关系。建筑通过多种途径与现有的教区中心联系起来，不仅限于通道，而是一种空间类型。

新建结构分为三层：地下室是服务区；一楼是主入口、行政区和咖啡厅；二楼是三个新建教室，并且直接与中心的其他教室连接起来。通道和连接系统保证了整体建筑结构的灵活性和个体结构的独立性。

一楼开敞的窗户朝向公园和街道，让新建筑呈现出一种开放的姿态，在繁华的街道和宁静的公园之间形成了过渡。二楼的墙壁更加封闭而紧凑，以内向型空间为宗教授课活动服务。

细部与材料

项目对混凝土的使用取决于建筑的雕塑感造型设计。建筑结构与混凝土完美的结合起来，外部的混凝土结构暗示出内部的空间构成。

带有自承和光滑模架的横梁及墙面模具（PERI模架Fin-Ply Maxi型）使建筑结构与建筑造型匹配起来，提升了混凝土的可塑性。

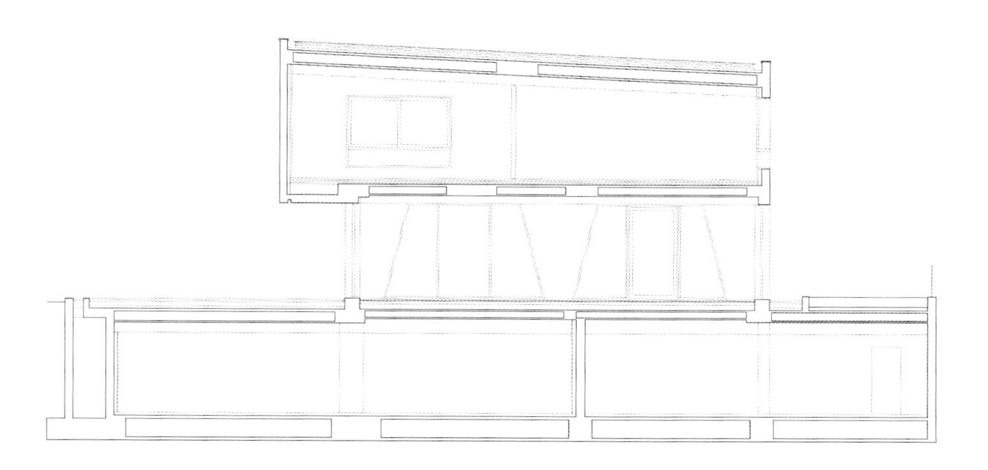

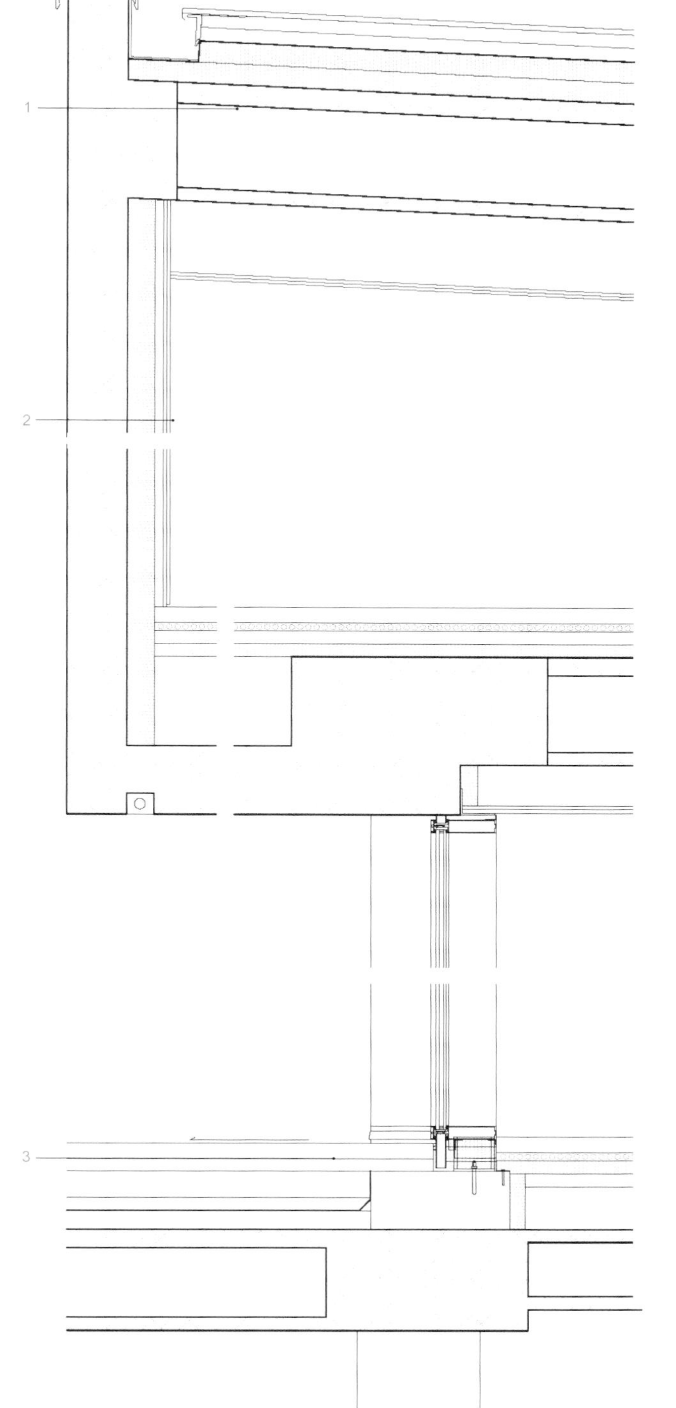

1. Roof Construction
   Zintek-70mm
   Ventilated cavity-50mm
   Thermal insulation-2x80mm
   Vapour barrier
   Concrete-400mm
   False ceiling
2. Wall Construction
   Concrete-220 mm
   Thermal insulation-100mm
   Plaster wall-60mm
3. Floor Construction
   Granolithic concrete- 60mm
   Screed with underfloor heating-80mm
   Polyethylene sheet-50mm
   Concrete-400mm
   False ceiling

1. **屋顶结构**
   Zintek涂层70mm
   通风腔50mm
   隔热层2x80mm
   隔汽层
   混凝土400mm
   假吊顶
2. **墙壁结构**
   混凝土220mm
   隔热层100mm
   石膏墙60mm
3. **地面结构**
   人造石混凝土60mm
   砂浆层+地热供暖80mm
   聚乙烯板50mm
   混凝土400mm
   假吊顶

New Theatre in Montalto di Castro

蒙塔尔托迪卡斯特罗新剧院

Location/地点: Montalto di Castro, Italy/意大利，蒙塔尔托迪卡斯特罗
Architect/建筑师: mdu architetti
Photos/摄影: Lorenzo Boddi, Valentina Muscedra, Pietro Savorelli
Site area/占地面积: 10,888m^2
Completion date/竣工时间: 2011
Key Materials: Façade – concrete, wood, alveolar polycarbonate
主要材料： 立面——混凝土、木材、齿槽聚碳酸酯

Overview

The design for the New Theatre in Montalto di Castro has a twofold objective: it is proposed as a conceptual model for measuring the territory and at the same time it attempts to express, through architecture, the magic of a theatrical event felt by the audience.

The territory of Montalto di Castro sinks its origins into Etruscan enthronisation whose ruins attest to architecture comprised of large stereometric masses in tufa; in the contemporary collective imagination Montalto di Castro evokes the world of the machines of the largest Italian power plant. The design proposes a temporal short circuit with respect to which the evolution of the territory is concentrated and expressed in a unique architectural moment: archaic Etruscan versus the aesthetics of the machine.

The new theatre is a large concrete monolith characterised by subtle variations in colour and texture, on which the fly tower appears to rest in an ethereal manner: an alveolar polycarbonate volume that dematerialises by day becoming indistinguishable from the sky, and lights up from within by night transforming into a large "lantern" on a territorial scale.

A new, extended, piazza in travertine and concrete, designed as a diversion of the road providing access to the historic centre, leads to the entrance of the New Theatre identified by an impressive overhanging roof. It introduces visitors to a continuous environment in which the foyer and the auditorium flow freely into one another. The wooden walls, with their

broken lines, create a space conceptually derived from the excavation of the concrete monolith. This morphological heaviness is contradicted by the vibration of the material that seems to envelope the space in a large curtain and introduces the spectator to the much awaited magical opening of the stage curtains.

The auditorium that seats 400 has its counterpart in the outdoor arena that seats 500, which can thus benefit from the theatre stage.

Detail and Materials

The main materials are concrete for the monolith, strips of wood covering the vertical structures creating a sequence of warm and vibrant curtains, and alveolar polycarbonate for the fly tower, which by day dematerialises becoming indistinguishable from the sky, while at night it lights up like a beacon on an urban scale.

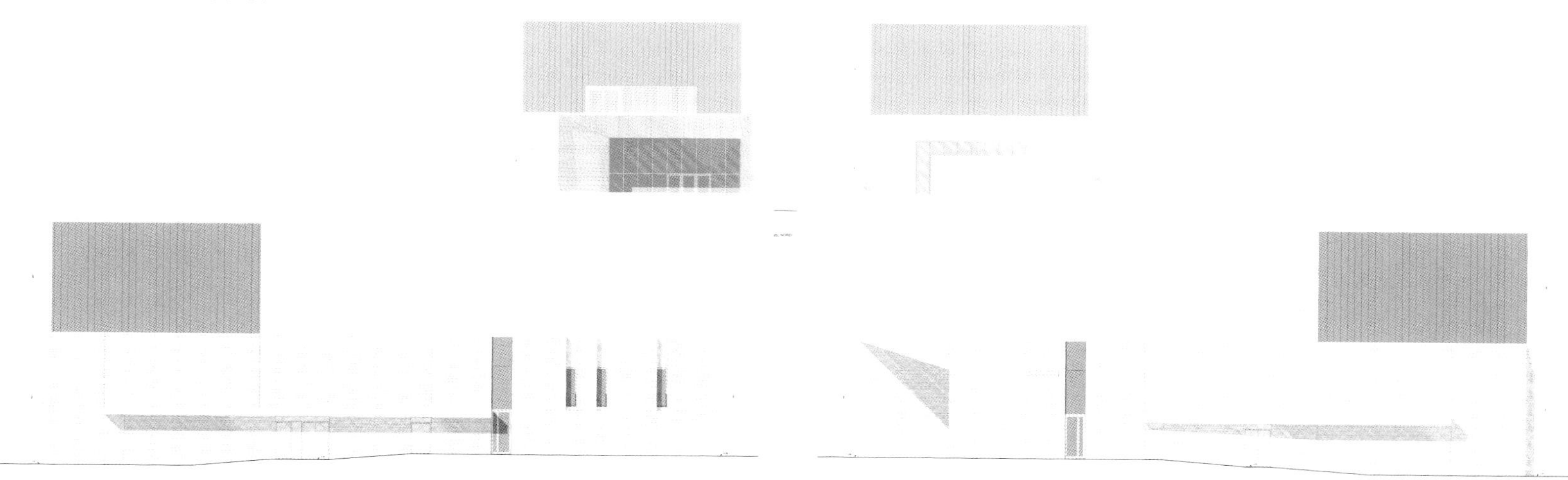

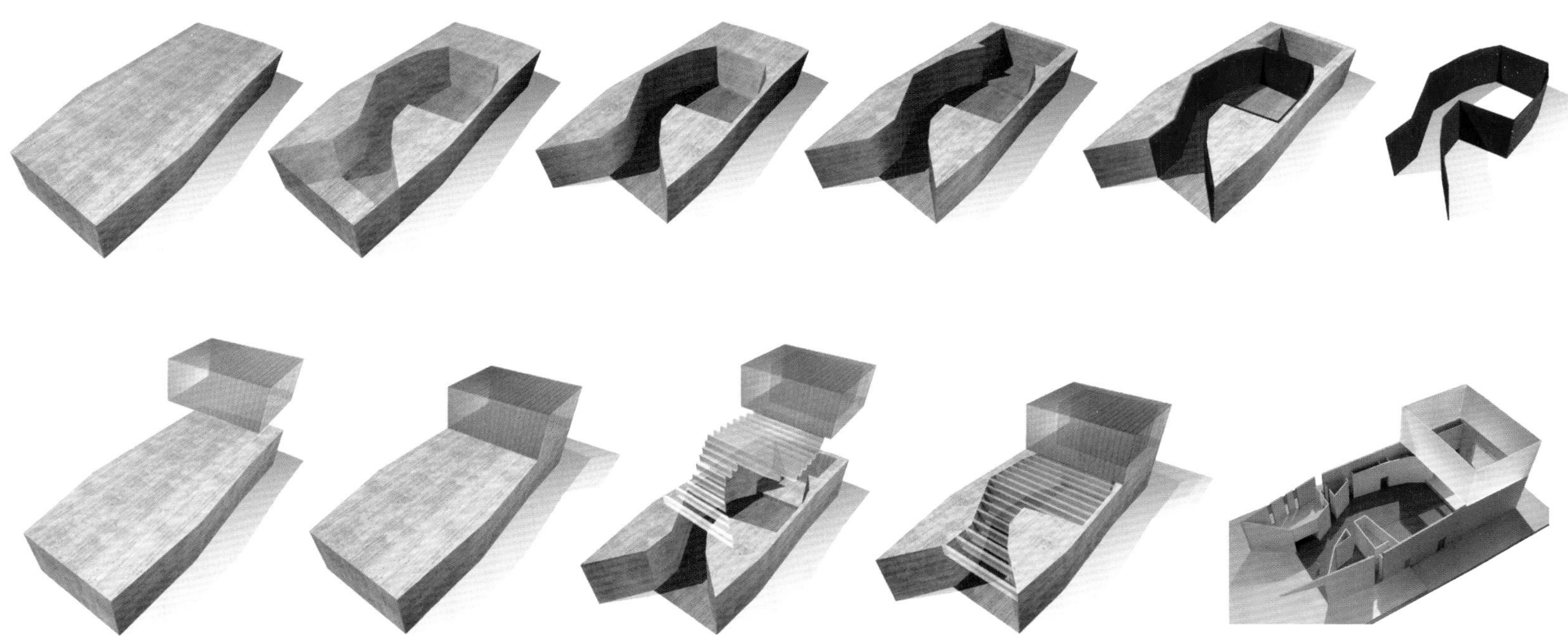

1. Fair faced concrete 28cm
   with water-repellent paint external treatment
2. Roofing system:
   waterproofing sheath 4mm
   rigid insulation with slope (2%) in light cellular concrete
3. Prefabricated concrete planks
4. Stalls wall:
   wooden fillets panels
5. Stalls floor: polished concrete floor 15cm

1. 琢面混凝土28cm
   防水涂料外部处理
2. 屋顶系统：
   防水外壳4mm
   刚性隔热层，坡度2%，覆于轻质多孔混凝土之上
3. 预制混凝土板
4. 剧场正前方墙壁：
   木嵌条板
5. 剧场正前方地面：
   抛光混凝土地面15cm

项目概况

蒙塔尔托迪卡斯特罗新剧院的设计有两个主要目标：一方面是在其所在地提出一个概念模型，另一方面是通过建筑向观众表现戏剧活动的魅力。

蒙塔尔托迪卡斯特罗的历史起源在伊特鲁里亚登基时被完全摧毁，他销毁了大量由石灰华构成的建筑体块；在现代人的印象中，蒙塔尔托迪卡斯特罗拥有意大利最大的发电厂，是一个机械的世界。剧院的设计既体现了对当地历史的尊重，又展示了独特的建筑形态，是古典的伊特鲁里亚王朝与现代的机械美感的融合体。

新剧院呈现为大型混凝土砌块结构，整体结构的纹理和色彩有着微妙的变化，而上方的舞台塔则呈现出轻盈的姿态：透明的齿槽聚碳酸酯结构在白天几乎可以消失在天空中，夜晚则会变成一盏明亮的灯笼。

由石灰华和混凝土铺设而成的露天广场将通往历史中心的城市街道和新剧院的入口连接起来，突出展示了新剧院引人注目的外伸屋顶。屋顶将访客引入流畅的内部环境，门厅与礼堂的连接天衣无缝。木制墙面通过上方的折线重现了外部混凝土的形态。这种厚重感与各种材料形成了一种对比。建筑内部的材料像一块巨幕将整个空间包裹起来，让观众更能全身心地等待舞台拉开神秘的帷幕。

礼堂有400个坐席，而露天剧场则能容纳500人，二者都能欣赏剧院舞台上的表演。

细部与材料

建筑的主要材料是混凝土（整体砌块）、木板条（覆盖垂直结构，形成温暖而活泼的幕墙）以及齿槽聚碳酸酯（舞台塔）。聚碳酸酯材料让舞台塔在白天几乎可以消失在天空中，夜晚则变成一盏明亮的灯笼。

Cuna de Tierra Winery

土地摇篮酒窖

Location/地点: Guanajuato, México/墨西哥，瓜纳华托
Architect/建筑师: CCA | Centro de Colaboración Arquitectónica
Photos/摄影: Estudio Urquiza
Built area/建筑面积: 1,800m^2
Completion date/竣工时间: 2013
Key materials: Façade – concrete, soil, wood and iron
主要材料：立面—— 混凝土、土、木材、铁

Overview

Cuna de Tierra Winery is a project located in the rural area of the state of Guanajuato in Mexico. The gross area of the whole construction contains 1,800 square metres, and it was open to the public on October 2013.

By using the wine's name as the concept driver (Cuna de Tierra / Soil Cradle), this project reaches an exploration in the relationship between a clean and functional wine development with the construction of a new place which tends to merge with its natural context.

In addition to the winery's development, the program required an observation platform from which a scenic view of the new construction is surrounded by the large extension of grape fields. With this tower, the project aims for a more complex solution by exploring the ways in which the experience of contemplating a view can be reshaped. The purpose of this space integrates the view, with a wine tasting pavilion and an excavated space, allowing the user to contemplate 3 different perspectives of the way an atmosphere of a winery can be sensed.

Detail and Materials

The mixture of concrete with the soil taken from the site becomes the main material for the construction of these spaces. This powerful element not only becomes the main element for the vine trees growth, but defines the project's motif.

The needs on mixing the thick soil walls with other native materials such as the iron and wood (used for passages, doors, ceilings, and lattices), gave the opportunity to deliver a high standard winery with the less-than-average number of high-tech constructive systems. These systems can be revealed on the way that natural lighting, humidity, and the air interacts within the closed and open spaces of the winery.

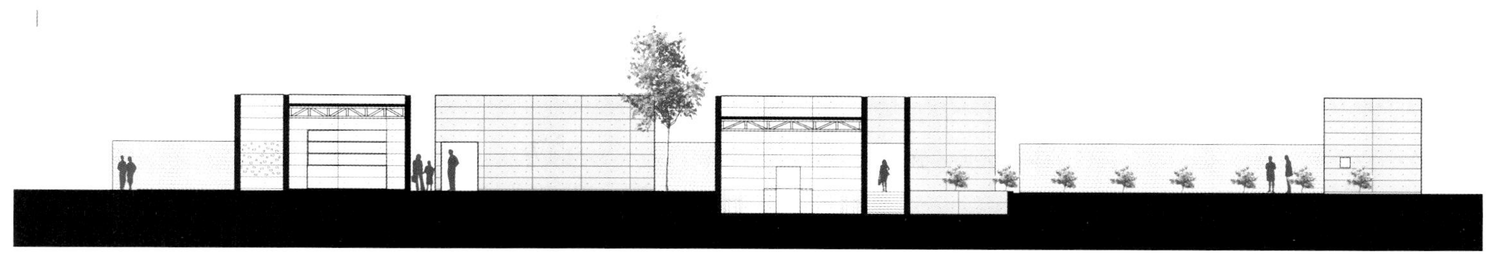

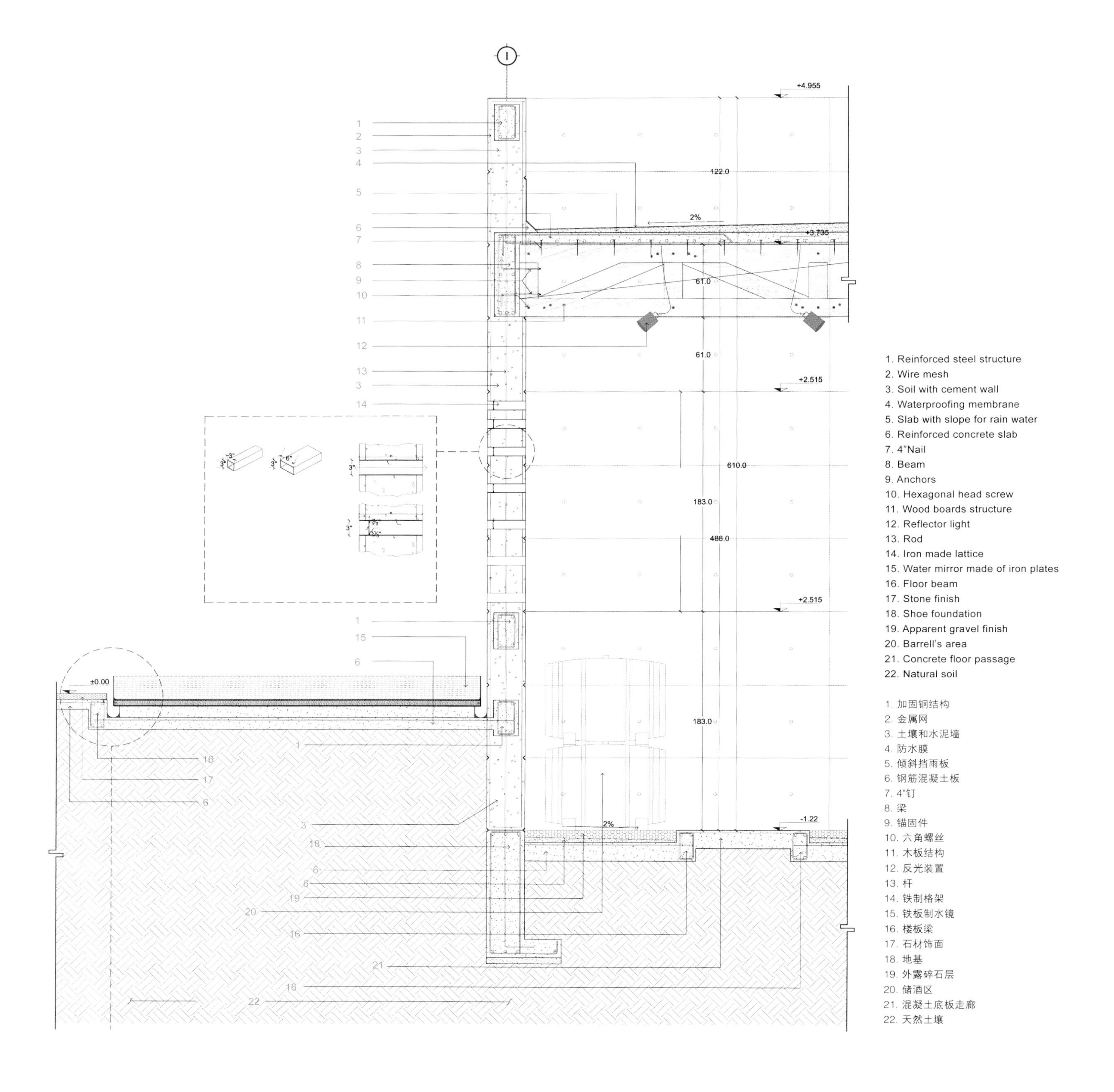

+4.955
122.0
2%
+3.735
61.0
61.0
+2.515
610.0
183.0
488.0
+2.515
183.0
±0.00
2%
-1.22
1. Reinforced steel structure
2. Wire mesh
3. Soil with cement wall
4. Waterproofing membrane
5. Slab with slope for rain water
6. Reinforced concrete slab
7. 4"Nail
8. Beam
9. Anchors
10. Hexagonal head screw
11. Wood boards structure
12. Reflector light
13. Rod
14. Iron made lattice
15. Water mirror made of iron plates
16. Floor beam
17. Stone finish
18. Shoe foundation
19. Apparent gravel finish
20. Barrell's area
21. Concrete floor passage
22. Natural soil
1. 加固钢结构
2. 金属网
3. 土壤和水泥墙
4. 防水膜
5. 倾斜挡雨板
6. 钢筋混凝土板
7. 4"钉
8. 梁
9. 锚固件
10. 六角螺丝
11. 木板结构
12. 反光装置
13. 杆
14. 铁制格架
15. 铁板制水镜
16. 楼板梁
17. 石材饰面
18. 地基
19. 外露碎石层
20. 储酒区
21. 混凝土底板走廊
22. 天然土壤

项目概况

土地摇篮酒窖项目位于墨西哥瓜纳华托州的一个乡村地区。整个项目的总面积为1,800平方米，于2013年10月正式开放。

项目设计以红酒的名字“土地摇篮”为出发点，探索并研究了干净实用的酿酒过程与融入自然的新建筑二者之间的关系。

处理酒窖的开发之外，项目还要求设计一个新的观景台，四周环绕着大片的葡萄园。观景台让项目显得更加复杂，为人们提供了全新的视角。这一空间的目标是整合视野，通过品酒亭和洞穴空间让人们在酒窖中体验三种截然不同的感觉。

细部与材料

混凝土与现场所取的土壤相混合，形成了施工的主要材料。这种材料不仅是辅助葡萄藤生长的主要元素，而且还明确了项目的主题。

厚重的土墙和铁、木材等其他本土材料（用于建造走廊、门、天花板、窗格）的混合，通过相对较少的高科技建造系统实现了高标准的酒窖设计。这些建造系统体现在自然采光、湿度控制以及封闭与开放空间的空气交换上。

Gilmosery

吉姆瑟里办公楼

Location/地点: Seoul, Korea/韩国，首尔
Architect/建筑师: Kim in cheurl + archium
Site Area/占地面积: 383m²
Construction Area/建筑面积: 1,028.97m²
Completion Date/竣工时间: 2012
Key materials: Façade – exposed concrete + THK24 Low-E pair glass
主要材料： 立面——素面混凝土、24厚低辐射双层玻璃

Overview

Gilmosery is office building for Settle Bank, the company that develop internet banking software. In order to demolish the existing residential home, and construct new building, architectural regulations(land ratio, floor space index, height, regulations on the road width and sunlight width) were treated with flexibility as basic conditions of the given site, rather than restrictions.

Building office on residential area is alike making formal dress on the regulation for making ordinary clothes. Social criterion to control urban density cannot help clashing with architectural will to get maximum utility. There is an easy solution, however, should the boundaries for a tall wider space be pushed to create more space rather than forcing them to comply with boundaries of limitations.

The correlation between floor area ratio 200% and building to land ratio 60% is very complicated in a typical residential area. Balcony is included building to land ratio but, not in floor area ratio. The double layered external wall will be constructed by setting the frame of structure top 60% and pushing back the wall that makes for the 40% of the interior space with the conditions to make balcony. This way, it can gain 20% balcony within structural frame. But it's outside space on regulations.

Securing both permitted building to land ratio and floor space ratio is giving up on a set pattern. The building form, due to setback regulations, was extended in its curvature to the bottom. While an atypical form might seem difficult to read, can be architectural suggestion on boring alley.

The frame of space and form is made by structure then invisible boundaries hide in it. The space is not closed and shut up but opens. As narrowness and openness and low and high are relative standards with expansion of the senses. The invisible space does not divide the outside and inside, but can create a sensation of unified space. The structure that is form and space that is contents gain void space on gap. The gap that is the role of bench and flowerpot don't belong to both outside and inside.

The goal of architecture is not making form but making space. Form is not essence in architecture but result of making space. Architectural structure is tool for define space. The space that set back is not surplus but very useful elements for changing relations between architecture and city as well as human and environment. In order to expression of architecture, if the framework that forms space can make form on its own without decorating. The ornamentation and exaggeration has already lost its reason.

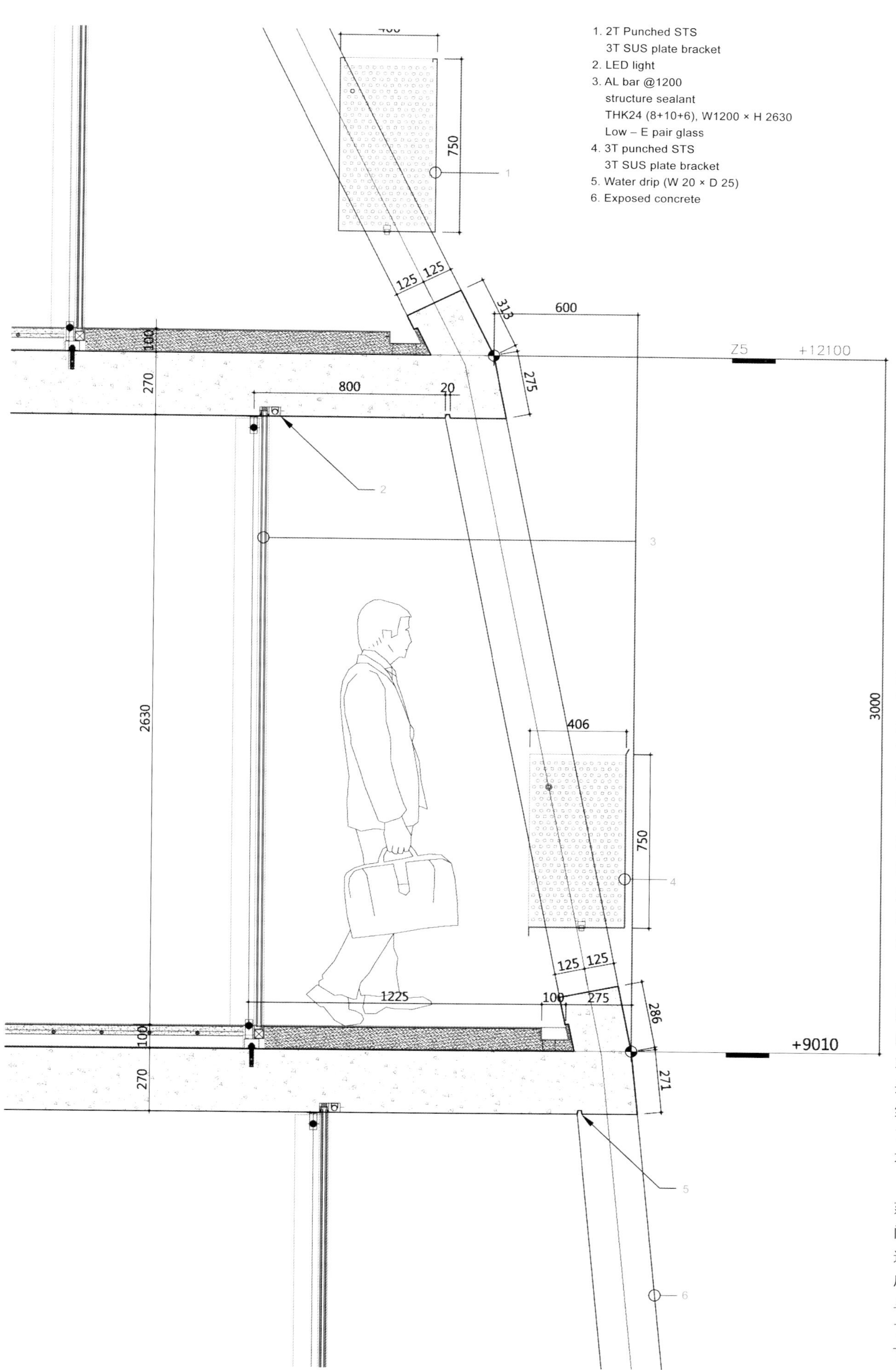

1. 2T穿孔STS
3T SUS板支架
2. LED灯
3. AL条@1200
结构密封胶
厚度24毫米（8+10+6），W1200 × H 2630
低辐射双层玻璃
4. 3T穿孔STS
3T SUS板支架
5. 滴水盘(W 20 × D 25)
6. 素面混凝土

项目概况

吉姆瑟里办公楼是为一家网上银行软件开发公司所设计的办公楼。为了拆除原有的住宅楼、建造新建筑，项目根据场地条件灵活运用了建筑规范（土地比例、建筑面积指数、高度、路宽、日照宽度等）。

在住宅区建造办公楼就像是用制作日常服装的规则设计礼服一样。城市密度控制标准与建筑功能性的最大化势必产生冲突。简单的解决方案是拓宽建筑的上部空间，而不是压迫有限的地面边界。

项目的建筑容积率可达200%，而建筑占地比率为60%，这在典型的住宅环境中是极难实现的。阳台被算在建筑占地比率中，而不是建筑容积率。双层外墙的设计让建筑结构占有60%的空间，而阳台的添加则能弥补剩余的40%室内空间。这样一来，建筑可以在结构框架内获得20%的阳台空间，而外部空间则满足了建筑法规要求。

项目通过打破常规造型而保证了建筑占地比率和建筑容积率。由于建筑必须从街道向后撤的规定，建筑呈现为弧形表面。非常规的建筑造型为小巷带来了独特的建筑景观。

建筑结构实现了空间框架和造型，内部则隐藏着看不见的边界。建筑空间不是封闭的，而是开放的。宽与窄、高与低都是相对的感官标准。无形的空间不分内外，而是形成了统一的整体空间。造型和空间的结构在缺口处获得了空间，缺口空间扮演了长凳和花槽的角色，既不属于外部，也不属于内部。

建筑的目标不是制造造型，而是制造空间。造型并不是建筑的基础，而是空间制造的结果。建筑结构是定义空间的工具。后撤的空间并不是过剩的，而是改变建筑与城市、人类与环境关系的重要元素。为了表现建筑，结构可以在没有装饰的前提下制造造型。装饰和夸张已经没有任何意义。

Dadong Art Centre

大东艺术中心

Location/地点: Kaohsiung, Taiwan, China/中国台湾，高雄
Architect/建筑师: MAYU architects + de Architekten Cie
Photos/摄影: Guei-Shiang Ke, Yu-lin Chen, Ya-Yun Wang
Built area/建筑面积: 36,470m^2
Completion date/竣工时间: 2012
Key materials: Façade – Exposed concrete, glass and membrane
主要材料： 立面 —— 素面混凝土，玻璃和薄膜

Overview

Dadong Art Centre is located in Kaohsiung City in the economic centre of southern Taiwan. Connecting the dense context of historic Feng-Shan District, a major park, and the Feng-Shan River, the Art Centre creates a sequence of public spaces by means of membrane-covered interstitial open space between four major volumes – theatre, exhibition hall, art library and the education centre.

The intensive use of public space by activities such as dance, Tai-Chi and various games characterises outdoor use in Taiwanese cities. A membrane roof creates a shaded condition for a wide range of activities. The roof shape prevents against extreme climatic conditions such as typhoon, periodically strong rain and high summer temperature. Wide holes in the roof guide the water to inner springs. Warm air rises to the top roof openings generating a fresh inner draft.

A 900-seat theatre and a small rehearsal hall, cladded in wood and perforated metal, are centre pieces of the programme. The theatre is designed as a wood volume contained inside the X-shaped concrete façade. The façade of the whole project is homogenous, yet implies the programme inside by subtle changes in density.

In order to achieve optimal acoustics in both music and theatre use, the ceiling of the auditorium is adjustable according to music and theatre modes. Sound absorption materials are also retractable behind the wood lamella.

Detail and Materials

The formal concept of Dadong Art Centre is

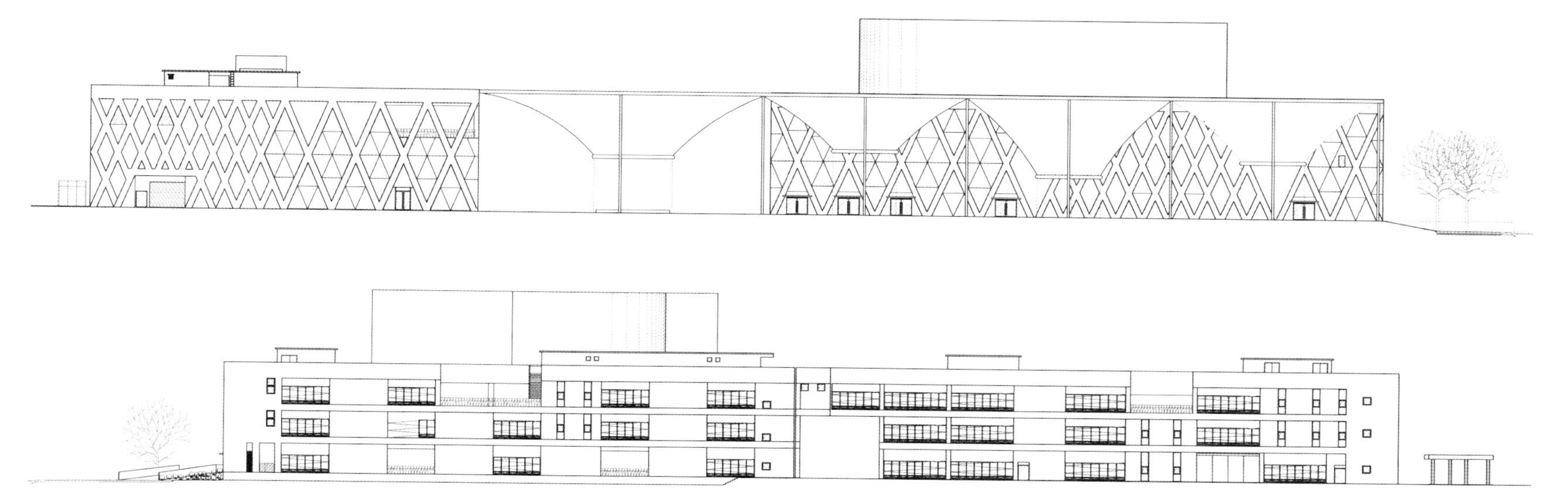

based on the duality of architectonics. The differentiation of "tectonic" and "stereotomic" exemplified in German architect and historian Gotterfield Semper's primitive hut is the guiding principles of structural, constructional and detailing expression of Dadong Art Centre.

While the membrane roof structure is the evidence of tectonic principle, the stereotomic concept is showed on the façade motif: tilted concrete columns, triangulated glass curtain walls and X-shape wall engraving. Such façade created horizontally extending tendency wrapping around the detached massing and amplifying the heaviness and monolithic quality of the architecture. Since theatre is a complicated building type, the structural system is the hybrid of reinforced concrete, steel frame and load bearing systems. The challenge here was to apply one principle over different structure systems while staying truthful to the tectonic logics.

Exposed concrete was chosen as primary material, and it was seen as antithesis to the soft membrane. The triangulation helped the façade to maintain an upward proportion and the lightweight membrane roof is supported by the convergent upper points. The glass curtain walls are set flush with the concrete surfaces. Effort was made to ensure high quality detailing especially at the building corners. It is essential to avoid any corner framing therefore the building façade can always be perceived as load bearing.

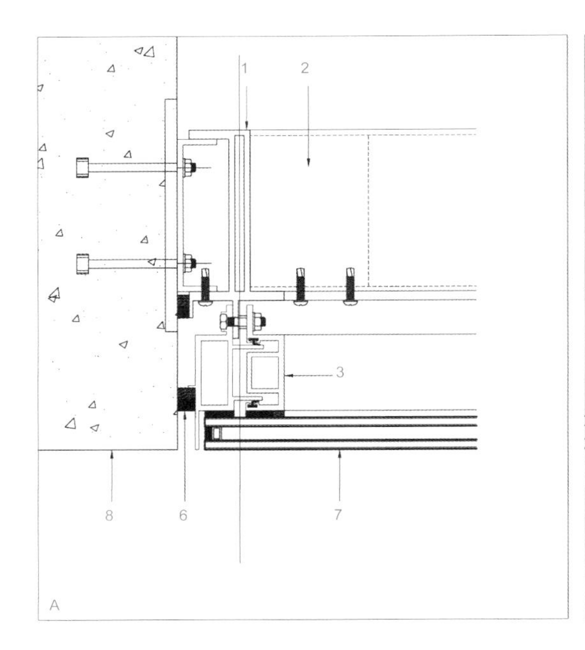

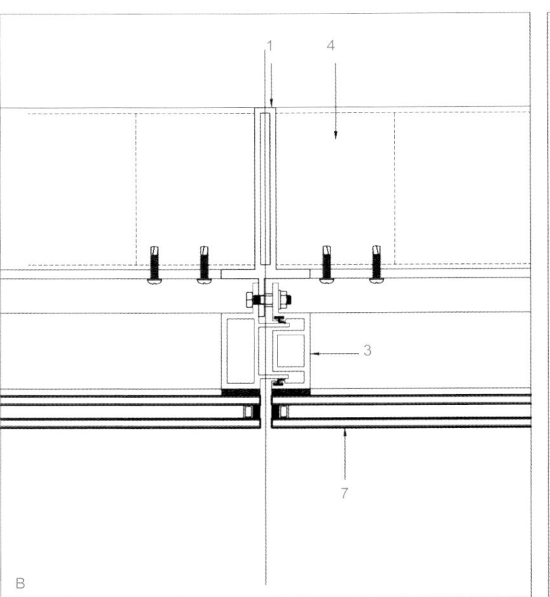

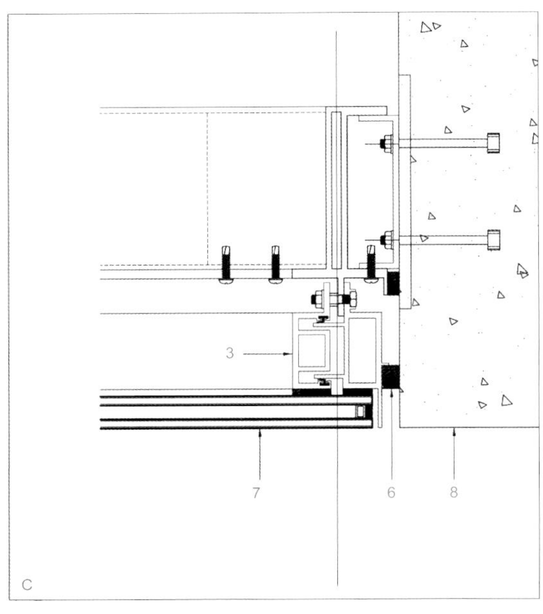

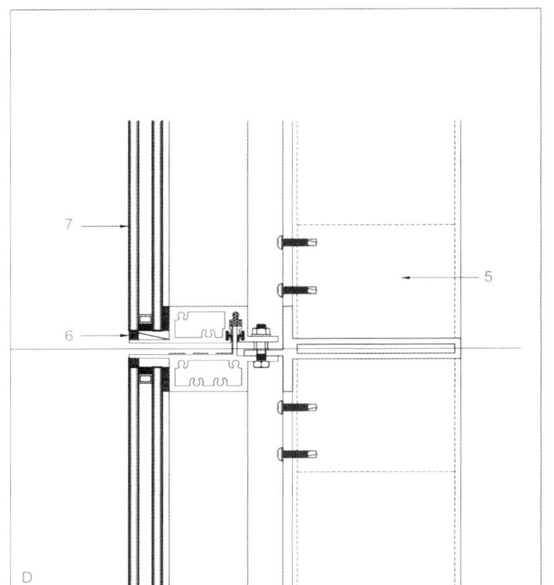

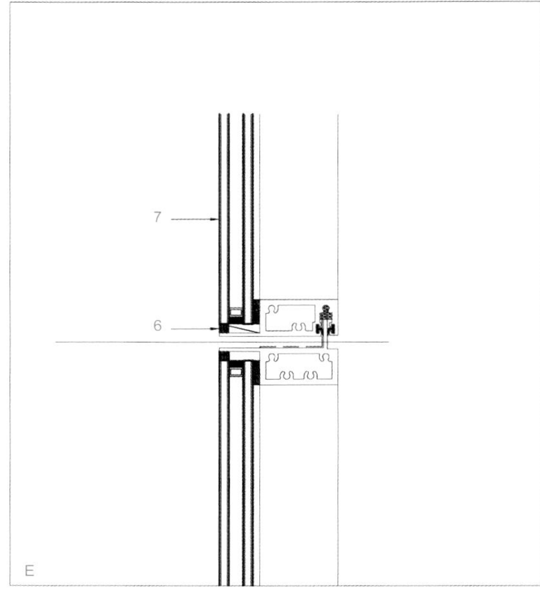

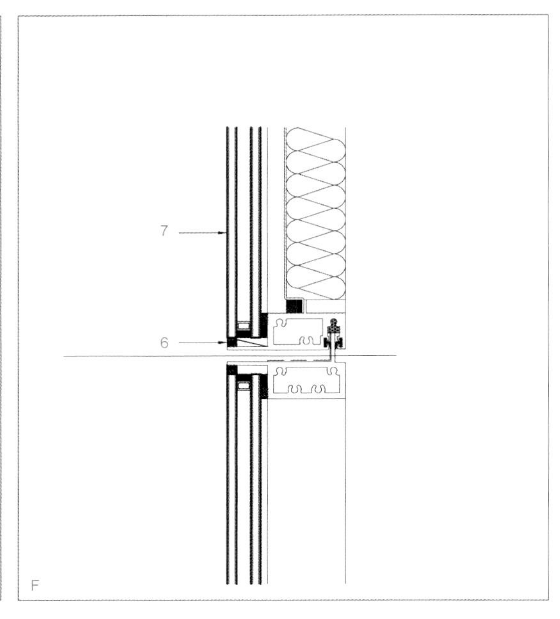

1. Aluminum T-profile vertical element
2. Cast aluminum T-profile connector
3. Aluminum frame
4. Cast aluminum six-way connector
5. Cast aluminum three-way connector
6. Structural silicone
7. Double-layered Low-E glass
8. Concrete wall

1. T形铝材垂直元件
2. T形铸铝连接件
3. 铝框架
4. 铸铝六通连接件
5. 铸铝三通连接件
6. 结构硅胶
7. 双层低辐射玻璃
8. 混凝土墙

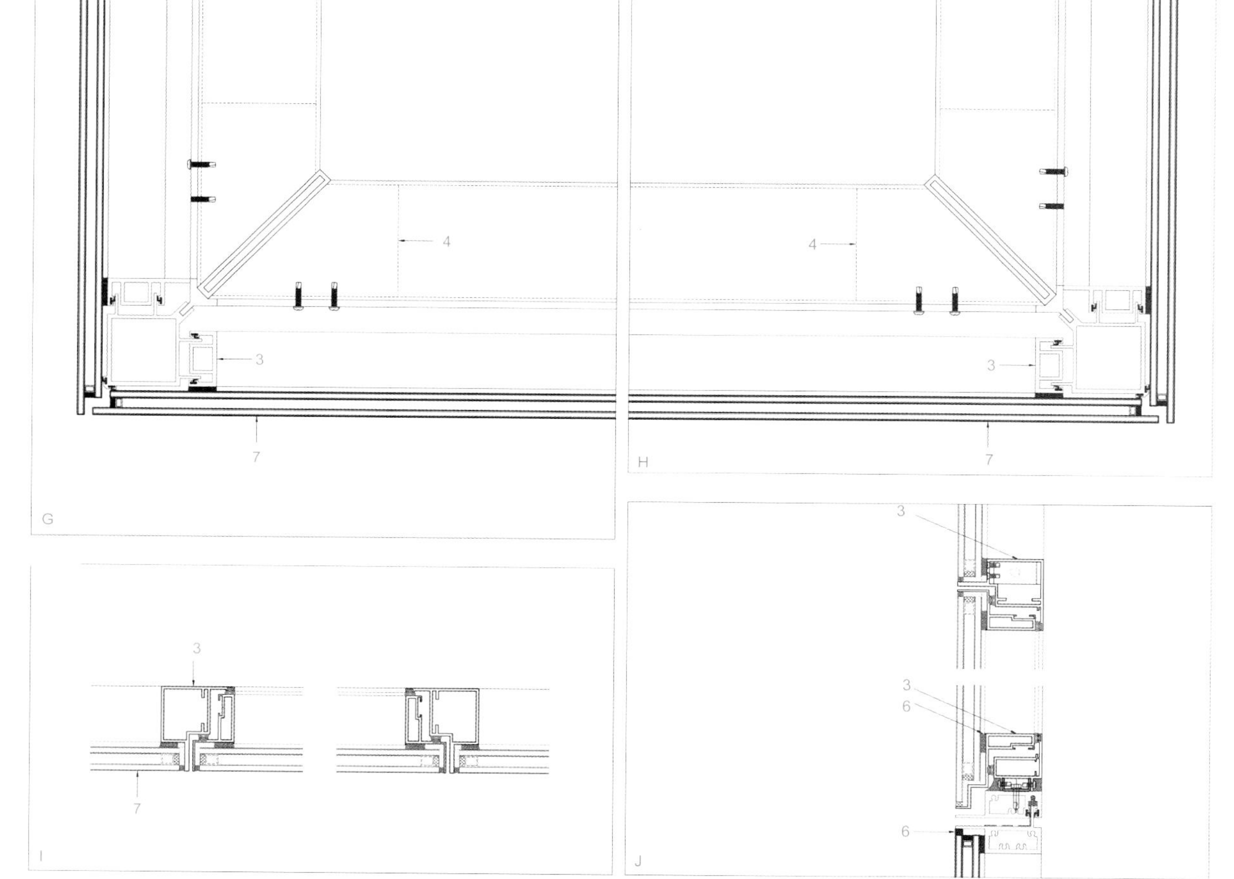

项目概况

大东艺术中心位于中国台湾南部的经济中心——高雄市。文化中心将凤山老城区、大东公园和凤山河连接起来，通过剧场、展览厅、艺术图书馆和教学中心之间的过渡空间形成了连续的公共空间。

公共空间被充分利用，进行了舞蹈、太极以及各种各样其他的运动形式，极具台湾特色。薄膜屋面为下面的活动空间提供了阴凉。屋顶造型能抵御台风、暴雨、高热等极端气候条件。屋顶上的大洞能让雨水流入内芯。热空气上升到屋顶开口，形成了新鲜的室内气流。900座位的剧场和小型排练场被木板和穿孔金属板包覆起来，是整个项目的核心。剧场外面是X形混凝土立面，内里则是木制空间。整个项目的立面和谐统一，同时又通过密集度的变化暗示了内部的空间设置。

为了实现更好的音响效果，礼堂的天花板可根据音乐和剧场模式进行调节。木板层后面的吸音材料同样是可伸缩的。

细部与材料

大东艺术中心的造型概念以构筑形式的二元性为基础。德国建筑师与历史学家哥特菲尔德·西姆珀的原始小屋所体现的“构造”与“分割”理论是大东艺术中心的结构、构造和细部设计的基本指导原则。

薄膜屋顶体现了“构造原则”，而立面图案则体现了“分割概念”，包括倾斜的混凝土柱、三角分割的玻璃幕墙和X形墙面雕刻。这种立面为分离的体块打造了水平延展的趋势，放大了建筑体量的厚重感和整体感。由于剧场是一种复杂的建筑类型，结构系统混合了钢筋混凝土、钢架和承重系统。设计所面临的挑战是如何在不同的结构系统中应用同一个原则，同时又认真贯彻构造逻辑。

素面混凝土是主要建筑材料，与软薄膜形成了对照。三角分割帮助建筑立面保持了向上的趋势，而轻质薄膜屋面则由趋向一点的结构支撑。玻璃幕墙与混凝土表面是平齐的。为了保证高品质的建筑细节，建筑师特别注重转角的开发，避免形成任何转角支架，一直以建筑立面作为承重结构。

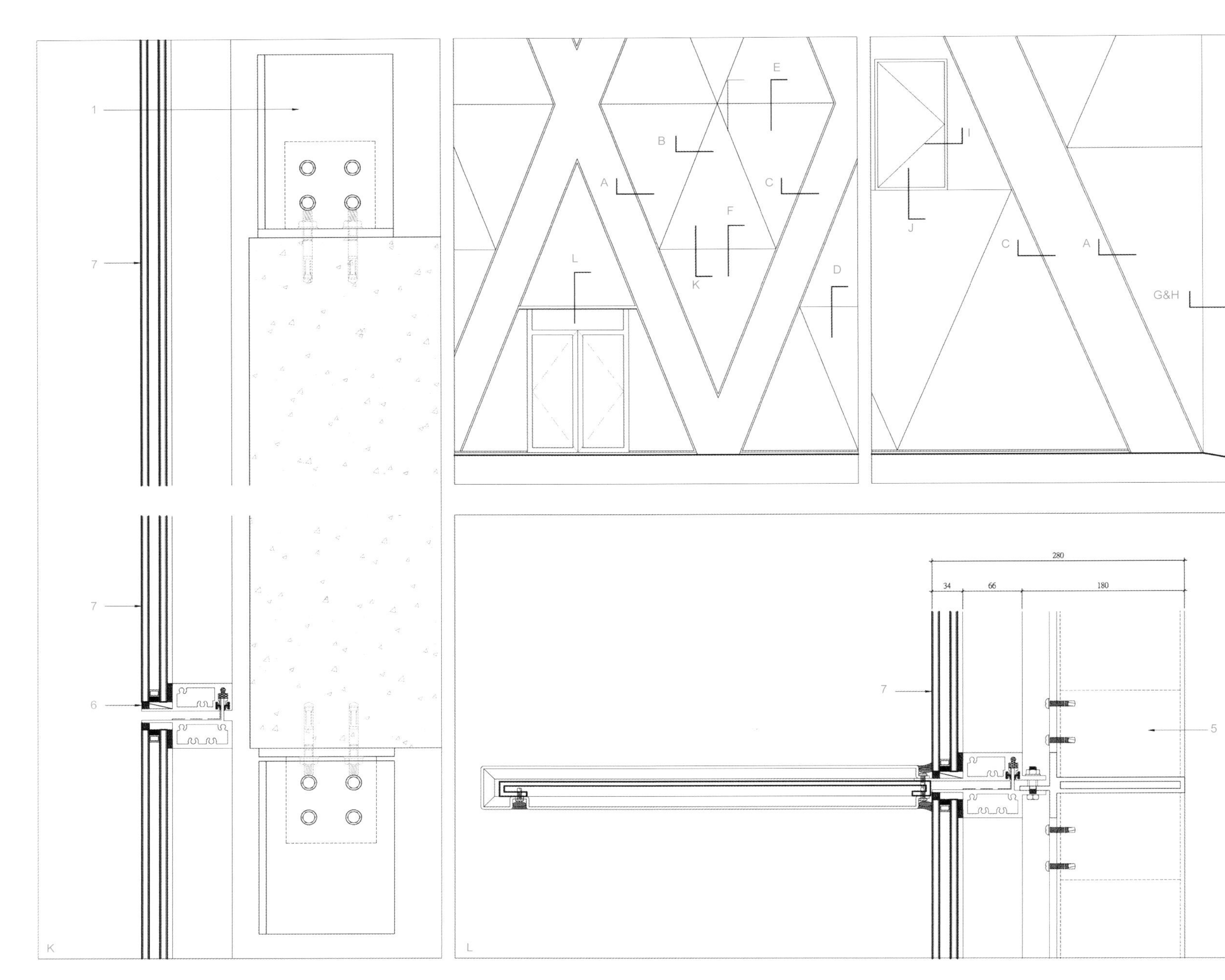

1. Aluminum T-profile vertical element
2. Cast aluminum T-profile connector
3. Aluminum frame
4. Cast aluminum six-way connector
5. Cast aluminum three-way connector
6. Structural silicone
7. Double-layered Low-E glass
8. Concrete wall

1. T形铝材垂直元件
2. T形铸铝连接件
3. 铝框架
4. 铸铝六通连接件
5. 铸铝三通连接件
6. 结构硅胶
7. 双层低辐射玻璃
8. 混凝土墙

Urban Hive

城市蜂箱

Location/地点: Seoul, Korea/韩国，首尔
Architect/建筑师: Kim in cheurl + archium
Site area/占地面积: 1,000.90m^2
Built area/建筑面积: 10,166.89m^2
Completion date/竣工时间: 2012
Key materials: Façade – cool gray colour exposed concrete, THK26 Low-E pair glass
主要材料: 立面——酷灰色素面混凝土、26厚低辐射双层玻璃

Overview

The essential component of architecture is structure. It is the absolute role of structure that overcoming gravity and creating space. As for a high-storeyed building with repeated typical floors, structure is the most important element. Structure is frame of space. Common sense says that the frame is first created and added with clothes later. Common sense is formal solution. But, it is often stereotyped. If clothes are first created and the frame is created later in a reversed way, the essence of architecture will be changed.

Rather than creating shape with distortions and controls, the architect wanted to make clear space with simplicity. Structure can be reveal with turning common sense that is the composition between interior and exterior over. Completing structure is completing of space. And space is the result of architecture.

Detail and Materials

Thick concrete wall that divides indoor and outdoor is just for bearing out weight, it is not closing space. So, it must be bored for connection between indoor and outdoor. Concrete that is bored is alike light sponge. The openings are for space. The openings are keys to architectural expression. However, how to look at the openings is more important than what the openings themselves looked like.

The openwork circular holes are not only suitable shape for structure but also maintain both solid and void conditions. The transparent glass screen that is into the structure is not windows but just maintains internal conditions.

项目概况

建筑的主要成分是结构。结构在克服重力、创造空间的过程中扮演着绝对重要的角色。对采用重复的标准楼层设计的高层建筑来说，结构是最重要的元素，它是空间的框架。从建筑常识上来讲，我们应当首先建造结构，然后再给建筑穿上衣服。但是这种常识具有思维定势，如果先给建筑穿上衣服，然后再建造结构，那么建筑的本质也将改变。

建筑师并不想打造扭曲的限制造型，而是想要一种简单干净的空间。如果打破常规，那么结构可以看作是室内外之间的构成物。完成结构就是完成空间，而空间正是建筑的结果。

细部与材料

厚重的混凝土墙将室内外分割开，仅起到承重作用，而不是封闭空间。因此，必须通过墙壁上的钻孔来连接室内外。混凝土像海绵一样被钻出圆孔。这些开口是建筑表达的关键元素。然而，如何看待这些开口比这些开口本身的外观更为重要。

圆形开口不仅在形状上与结构相称，还保持了虚实结合的空间环境。嵌入结构的透明玻璃幕墙不是窗户，而是为了保持室内环境的舒适度。

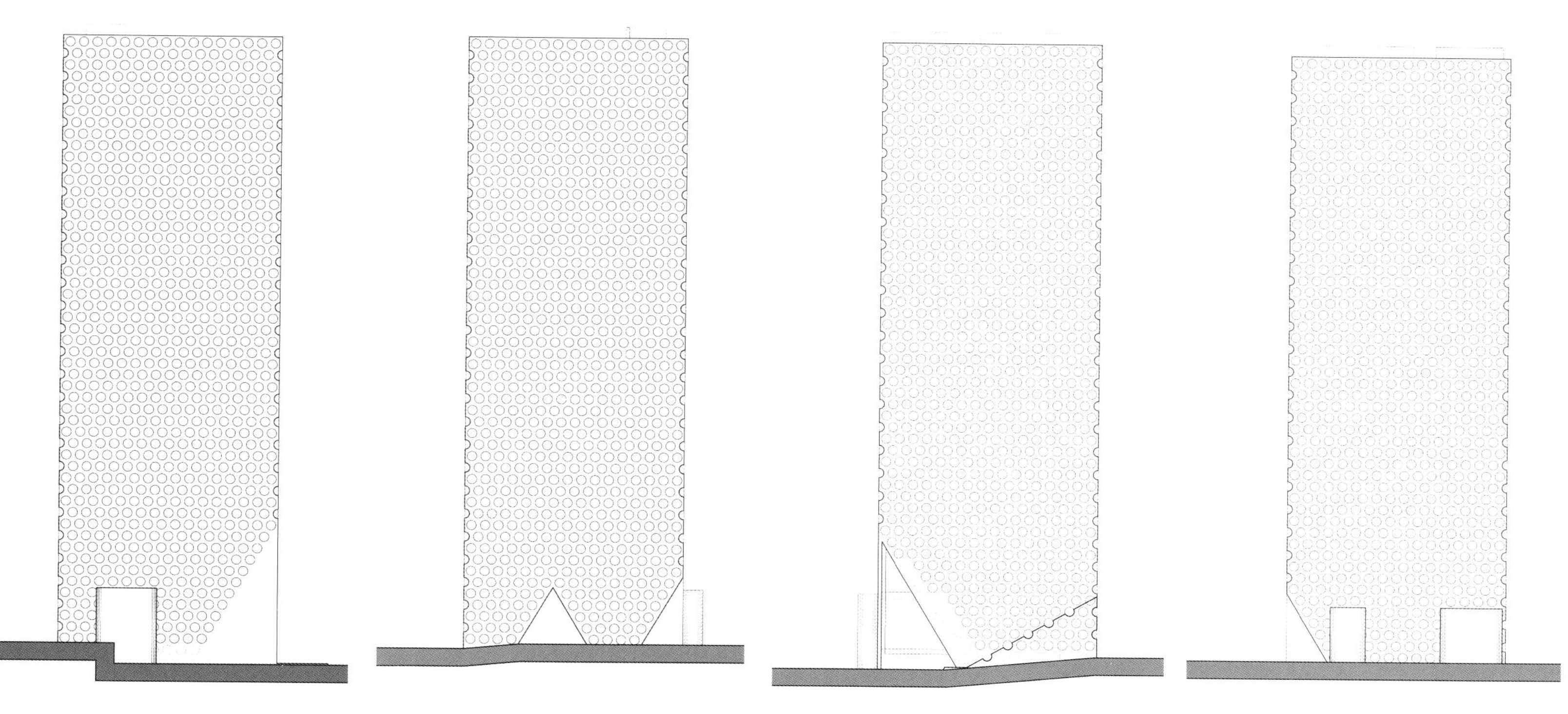

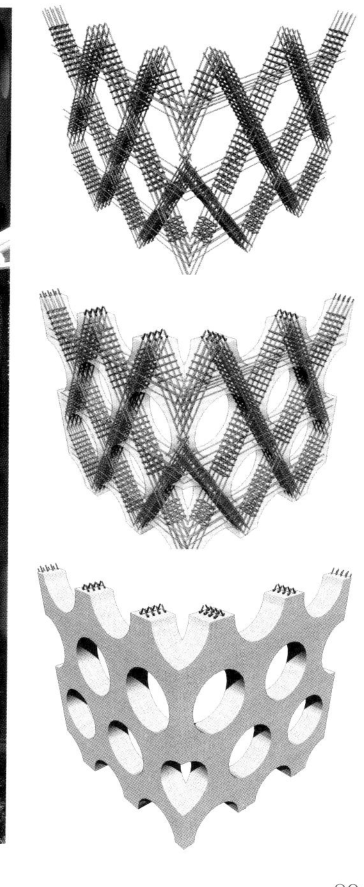

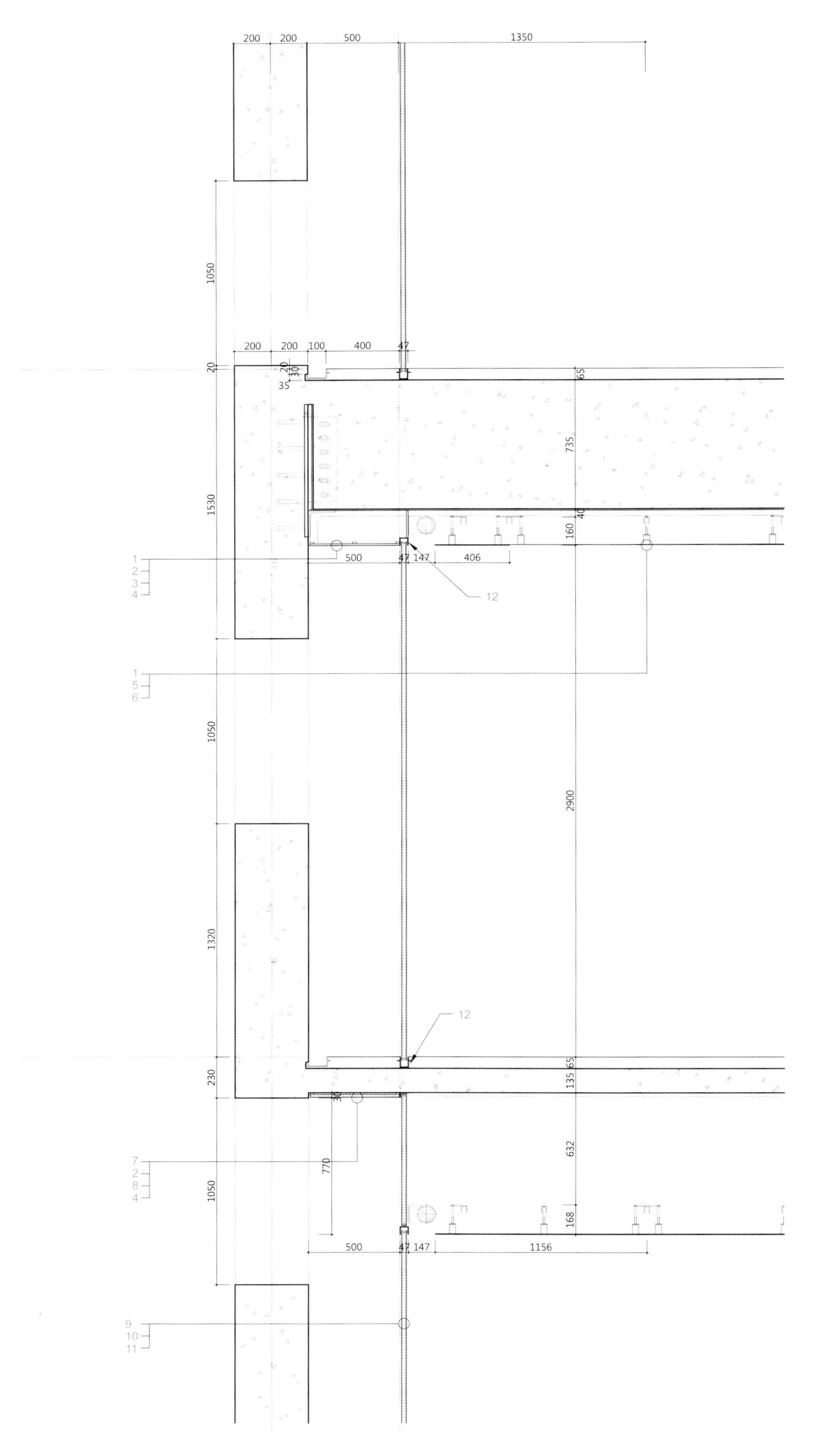

1. Concrete, deck slab & TSC beam
2. STS (Stainless steel) square pipe 10×20×1@200
3. THK6 N.F. board/THK9 CRC board
4. Exterior indicated color paint
5. Light weight steel ceiling frame
6. THK3 AL punched metal
7. THK38 new HICOTE TP spray coating
8. THK9 CRC board
9. STS (Stainless steel) T-bar @1500
10. Structural sealant
11. THK26 (8+12+6, W1500×H2950) low-e pair glass
12. THK3 glass shoe + SST T-bar @1500

1. 混凝土，楼板 & TSC梁
2. STS (不锈钢) 方管10 × 20 × 1@200
3. 厚度6毫米N.F.板/厚度9毫米CRC板
4. 外部指定颜色涂漆
5. 轻质钢材天花板结构
6. 厚度3毫米AL穿孔金属
7. 厚度38毫米新 HICOTE TP喷漆涂层
8. 厚度9毫米CRC板
9. STS (不锈钢) T形条@1500
10. 结构密封胶
11. 厚度26毫米(8+12+6, W1500 × H2950) 低辐射双层玻璃
12. 厚度3毫米玻璃底板+SST T形条@1500

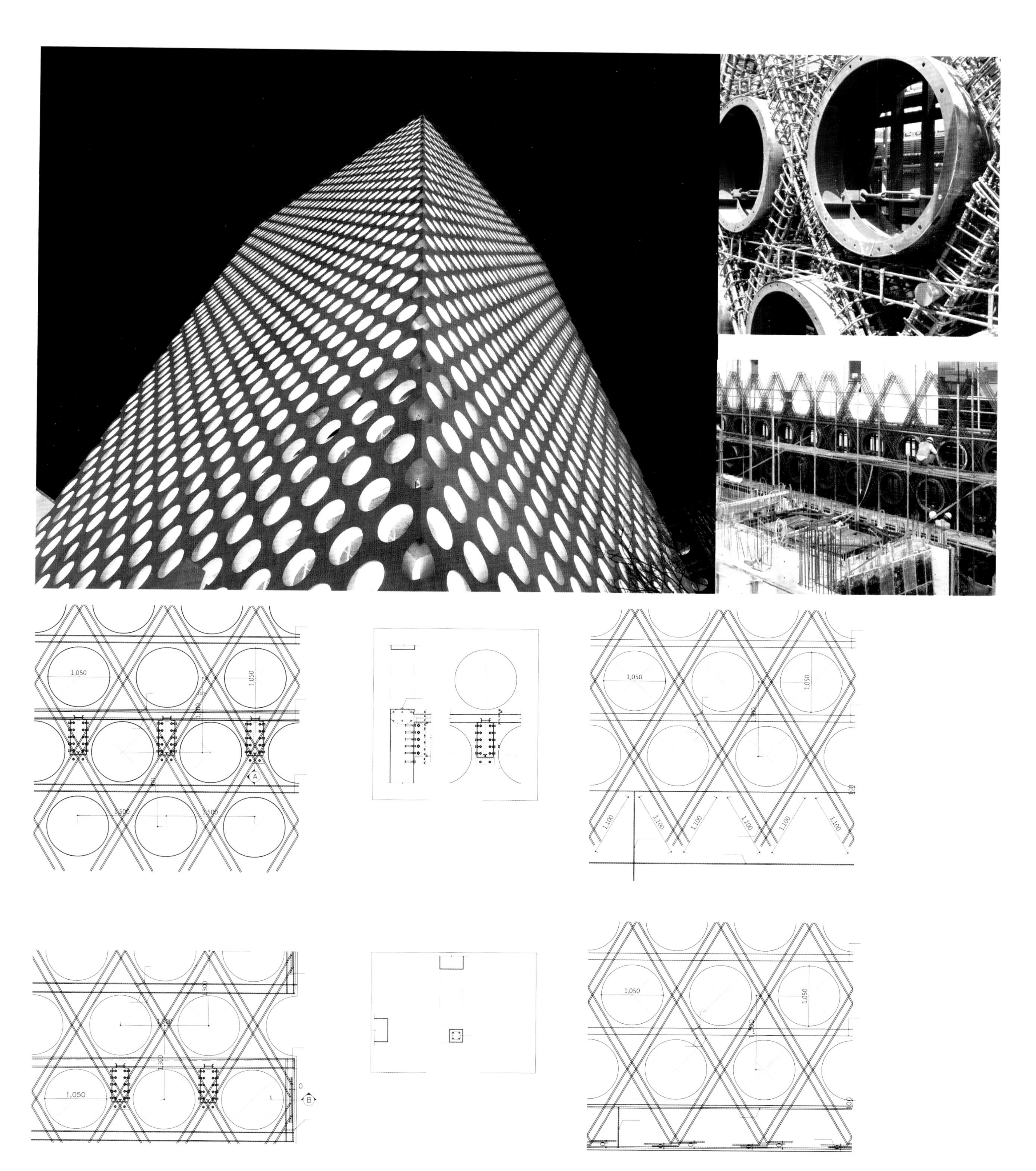
1,050
1,050
1,500
1,500
A
1,050
1,050
100
1,100
1,100
1,100
1,100
1,100
1,100
1,300
1,300
1,050
B
1,050
1,050
100

Can Felic Nursery

坎菲里克幼儿园

Location/地点: Benicàssim, Spain/西班牙，贝尼卡西姆
Architect/建筑师: Estudio Fernández-Vivancos, Abalosllopis Arquitectos
Photos/摄影: José Manuel Cutillas
Built area/建筑面积: 2,340m²
Key materials: Façade – sandblasted reinforced concrete
主要材料： 立面—— 喷砂面混凝土

Overview

The design of Can Feliç offers the opportunity to investigate the concepts of the one and the multiple, the same and the different, as a way to deepen the understanding of the human relationships that are established between the individual and society.

Given that a nursery education programme is being developed for children aged 0-3, the objective of this project is not only to fit out and equip the necessary space for learning, but also to ensure that a tranquil and protected atmosphere is created, suitable for both recreation and teaching.

The composition cell is defined as the unit made up of two rooms grouped around an area made up of service and communal installations, optimising the running of the nursery. These cells are located around a central space, a covered plaza, surrounded by separate small pavilions that open onto an exterior patio, that combine to form a play area that complements the patio.

The fan-like orientation of the pavilions around the covered plaza is a liberation from the strict logic of parcelling out, and seeks a link with the passing of the sun from east to west, the presence of trees and the oblique view to the patios to achieve the largest possible focal depth in the available space.

Each group of eight children form a small family and to each one is assigned an independent place, a home. All are equal and at the same time are different according to their position relative to the sun, the view and the relationship with their neighbours. Together they form a community that gathers around the central space has continuity in a grove of trees where the units are sheltered, a courtyards house in the woods. The grove grows and merges with the urban space forming a publicly accessible garden overlooking the Desierto de las Palmas, the fundamental landscape of the common identity of Benicàssim.

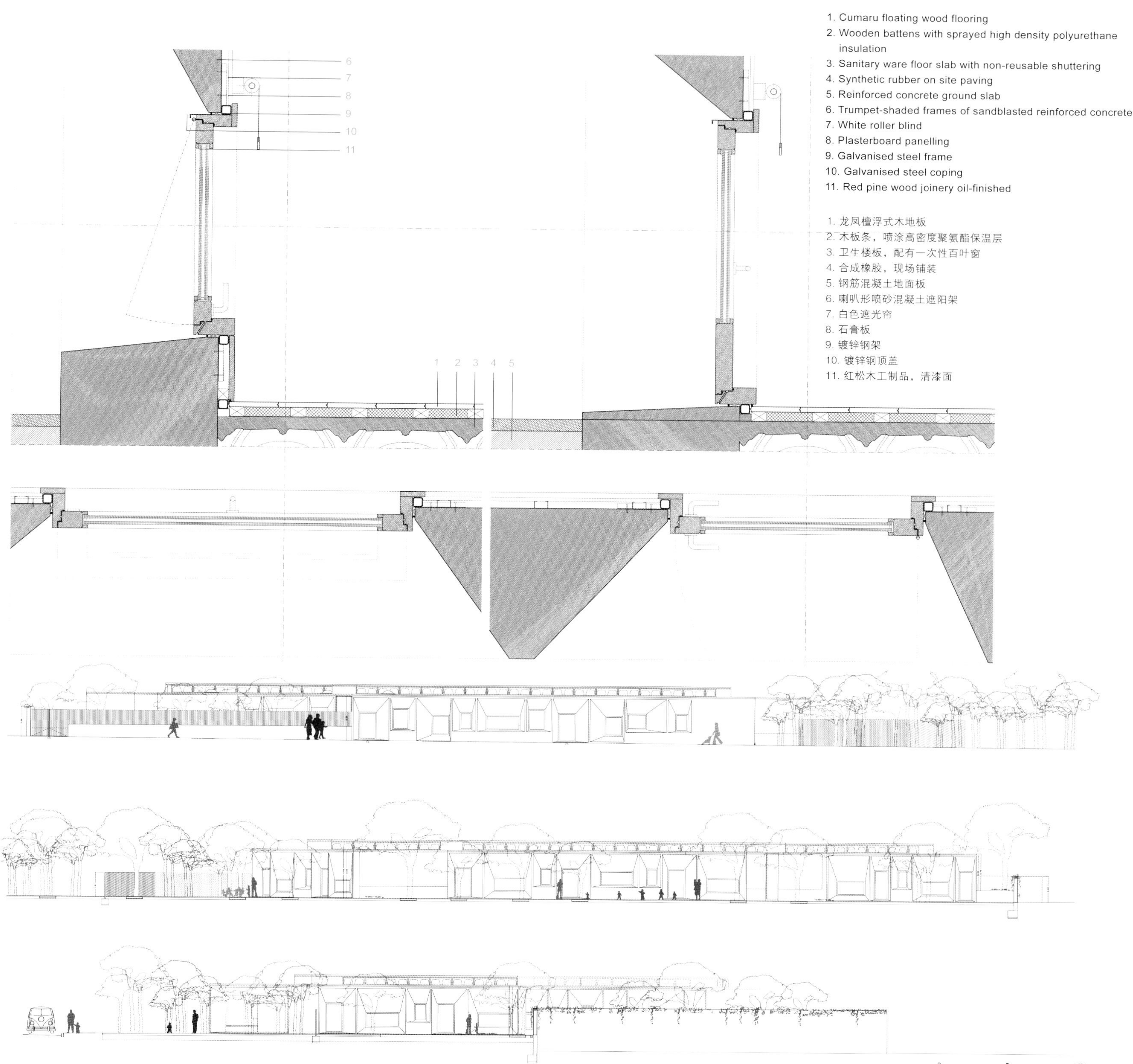

项目概况

坎菲里克幼儿园的设计让建筑师有机会探索了单一与多样、相同与差异的概念，使建筑师进一步了解了个人与社会之间的关系。

幼儿园专为0至3岁的幼儿提供教学看护服务，项目的目标不仅是配置必要的教学空间，还要营造出一种安静祥和的氛围，既适合娱乐，又适合教学。

建筑以两个房间为一组，围绕着一个服务和公共设施区展开，优化了幼儿园的交通流线。这些单元围绕着中央空间展开，中央广场四周环绕着与露天天井相连的独立小屋，小屋与天井共同组成了游戏区。

小屋环绕广场的扇形布局打破了严格的打包式布局，试图在由东至西的太阳轨迹、树木以及天井的侧面视野之间实现一种联系，在有限的空间内实现焦点深度的最大化。

每8个孩子组成一个小群体，每个群体拥有一个独立的小家。所有小家的空间大小都相同，但是它们与太阳的相对位置、视野、与周边建筑的联系都是不同的。它们共同组成了一个社区，通过中央广场和小树林联系起来。随着树林的生长，它将融入城市，成为一个公共花园，俯瞰拉丝帕尔马斯沙漠——贝尼卡西姆的标志性景观。

Refurbishment and Extension "Balainen School"

巴来恩学校翻修及扩建

Location/地点: Nidau, Switzerland/瑞士，尼道
Architect/建筑师: Wildrich Hien Architekten
Photos/摄影: Johannes Marburg
Site area/占地面积: 10,900m²
Built area/建筑面积: 1,400 m²
Completion date/竣工时间: 2012
Key materials: Façade – sandblasted concrete
主要材料: 立面——喷砂面混凝土

Overview

A solitaire that is inspired by the existing buildings, but who re-interprets their structure in a contemporary way complements the historical school complex with its main building and gymnasium. The extension approaches the 1918-built ensemble respectfully, yet takes a conscious role within the complex. The extension reacts to the scale of the surrounding one-family dwellings and thus forms a transition between the quarters and the school's dominant main building. The oversized dormer window contributes to the extension's independence and creates a visual link with the shape of the opposing gymnasium. The sandblasted concrete shell was carefully tuned to fit the colours of the existing buildings' plaster façades.

While the existing school building contains the main teaching rooms, the rooms for handcraft-education and the teachers' facilities, the extension provides room for the library, natural science classes and a large auditorium. Around the lofty and generously dimensioned staircase-hall, the classrooms are organised in enfilade in L-shaped layers. White plaster walls and colourful screed floors enclose the hall and staircase, while the classrooms are equipped with wooden floors and colourful wooden cabinets that contain the blackboards, fountains and which divide the classrooms from one another. The auditorium is situated in the extension's upper floor and can easily be identified from the outside due to its large dormer window. The stage acts as a black box within the auditorium.

A generous plateau, elevated above ground, links the existing buildings barrier-free with the extension and creates a distinctive outdoor "classroom" with tiered seats towards the schoolyard and river. Filigree parasol-like sculptures cover the plateau and form a signature element that is implemented in various places across the complex.

Detail and Materials

The extension's structural system comprises of load-bearing outer walls and concrete ceilings whose reinforcements are linked together. Isolating layers on the inside enable an effective, low-cost structural approach.

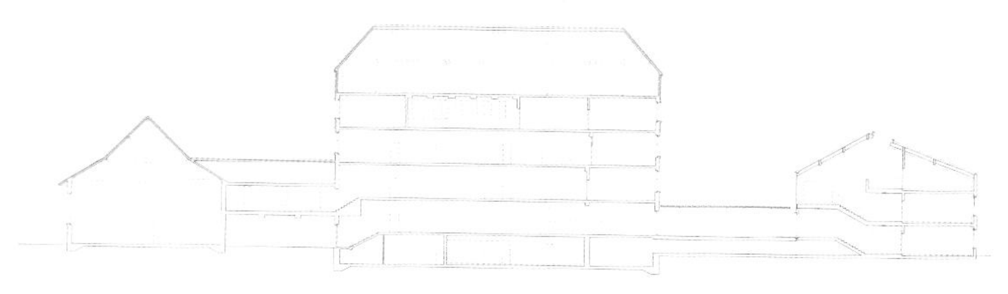

The outer shell contains a mixture of concrete and colour-giving mineral additives and was sand blasted on the outside, while the window reveals where left smooth.

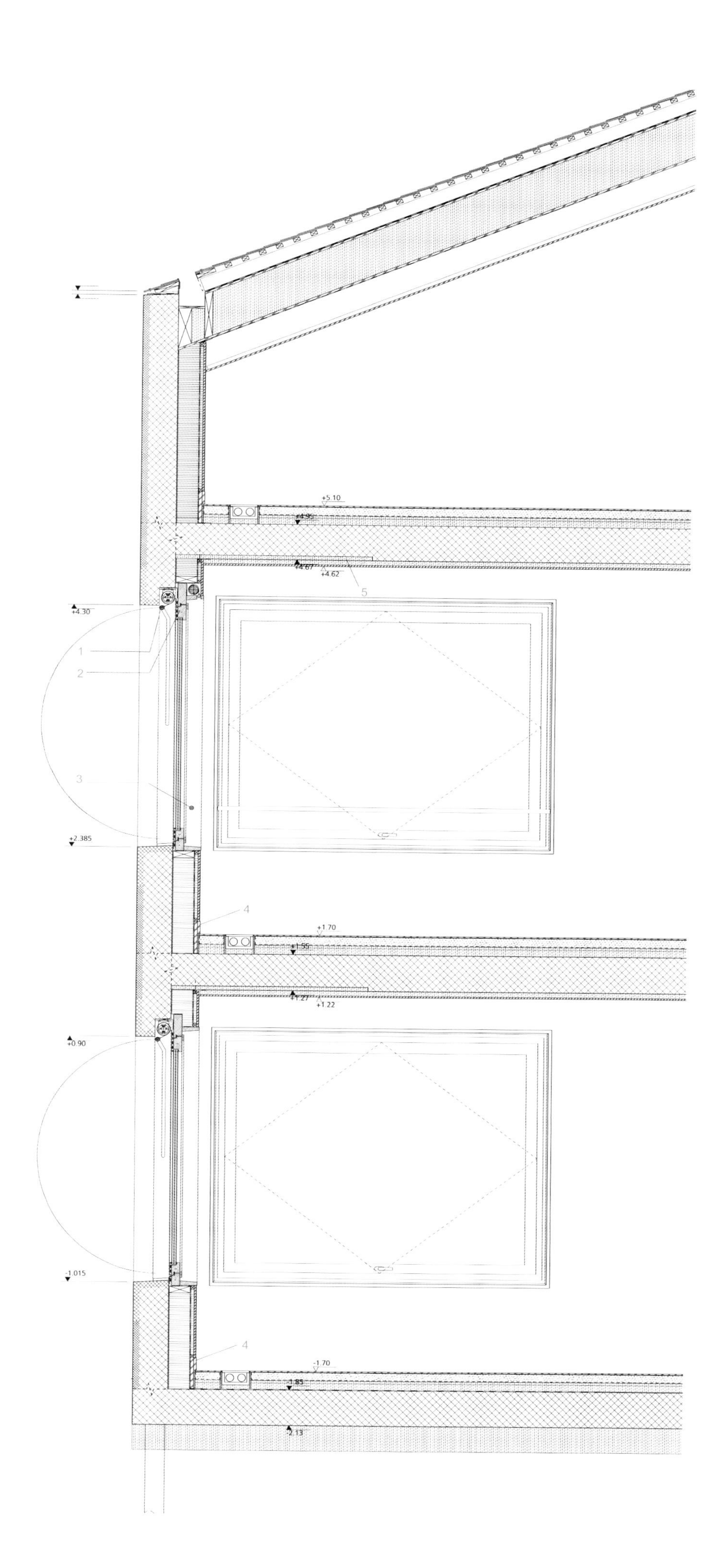

项目概况

新建筑的设计从原有建筑中获得了灵感，但是以现代的方式重新诠释了它们的结构，与历史悠久的校园整合了起来。扩建项目充分尊重了建于1918年的历史建筑，同时也呈现出鲜明的特色。它反映了周边独栋住宅的规模，在住宅区与学校主楼之间形成了过渡。超大的屋顶窗保证了新建筑的独立性，使其与对面的体育馆在视觉上连接起来。通过精心调制，喷砂面混凝土外壳的色彩与原有建筑的石膏外墙的色彩几乎一致。

原有的教学楼主要包含教室、手工培训室和教师办公室；扩建楼内是图书馆、自然科学教室和大型礼堂。教室环绕着宽敞的走廊大厅，呈L形分层纵向排列。白色石膏墙面和彩色砂浆地面将大厅和楼梯间包围起来。教室则铺设着木地板，黑板配有彩色木框，将不同教室区分开。礼堂位于扩建楼的顶部，拥有巨大的屋顶窗。舞台则作为一个黑盒子结构设置在礼堂内。

新旧建筑通过一块高出地面的平台无障碍连接起来，形成了独特的露天教室，台阶座位面向校园和河流。平台上布满了金银丝工艺的阳伞状雕塑结构，构成了整个空间的标志性元素。

细部与材料

扩建楼的结构系统由承重外墙和混凝土天花板组成，二者的钢筋加固件连接在一起。内部的隔离层保证了高效经济的结构设计。

建筑外壳由混凝土和彩色矿物添加剂的混合物构成，外表面采用喷砂处理，窗侧部分则是光滑的表面。

1. Drop arm awning with cranked steel frame
2. Wood-Metal window
3. Steel rod φ 33mm as fall protection
4. Baseboard flush with the plaster
5. Sound insulation insert 30mm XPS in formwork

1. 转向轴遮阳篷，配弯曲钢框
2. 木制/金属窗
3. 防跌落保护钢杆，φ 33mm
4. 护壁板，与石膏抹面平齐
5. 隔音层，在模架中嵌入30mm挤塑聚苯板

Lar Casa de Magalhães

麦哲伦之家

Location/地点: Ponte de Lima, Portugal/葡萄牙，朋特利马
Architect/建筑师: José Manuel Carvalho Araújo
Photos/摄影: Hugo Carvalho Araújo
Site area/占地面积: 18,000m²
Completion date/竣工时间: 2010
Key materials: Façade – concrete
主要材料: 立面——混凝土

Overview
Inside a white house lives a benefactress lady. She donates the land, home and attachments to build a home for the elderly. She only requires that the construction be done prior to her death. It all starts from the house; it is the center and the symbol.

The building of the home for the elderly results from the expansion of the platform floor of the existing house. In the center are built two courtyards, one social and one for services, dematerializing the built mass, as if a part of the building had been removed, exposing its interior. You realize there's life and movement.

Arranged around the courtyard the social areas and the 27 rooms invite the fellowship of a shared life, a sense of connection and security. The building is reduced to the scale of a house. Everything is connected in one floor, in ground level. There is a certain feeling of familiarity and a scale possible of controling.

The circulations are simple, direct, wide and bright. The lobby intersects the building and creates two opposing entrances, the main and service. And the circulation ring that embraces the two patios.

Detail and Materials
Outside, the rhythm of the facade refers to the tree trunks, a blending option, highlighting the existing houses, white-washed.

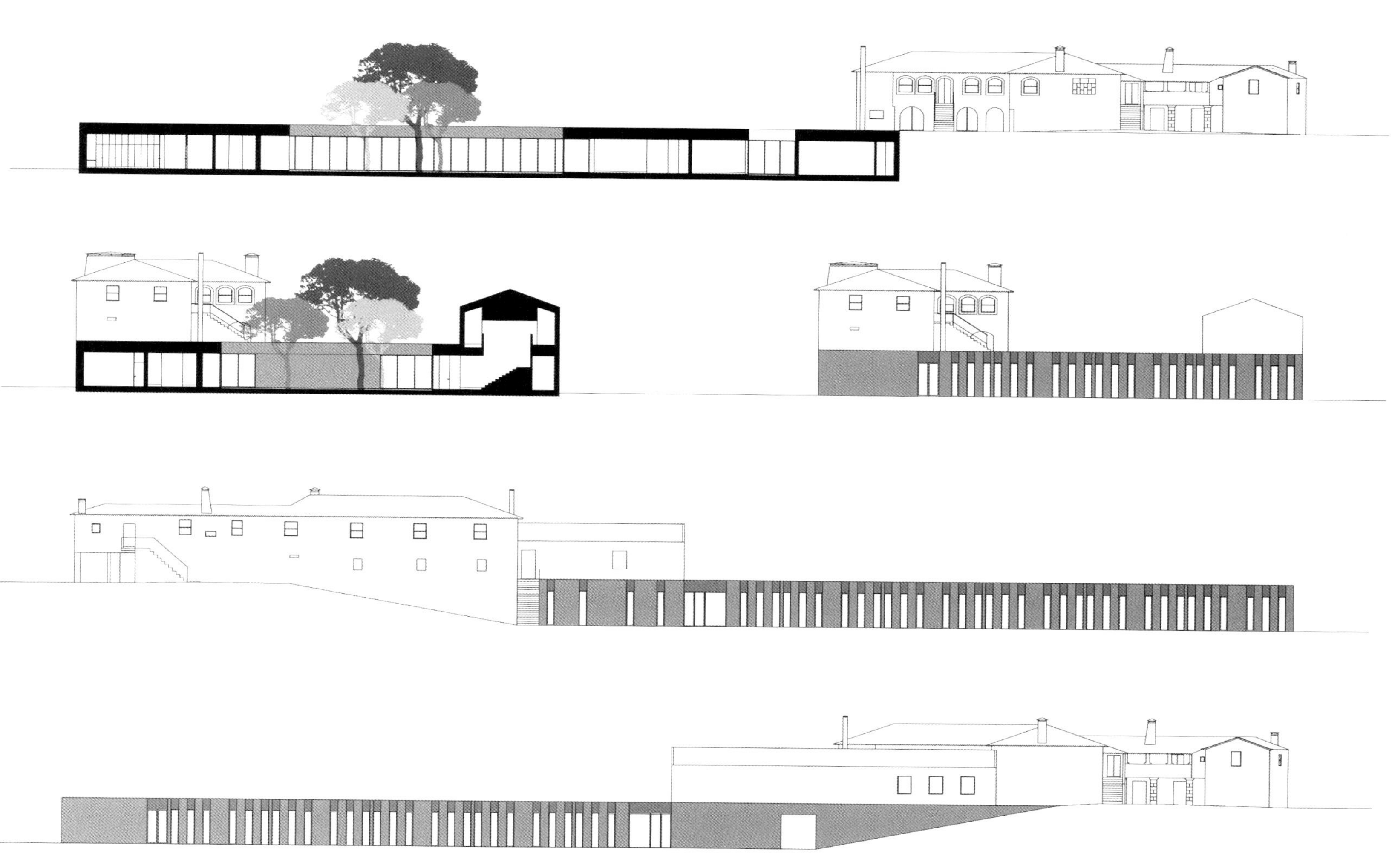

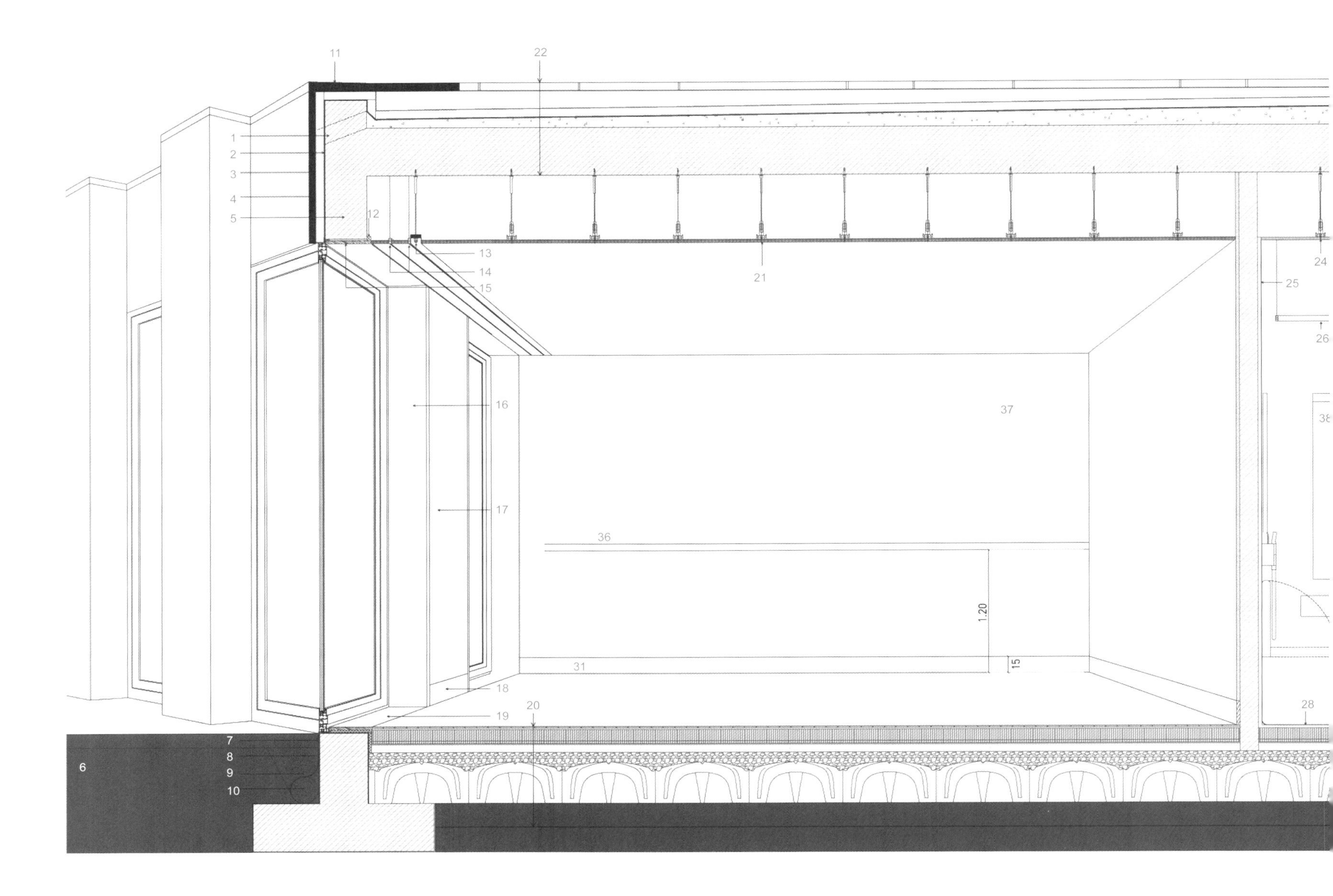

1. Hidden drain pipe w/pine top;
2. Waterproofing;
3. Concrete paving flag with incorporated thermal insulation mechanically fixed to the support with adequate texture to fix monomass;
4. Covering in monomass w/ inerts;
5. Structure in waterproofed reinforced concrete;
6. Soil w/ draining layer;
7. Metallic piece to finish membranes;
8. Geotextile draining felt (covers drain) draining membrane, "DELTA-DRAIN"; type, waterproofing w/ asphalt membranes;
9. Rock fill with variable gravel;
10. Drain covered with geotextile;
11. Finishing slab w/ slope towards the interior, in pigmented concrete w/inerts similar to those in the façade (made to measure);
12. Batten to fix board;
13. Steel track with ABA for fluorescent light fitting;
14. Double track for curtains, embedded into ceiling, type Kirsck – cabinskena;
15. Trim of the bedroom frames with embedded pine wood;
16. Posts in pine wood on same plane as wall;
17. Wall, plastered and painted with matt plastic paint, 15cm ceramic brick;
18. Skirting in pine wood embedded into wall w/ 15cm height;
19. Board in pine wood to level with floor;
20. Bedroom slab – bedroom pavement in glued pine lamparket, levelling mscreed, 0.10m thickness w/mesh or fibres and radiant floor type heating system, thermal insulation w/ mechanical protection / insulating membrane, levelling (6cm min.), "cupolex"

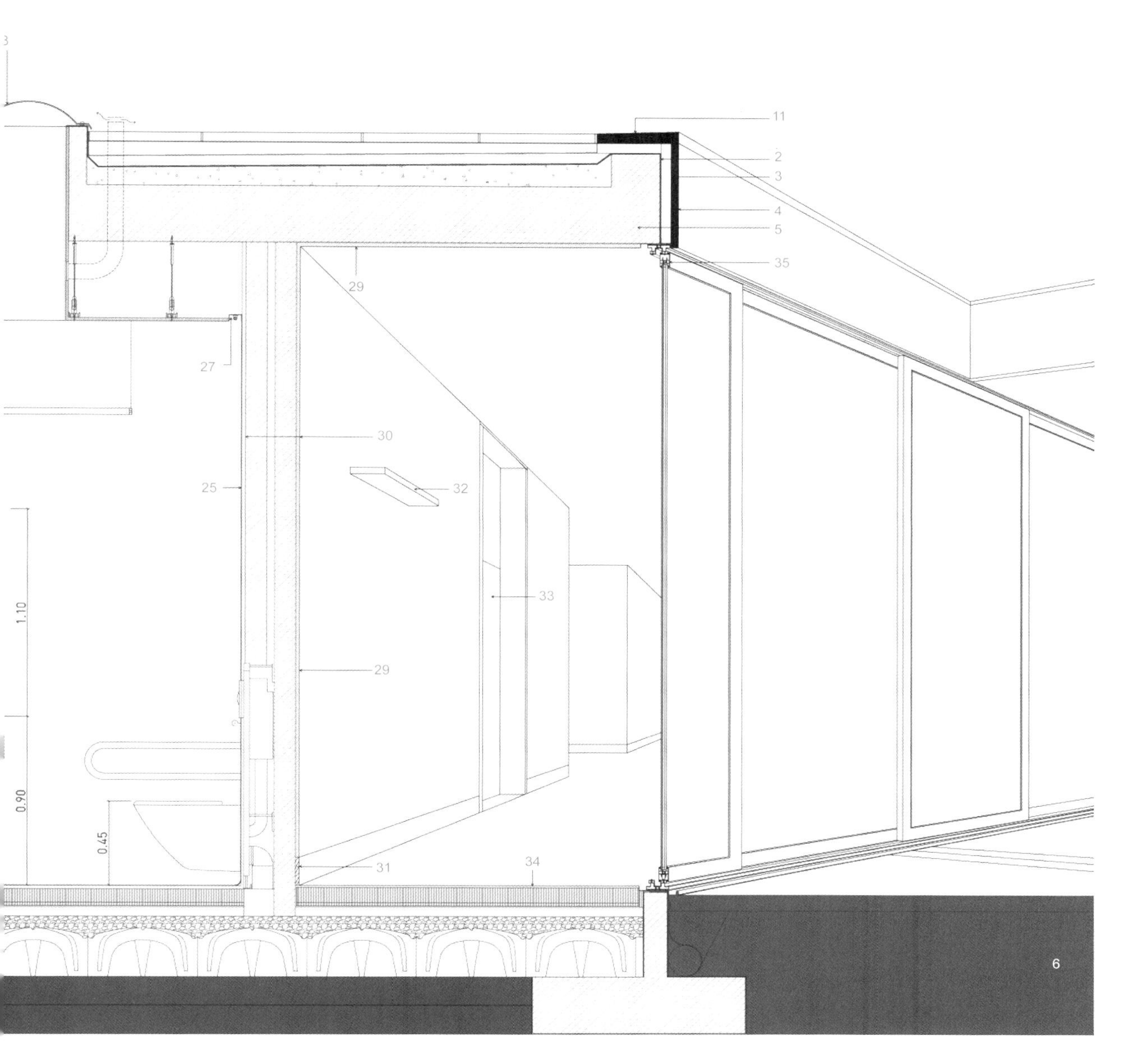

项目概况

本项目的赞助人居住在一座白色的住宅内。她为这个老年之家捐赠了土地、住宅和附属设施。她的唯一要求是项目必须在她有生之年完工。所有的一切都围绕着这座住宅展开，以它为中心和象征。

老年住宅的建造始于对原有住宅平台层的扩建。项目中央是两个庭院，一个用于社交，一个用于服务。它们弱化了建筑的集中感，就像是拆除了建筑的一部分，露出了内部。你能从中感受到生命与活力。

围绕着庭院展开的社交区和27间客房吸引着老年人前来体验共同生活，给人以团结感和安全感。建筑的规模与住宅相似，所有设施都在同一层楼连接起来，从而营造出一种居家感，让人感到一切尽在掌控之中。

建筑的内部交通动线简单、直接、宽敞、明亮。大厅贯穿建筑，形成了两个相对的入口——正门和服务门。交通动线连接了两个天井。

细部与材料

参考了树干的感觉，协调统一，突出了原有住宅的白色粉刷外墙。

system w/ ventilation of the airchamber, cleaning concrete;
21. False ceiling of the bedroom in "scots" type pine wood panelling w/ adequate structure to support the panelling;
22. Concrete slabs, draining layer in porous concrete, geotextile felt, thermal insulation – extruded polystyrene, double layered asphalt membranes, slope creating layer w/ levelling – 2% slope, structures in waterproofed reinforced concrete;
23. Standard fixed circular skylight;
24. False ceiling of the toilet in hydrofuge plasterboard or similar, painted w/ adequate structure to support false ceiling;
25. Covering in monochromatic vinyl over parget;
26. Rail suspended at 2.50m for shower curtain, type Kirsck – cabinskena;
27. Steel track with tab for fluorescent light fitting;
28. Pavement in monochromatic vinyl w/ reinforcement at edges, levelling layer;
29. Plaster parget, painted;
30. Two walls in 11cm brick;
31. Skirting in pine wood embedded into wall w/15cm height;
32. Corridor lighting fixed to wall, type Level by Regent c/600mm (2x55w);
33. Bedroom entrance door in pine w/ frame in pine boards in flush with walls;
34. Pavement of the corridor in pigmented concrete w/ adequate finish;
35. Sliding, matt thermal lacquered aluminium frame (type Navarra N24 200), placing of the pre-frame to fix the fixed frame;
36. Electrified rail on wall for sockets, switches and bed head lighting;
37. Wall parged and painted with matt plastic paint;
38. Mirror fixed to wall;

1. 隐藏式排水管道
2. 防水层
3. 混凝土铺路石板，配有保温层，固定在支架上
4. 覆盖层
5. 防水钢筋混凝土结构
6. 土壤，配排水层
7. 金属件，用于装饰薄膜
8. 土工布排水毡排水膜，DELTA-DRAIN型，防水，配沥青膜
9. 碎石填充
10. 排水沟，土工布覆盖
11. 装饰板，斜坡朝向室内，染色混凝土，纹理与外墙类似
12. 板条，用于固定板材
13. 钢轨，固定荧光灯装置
14. 双层窗帘轨道，嵌入天花板，Kirsck – cabinskena型
15. 卧室边框修整，嵌入松木
16. 松木杆，与墙面平齐
17. 墙面，石膏抹面+压光塑料漆，15cm瓷砖
18. 松木壁脚板，嵌入墙壁，15cm高
19. 松木板与地面平齐
20. 卧室板——卧室铺装胶合松木地板、找平砂浆（0.10m厚，配网）、地热系统、隔热层（机械保护、隔热膜）、找平层（最薄6cm）、cupolex系统（配通风气腔）、清水混凝土
21. 卧室假吊顶，松木镶板，配支撑结构
22. 混凝土板，多孔混凝土排水层、土工布毡、隔热层（挤塑聚苯乙烯）、双层沥青膜、斜坡层（带找平，坡度2%）、防水混凝土结构
23. 标准固定圆形天窗
24. 洗手间假吊顶，防水石膏板（涂漆），配支撑结构
25. 单色乙烯基覆盖层
26. 2.50m高轨道，用于悬挂浴帘，Kirsck – cabinskena型
27. 钢轨，用于安装荧光灯
28. 单色乙烯基铺装，边缘加固，配找平层
29. 石膏灰泥
30. 两面墙壁，11cm砖块砌成
31. 松木壁脚板，嵌入墙壁，15cm高
32. 走廊墙面照明，Level by Regent c/600mm（2x55w）型
33. 松木卧室门，松木板框与墙壁平齐
34. 走廊铺装，染色混凝土
35. 滑动亚光热漆铝框（Navarra N24 200型），安装在预制框架上
36. 电线轨道，用于在墙壁上安装插座、开关和床头灯
37. 灰泥抹面墙，亚光塑料漆
38. 镜子，固定在墙面上

Expansion of Cemetery of Ponte Buggianese

蓬泰布贾内塞公墓扩建

Location/地点: Ponte Buggianese, Italy/意大利，蓬泰布贾内塞
Architect/建筑师: Massimo Mariani
Photos/摄影: Massimo Mariani, Alessandro Ciampi
Area/面积: 4,500m^2
Completion date/竣工时间: 2012
Key materials: Façade – concrete and acrylic coating
主要材料：立面——混凝土与丙烯涂层

Overview

The design concerns an area of about 4,500 square metres, which is placed next to the already existing cemetery (on the west side of it). The plan, which also involves the south side, where is placed the current side entrance, consists of the building of about 1,230 loculi, 720 ossuaries/cinerary urns and 12 private chapels, thus the complex will cover the citizen's needs for the next 15 years.

From the urban point of view the cemetery is divided into 2 parts: the first-one reveals a clear 19th century influence and looks like a lot of other places of the Tuscan surroundings; the second has been built later and dates from the 60's/80's. Therefore the different languages of the two planners create a typological confusion producing a loss of identity.

The new plan tries to give up the cemetery with its own identity both of worship and respect. The old church has been demolished to free the access to the expansion area, and a new church has been built as to complete the central perspective axis. By this way the ensemble is crossed by two longitudinal axes: the first one is connecting the main entrance to the church, the second starts from the side entrance and ends in front of the loculi building.

The new church is characterised by four large crosses, one per side of the building, they solve both structural and iconographic function. The rest of the church is made of glass and it comes out from the stone-coated private chapels. On both sides of the church there are two buildings in white reinforced concrete, featured by the large grating, which creates a shaded zone behind.

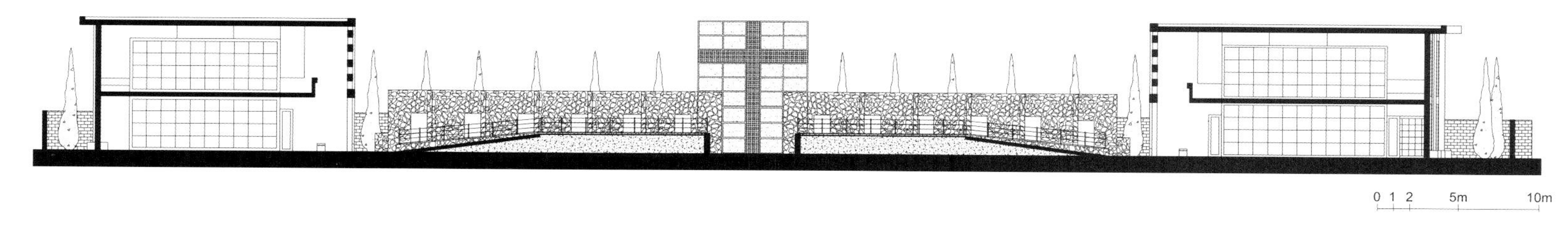

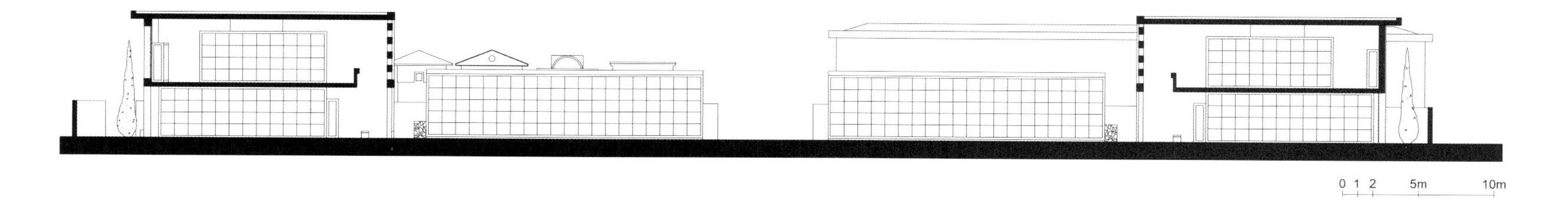

The buildings are developed on two floors: on the ground floor there are some containers of various dimensions in order to host the loculi, the cinerary urns and the ossuaries. On the upper floor, the loculis occupy the central area and look out the balcony, which runs behind the main façade.

The buildings which give shape to this space are set on the perimeter of the site in order to make up a yard. They are detached from the walls by 3 metres. These empty spaces are lined with some rows of cypress trees, which give this area an immediate sense of identity as "gardens of prayer". Finally the parking area has been doubled up in size.

Detail and Materials

The building have been constructed with no particular concrete; the fencing façade is made of common concrete with thinner, and painted in white acrylic protective coating, called Beton Cryll.

The architects decided to adopt this kind of building technology, to keep down construction costs; the common concrete also give the appeal of popular culture. The building grew up by the work of non-professional manpower, and by this way it looks something like Brutalistic architecture.

1. Concrete
   - Resistance class RCK 350
   - Exposition class XC3
   - Solidity class S3 (stairs), S4 (pillars), S5 (other parts)
   - Max aggregates diameter 10mm
     dept castings<12cm
     30mm (pillar castings),
     20mm (other part castings)

2. Concrete
   - Resistance class RCK 400
   - Exposition class XC4
   - Solidity class SCC
   - Max aggregates diameter 20mm

1. 混凝土
   – 电阻等级 RCK 350
   – 暴露等级 XC3
   – 固体等级S3（楼梯）、S4（柱）、S5（其他）
   – 最大颗粒直径10mm
     铸件深度＜12cm
     30mm（柱铸件）
     20mm（其他铸件）

2. 混凝土
   – 电阻等级RCK 400
   – 暴露等级XC4
   – 固体等级SCC
   – 最大颗粒直径20mm

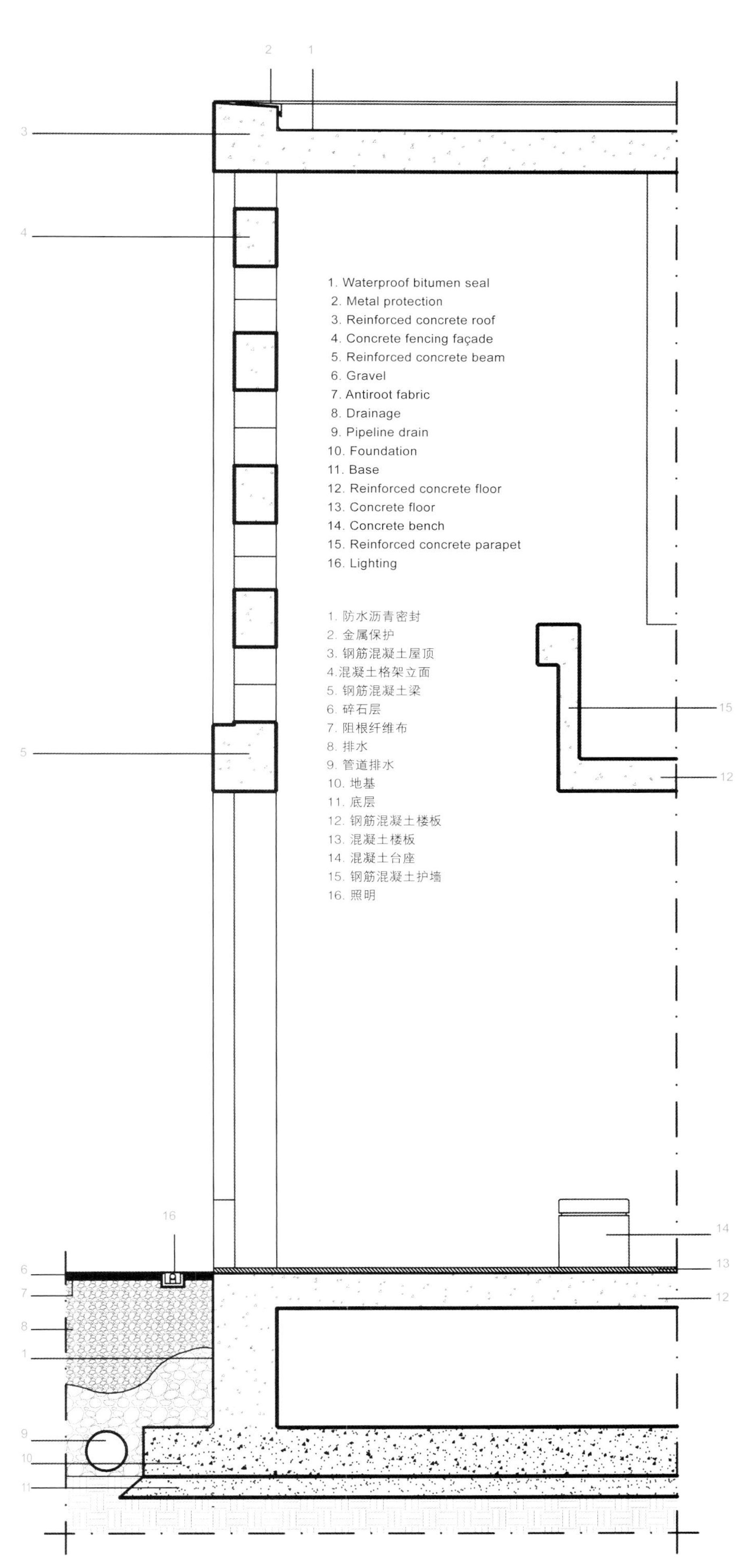

项目概况

项目的占地面积约4,500平方米，紧邻西侧已有的墓地。项目还涉及场地南面的入口区域，包含一座拥有1,230个小隔间、720个藏骨堂/骨灰罐和12个私人小礼拜堂，整个项目能够满足未来15年内市民的丧葬需求。

从城市规划角度来说，墓地分为两个部分：第一部分明显受到19世纪的影响，看起来与托斯卡纳周边的其他地区十分相似；第二部分建造时间较晚，大概是20世纪60至80年代。因此两种不同的设计风格造成了类型的混乱，使整体空间缺乏身份特征。

新规划试图摆脱墓地的崇拜感和尊重感。旧教堂的拆除解放了通往扩建区域的通道，新建的教堂则让中央透视轴变得更为完整。这样一来，整个墓地有两条纵轴：第一条连接主入口和教堂；第二条从侧门一直到殡葬楼的前方。

新教堂拥有四个大型十字架，分别位于建筑的四面，它们兼具结构及象征功能。教堂的其他部分由玻璃构成，四周环绕着石砌的私人礼拜堂。教堂两侧是两座白色钢筋混凝土建筑，巨大的格架为后方的区域提供了遮阳。

建筑有两层楼高：一楼是一些大小不一的空间，用于安置隔间、骨灰罐和藏骨堂；二楼的隔间占据了中央区域，面朝主立面后方的阳台。

建筑塑造了空间，被设置在场地外围，以构成一个庭院。它们距离围墙3米，空白空间沿途种植着几排柏树，赋予了整个区域“祈祷者花园”的感觉。最后，建筑师还将停车场扩大了一倍。

细部与材料

建筑并没有采用特殊的混凝土，格架立面的普通混凝土上涂有一层薄薄的白色丙烯酸保护涂层（Beton Cryll）。

建筑师利用这种建筑技术来控制建造成本，普通混凝土同时也彰显了大众文化。建筑由非专业人力建造而成，因此看起来有点像野兽派建筑。

Doninpark

多宁大厦

Location/地点: Vienna,Austria/奥地利，维也纳
Architect/建筑师: Love Architecture
Photos/摄影: Jasmin Schuller
Built area/建筑面积: 15,000m²
Completion date/竣工时间: 2013
Key Materials: Façade – concrete and acrylic coating
主要材料：立面——混凝土与丙烯涂层

Overview

The Doninpark project was developed as an eight-storey residential, office and retail building, located in the 22nd district in Vienna, directly behind the "Kagraner Platz" subway stop. In terms of urban planning, this location is characterised be enormous leaps in scale: to the east lies a dense, urban area with extensive infrastructure, while the area to the west has a more suburban feel, with numerous single-family and multi-family dwellings and sports fields.

For this reason, the leap in scale also became the central design idea. The window openings and alcoves, which seem to be almost randomly placed, create a façade that is "scaleless", just like the surrounding area. This makes it almost impossible for the viewer to capture the true dimensions of the building at first sight and effectively masks the true size and expanse of the building.

However, the size and expanse were essentially imposed, since the building exactly meets the requirements of Vienna's master urban plan. One could thus say that the city of Vienna designed the building, and the building thus displays a kind of radical pragmatism (i.e. one does exactly what one is allowed to do).

The ground floor of the building houses a shopping zone that faces the subway stop, while the first and second floors feature office space and a gym. The third floor upwards is

residential quarters. The apartments are oriented towards the east or west and can be accessed via a central aisle. Each residential unit is equipped with a balcony, loggia or alcove. The loggias are on the east side, thereby creating a distance to the street space.

Detail and Materials

Concrete is one of the most durable building materials. It provides superior fire resistance compared with wooden construction and gains strength over time. Façade made of concrete can have a long service life. Concrete is used more than any other manmade material in the world. Reinforced concrete, pre-stressed concrete and precast concrete are the most widely used types of concrete functional extensions in modern days.

Energy requirements for transportation of concrete are low because it is produced locally from local resources, typically manufactured within 100 kilometres of the job site. Similarly, relatively little energy is used in producing and combining the raw materials (although large amounts of CO_2 are produced by the chemical reactions in cement manufacture). The overall embodied energy of concrete is therefore lower than for most structural materials other than wood.

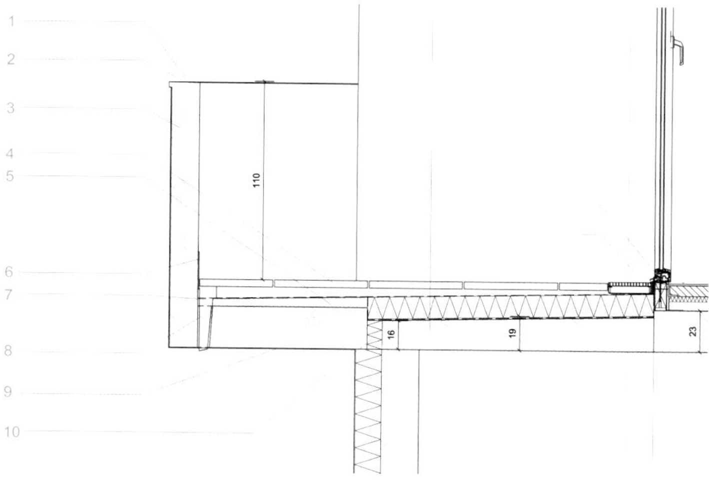

1. Inside buffer plate
2. Acrylic painted STOcryl V100
3. STO SUPERLIT directly on concrete
4. Concrete slab 4cm
5. Rack or shade
6. Sealing foil connected with joining plate
7. Sealing
8. Gargoyle stainless steel
9. STO SUPERLIT directly on concrete
10. Isokorb thermal materials 8cm

1. 内置缓冲板
2. 带有丙烯酸涂料涂层的STOcryl V100
3. 混凝土表面上的STO SUPERLIT涂层
4. 混凝土板4cm
5. 支架或遮阳板
6. 与链接板相连的密封膜
7. 密封
8. 不锈钢排水槽
9. 混凝土表面上的STO SUPERLIT涂层
10. Isokorb隔热材料8cm

项目概况

多宁大厦项目被开发成一个8层高的大楼，集住宅、办公和零售于一身，位于维也纳22区，就在卡格拉纳广场地铁站后面。在城市规划的层面上，这个位置的建设比例十分跳跃：南面是大面积的高密度基础设施，而西面则更具有城郊感，是大量独栋住宅、多层住宅楼和运动场地。

因此，在比例上的跳跃也成为了设计的核心主题。错落的窗口和凸出的阳台看起来几乎毫无规律，构成了与周边环境相似的无比例空间。从外面几乎无法看出建筑的真正规模，有效地将建筑的真实规模和跨度隐藏了起来。

事实上，建筑的规模和跨度确实符合维也纳的城市规划要求。可以说是维也纳设计了这座建筑，在可允许范围内，建筑彻底贯彻了实用主义。

建筑的一楼是一个购物中心，正对地铁站；二、三楼是办公区和一个健身房；四楼以上是住宅。公寓全部采用东西朝向，可由中央走廊进入。每个住宅单元都配有阳台、凉廊或凹室。凉廊位于东侧，与街道保持了一定的距离。

细部与材料

混凝土是最耐用的建筑材料之一。与木结构相比，它具有出色的耐火性，并且会随着时间变得更强。混凝土立面的使用寿命很长。混凝土可以说是世界上使用最多的人造材料。钢筋混凝土、预应力混凝土和预制混凝土是当前应用最广泛的混凝土功能扩展类型。

混凝土运输所需的能源很少，因为它是本地生产的材料，基本都在施工场地100千米以内的地方加工而成。同样地，加工和混合原材料的过程所消耗的能源也很少（但是在实施水泥制造的化学反应中会产生大量的二氧化碳）。因此，混凝土的整体物化能比大多数建筑材料（木材除外）都要低。

Pago de Carraovejas Winery

帕谷卡拉欧贝哈酒庄

Lcoation/地点: Valladolid, Spain/西班牙，巴利亚多利德
Architect/建筑师: Amas4arquitectura – Javier López de Uribe, Fernando Zaparaín, Fermín Antuña & Eduardo García
Photos/摄影: José María Díez Laplaza & Amas4arquitectura
Built area/建筑面积: 10,112.09m^2
Completion date/竣工时间: 2011
Key materials: Façade – coloured Concrete, stone and glass
主要材料: 立面—— 彩色混凝土、石材、玻璃

Overview

The whole proposal for the extension of Pago de Carraovejas sought to keep the old premises where the winery was born and envelop them with the new constructions. Its hillside location allows a gravity-flow winery. Grapes are received at the upper floor, fermentation takes place at the intermediate level and both aging and shipment are located at the lower one. Thus, premises for wine aging, in white concrete, are buried in the slope, while the representative ones, in grape colour concrete, spring up to catch the stunning landscape dominated by Peñafiel Castle, just in the middle of winery's vineyards.

Visitors and administration buildings constitute the external image of the winery by using the topography of the site to articulate a set of outdoor spaces at different levels. This way it is possible to differentiate accesses according to production necessities.

Wide terraces and cantilevers placed between two parallel planes, which form the façades, give shade to the glass enclosure. Views turn into the chief feature of the building, as they are lead and frame from these viewing decks, showing a surrounding landscape delimited by architecture.

The project incorporates simple and effective energy saving resources such as sheets of water, ecological roofs, natural materials, window overhangs, operable external solar shades, earth sheltering and occupied basements, thick thermal walls, natural cross ventilation and ventilated ceilings.

Detail and Materials

This project is part of a series of works where the architects tried to look into a massiveness which liberated space and turned what isn't done into the main focus. The whole building is articulated by this blank space in the shape of patios, subtractions or voids. Façade design follows these ideas: daylight is distributed through indirect and singular openings. Structure is resolved through big superficial components in the form of enclosure. This bearing mass accumulation gives some freedom on façade to arrange shades and transparencies of a singular plastic value.

Concrete, makes possible that wished continuity on which to open significant voids. A homogeneous finishing, concrete coloured in red-wine, unifies different uses and gives some warmth both inside and outside. A pigment was added to a prescribed concrete mix. Mixing water was purified by osmosis in order to avoid efflorescences.

Stone taken from the site surroundings forms gabions walls and is used in façades with high performance thermal isolation requirements.

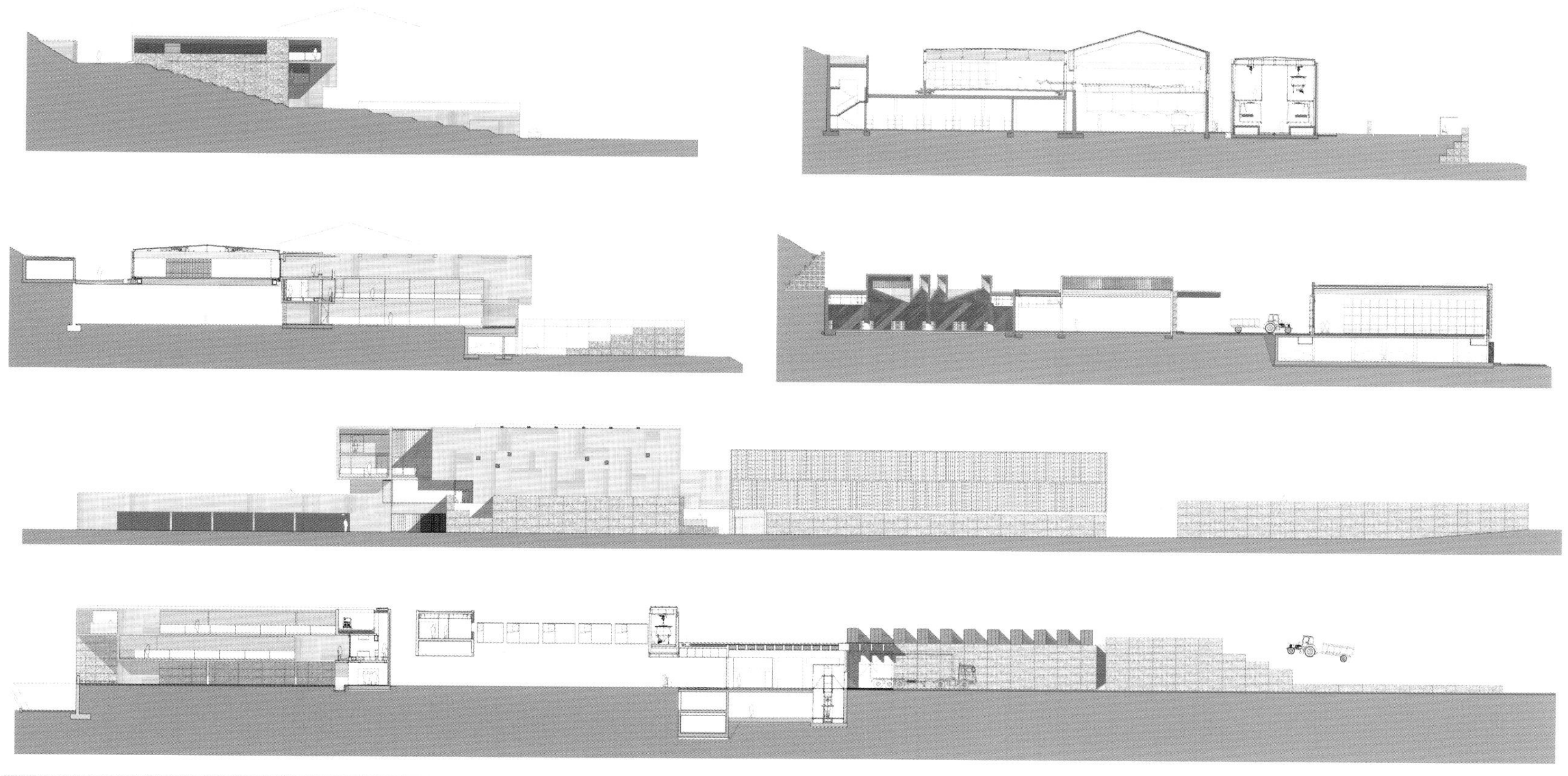

项目概况

帕谷卡拉欧贝哈酒庄的扩建方案是保留酒庄原址，然后在外围建造一圈新建筑。酒庄所在的山坡地势实现了酒庄的重力流酿酒方式。顶楼接收葡萄，发酵在中间层进行，酿造和装货则位于最底层。因此，由白色混凝土建造的酿酒区埋在山坡里，而具有代表性的葡萄色混凝土则向外凸出，与位于酒庄葡萄园中央的卡拉欧贝哈城堡的极致景观遥相呼应。

参观楼和行政楼构成了酒庄的外部形象，它们利用场地地形将一系列不同层次的户外空间连接起来。这种设计使参观和行政区与生产区简单地区分开。

宽敞的露台和悬臂式结构组成了建筑立面，为玻璃墙面提供了遮阳。良好的视野是建筑的主要特征，这些观景台为建筑提供了各种各样的美景。

项目采纳了简单高效的节能资源，例如生态屋顶、天然材料、悬臂窗、可控式外部遮阳、原土掩体和地下室、厚保温墙、自然交叉通风、通风天花板等。

细部与材料

事实上，项目是一系列建筑结构的一部分，建筑师试图呈现一种厚重感，将空间释放出来，使留白成为焦点。整座建筑通过各种形式的天井、缩进空间和露台连接起来。立面设计实现了这种概念：阳光通过间接而单一的窗口进入室内。建筑结构通过各种封闭的表面构件组合而成。承重体块的累积释放了部分立面，形成了独特的阴影和透明效果。

混凝土的应用实现了空间的开口设计。带有均匀饰面的酒红色混凝土将不同的功能区统一起来，为室内外都带来了温馨之感。建筑师在预制混凝土拌合料内添加了染料，并且通过渗透作用对拌合水进行了净化，避免了混凝土的风化。

从场地周边采集的石材构成了石笼墙，并且被应用在高性能保温立面上。

1. 聚乙烯层
2. 地基板，与墙壁的接缝由膨润土土工复合材料密封
3. 彩色钢筋混凝土，用于承重墙和楼板；模架由交错方向的厚木板制成
4. 钢筋混凝土板
5. 碎石层
6. 高密度膨胀聚苯乙烯泡沫
7. 防水系统，由高密度聚苯乙烯结节板、聚丙烯土工布（用于地下结构排水）和双层沥青薄板构成
8. 坡面，找平砂浆层修饰
9. 10mm柯尔顿扁钢条
10. 防水混凝土地面，位于碎石层上方
11. 10mm柯尔顿钢槽，与EPDM排水槽相连
12. 复合木板
13. 橡木地板
14. 高密度饰面隔热层，环绕建筑外围，H=100cm
15. 沟槽和水平玻璃装配结构，由扁钢条和氯丁橡胶板构成，采用RAL 9006金属色饰面
16. 夹层安全玻璃8+8，透明硅树脂密封
17. 夹层安全玻璃10+10，玻璃装配结构由10mm扁钢条附于混凝土板上构成；4mm开口接缝，采用RAL 9006金属色饰面
18. 低辐射玻璃8+8/16/8+8，垂直接缝由结构硅树脂密封
19. 夹层安全玻璃，内部隔断，Trebe (Sitab)系统
20. 隔断组合家具，Trebe (Sitab)系统
21. 白色层压板
22. 橡木饰面10mm胶合板
23. 空调管道
24. 隔热石膏板包层
25. 隔热层：石棉+隔汽层
26. 伸缩接缝：5cm挤塑聚苯乙烯板
外墙上8cm聚苯乙烯板隔热层放置在墙壁层之间
27. 防火门，RAL 9006金属色饰面
28. 覆盖层/屋面PUR夹心板
29. 3mm镀锌板槽
30. 砖砌结构上设备层
31. 找平砂浆层
32. 基础垫层混凝土H_{min}=10cm
33. 彩色裂纹桥联屋顶防水：两层可喷涂型聚脲薄膜
34. 防水系统：聚丙烯土工布+双层沥青板+聚丙烯土工布
35. 隔热层：高密度挤塑聚苯乙烯板
36. 钢筋混凝土，带有外围伸缩接缝
37. 混凝土板，嵌入射灯
38. 线形吊灯
39. 节能荧光射灯，固定在天花板上
40. 吊顶：100x50mm橡木条（间距：75mm）
41. 吊顶：15mm石膏板
42. 遮阳：外部遮阳卷帘

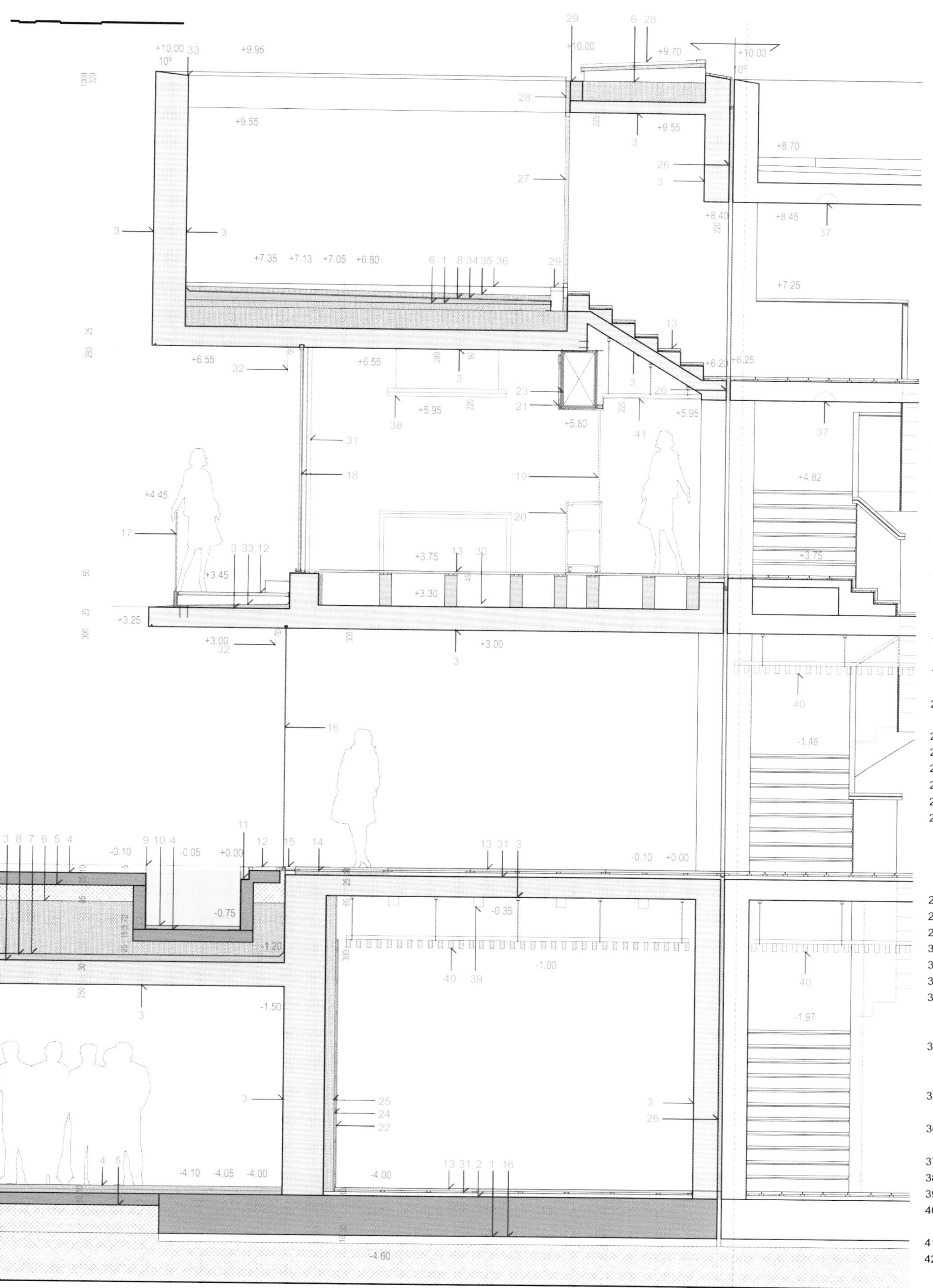

1. Polythene gauge
2. Foundation slab, according to structure details. Joints with walls sealed with string form bentonite geocomposite
3. Coloured reinforced concrete in load bearing walls and slabs, according to structure details. Formwork made with wooden plank in alternate directions
4. Reinforced concrete slab, according to structure details
5. Gravel bedding
6. High density expanded polystyrene foam (EPS). Thickness according to drawing-annotations
7. Waterproofing system formed by: High-density polyethylene nodular sheet (HDPE) with polypropylene geotextile for underground structure drainage and double bituminous sheet
8. Slope finished by regulating mortar layer
9. 10mm Corten steel flat bar
10. Water-repellent concrete floor on selected gravel bedding
11. 10mm Corten steel gutter. Overflowing to EPDM drains
12. Wood composite (Geolam) planking
13. Solid oak wooden plank flooring
14. High density rock wool thermal insulation in building's perimeter. H= 100cm
15. Gutter and horizontal glazing structure made with steel flat bars on neoprene sheet. Finished in Metalic RAL 9006
16. Laminated safety glass 8+8 sealed with transparent silicone.
17. Laminated safety glass 10+10. Glazing structure made with 10mm steel flat bars attached to concrete slab. 4mm opened joints. Finished in Metalic RAL 9006
18. Low-E glazing 8+8/16/8+8. Vertical joints sealed with structural silicone
19. Laminated safety glass in interior partitions. Trebe (Sitab) system.
20. Partitions integrated furniture. Trebe (Sitab) system
21. White laminate board
22. Oak finished 10mm plywood board
23. AC duct
24. Thermal isolated gypsum board coating
25. Thermal-Isolation: Mineral wool with vapour barrier
26. Expansion joints: 5cm extruded polystyrene board (XPS) In façade walls, 8cm polystyrene board (XPS) thermal insulation is placed as lost-workform between wall layers.
27. Fire door. Finished in Metalic RAL 9006
28. Cladding/roofing PUR sandwich panel
29. 3mm galvanised sheet gutter
30. Technical flooring on brickwork
31. Regulating mortar layer
32. Blinding concrete. Hmin=10cm
33. Coloured crack-bridging roof waterproofing: two component sprayable membrane of polyurea
34. Waterproofing system formed by: polypropylene geotextile, double bituminous sheet and polypropylene geotextile
35. Thermal-Isolation: high density extruded polystyrene board (xps)
36. Reinforced concrete with perimeter strechable joint
37. Concrete slab built in Downlights
38. Linear suspended luminaire
39. Downlight for compact fluorescent fixed to ceiling
40. Suspended ceiling: 100x50mm solid oak bars(distanced: 75mm)
41. Suspended ceiling: gypsum board 15mm
42. Sun protections: exterior roller blind

Centre for Interpretation of the Battle of Atoleiros

艾托莱罗斯战役解读中心

Location/地点: Fronteira, Portugal/葡萄牙，弗龙泰拉
Architect/建筑师: Arch. Gonçalo Byrne (Gonçalo Byrne Arquitectos, Lda.), Arch. José Laranjeira (Oficina Ideias em linha – Arquitectura e Design, Lda.)
Project team/ 项目团队: Doriana Reino, Ana Abrantes, Arq.° Tiago Oliveira
Photos/摄影: Fernando Guerra
Built area/建筑面积: 935m^2
Completion date/竣工时间: 2012
Key materials: Façade – coloured concrete
主要材料：立面——彩色混凝土

Overview

The Centre for Interpretation of the Battle of Atoleiros, in Fronteira, is a cultural equipment intended to raise social awareness on the several perspectives over the battle occurred on April 6th 1384, and its importance in the context of the dynastic disputes between the kingdoms of Portugal and Castela, by the end of the XIV Century.

Given the impossibility on plotting the Interpretation Centre on-site, in the battlefield area, the City Council approved its plot in the town core, on a location with high visibility and inserted in an urban park system that simulates and evokes the old battlefield. During the visit to the Interpretation Centre, visitors will experience different visual perspectives of the battle field, but also about the history, through its protagonists and authors, led by the hand of the painter Martins Barata.

A large bench, at the end of the exhibition circuit, presents urban park in all its dimensions, rehearsing another exhibition discourse, this made of vegetables and inert elements, a sculptural dimension that simulates the plains and the imagination refers to the Battle of Atoleiros.

Portuguese southern landscape has a golden/ reddish tone. The reddish wash of the building tries to emulate those colours and patterns, therefore reinforcing a sense of belonging. Colour and textures are also enhanced by the usage of the same pitch used on the urban park paving system, serving as an essential framework for the Interpretation Centre, yet reinterpreting the battlefield original landscape.

Detail and Materials

The body of the building recalls the tactility of the traditional medieval construction, pre-

senting rough textured surfaces, achieved by the use of pigmented concrete with raw and irregular expression, very close to the primal textures achieved by human hand. This texture is enhanced by interposing lines of schist slabs in the horizontal joints of the building.

As a whole, the building generates a gravitational presence; almost an earth sculpture dyed in its own tonalities, evoking time in the spontaneous patina patterns, resembling a stained vertical battlefield, between a small and a larger body, like the two armies in conflict.

The combined use of concrete walls and a structure formed by a concrete column/beam/slab system, allowed maximal area exploitation and the display of generous exhibition areas.

Through the completion of consoles the structure had acquired more complexity, allowing lateral glazing and motivating an open relationship between inner and outside areas, between exhibition and urban park.

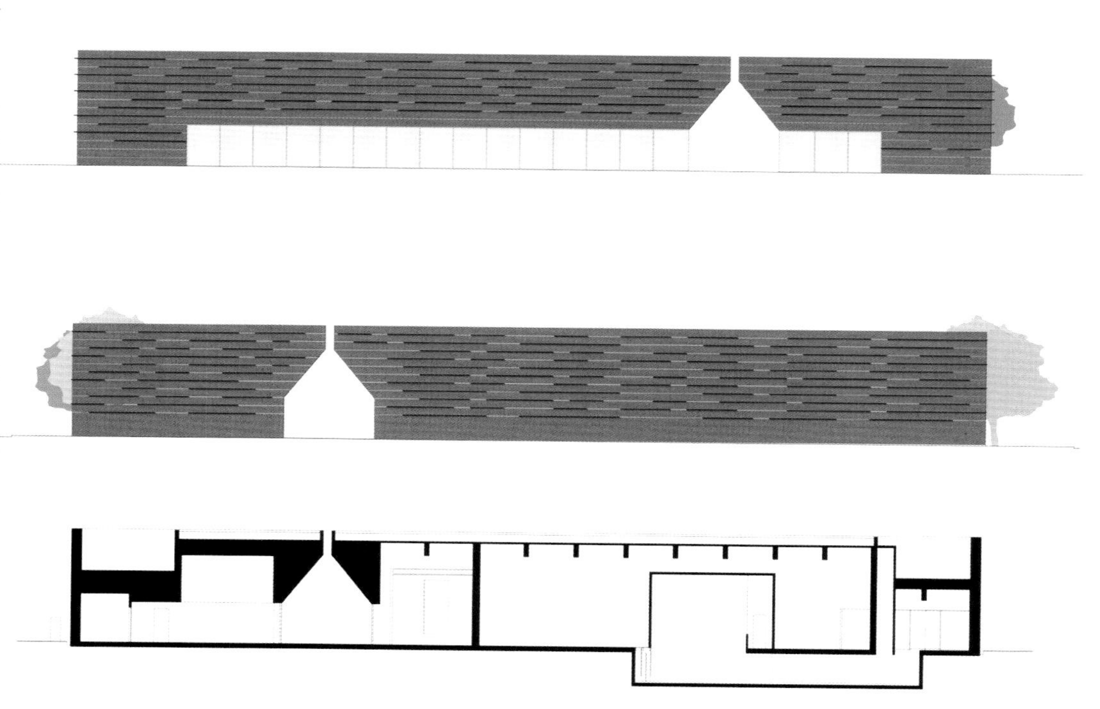

项目概况

作为一座文化设施，艾托莱罗斯战役解读中心的目标是让公众更多地了解1384年4月6日所发生的这场战役以及它在葡萄牙王国和卡斯蒂拉王国在14世纪末的王朝争端中的重要意义。

由于无法将解读中心建在战场原址，弗龙泰拉市议会批准将其建在市中心十分显著的地理位置，通过嵌入城市公园的方式再现古战场。在参观解读中心的过程中，参观者不仅能体验各种各样的战场效果，还能在画家马丁斯•巴拉塔的引导下，通过主人公和描述者了解这段历史。

展览路线的终点有一张巨大的长椅。坐在长椅上，可以将城市公园尽收眼底。公园内的植物和静物能够将人们的想象带到艾托莱罗斯战役的现场。

葡萄牙南部景观以金色和红色色调为主。建筑的红色水洗外观试图模仿那些色彩和图案，增加一种归属感。城市公园的地面铺装系统采用了同种色系的沥青，进一步突出了色彩和纹理，相当于解读中心的基本框架，重新诠释了战场的原始景观。

细部与材料

建筑主体令人想起中世纪传统建筑的触感，染色混凝土以及原始而不规则的表达让表面纹理显得粗糙，与手工纹理效果十分相似。水平接缝处的片岩板条则进一步凸显了这些纹理效果。

整体来看，建筑显得厚重沉稳，就像是一块染色的大地雕塑，以其沧桑的质感模仿着垂直战场。大小两个结构就像是交战中的两支军队。

混凝土墙壁和混凝土梁、柱、楼板系统的组合实现了空间开发的最大化，呈现出宽阔的展览区域。

支柱的设计让建筑结构变得更加复杂，实现了横向玻璃墙壁和室内外区域的开放关系，将展览区与城市公园连接起来。

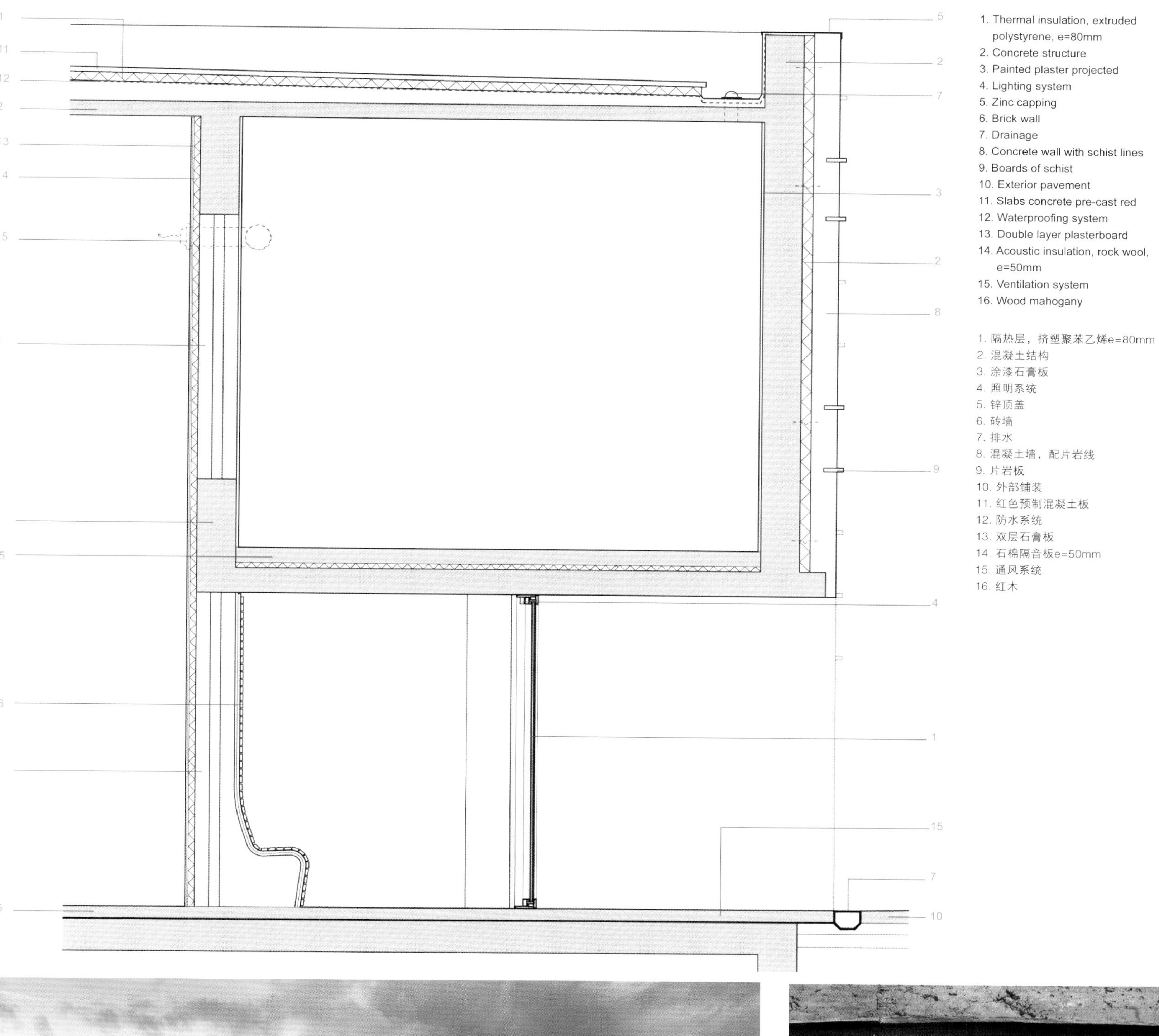
1. Thermal insulation, extruded polystyrene, e=80mm
2. Concrete structure
3. Painted plaster projected
4. Lighting system
5. Zinc capping
6. Brick wall
7. Drainage
8. Concrete wall with schist lines
9. Boards of schist
10. Exterior pavement
11. Slabs concrete pre-cast red
12. Waterproofing system
13. Double layer plasterboard
14. Acoustic insulation, rock wool, e=50mm
15. Ventilation system
16. Wood mahogany
1. 隔热层，挤塑聚苯乙烯e=80mm
2. 混凝土结构
3. 涂漆石膏板
4. 照明系统
5. 锌顶盖
6. 砖墙
7. 排水
8. 混凝土墙，配片岩线
9. 片岩板
10. 外部铺装
11. 红色预制混凝土板
12. 防水系统
13. 双层石膏板
14. 石棉隔音板e=50mm
15. 通风系统
16. 红木

Praça das Artes

艺术广场

Location/地点: São Paulo, Brasil/巴西，圣保罗
Architect/建筑师: Brasil Arquitetura
Photos/摄影: Nelson Kon
Built area/建筑面积: 28,500m²
Completion date/竣工时间: 2012
Key materials: Façade – coloured concrete
主要材料： 立面——彩色混凝土

Overview

Since the initial site study, the former Conservatory, restored and converted into a concert hall and an exhibition space, represented the anchor for the project. The new buildings are mainly positioned along the boundaries of the site and, to a large degree, lifted off the ground. Thus, it was possible to create open spaces and generous circulation areas, resulting in the plaza which gives the project its name. This paved plaza can be accessed from rua Conselheiro Crispiniano, avenida São João and, in the next construction phase, also from rua Formosa (Anhangabaú) via a flight of stairs, which connect the different levels of the streets.

The new volumes reach from the centre outwards towards the three adjacent streets. A series of interconnected buildings in exposed concrete, with ochre pigments, accommodate the various functions and is the main element establishing a new dialogue with the neighbourhood and with the remaining constructions that will be incorporated into the project, the former Conservatory and the façade and foyer of the former Cairo Cinema.

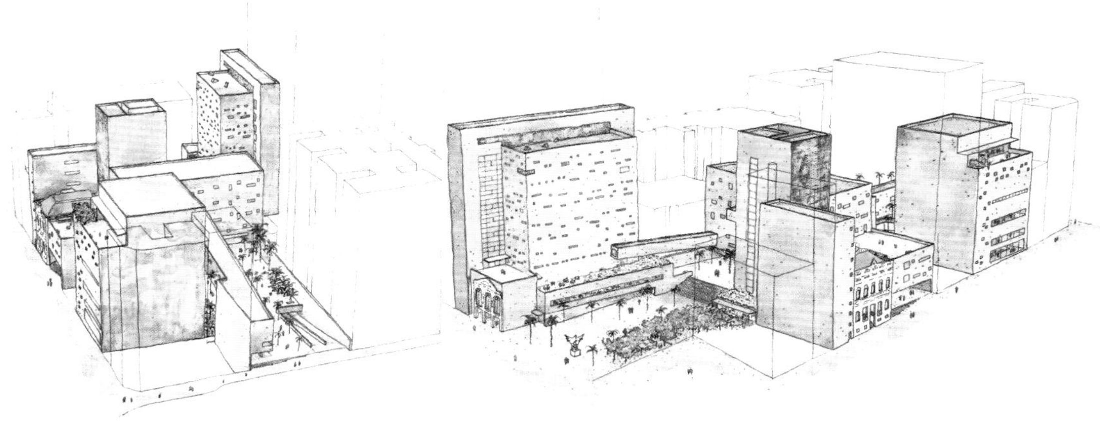

These historic buildings are physical and symbolic records, remains of the city of the end of the 19th and the beginning of the 20th century. Restored in all aspects and converted for new uses, they will sustain a life to be invented. Incorporated them into the project, they became unconfined from neighbouring constructions and gained new meanings. The historic building on Avenida São João came into being as a commercial exhibition space for pianos in 1986, then gained an extension to become a hotel and shortly after was transformed into a musical conservatory, even before the creation of the Municipal Theatre, for which it is a precursor, in certain ways, and the training centre for the musicians who would then make up its orchestra.

In order to satisfy the high requirements in preventing the propagation of noise and vibrations, specific details were used, such as floating slabs, acoustic walls and ceilings made of gypsum panels and rock wool, a system called acoustic isolation. The administrative office areas of the extension to the Conservatory are equipped with a raised floor, which means that electrical, logistical, and communication installations can be adapted freely for allowing a greater flexibility in the arrangements of the work spaces.

It was possible to achieve large spans without intermediate columns by using shear walls, thus guaranteeing complete flexibility of the internal spaces and unobstructed external spaces on the plaza level.

Detail and Materials

A new addition to the Conservatory was built, which works as a pivot point for all departments and sectors within the complex. All administrative offices, vertical circulation (stairs and lifts), entrance and distribution halls, toilets, changing rooms, and shafts for building services are concentrated in this building, which is the only one coloured with red pigments. Towards the plaza, a sculptural triangular staircase built in concrete and glass allows for a direct access between the level of the plaza and the first floor, where the concert hall is located.

Besides the coloured concrete, the windows represent an important part of project. They are either externally attached or placed within the opening. In the rooms with special acoustic requirements, the windows are fixed and attached to the building from the outside with 16mm-thick glass; in other spaces awning windows are used.

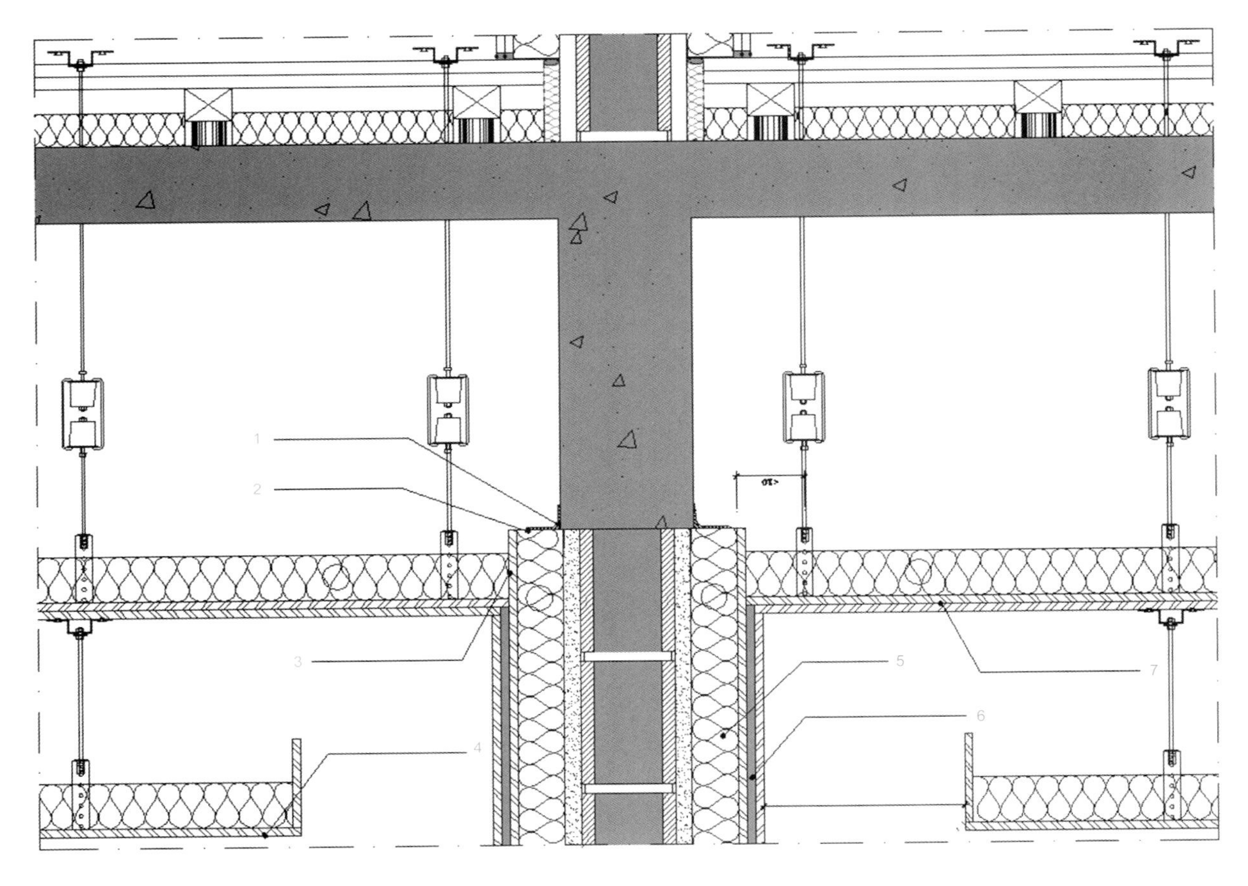

1. Metal bracket to support against wall gypsum
2. Metal guide 70x28mm
3. Sheet of plasterboard, 12.5Mm each Fixation at double amount 70x35mm
4. Perforated plaster lining
5. Double amount, 70x35mm
6. Plywood sheet 12mm Clamped between two plasterboard Gypsum each with 12.5Mm Fixation in double amount, 70x35mm
7. 2 Sheets of plasterboard, 12.5Mm each Fixation double amount 70x35mm

1. 金属支架，背靠石膏墙面
2. 金属导轨70x28mm
3. 石膏板12.5mm 双倍固定70x35mm
4. 穿石膏内衬
5. 双倍固定70x35mm
6. 胶合板12mm 夹在两层石膏之间 石膏板12.5mm 双倍固定70x35mm
7. 双层石膏板12.5mm 双倍固定70x35mm

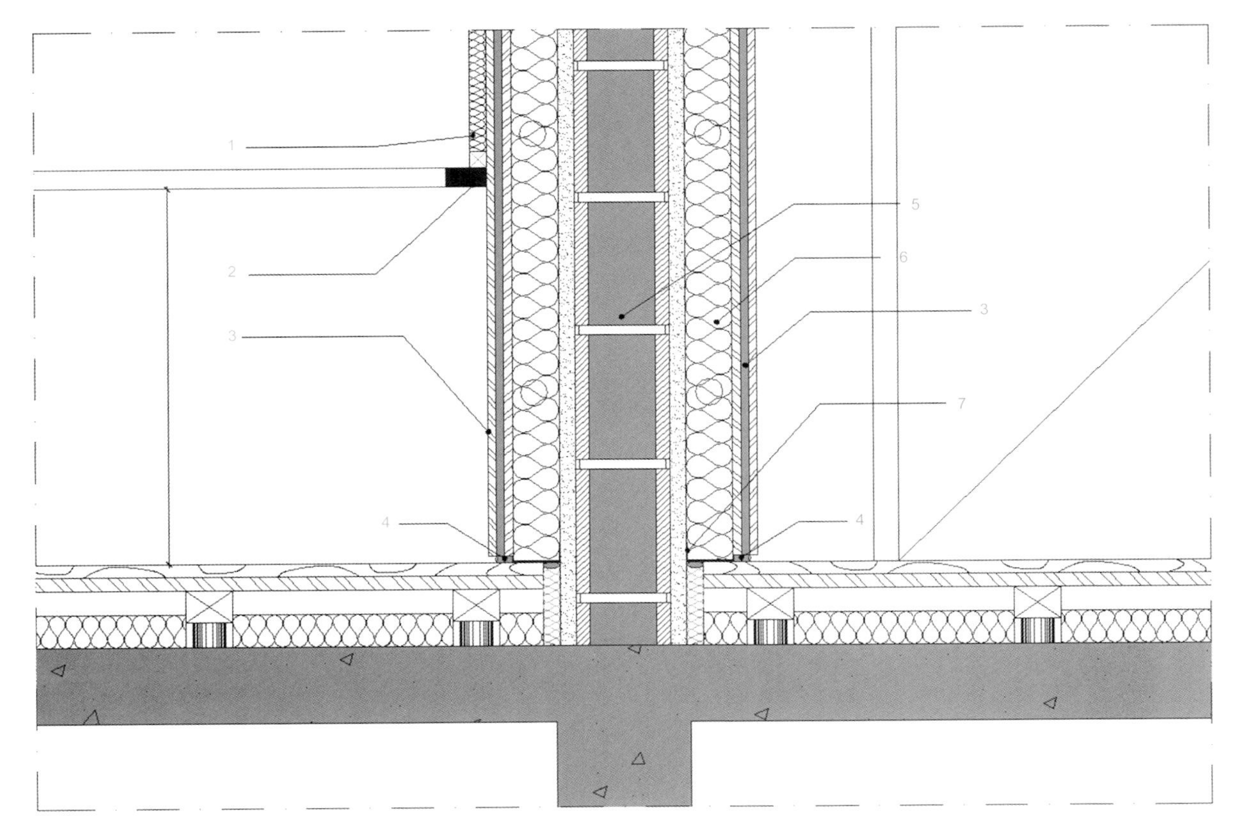

1. Absorbent panel with coated fabric Glass 25mm
2. Plywood frame 30x60mm
3. Plywood sheet 12mm Clamped between two plasterboard Gypsum 12.5Mm each with Fixation in double amount, 70x35mm
4. Railing
5. Concrete blocks
6. Double amount, 70x35mm
7. Metal guide 70x38mm

1. 吸音板，配涂层布玻璃25mm
2. 胶合板框30x60mm
3. 胶合板12mm 夹在两层石膏板之间 石膏厚度12.5mm 双倍固定70x35mm
4. 栏杆
5. 混凝土砌块
6. 双倍固定70x35mm
7. 金属导轨70x38mm

项目概况

经过初期场地调研，前音乐学校经过修复被改造成了音乐厅和展览空间，建筑师将其呈现为项目的核心。新建的建筑主要围绕着项目场地的外围建造，因此在场地中央形成了开放宽敞的空间，项目的名称“文化广场”正是由此得名。行人可经由一段楼梯从克里斯皮亚诺街、圣约翰大道以及福莫萨街进入广场。

新的建筑空间由中央向外扩展到三条邻近的街道。赭石色的清水混凝土建筑相互连接，内部设置着各种文化功能，与周边的社区、前音乐学院和前开罗电影院等设施建立起全新的对话。

这些历史建筑是富有象征意义的记录，记录着19世纪末、20世纪初的城市轨迹。经过了全方面的改造并被赋予了新的功能，它们重新焕发出活力。通过融入项目，它们与周边的建筑结构区分开来，获得了新的意义。1986年，圣约翰大道上的历史建筑被改造成了商业化的钢琴展览空间，后来经过改造形成了酒店，不久之后又被改造成了音乐学校。它的音乐历史甚至比市政剧院还要长，是音乐人的培训摇篮。

为了满足降低噪声和共鸣的高标准要求，项目采用了特别的细节，例如浮式板材、由石膏板和石棉构成的隔音墙和吊顶等隔音系统。音乐学院扩建部分的行政办公区配有活动地板，所有电气、后勤、通信装置都能在地板下方进行灵活的配置。

剪力墙的应用实现了无中柱的大跨度楼面，保证了内部空间的灵活运用，同时也在广场层实现了无障碍的交通。

细部与材料

音乐学校的扩建部分是连接各个院系部门的枢纽。所有行政办公室、垂直交通（楼梯和电梯）、入口、分流厅、洗手间、更衣室、服务维修井都集中在这幢楼内。它也是整个项目中唯一一座红色的建筑。

除了彩色混凝土之外，窗户的设计也是项目的一大特色。有的窗户附在外面，有的则安装在窗口内。在具有特殊音响要求的房间内，窗户被固定在建筑之外，配有16毫米厚的玻璃；其他空间则采用了篷式天窗。

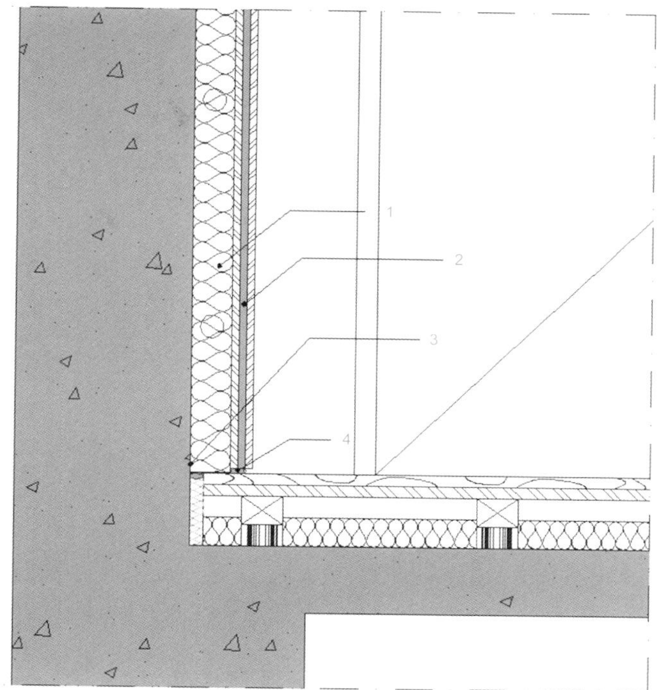

1. Double amount, 75x35mm
2. Plywood sheet 12mm
   Clamped between two plasterboard
   Gypsum each with 12.5Mm
   Fixation in double amount, 70x35mm
3. Metal guide 70x28mm
4. Railing

1. 双倍固定75x35mm
2. 胶合板12mm
   夹在两层石膏板之间
   石膏厚度12.5mm
   双倍固定70x35mm
3. 金属导轨70x28mm
4. 栏杆

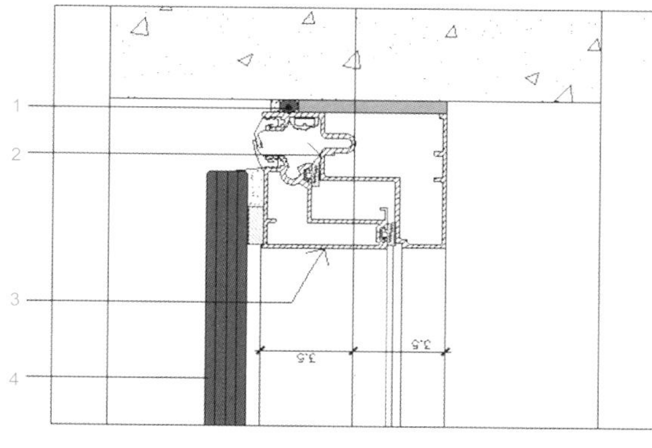

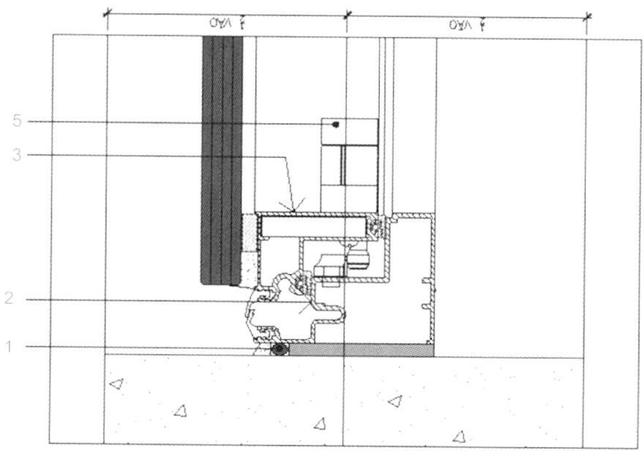

1. Joint bottom and silicone
2. Amount of fixed aluminum
3. Amount of aluminum dump
4. Glass e=16mm
5. Closure

1. 底部硅胶接缝
2. 铝件固定
3. 铝型材
4. 玻璃e=16mm
5. 包围结构

The MuCEM Conservation and Resources Centre

欧洲与地中海文明博物馆收藏与资源中心

Location/地点: Marseille, France/法国，马赛
Architect/建筑师: Corinne Vezzoni et associés
Photos/摄影: David Huguenin
Built area/建筑面积: 13,033m^2
Completion date/竣工时间: 2012
Key materials: Façade – coloured concrete
主要材料：立面—— 彩色混凝土

Overview

In 2004 the French Ministry of Culture held a competition for the MuCEM Conservation and Resources Centre, to be built on industrial wasteland on the La belle de mai site near the Saint-Charles Railway Station in Marseille. The competition winners announced in September 2004 were the architectural team of Corinne Vezzoni et associés/AURA.

Designed for the storage and study of the collections, the 13,000 m2 Conservation and Resources Centre (CCR) is situated on the former 1.2 hectare Bugeaud military terrain housing the Le muy barracks. The site is located in Marseille's La belle de mai district. It has room for a second phase extension for the reserve collections. The Centre houses all the museum's reserve collections, its documentation resources, library and scientific archives.

The MuCEM CCR carries out the museum's primary missions, namely to store, conserve, study, document, maintain and develop the collections. Another function of the Centre is to be a living instrument for promoting and disseminating the collections through a vigorous loan and deposit policy. The Centre must also help heritage professionals, researchers and students by making all the MuCEM collections and resources accessible to them. Part of the reserve collections and the temporary exhibitions are open to the public by appointment.

Detail and Materials

Corinne Vezzoni installed the MuCEM collections conservation building in a radical and compact manner on the site, selecting the largest possible size to reflect the nearby

industrial structures and base of the artillery barracks. The intent of the building's unique volume – easily identifiable in a fleeting glimpse by train passengers – is to send out an urban message and to echo the MuCEM.

With the same 72x72 metre footprint as the MuCEM, the Centre sends a fraternal message to the museum. On the site the architect has implanted a large rough-finished concrete monolith, akin to a block of rock, scooped out to let the light in: a direct reference to the works of the Spanish sculptor Eduardo Chillida. The shell is made of rough wood-textured concrete (mottled and poured in situ) that is carved out and hollowed to highlight the luminosity of the smooth white reflective concrete in the heart of the building.

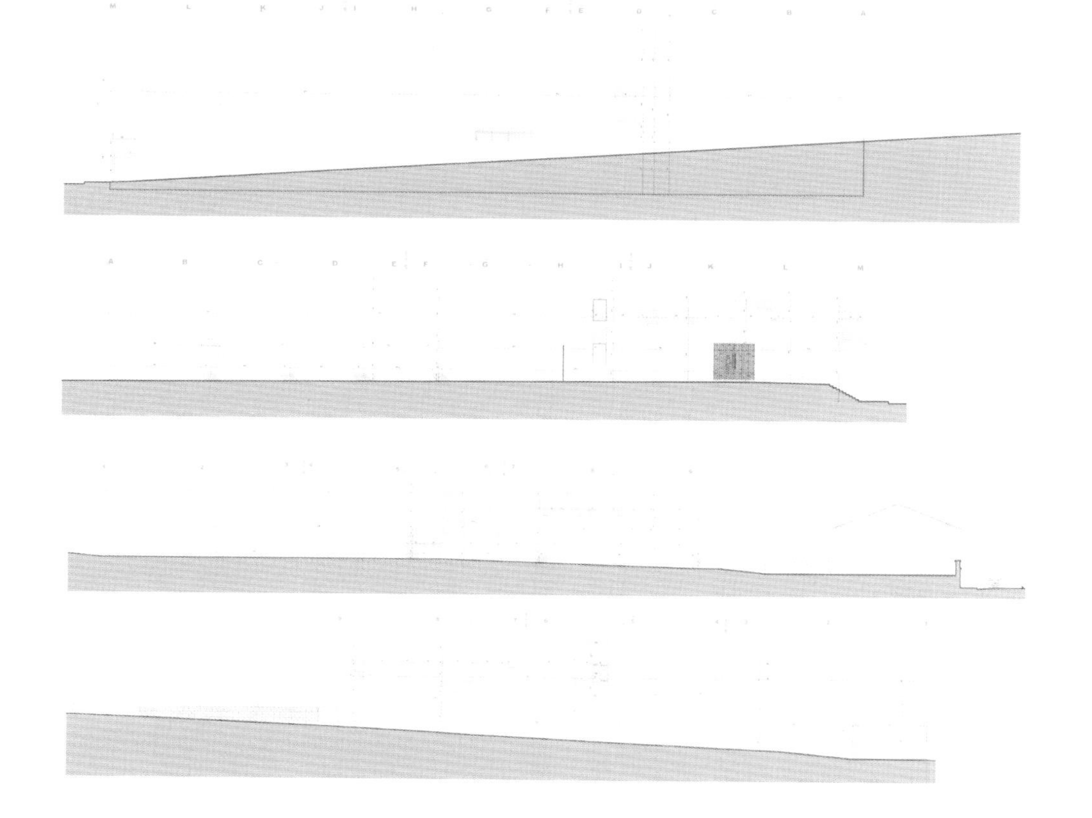

From the street, the Centre matches the scale of the nearby industrial structures and emphasises the lines of buildings. Although passers-by only see a rough-finished, ochre-coloured concrete façade, this apparent impenetrability is offset by the interior, structured around bright south-facing patios. An inner world is created around these hollow spaces. The offices are sheltered from city noise and look out over the park, creating a peaceful atmosphere. As much as possible of the existing vegetation has been kept and the presence of the interior garden is sensed from the street. Pedestrians enter from the original small square.

The reserve collection building is naturally more introverted as it tells the story of the hidden works and develops what is not visible – unlike the MuCEM, which represents the visible part of the institution. For the architect, "the reserve collections are the other side of the coin, what is behind the scenes".

The roof is covered by large slabs of coloured concrete, emphasising the shape of the monolith and creating a fifth façade, viewed from the Le Muy barracks and railway tracks. Two of the existing buildings have been kept and renovated: a hangar on the street to welcome transiting collections and a building of living quarters for foreign research workers.

项目概况

2004年，法国文化部为建造欧洲与地中海文明博物馆的收藏与资源中心而举办了一次设计竞赛，项目场地就设在马赛圣查尔斯火车站附近的一块工业废地上。柯林•维佐尼建筑事务所和AURA所组成的团队获得了竞赛的优胜。

收藏与资源中心用于储藏和研究博物馆的藏品，总面积13,000平方米。项目场地坐落在占地1.2公顷的前毕若军事区内，属于马赛市的拉贝勒区，还留有空间进行二期扩建。中心收藏着博物馆的所有储备藏品、文件资源、图书馆和科学档案。

欧洲与地中海文明博物馆收藏与资源中心贯彻了博物馆的主要任务，即储藏、保存、研究、归档、维护和开发藏品。中心的另一个功能是通过有力的出借和储藏政策来宣传藏品。它还必须通过开放博物馆的藏品资源来帮助专业的文物工作者、研究人员和学生。部分藏品和临时展览可通过预约参观的方式向公众开放。

细部与材料

建筑师将欧洲与地中海文明博物馆的收藏楼呈紧凑的造型展开，以最大的尺寸来呼应附近工业结构与军事基地的规模。建筑独特的造型让火车上的乘客过目不忘，发出了独特的城市讯息，并且与欧洲与地中海文明博物馆遥相呼应。

收藏与资源中心的地基面积与博物馆相同，都是72x72米，有一种兄弟楼的感觉。建筑师采用了大型糙面混凝土砌块结构，使建筑看起来像一块岩石。建筑外壳挖出一些开口让光线进入，这直接参考了西班牙雕刻家爱德华多•奇里达的作品。外壳由糙面木纹混凝土（现场制作）构成，切口处露出了建筑内部的白色反光混凝土。

从街面上看，中心与附近工业建筑规模相似，突出了建筑群的线条。尽管过往的行人只能看到粗糙的赭石色混凝土立面，环绕着明亮天井的内部空间还是弥补了这种封闭感。建筑内部围绕着若干个天井展开。办公区远离尘世的喧嚣，朝向公园，营造出平和的氛围。场地上原有的植物被尽量保留，从街道上就能感受到美丽的内部花园。行人可经由小广场进入。

收藏楼的设计比较内向，因为它的功能与博物馆不同，更倾向于收藏和保护。用建筑师的话来说："保留藏品是硬币的另一面，它们隐藏在舞台布景后面。"

屋顶由大块的彩色混凝土板覆盖，进一步突出了建筑的体量感，形成了第五个立面。两座原有的建筑结构被得以保留和翻修：一座临街用于接收藏品的运输，另一座是外国研究人员的宿舍。

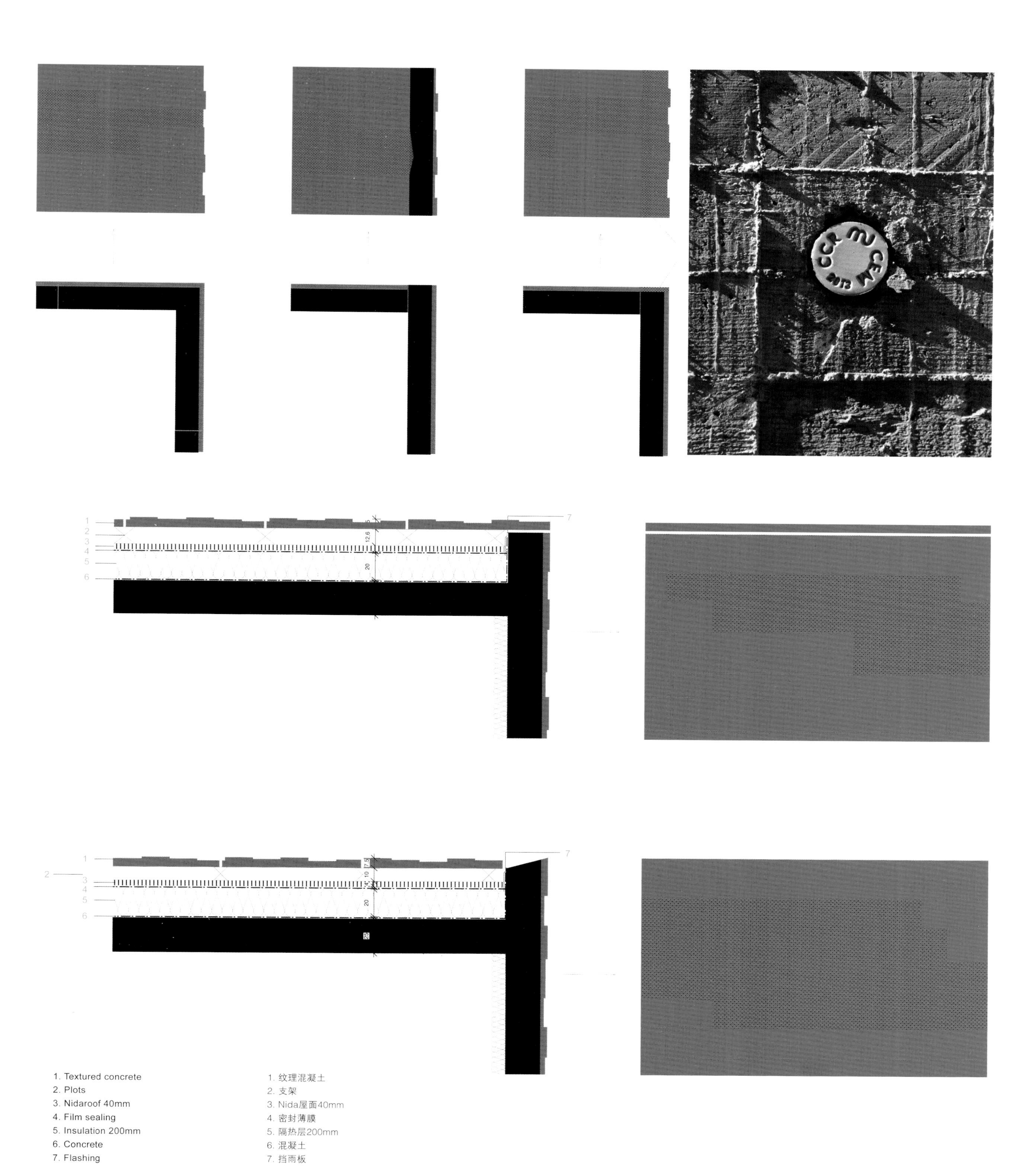
1. Textured concrete
2. Plots
3. Nidaroof 40mm
4. Film sealing
5. Insulation 200mm
6. Concrete
7. Flashing
1. 纹理混凝土
2. 支架
3. Nida屋面40mm
4. 密封薄膜
5. 隔热层200mm
6. 混凝土
7. 挡雨板

Centre for Persons with Disabilities "Aspaym"

阿斯帕伊姆残障人士中心

Location/地点: San Juan de Sahagún, León, Spain/西班牙，莱昂
Architect/建筑师: Amas4arquitectura
Photos/摄影: Amas4arquitectura
Built area/建筑面积: 1,035m²
Completion date/竣工时间: 2011
Key materials: Façade – coloured concrete, polycarbonate and glass
主要材料: 立面——彩色混凝土、聚碳酸酯、玻璃

Overview

This project belongs to a series which explores massing that liberates space, and that renders protagonist that which "is not". Such "blank space", in the form of courtyards, subtractions, or voids, articulates the building. Light is given by indirect and unique openings. The structure is solved with large surface elements, which are manifested as enclosure. This accumulation of load-bearing mass yields great façade panes, with generate shadows and transparencies of unique artistic value. The concrete, coloured black, allows us to obtain a continuity in which significant gaps open.

The building occupies a triangular site with a very pronounced geometry, in an environment of large residential blocks. An orthogonal, single floor geometry is implanted in the lot, articulating a set with different traces and heights, in which interstitial spaces refer to the geometry of the site.

The combination of built volumes and open spaces – in the shape of courtyards and plazas – takes advantage of the existing oriental planes and acacias on the roads nearby, generating different degrees of continuity between the interior and exterior spaces. The guarding of the centre prevails, protecting the classrooms from the traffic noise, taking advantage of the shade generated by the yearly cycles of foliage. Compact exteriors contrast with a transparent and dense interior, fractured by diagonal and horizontal lights.

The entrance, lobby, cafeteria and multipurpose spaces are linked through a large porch pierced on the north side of the building. The classrooms, the space for physiotherapy and the administration area are organised around it, so that their solar exposure is optimised in the extreme weather of Leon.

Detail and Materials

The project proposes a material economy that combines exposed concrete, polycarbonate, and glass, arranged constructively so as to enhance passive energy efficiency. The chromatic contrast of the textured con-

crete stained black and the fuchsia plastic materials, as the hallmark of the building, contributes to a unique urban image while screening natural light, creating different effects inside.

项目概况

项目力求解放空间，让建筑结构之外的空间成为设计的主角。建筑通过庭院、天井等“空白空间”清晰地表现出来。间接和独特的开口为室内提供了光线。建筑结构通过大片的表面元素（即围墙）表现出来。承重结构累

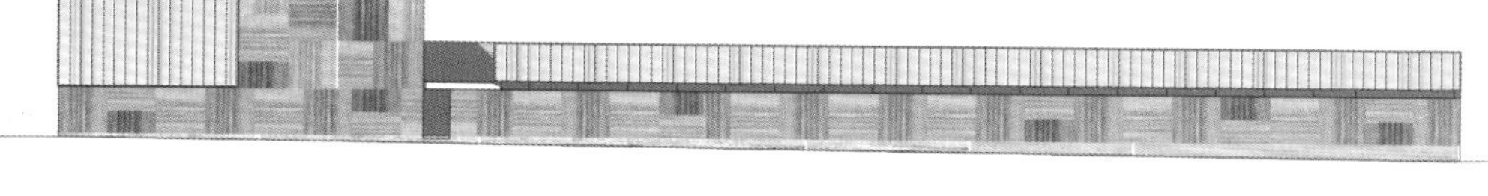

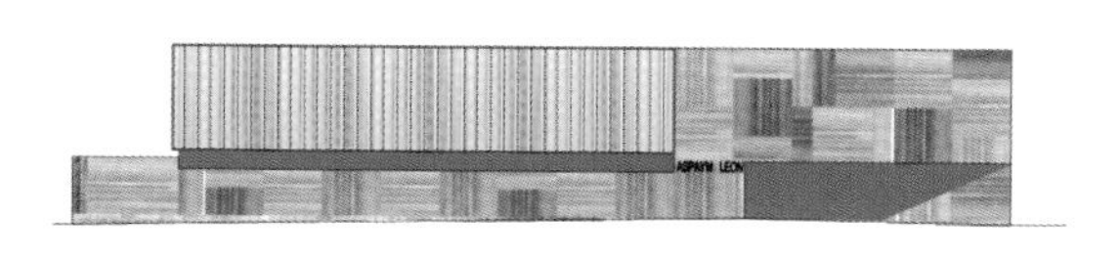

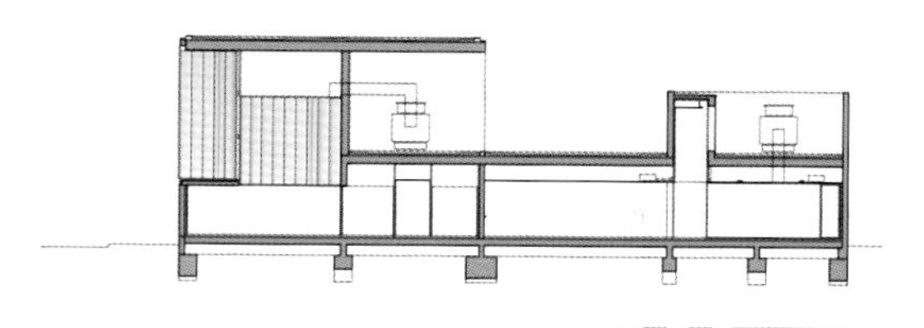

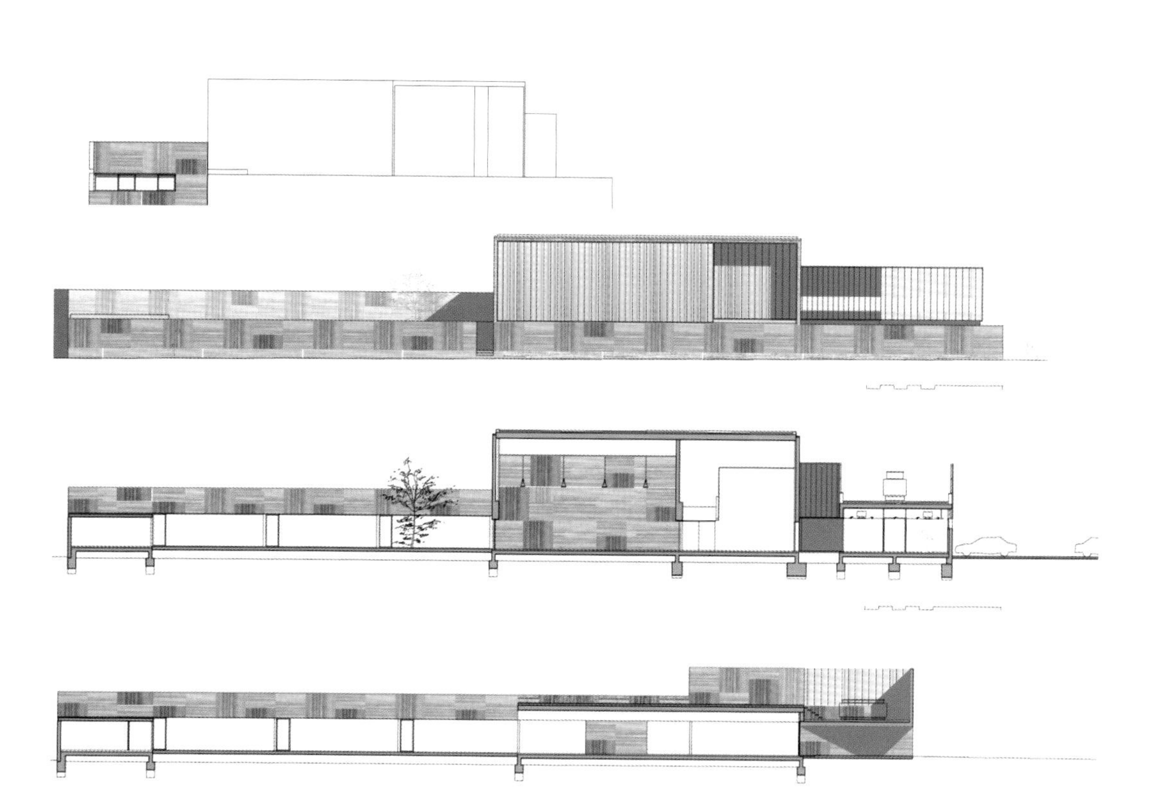

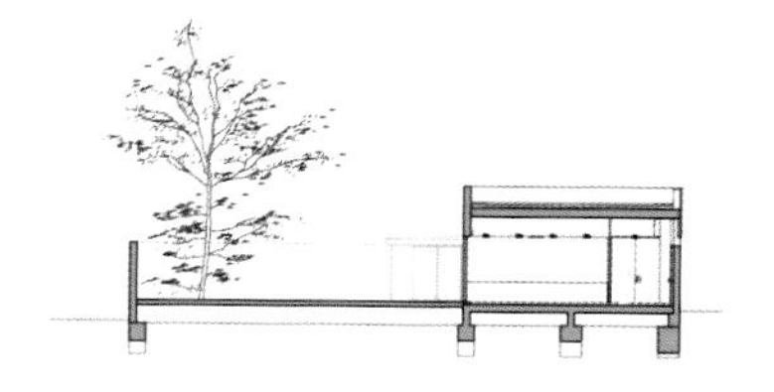

积成为巨大的外墙平面，同时也形成了具有独特艺术价值的阴影和透明效果。黑色混凝土让墙面的缺口处都能与墙面融为一体。

建筑位于居民楼林立的街区之中，占据一块三角形场地，采用直角单层建筑结构，以高低错落的建筑结构为特色，其间隙空间参考了场地的地形。

建成空间与开放空间（体现为庭院和广场的形式）的组合充分利用了附近街道旁的法国梧桐和洋槐，在室内外空间实现了不同程度的连续感。教室被设在中央，远离交通噪声，同时也充分利用树木的阔叶周期实现了遮阳。简洁的外观与通透密集的内部结构形成了鲜明的对比。

入口、大厅、自助餐厅和多功能空间通过一个贯穿建筑北面的大型门廊连接起来。教室理疗室和行政区都围绕着它而展开，即使在寒冷的冬季也能得到充分的阳光辐射。

细部与材料

项目以素面混凝土、聚碳酸酯和玻璃为主要材料，保证了建筑的被动节能性能。黑色纹理混凝土和紫红色塑料材质的色彩对比是建筑的主要特色，形成了独特的城市形象。同时，外墙设计还能过滤自然光线，在室内形成丰富的效果。

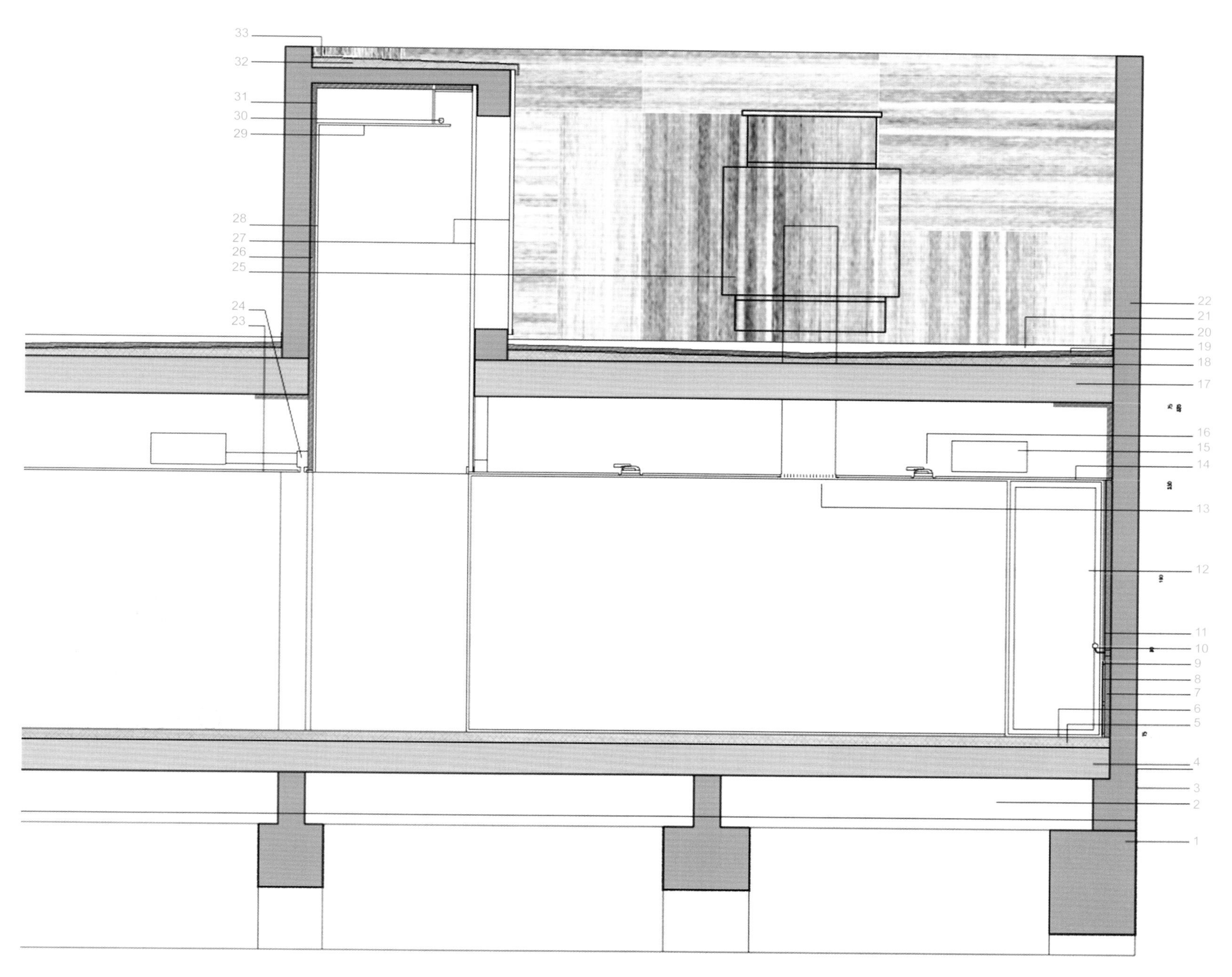

1. Reinforced concrete slab
2. Air cavity
3. Waterproofing system formed by: high-density polyethylene nodular sheet (HDPE) with polypropylene geotextile for underground structure drainage and double bituminous sheet
4. Reinforced concrete slab
5. Regulating mortar layer
6. PVC flooring
7. Thermal-isolation: mineral wool with vapour barrier
8. Laminate board
9. 22x40mm wooden structure
10. Handrail: Ø50mm inox-steel pipe
11. Thermal isolated gypsum board coating
12. Laminated safety glass 4+4 in interior partitions
13. Linear AC diffuser connected to plenum: Trox AH
14. Suspended ceiling: gypsum Board 15mm
15. AC duct
16. Downlight for compact fluorescent fixed to ceiling
17. Reinforced concrete slab
18. 3% slope finished by regulating mortar layer
19. Thermal-isolation: high density extruded polystyrene board (XPS)
20. Waterproofing system formed by: polypropylene geotextile, double bituminous sheet and polypropylene geotextile
21. Gravel bedding
22. Black reinforced concrete in load bearing walls and slabs. Formwork made with wooden plank in alternate directions
23. Suspended ceiling: gypsum board 15mm
24. Radial AC diffuser: Trox VSD35
25. AC unit
26. Thermal-isolation: Mineral wool with vapour barrier
27. Cellular polycarbonate wall
28. Thermal isolated gypsum board coating
29. Suspended ceiling: gypsum board 15mm
30. Linear luminaire
31. Thermal-isolation: high density extruded polystyrene board (XPS)
32. 3% slope finished by regulating mortar layer
33. Waterproofing system formed by: polypropylene geotextile, double bituminous sheet with self-protected sheet

1. 钢筋混凝土板
2. 空气腔
3. 防水系统：高密度聚乙烯板、聚丙烯土工布（地下结构排水）、双层沥青板
4. 钢筋混凝土板
5. 找平砂浆层
6. PVC地面
7. 隔热层：石棉+隔汽层
8. 层压板
9. 22x40mm木结构
10. 栏杆：Ø50mm inox不锈钢管
11. 隔热石膏板包层
12. 夹层安全玻璃4+4，拥有室内隔断
13. 条形空调散流器，与增压室相连
14. 吊顶：石膏板15mm
15. 空调管道
16. 射灯：节能灯，固定在天花板上
17. 钢筋混凝土板
18. 3%斜坡，找平砂浆层饰面
19. 隔热层：高密度挤塑聚苯乙烯板（XPS）
20. 防水系统：聚丙烯土工布、双层沥青板、聚丙烯土工布
21. 碎石层
22. 黑色钢筋混凝土，用于承重墙和底板；模架由木板制成
23. 吊顶：石膏板15mm
24. 放射状空调散流器：Trox VSD35
25. 空调机组
26. 隔热层：石棉+隔汽层
27. 多孔聚碳酸酯墙
28. 隔热石膏板包层
29. 吊顶：石膏板15mm
30. 条形照明
31. 高密度挤塑聚苯乙烯板（XPS）
32. 3%斜坡，找平砂浆层饰面
33. 防水系统：聚丙烯土工布、自带保护层的双层沥青板

Construction of a Performance Hall and Development of Its Surroundings in Marciac

拉斯特拉达音乐厅及其周边开发

Location/地点: Marciac, France/法国，马尔西亚克
Architect/建筑师: atelier d'architecture king kong
Photos/摄影: Arthur Péquin
Site area/占地面积:1,500m^2
Built area/建筑面积: 1,450m^2
Completion date/竣工时间: 2011
Key materials: Façade – concrete and wood
主要材料：立面——混凝土和木材

Overview

Officially opened in the spring of 2011, L'Astrada is a music venue and home to the renown Marciac jazz festival which prior to its construction had no permanent quarters to stage the international artists who come to perform each year. Marciac is a hive of activity during the festival, and an ever more numerous public of jazz lovers flock from home and abroad to enjoy the top notch facilities and finely tuned spatial arrangement of the new venue. L'Astrada has received unanimous acclaim from architecture critics and historians alike.

The decision to set the new concert hall within Marciac's fortified Medieval bastide was part of a wider project for the town as a cultural pole. It triggered the redevelopment of a portion of the urban fabric to the north-east of the town centre, obliging the architects at King Kong to give careful consideration not only to the practicalities of the new building and its architectural context, but also to ways in which it could serve to revitalise this somewhat isolated neighbourhood as part of an urban design project still in the process of becoming. Their solution was to deftly graft the new building onto the existing grid layout and the result, while it refuses to unsettle the order of the town's urban composition, does, like so many of King Kong's designs, offer elements of surprise.

Detail and Materials

L'Astrada dips unrestrainedly into the architectural idiom of its urban context, reworking its main characteristics in contemporary vein, in deference to the past but refusing hollow pastiche. It echoes back the straight lines and regularly recurring rhythms which shape the essence of the bastide's soul. The authenticity oozing from the town's low-lying houses with their simple, unadulterated walls is inspiration for the use of stamped or exposed-aggregate concrete, deliberately left untreated in the main surfacing, while the glazing, however functional, is reminiscent of Gothic ogee arching. The rustic wood panelling evokes the decorative trimming still visible on the exterior of many of the bastide's historical abodes.

The building's exterior cladding is at once smooth and forceful, voicing with unaffected simplicity the distilled embodiment of the fertile substance deposited through the centuries of the town's architectural past. Refusing all sense of pretence and artifice, it echoes back the homogeneous plurality of Marciac's rich identity. There is no misguided nostalgia, no backward-looking regret, nor does it imprison the town in its preordained role as a centre for tourism. Rather, L'Astrada has equipped the town with a first rate facility on a par with the extraordinary festival it hosts each year, a fitting reflection of its vitality and reputation. The wholesome purity of its architecture bears witness to the building's filiation with Marciac's rich historical legacy, while offering exciting new horizons for days to come.

The concrete casing creates both forceful and gentle contrasts with the natural wood also used. The latter's shades vary with the shifting atmospheric conditions, creating a rich dialogue between yesterday and tomorrow, redolent with unceasingly renewed sensations. Past and future come together here, each with their own specific part to play, in a mutually respectful union of rich and subtle complicity.

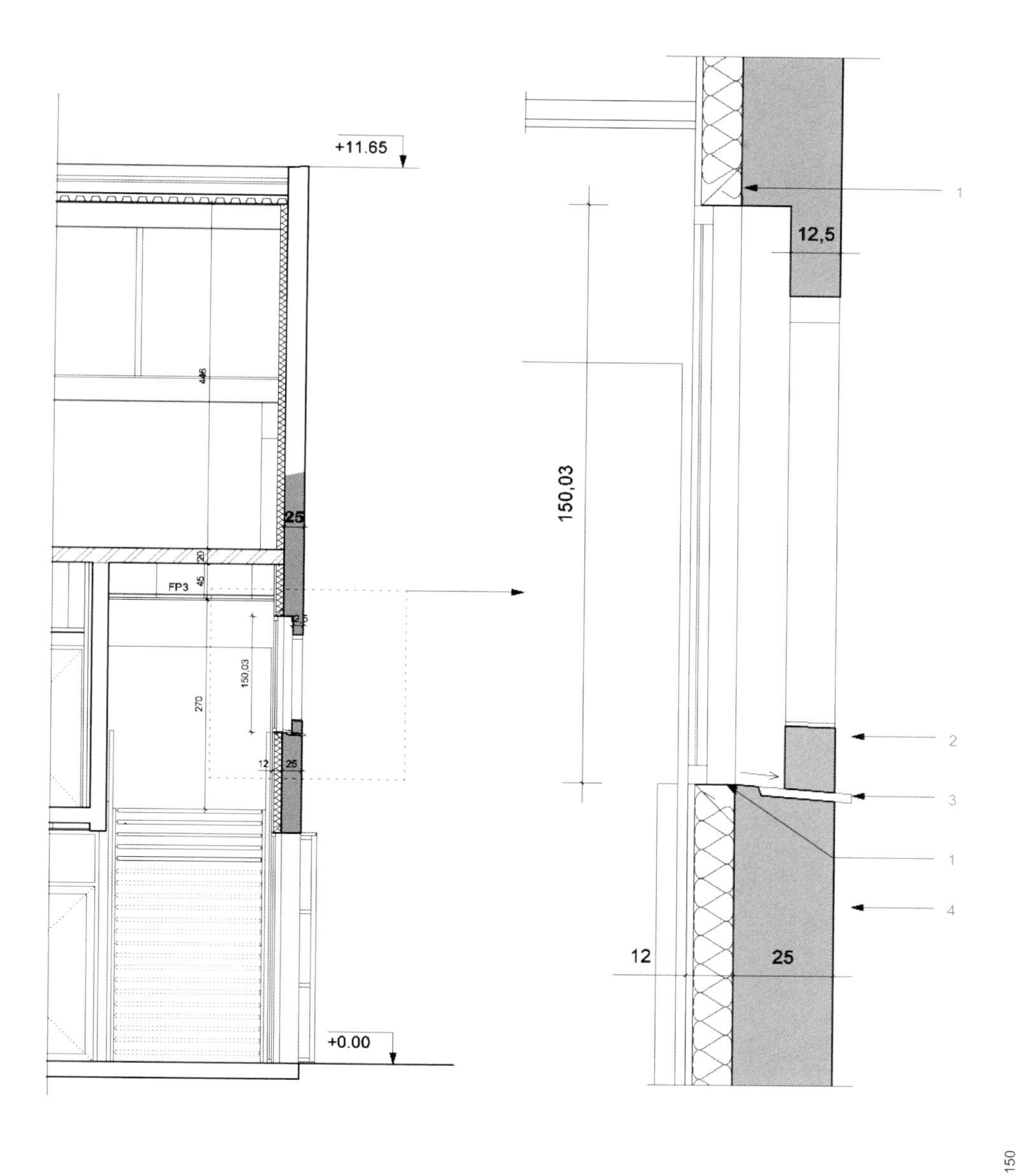

1. Pre-frame
2. Carving sheet thickness 12.5 cm
3. Stormwater drain
4. Concrete sheet thickness 25 cm

1. 预制框架
2. 雕刻板，12.5cm厚
3. 暴雨排水槽
4. 混凝土板，25cm厚

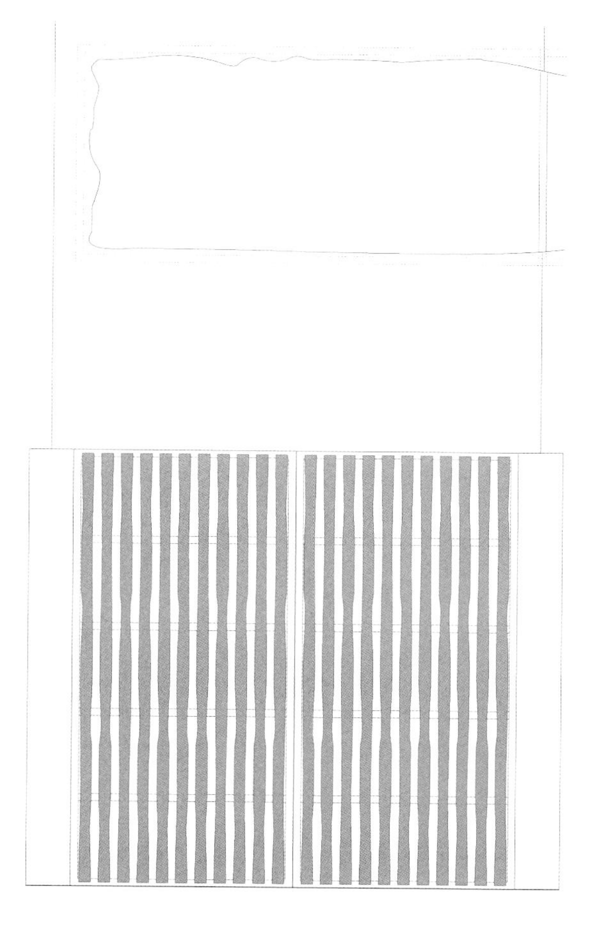

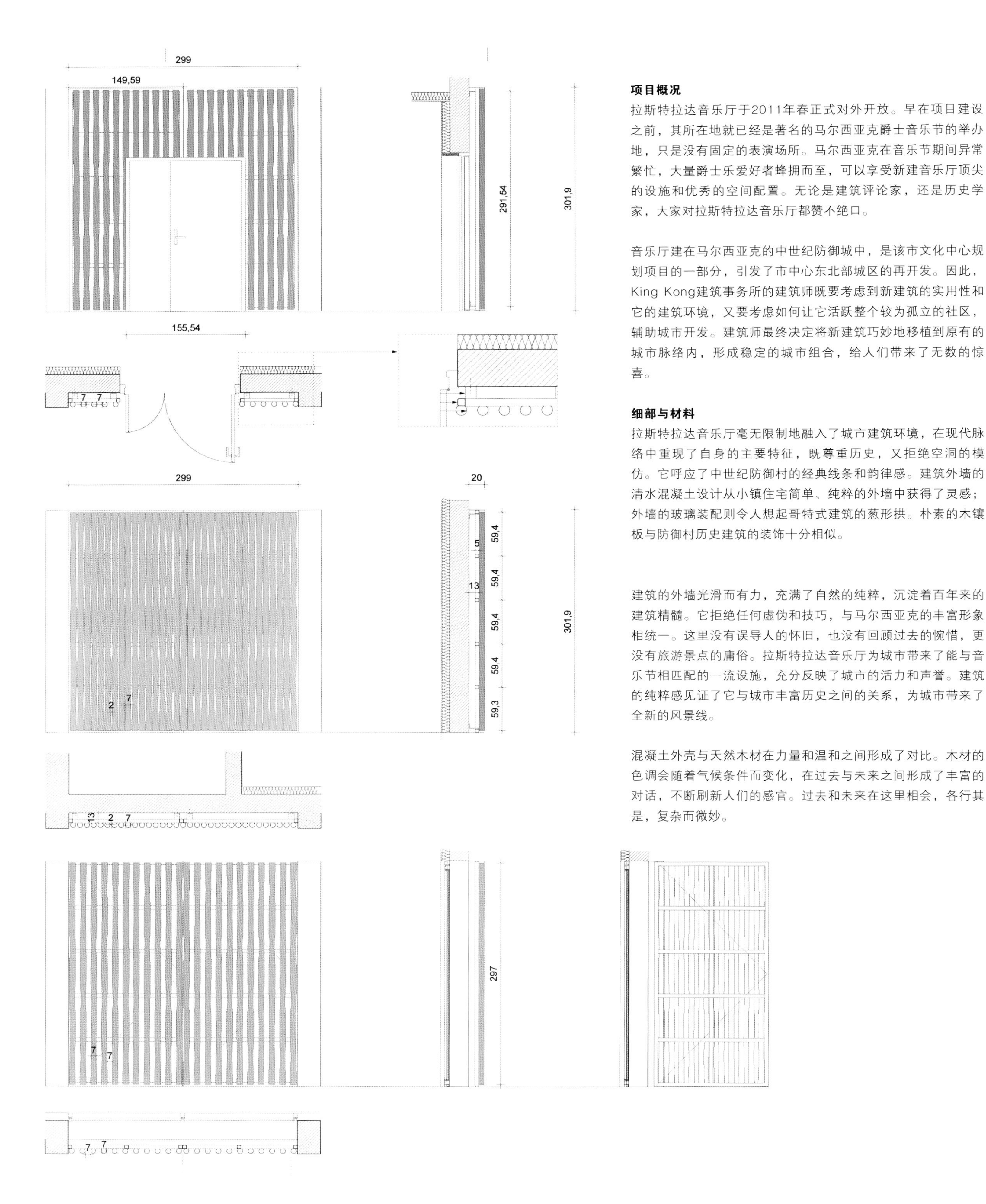

项目概况

拉斯特拉达音乐厅于2011年春正式对外开放。早在项目建设之前，其所在地就已经是著名的马尔西亚克爵士音乐节的举办地，只是没有固定的表演场所。马尔西亚克在音乐节期间异常繁忙，大量爵士乐爱好者蜂拥而至，可以享受新建音乐厅顶尖的设施和优秀的空间配置。无论是建筑评论家，还是历史学家，大家对拉斯特拉达音乐厅都赞不绝口。

音乐厅建在马尔西亚克的中世纪防御城中，是该市文化中心规划项目的一部分，引发了市中心东北部城区的再开发。因此，King Kong建筑事务所的建筑师既要考虑到新建筑的实用性和它的建筑环境，又要考虑如何让它活跃整个较为孤立的社区，辅助城市开发。建筑师最终决定将新建筑巧妙地移植到原有的城市脉络内，形成稳定的城市组合，给人们带来了无数的惊喜。

细部与材料

拉斯特拉达音乐厅毫无限制地融入了城市建筑环境，在现代脉络中重现了自身的主要特征，既尊重历史，又拒绝空洞的模仿。它呼应了中世纪防御村的经典线条和韵律感。建筑外墙的清水混凝土设计从小镇住宅简单、纯粹的外墙中获得了灵感；外墙的玻璃装配则令人想起哥特式建筑的葱形拱。朴素的木镶板与防御村历史建筑的装饰十分相似。

建筑的外墙光滑而有力，充满了自然的纯粹，沉淀着百年来的建筑精髓。它拒绝任何虚伪和技巧，与马尔西亚克的丰富形象相统一。这里没有误导人的怀旧，也没有回顾过去的惋惜，更没有旅游景点的庸俗。拉斯特拉达音乐厅为城市带来了能与音乐节相匹配的一流设施，充分反映了城市的活力和声誉。建筑的纯粹感见证了它与城市丰富历史之间的关系，为城市带来了全新的风景线。

混凝土外壳与天然木材在力量和温和之间形成了对比。木材的色调会随着气候条件而变化，在过去与未来之间形成了丰富的对话，不断刷新人们的感官。过去和未来在这里相会，各行其是，复杂而微妙。

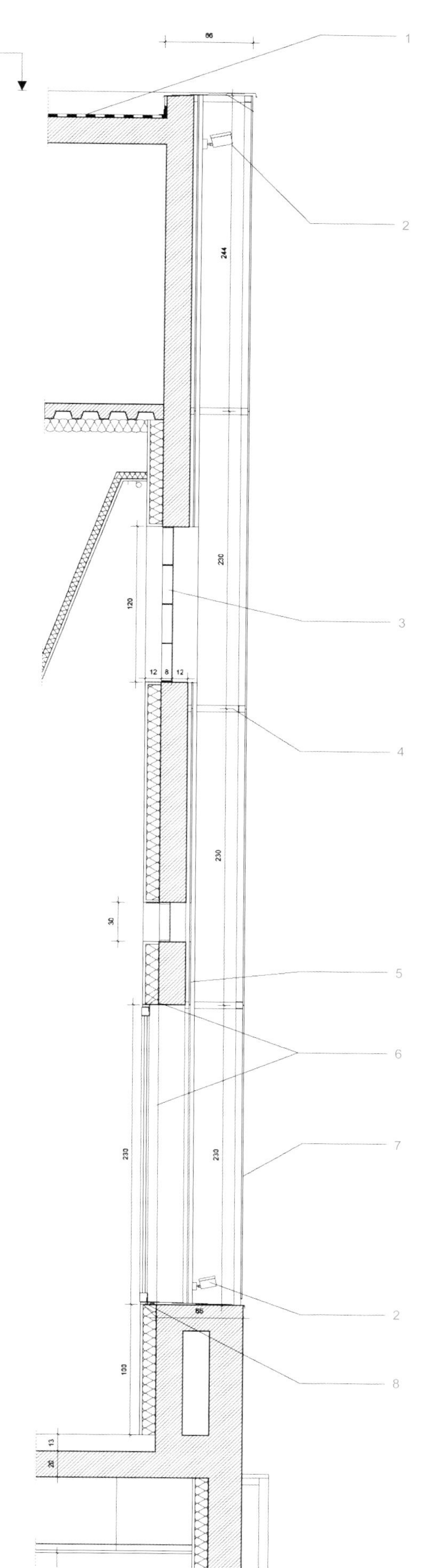

1. Multilayer sealing
2. Searchlight
3. Paved with glass
4. Metal tube support fixed to structural work
5. Powder coated aluminium cladding fixed on wood bracket 5 cm
6. Pre-frame
7. Expanded metal welded to the metal support
8. Lacquered folded sheet on supporting and spinning drop

1. 多层密封
2. 探照灯
3. 玻璃铺装
4. 金属管支撑，固定在结构框架上
5. 粉末涂层铝板，固定在木支架上5cm
6. 预制框架
7. 金属网，焊接在金属支架上
8. 涂漆折叠板

The San Jorge Church and Parish Centre

圣乔治教堂及教区中心

Location/地点: Pamplona, Spain/西班牙，潘普洛纳
Architect/建筑师: Tabuenca & Leache, Arquitectos
Photos/摄影: José Manuel Cutillas, Pedro Pegenaute
Built area/建筑面积: 2,746.24m^2
Key materials: Façade – concrete in situ
主要材料： 立面——现浇混凝土

Overview

For this project, a church-building had to be designed and built for 400 people, including a chapel which would cater for about 100 people to use it on a daily basis. A parish centre also included offices, multipurpose rooms, classrooms for catechesis, two houses for priests and a guest room.

An urban planning study placed the building in the centre of a space which was surrounded by buildings as high as eight storeys: a difficult and somewhat bland situation to deal with. In addition, the site spilled out onto 2 squares on 2 of its sides. In the building's design process, the architects realised that the relationship between these 2 squares and how they interconnect, would become the pivot point for the complex as a whole.

The church stands perpendicular to the neighbourhood's main street. The architects positioned it so that it would conform to the sequence of the existing row of buildings, by having it follow the same line as the others. This helps the building occupy its place in the neighbourhood, seamlessly and with discretion. A large atrium acts as an exterior entrance hall, linking the 2 squares previously mentioned and creating an urban scenario for those people who simply pass through it, while at the same time, it serves as a meeting and gathering point for congregants just before they enter the temple. The atrium also connects the church to the parish centre, which contains the apartments for the priests on the top floor. A raised patio that runs the full length of the façade allows for illumination without disturbing the privacy of others.

This trinity of its main parts (the atrium, temple and parish centre) combine under a unified appearance, like that of a church-fortress that protects all that is within and around her, but also serves as a neutral backdrop to the built environment.

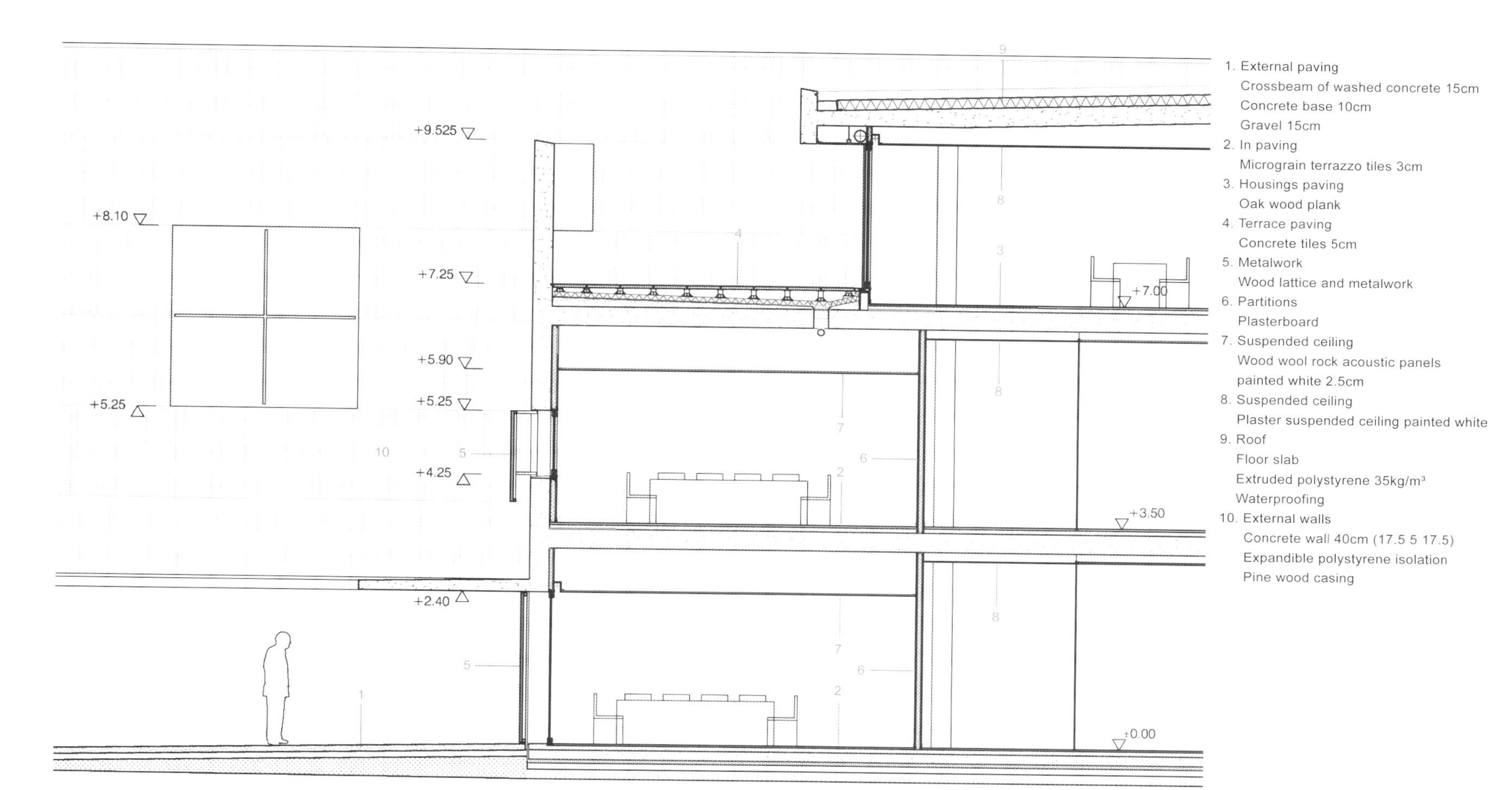

1. External paving
   Crossbeam of washed concrete 15cm
   Concrete base 10cm
   Gravel 15cm
2. In paving
   Micrograin terrazzo tiles 3cm
3. Housings paving
   Oak wood plank
4. Terrace paving
   Concrete tiles 5cm
5. Metalwork
   Wood lattice and metalwork
6. Partitions
   Plasterboard
7. Suspended ceiling
   Wood wool rock acoustic panels painted white 2.5cm
8. Suspended ceiling
   Plaster suspended ceiling painted white
9. Roof
   Floor slab
   Extruded polystyrene 35kg/m³
   Waterproofing
10. External walls
    Concrete wall 40cm (17.5 5 17.5)
    Expandible polystyrene isolation
    Pine wood casing

1. 外部铺装
   水洗混凝土横梁15cm
   混凝土底座10cm
   碎石层15cm
2. 地面铺装
   微粒水磨石砖3cm
   地热供暖7+1cm
   混凝土底座10cm
   碎石层15cm
3. 房屋铺装
   橡木地板
4. 平台铺装
   混凝土砖5cm
5. 金属件
   木格架+金属件
6. 隔断
   石膏板
7. 吊顶
   石膏吊顶，涂有白色涂料
9. 屋顶
   楼板
   挤塑聚苯乙烯35kg/m³
   防水层
10. 外墙
    混凝土墙40cm（17.5 5 17.5）
    膨胀聚苯乙烯隔热层
    松木槽板

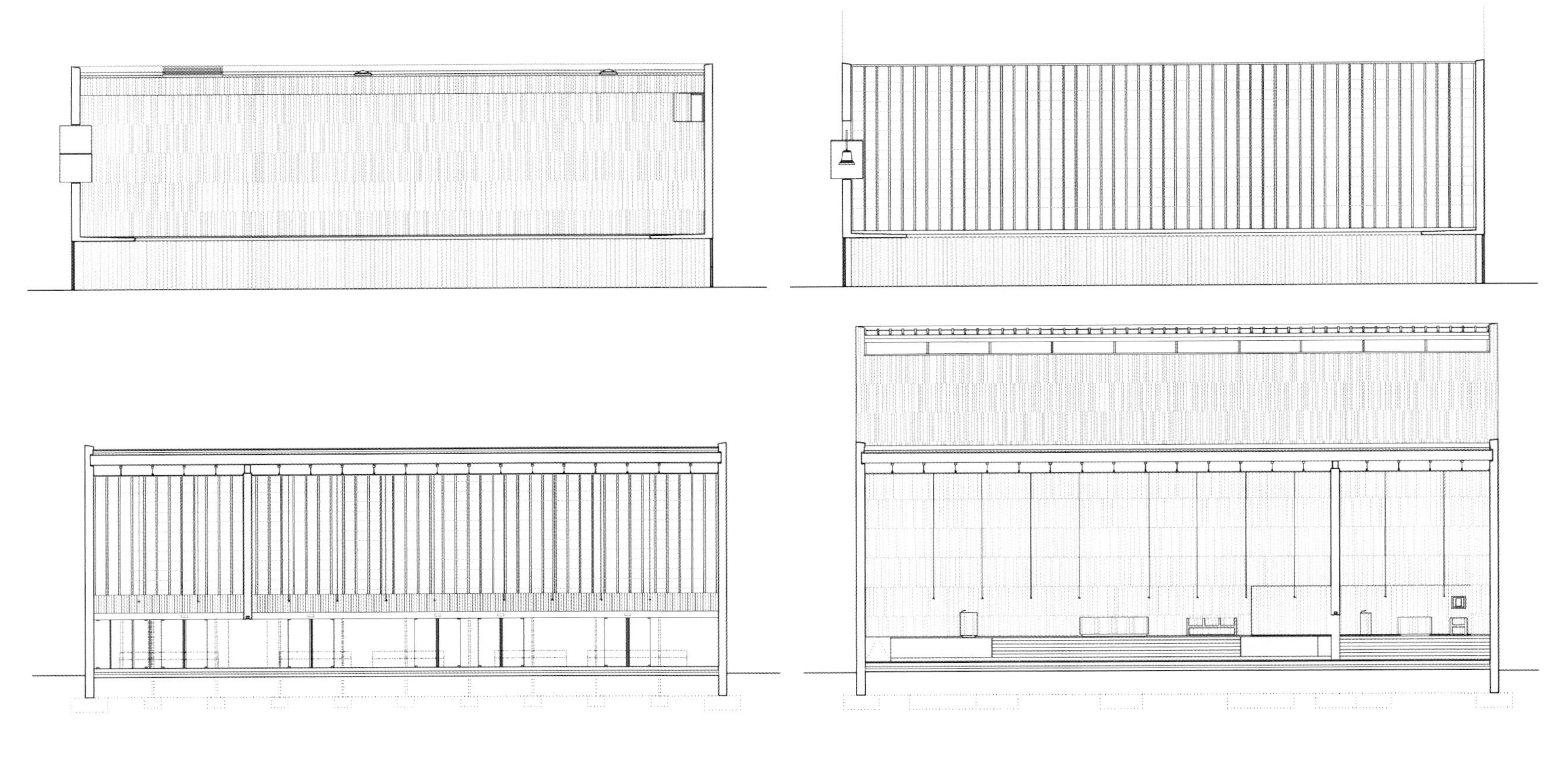

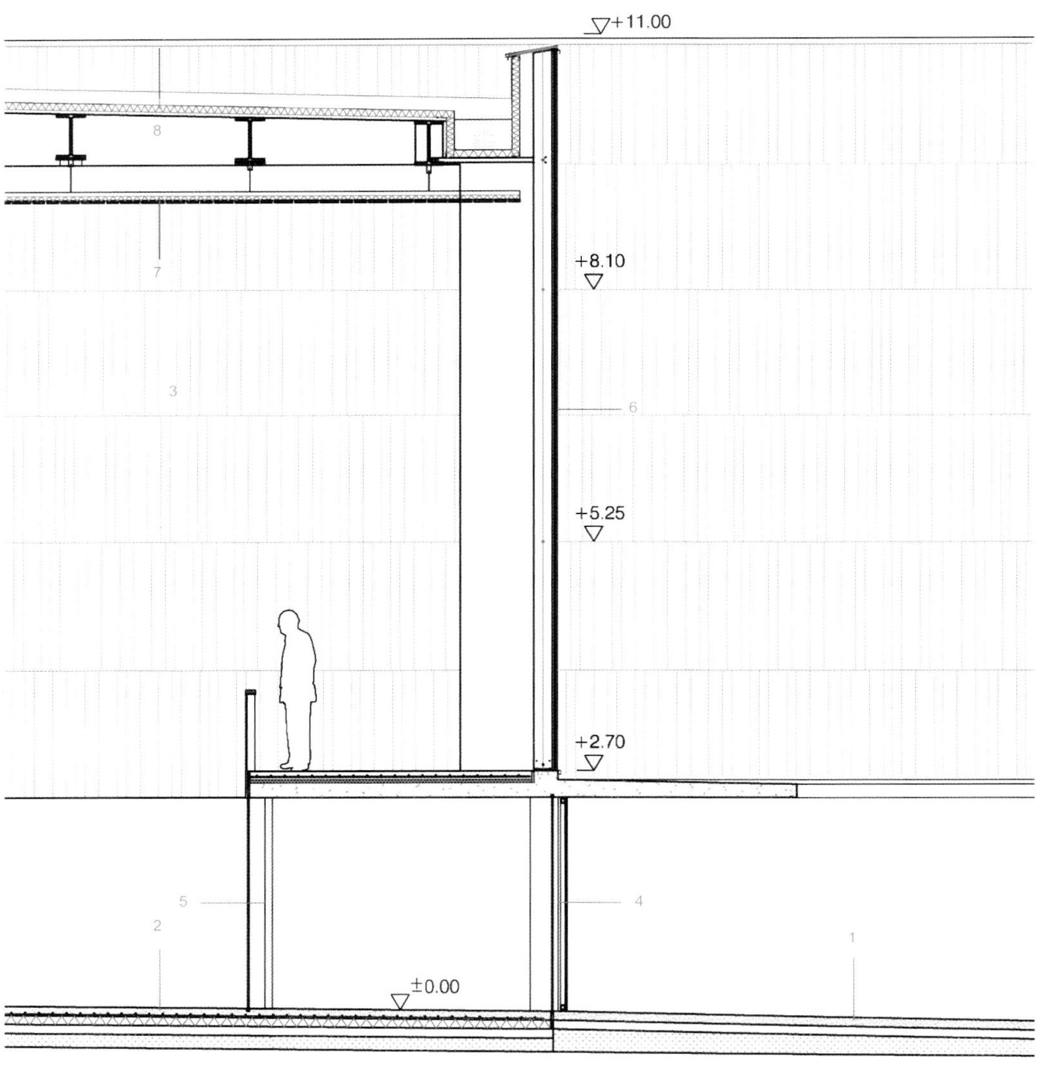

1. External paving
   Crossbeam of washed concrete 15cm
   Concrete base 10cm
   Gravel 15cm
2. In paving
   Micrograin terrazzo tiles 3cm
   Underfloor heating 7+1cm
   Concrete base 10cm
   Gravel 15cm
3. External walls
   Concrete wall 40cm (l7.5 5 17.5)
   Expandible polystyrene isolation
   Pine wood casing
4. Metalwork
   Wood lattice and metalwork
5. Choir structure
   Square metal pillar # 80.80.5
6. Window
   Microlaminated wood beams
   Alabaster 20mm
7. Ceiling
   Pine wood suspended ceiling
8. Roof
   IPE 250 beams and roof decking

1. 外部铺装
   水洗混凝土横梁15cm
   混凝土底座10cm
   碎石层15cm
2. 地面铺装
   微粒水磨石砖3cm
   地热供暖7+1cm
   混凝土底座10cm
   碎石层15cm
3. 外墙
   混凝土墙40cm（17.5 5 17.5）
   膨胀聚苯乙烯隔热层
   松木槽板
4. 金属件
   木格架+金属件
5. 唱诗班结构
   方形金属柱# 80.80.5
6. 窗户
   微胶合木板梁
   雪花石膏2mm
7. 天花板
   松木吊顶
8. 屋顶
   IPE 250梁+屋顶板

1. In paving
   Micrograin terrazzo tiles 3cm
   Underfloor heating 7+10cm
   Concrete base 10cm
   Gravel 15cm
2. External walls
   Concrete wall 40 cm(17.5 5 17.5)
   Expandible polystyrene isolation
   Pine wood casing
3. Ceiling
   Pine wood suspended ceiling
4. Roof
   IPE 250 beams and roof decking
5. Skylight
   Glass 6+12+12 over wood beams

1. 地面铺装
   微粒水磨石砖3cm
   地热供暖7+10cm
   混凝土底座10cm
   碎石层1cm
2. 外墙
   混凝土墙40cm（17.5 5 17.5）
   膨胀聚苯乙烯隔热层
   松木槽板
3. 天花板
   松木吊顶
4. 屋顶
   IPE 250梁+屋顶板
5. 天窗
   玻璃6+12+12，安装在木梁之上

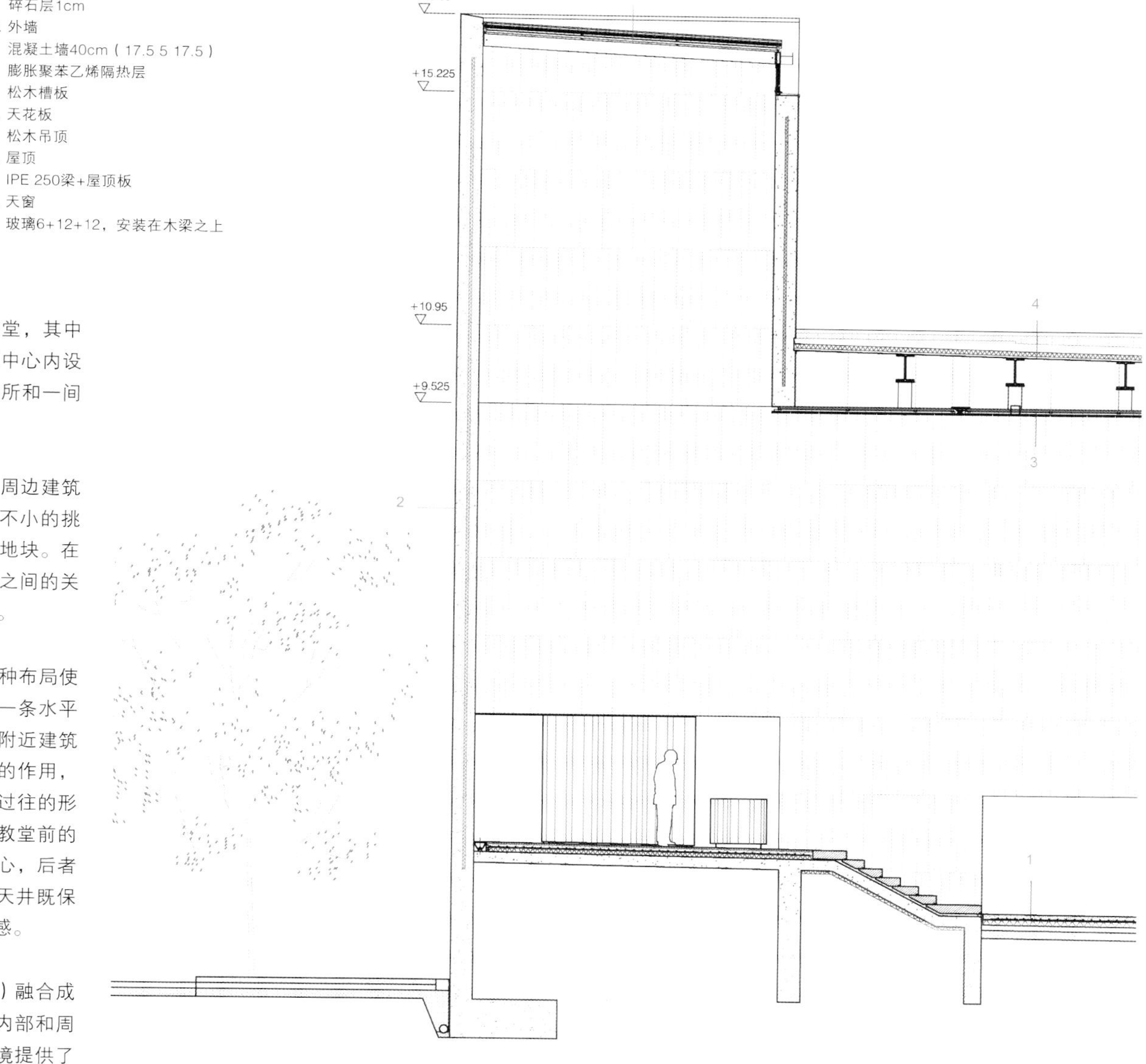

项目概况

本项目要求设计并建造一个可容纳400人的教堂，其中包含一个可供100人日常使用的礼拜堂。教区中心内设有办公室、多功能室、教理教室、两个牧师住所和一间客房。

城市规划研究将建筑设置在一个区域的中心，周边建筑的高度均不高于8层；这对建筑师来说是一个不小的挑战。此外，项目场地在两面还伸出了两个方形地块。在设计过程中，建筑师认识到了这两个方形地块之间的关系及它们的互联方式将是整个项目设计的核心。

教堂与相邻的主要街道相垂直。建筑师通过这种布局使教堂与已有的建筑序列相一致，使它们处在同一条水平线上。这有助于建筑在所在地区站稳脚跟，与附近建筑紧密联系起来。大型中庭起到了室外入口大厅的作用，将上面提到的两个方形地块连接起来。它既为过往的形成提供了独特的城市景观，同时又可作为进入教堂前的会面和集会地点。中庭还连接了教堂与教区中心，后者在顶楼设有牧师的公寓。贯穿建筑立面的架高天井既保证了室内空间的采光，又不会影响它们的私密感。

建筑的三个主要部分（中庭、教堂和教区中心）融合成一个统一的外观，就像是一个教堂堡垒保护着内部和周边的一切。与此同时，建筑还为周边的建成环境提供了统一的背景。

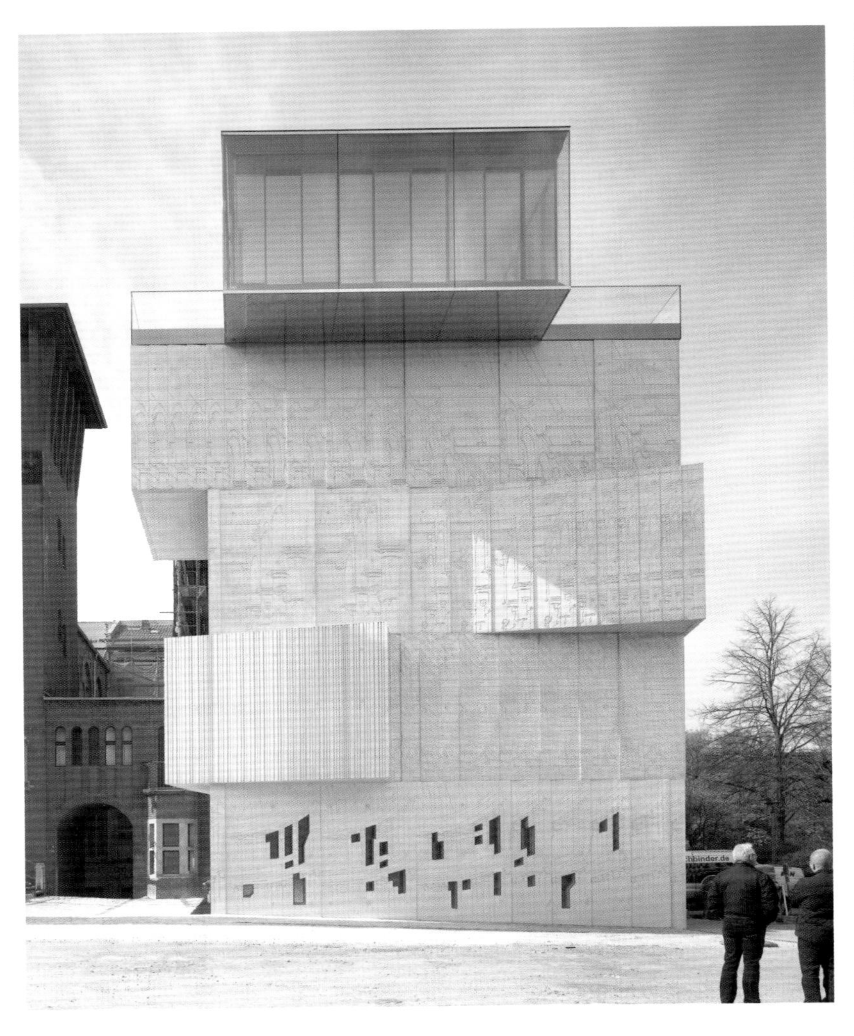

Museum for Architectural Drawing

建筑绘图博物馆

Location/地点: Berlin, Germany/德国，柏林
Architect/建筑师: nps tchoban voss
Photos/摄影: Roland Halbe
Site area/占地面积: 498m^2
Completion date/竣工时间: 2013
Key materials: Façade – coloured concrete and glass
主要材料：立面——彩色混凝土、玻璃

Overview

The Architectural Graphics Museum is meant for placing and exposing the collections of Sergey Choban's Fund founded in 2009 for the purpose of architectural graphics art popularisation as well as for interim exhibitions from different institutions including such famous as Sir John Soane's Museum in London or Art school in Paris.

For the construction of the Museum, the Foundation purchased a small lot on the territory of the former factory complex Pfefferberg, where the art-cluster is formed. Here are already located the famous architecture gallery AEDES, modern art gallery and artists' workshops. The Architectural Graphics Museum that is being constructed became a logical continuation to the development of the new cultural centre in a district Prenzlauer Berg that is very popular among Berlin residents.

The new Museum building flanks the firewall of the adjacent four-storey residential house. Such neighbourhood and the location under the conditions of the current development implied the irregular space-planning arrangement of the Museum.

On the ground and second floors from the side of Christinenstrasse, the flat surfaces of the massive concrete walls are alternate with large glass paintings accentuating the main building entrance and recreation area in front of one of the graphic cabinets. On the ground floor there will be the entrance hall – library. Two cabinets for drawings exposition and archive are located on the upper floors. The levels are connected by an elevator and stairs.

Detail and Materials

The volume that is compact in terms of design rises up to the mark of the neighbouring

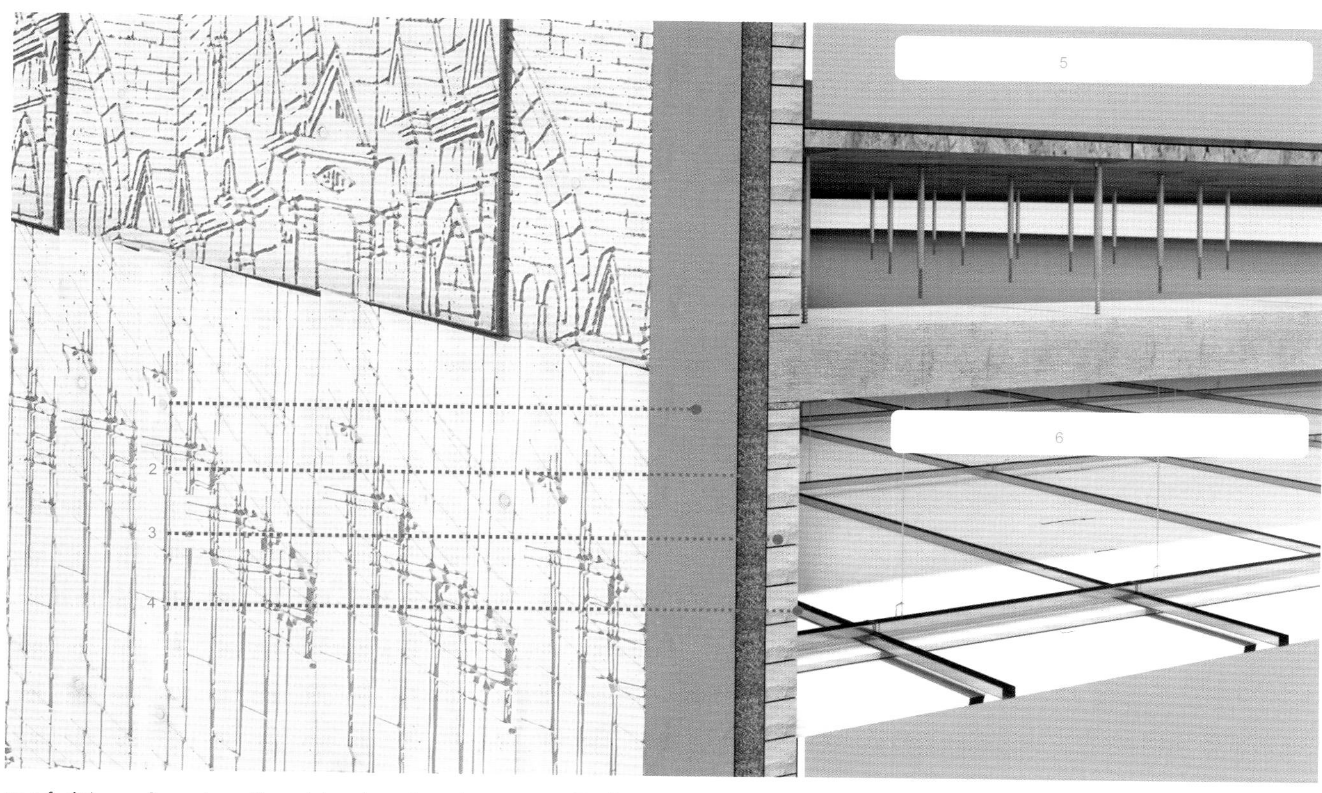

1. Textured concrete
2. Thermal insulation
3. Cladding
4. Plaster
5. Double floor
6. Suspended ceiling

1. 纹理混凝土
2. 隔热层
3. 外墙覆层
4. 石膏
5. 双层地板
6. 吊顶

roof ridge, forming five blocks clearly cut in the building carcass and offset in relation to each other. The upper block, made of glass, hang over the whole volume of the building in cantilever. The façades of the four lower blocks are made of concrete and its surfaces are covered with relief drawings with architectural motives, repeating on every level and overlapping each other as sheets of paper. This artistic touch is supposed to emphasise the function and contents of the exposition in the Museum architectural look.

项目概况

建筑绘图博物馆主要存放并展示了成立于2009年的谢尔盖•科班基金会的建筑绘图藏品，旨在提升建筑绘图艺术的认知度，同时也举办一些来自各大机构的临时展览，例如，伦敦约翰•索恩爵士博物馆或巴黎艺术学院等。

为了建造博物馆，基金会购买了前普费佛酒厂厂区的一小块地块，如今这里已经形成了一个艺术聚集地，拥有著名的AEDES建筑画廊、现代艺术画廊和众多艺术家的工作室。建筑绘图博物馆将延续这个文化中心的开发，成为柏林居民的新去处。

新建的建筑位于旁边四层高的住宅楼的防火墙一侧。这种周边环境和地理位置决定了博物馆将采用不规则的空间规划布局。

在克里斯丁街一侧的一楼和三楼，平整的混凝土墙面上点缀着大块玻璃窗，标志出建筑的主入口和展览厅前方的休闲区。一楼的入口大厅是一个图书馆。两个绘图展厅和档案馆位于上方的楼层，各个楼层之间由电梯和楼梯相连。

细部与材料

建筑的空间设计十分紧凑，高处周边建筑的屋脊，形成了五个界限分明、相互错落的建筑体块。最顶上的一块由玻璃制成，突出于下方的体块。下面四层的建筑立面由混凝土制成，表面覆盖着建筑浮雕图案，它们像纸片一样重复叠加。这种艺术风格彰显了博物馆建筑的功能和内涵。

Chapter 2
Precast Concrete
第二章 预制混凝土

Precast concrete is a construction product produced by casting concrete in a reusable mold or "form" which is then cured in a controlled environment, transported to the construction site and lifted into place. In contrast, standard concrete is poured into site-specific forms and cured on site. Precast stone is distinguished from precast concrete by using a fine aggregate in the mixture, so the final product approaches the appearance of naturally occurring rock or stone.

预制混凝土是一种在可再用模具中成形、在受控环境中固化、运输到施工现场进行安装的建筑产品。相反，标准混凝土是被浇注在现场的模板中并现场固化的。预制石材与预制混凝土的区别在于它使用了更精细的骨料，因此最终产品的外观接近于天然石材。

2.1 Advantages of Precast Concrete

(See Figure 2.1 to Figure 2.3)

During Construction:

Waste Minimisation. Less materials are required because precise mixture proportions and tighter tolerances are achievable. Less concrete waste is created due to tight control of quantities of constituent materials. Waste materials are more readily recycled because concrete production is in one location. Sand and acids for finishing surfaces are reused. Steel forms and other materials are reused.

Recycled Content. Recycled materials such as fly ash, slag cement, silica fume, and recycled aggregates can be incorporated into concrete, thereby diverting materials from the landfill and reducing use of virgin materials. Hardened concrete is recycled (about 5% to 20% of aggregate in precast concrete can be recycled concrete). Gray water is often recycled into future mixtures. All above may contribute to LEED Credit M 4.

Less Community Disturbance. Less dust and waste is created at construction site because only needed precast concrete elements are delivered; there is no debris from formwork and associated fasteners. Fewer trucks and less time are required for construction because concrete is made offsite; particularly beneficial in urban areas where minimal traffic disruption is critical. Precast concrete units are normally large components, so greater portions of the building are completed with each activity, creating less disruption overall. Less noise at construction sites because concrete is made offsite.

During the Life of the Structure:

Energy Performance. Energy savings are achieved in buildings by combining the thermal mass of concrete with the optimal amount of insulation in precast concrete walls. Precast concrete acts as an air barrier, reducing air infiltration, and saving more energy. This may contribute to LEED Credit EA 1.

Disaster Resistant. Precast concrete structures are resistant to fires, wind, hurricanes, floods, earthquakes, wind-driven rain, and moisture damage.

Cool. Light- or natural-coloured concrete reduces heat islands, thereby reducing outdoor temperatures, saving energy, and reducing smog.

Indoor Air Quality. Precast concrete has low VOC emittance and does not degrade indoor air quality.

Recyclable. Precast concrete structures in urban areas can be recycled into fill and road base material at the end of their useful life.

2.2 Productions of Precast Concrete

(See Figure 2.4 to Figure 2.9)

Standard precast products such as beams, decks, and railroad ties are shaped in one type of form that is used repeatedly. Specialty precast products are designed for the particular building, bridge, or other structure. Most precast companies have their own carpentry shops where skilled workers create forms

2.1 预制混凝土的优势（见图2.1 ~ 图2.3）

施工过程中：

减少废物排放。精确的混合比例和更严格的容差有效减少了原材料的浪费。严格的组分材料控制减少了混凝土废料的产生。因为混凝土在同一地点生产，废料更容易回收。修整表面所用的沙子和酸类物质、钢模板以及其他材料都可以重复利用。

回收物质含量。粉煤灰、矿渣水泥、硅粉等回收材料以及回收骨料都可以成为混凝土的一部分，从而有效地将材料从填埋场转移出来进行再利用。硬化混凝土也可以回收利用（预制混凝土中，约5%至20%的骨料可以制成再生混凝土）。灰水通常被回收用作再次搅拌。以上做法都有利于项目在绿色建筑认证（LEED）M 4标准中获得更高的评分。

减少对社区的干扰。因为只需要运输预制混凝土构件，在施工现场可以减少粉尘和废物的产生，并且不会产生来自模架及相关固定配件的碎片。由于混凝土在场外制作，施工所需的货车和时间更少；这点对城市地区特别有力，能有效减少交通拥堵。预制混凝土构件通常为大体块，因此每个阶段都能完成大部分的建造过程，从整体上减少了破坏的数量。预制混凝土的使用还能减少施工现场的噪声。

使用过程中：

能源绩效。在预制混凝土墙面上，将混凝土的热容量与优化的隔热材料相结合，能够实现建筑的高效节能。作为气密层，预制混凝土能减少空气渗透，节约更多能源。这有利于项目在绿色建筑认证（LEED）EA1标准中获得更高的评分。

抵御灾害。预制混凝土结构能防火、防风、防飓风、防洪水、防地震、防暴风雨、防潮。

冷却效果。浅色或自然色混凝土能有效减少热岛效应，从而降低室外温度、节约能源、减少雾气。

室内空气质量。预制混凝土的有机挥发物成分低，不会影响室内空气质量。

可回收性。城市地区的预制混凝土结构废弃之后可以用作填埋材料和路基材料。

2.2 预制混凝土的制作（见图2.4 ~ 图2.9）

建筑梁、板面、铁轨轨枕等标准预制混凝土产品采用

2.1

2.2

2.3

2.4

2.5

2.6

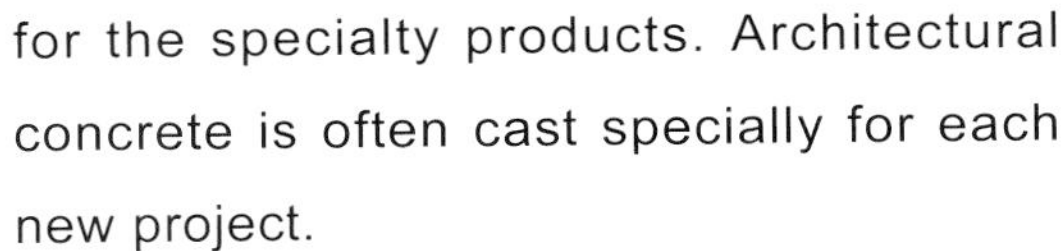

for the specialty products. Architectural concrete is often cast specially for each new project.

During the production process, forms for concrete are well lubricated. Concrete is placed in the forms and allowed to cure. After curing, the product is carefully lifted from the form and taken to a yard for further curing before it is shipped to the project site. The form is cleaned and prepared for the next batch of concrete. Many precasters can reuse their forms every one to two days.

Exterior finishes for architectural precast concrete can incorporate a full range of colours and textures. Textures are achieved by acid-etching, retarders, or sandblasting.

2.3 Applications in Building Façade

(See Figure 2.10 to Figure 2.14)

Precast Concrete Sandwich Wall (or Insulated Double-Wall) Panels

Origin

The precast concrete double-wall panel has been in use in Europe for decades. The original double-wall design consisted of two wythes of reinforced concrete separated by an interior void, held together with embedded steel trusses. With recent concerns about energy use, it is recognised that using steel trusses creates a "thermal bridge" that degrades thermal performance. Also, since steel does not have the same thermal expansion coefficient as concrete, as the wall heats and cools any steel that is not embedded in the concrete can create thermal stresses that cause cracking and spalling.

Development

To achieve better thermal performance, insulation was added in the void, and in many applications today the steel trusses have been replaced by composite (fibreglass, plastic, etc.) connection systems. These systems, which are specially developed for this purpose, also eliminate the differential thermal

统一的模板成形，可以反复利用。特制预制混凝土产品专为特定建筑、桥梁或其他结构设计。大多数预制混凝土公司都有自己的木工车间，由专业的工人为特制产品制作模板。建筑混凝土大多专门为新项目所定制。

在生产过程，混凝土的模板经过了良好的润滑。混凝土被放置在模板中进行固化。固化结束后，产品被小心地从模板中取出，运往场院进行进一步的固化，最后再运往项目现场。模板经清洁后为下一批次混凝土产品所使用。许多预制制造商每隔一到两天重复利用一次模板。

建筑预制混凝土的外表面可以结合各种色彩和质感。酸蚀刻、混凝剂或喷砂等处理方式都能实现多变的质感。

2.3 建筑表皮中应用（见图2.10 ~ 图2.14）

预制混凝土夹层墙板（隔热双层墙板）

起源

预制混凝土双层墙板在欧洲已经有几十年的应用。最初的的双层墙设计由两层钢筋混凝土和中空层构成，二者由埋置钢桁架连接起来。近年来，考虑到能源利用，人们意识到钢桁架会产生“热桥效应”，影响热性能。此外，由于钢的热膨胀系数与混凝土不一致，随着墙壁的升温和冷却，未埋入混凝土的钢材会产生热应力，造成混凝土的开裂和剥落。

发展

为了实现更好的热性能，建筑师开始在中空层中加入

expansion problem. The best thermal performance is achieved when the insulation is continuous throughout the wall section, i.e., the wythes are thermally separated completely to the ends of the panel. Using continuous insulation and modern composite connection systems, R-values exceeding R-22 can be achieved.

Characteristics

The overall thickness of sandwich wall panels in commercial applications is typically 8 inches, but their designs are often customised to the application. In a typical 8-inch wall panel the concrete wythes are each 2-3/8 inches thick, sandwiching 3-1/4 inches of high R-value insulating foam. The interior and exterior wythes of concrete are held together (through the insulation) with some form of connecting system that is able to provide the needed structural integrity. Sandwich wall panels can be fabricated to length and width desired, within practical limits dictated by the fabrication system, the stresses of lifting and handling, and shipping constraints. Panels of 9-foot clear height are common, but heights up to 12 feet can be found.

The fabrication process for precast concrete sandwich wall panels allows them to be produced with finished surfaces on both sides. Such finishes can be very smooth, with the surfaces painted, stained, or left natural; for interior surfaces, the finish is comparable to drywall in smoothness and can be finished using the same prime and paint procedure as is common for conventional drywall construction. If desired, the concrete can be given an architectural finish, where the concrete itself is coloured and/or textured. Colours and textures can provide the appearance of brick, stone, wood, or other patterns through the use of reusable formliners, or, in the most sophisticated applications, actual brick, stone, glass, or other materials can be cast into the concrete surface.

隔热材料。在当前许多应用中，钢桁架已经被复合连接系统（纤维玻璃、塑料等）所取代。这些系统专为此而开发，排除了热膨胀系数差异的问题。当隔热材料遍布整个墙壁截面时，热性能会达到最佳表现，即两层混凝土板完全被隔热层隔开。连续的隔热层和现代复合连接系统可以让墙壁的热阻值超过R-22。

特色

商用夹层墙板的标准厚度为8英寸（约20厘米），但是大多数设计都是特别定制的。在典型的8英寸墙板中，两层混凝土层分别为2.375英寸（约6厘米），高热阻

2.7

2.8

2.9

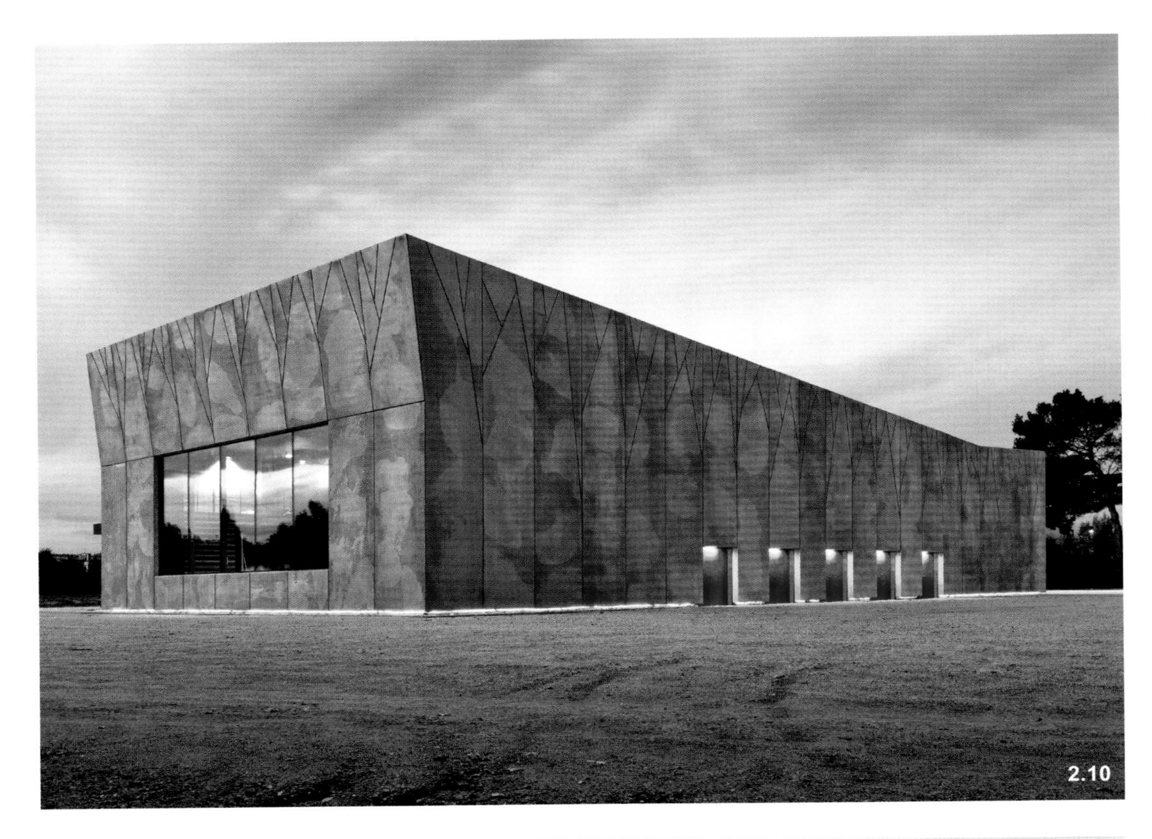
2.10

2.11

2.12

2.13

Window and door openings are cast into the walls at the manufacturing plant as part of the fabrication process. In many applications, electrical and telecommunications conduit and boxes are cast directly into the panels in the specified locations. In some applications, utilities, plumbing, and even heating components have been cast into the panels to reduce on-site construction time. The carpenters, electricians, and plumbers do need to make some slight adjustments when first becoming familiar with some of the unique aspects of the wall panels. However, they still perform most of their job duties in the manner to which they are accustomed.

Applications and Benefits

Precast concrete sandwich wall panels have been used on virtually every type of building, including schools, office buildings, apartment buildings, townhouses, condominiums, hotels, motels, dormitories, and single-family homes. Although typically considered part of a building's enclosure or "envelope", they can be designed to additionally serve as part of the building's structural system, eliminating the need for beams and columns on the building perimeter. Besides their energy efficiency and aesthetic versatility, they also provide excellent noise attenuation, outstanding durability (resistant to rot, mold, etc.), and rapid construction.

2.14

隔热泡沫夹层为3.25英寸（约8厘米）。内外两层混凝土通过连接系统（穿过隔热层）连接起来，提供必要的结构完整性。在制造系统的实践限制、起重压力和装运限制的范围内，夹层墙板的长度和宽度都可以定制。净高9英尺（约2.7米）的板材最为常见，但是最高的板材也可达12英尺（约3.66米）。

预制混凝土夹层墙板的制作过程允许内外两面都能配有装饰表面。它的饰面可以十分光滑，可以喷涂、染色或保持自然的状态；内表面的光滑度堪比干式墙，也可以使用与传统干式墙相同的打底和喷涂工艺。如有需要，可以赋予混凝土建筑饰面，对其进行染色或添加纹理。可再用模板形成的色彩与纹理赋予混凝土类似于砖、石材、木材的外观。在复杂的应用中，设置可以在混凝土表面上浇铸真正的砖、石材、玻璃或其他材料。

工厂制作过程中就能把门窗开口浇铸在墙面上。在许多应用中，电线和通信线导管和线路盒都被直接浇铸在板材的特定位置。在一些应用中，公共事业设备、水管装置乃至供暖组件都可以被浇注在板材之中，以减少现场施工的时间。木工、电工和管道工在初次面对预制墙板时确实需要进行调整，但是一旦习惯之后，他们就能快速高效地完成自己的工作职责。

应用与优势

预制混凝土夹层板已经被应用在各种类型的建筑中，包括学校、办公楼、公寓楼、联排别墅、托管公寓、酒店、汽车旅馆、宿舍、独户住宅等。尽管人们通常将它们看作是建筑的表皮或外壳，它们还可以被设计成建筑的结构系统，免除了建筑外围结构所需的梁柱支撑。除了良好的能源效率和美学价值外，它们还具有出色的隔音性、耐久性（防腐、防霉），可以快速地进行施工。

Braamcamp Freire Secondary School

布拉姆坎博•弗莱德中学

Location/地点: Lisbon, Portugal/葡萄牙，里斯本
Architect/建筑师: CVDB Arquitectos
Photos/摄影: invisiblegentleman.com
Built area/建筑面积: 15,800m^2
Completion date/竣工时间: 2012
Key materials: Façade – situ concrete and precast concrete panels
主要材料：立面—— 现场浇筑混凝土、预制混凝土板

Overview

The Braamcamp Freire Secondary School is located at the edge of the historical centre of Pontinha, Lisbon. The site has approximately 17,380 square metres and borders an accentuated topography. The school is part of Pontinha's urban fabric with the exception of its north boundary that faces an unconstructed valley. The School was originally built in 1986, with 5 standardised prefabricated pavilions – a central one with a single storey and four two storey pavilions. These pavilions were organised along an east-west axis, connected by covered walkways. The existing school included a gym as well as an outside playground at a lower level and very disconnected from the buildings.

The rehabilitation project of the building was part of the Portuguese "Modernisation of Secondary Schools Programme", which has been implemented by the Parque Escolar E.P.E., since 2007. The Programme's objective is to reorganise schools spaces, to articulate their different functional areas and to open these schools to their local communities.

The project proposes to restructure the dispersed pavilion typology into one single building, to connect all the pavilions through interior circulation spaces. The new buildings are built to work as a link in between the existing pavilions. The programme is structured as a learning street and a continuous path throughout the various building levels and floors. These pathways consist in a succession of several interior spaces, offering different informal learning opportunities. The learning street therefore articulates the various programmes of the school. The pathways are punctuated with social areas which actively contribute to interactions between students, the various educational programmes and the school community.

The school is structured around a central open space, a "learning square" that expands the "learning street" as an outside social central

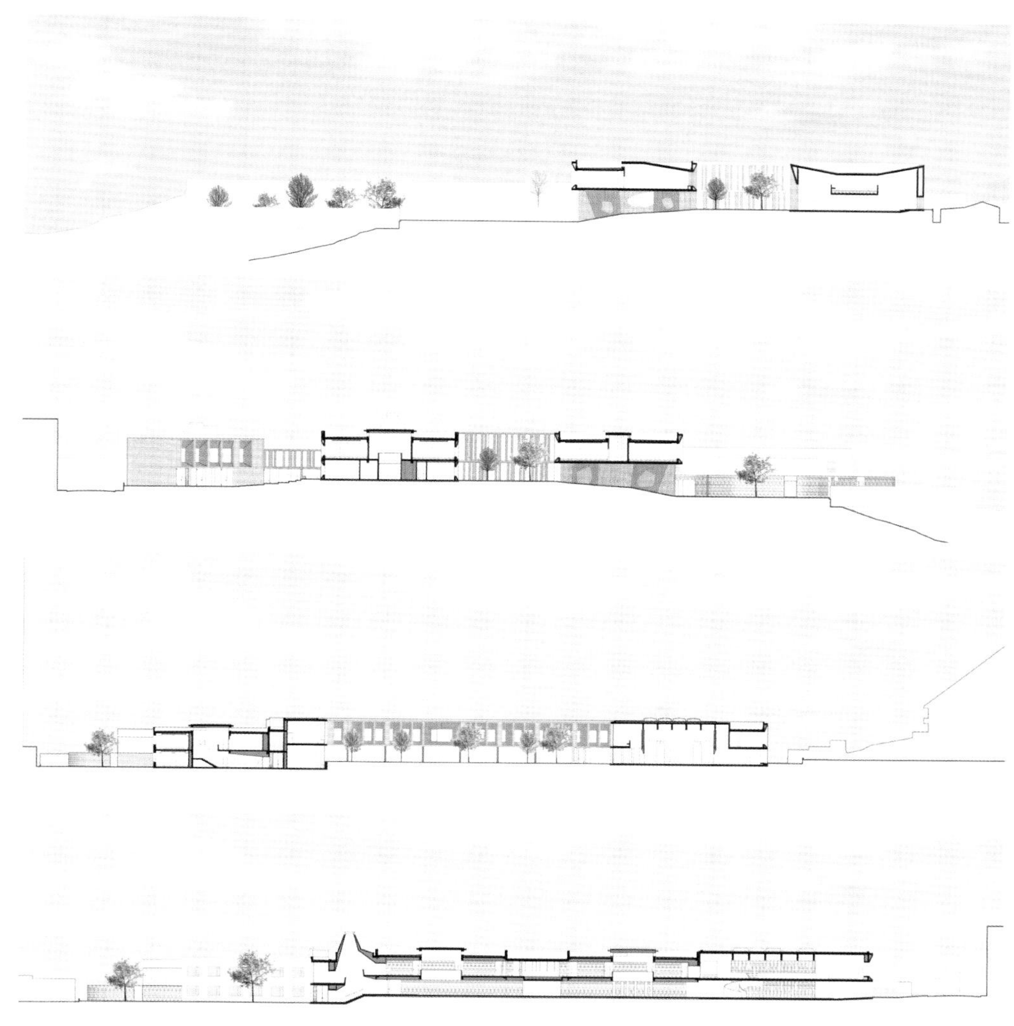

space of the school. The square's relationship with the playground areas provides a strong relationship with the existing natural landscape and topography. The Square is open as an amphitheater connecting it to the playgrounds in the northern part of the school grounds. This amphitheater is below the new classrooms building supported by a series of punctured concrete walls allowing students either to walk through them or to use them as places to sit, talking and playing.

In the interior spaces, adequate resistant materials were chosen for an intensive use and very low maintenance costs. The multipurpose hall has timber studs and acoustic panels. The circulation spaces walls are mainly done with concrete acoustic blocks. The social spaces present themselves as niches in bright colours.

Detail and Materials

The façades of the school are essentially constituted in exposed in situ concrete and prefabricated concrete panels, to minimise maintenance costs. The concrete panels were carefully designed to respond adequately to each façade's solar orientation.

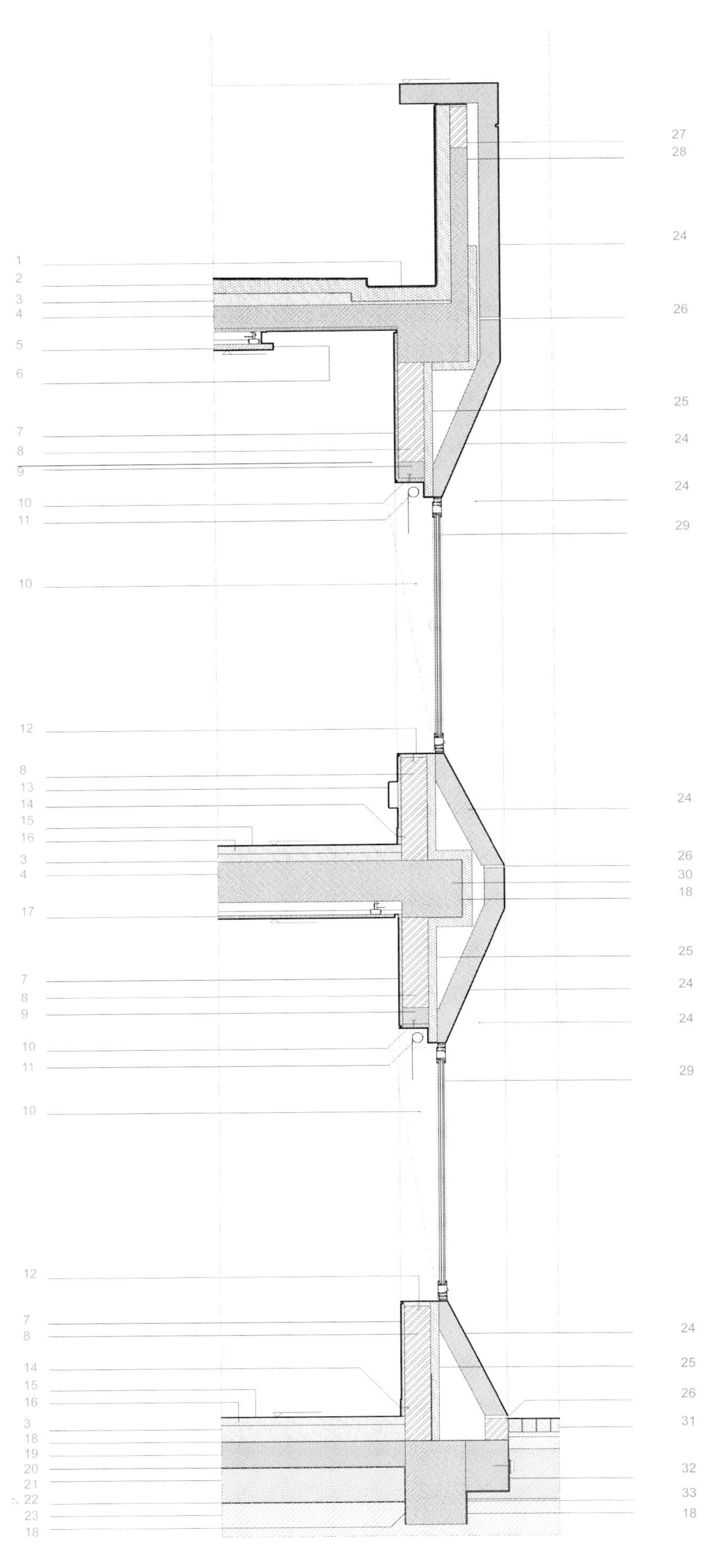

1. Gutter
2. Roof system imperalum "imperkote" with 80mm of thermal insulation
3. Lightweight concrete filling
4. Existing building reinforced concrete slab
5. 35mm acoustic panel Celenit ABE
6. Metal finishing profiles
7. Painted plaster
8. Brick wall
9. Concrete lintel
10. Aluminium frame
11. Shading roller
12. Aluminium sill
13. Device installation trunking
14. Metal skirting (100×5mm)
15. Self leveling floor with elastic and polyurethane based finishing
16. 47mm reinforced cement screed
17. Gypsum board acoustic ceiling
18. Waterproofing
19. Ground floor slab
20. Vapour barrier
21. Rockfill
22. Geotextile fabric (200gr/m^2)
23. Compacted natural soil
24. Precast concrete panel with water repellent coat
25. 60mm thermal insulation
26. Mastic and neoprene sealant
27. Brick work
28. Existing building reinforced concrete roof beam
29. Double glazing aluminium frame
30. Existent reinforced concrete beam
31. Concrete pavement blocks
32. Precast concrete panels support foundation
33. Existing building foundation

1. 排水槽
2. 屋顶系统imperkote，配有80mm隔热层
3. 轻质混凝土填充
4. 原有建筑钢筋混凝土结构板
5. 35mm隔音板Celenit ABE
6. 金属饰面型材
7. 涂漆石膏
8. 砖墙
9. 混凝土过梁
10. 铝框
11. 遮阳卷帘
12. 铝窗台
13. 设备安装轨道
14. 金属踢脚板（100x5mm）
15. 自平层，弹性聚氨酯基饰面
16. 47mm增强水泥砂浆
17. 石膏板隔音天花板
18. 防水层
19. 地面楼板
20. 隔汽层
21. 碎石填充
22. 土工布（200gr/m^2）
23. 压实天然土
24. 预制混凝土板，带防水涂层
25. 60mm隔热层
26. 胶泥和氯丁橡胶密封
27. 砖砌结构
28. 原有建筑钢筋混凝土屋梁
29. 双层玻璃铝框
30. 原有钢筋混凝土梁
31. 混凝土铺装块
32. 预制混凝土板支撑地基
33. 原有建筑地基

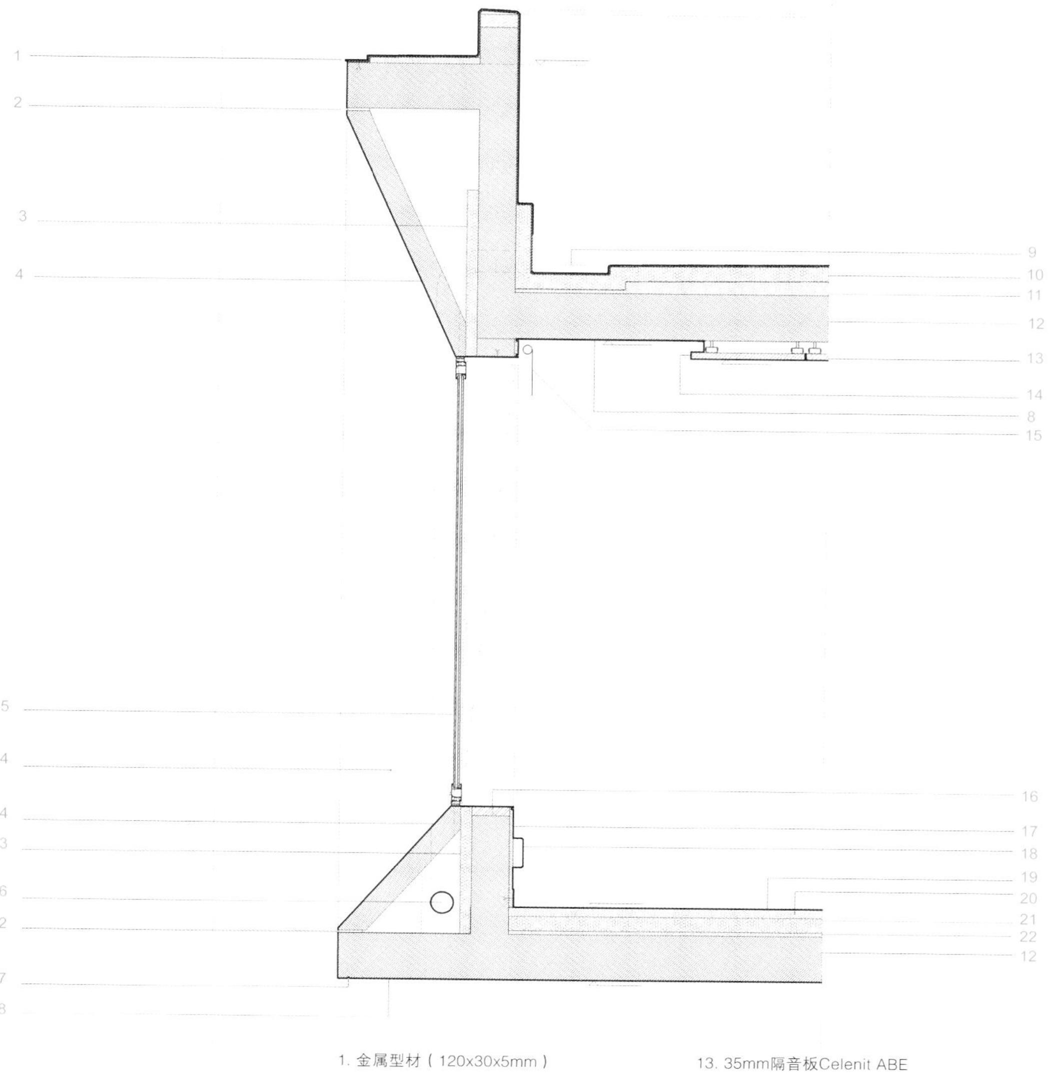

1. Metallic profile (120×30×5mm)
2. Mastic and neoprene sealant
3. 60mm thermal insulation
4. Precast concrete panel with water repellent coat
5. Double glazing aluminium frame
6. Rainwater pipe – roof drainage system
7. Flashing drip detail
8. Fair faced concrete
9. Gutter
10. Roof system imperalum "imperkote" with 80mm of thermal insulation
11. Lightweight concrete filling
12. Reinforced concrete structural slab
13. 35mm acoustic panel Celenit ABE
14. Metal finishing profiles
15. Shading roller
16. Aluminium sill
17. Painted plaster
18. Device installation trucking
19. Self leveling floor with elastic and polyurethane based finishing
20. 47mm reinforced cement screed
21. 80mm thermal insulation
22. 20mm leveling cement screed
23. Concrete pavement blocks

1. 金属型材（120x30x5mm）
2. 胶泥和氯丁橡胶密封
3. 60mm隔热层
4. 预制混凝土板，带有防水涂层
5. 双层玻璃铝框
6. 雨水管——屋顶排水系统
7. 防水板滴水槽
8. 琢面混凝土
9. 排水槽
10. 屋顶系统imperkote，配有80mm隔热层
11. 轻质混凝土填充
12. 钢筋混凝土结构板
13. 35mm隔音板Celenit ABE
14. 金属饰面型材
15. 遮阳卷帘
16. 铝窗台
17. 涂漆石膏
18. 设备安装轨道
19. 自平层，弹性聚氨酯基饰面
20. 47mm增强水泥砂浆
21. 80mm隔热层
22. 20mm找平水泥砂浆
23. 混凝土铺装块

23

项目概况

布拉姆坎博•弗莱德中学位于里斯本庞蒂尼亚历史中心的边缘，占地总面积约17,380平方米，地理位置比较突出。学校是庞蒂尼亚城市网格的一部分，只有北部边缘朝向未建设的山谷。学校初建于1986年，拥有五座标准教学楼——中央楼为单层，其他四座为双层。这些教学楼沿着校园的东西向中轴展开，由铺装的走道相连接。学校还在低处设有一个体育馆和户外操场，它们与教学楼的连接十分分散。

这个重建项目是葡萄牙“中学现代化改造计划”的一部分，该计划开始于2007年，由Parque Escolar E.P.E.负责实施，目的是重新组织学校空间，使不同的功能区域更加清晰，并且让这些学校面向当地社区开放。

该项目的方案是对分散的结构进行重组，将其连成一座建筑，通过室内交通流线空间连接起所有的结构。原有的结构之间建造了新的建筑作为连接元素。项目设计了一条学习街和一条穿过所有建筑楼层的连续通道。这些通道包括一系列室内空间，为学生们提供了不同的非正式学习的机会。学习街因此连接起学校不同的功能空间。这些通道上设置了一些社交区域，可以促进学生之间的互动、各种教育计划的开展和学校社团活动的举行。

学校围绕一个中央开放空间而展开。作为一个 “学习广场”，这个开放空间扩展了“学习街”，是学校的室外中央社交空间。广场和操场的关系与原有的自然景观和地形关系密切。该广场是开放式的，像是一个露天剧场，与校园北部的操场相连，位于由一系列穿孔混凝土墙支撑的新教室建筑下方，学生们可以在这里穿行或是休息、谈话、玩耍。

室内空间选用了非常耐用的材料，以满足密集的使用需求和较低的维护成本需求。多功能大厅采用了木龙骨和吸音板。交通流线空间的墙体主要安装了混凝土吸音块。社交空间表现为色彩亮丽的小空间。

细部与材料

学校的外墙基本上由现场浇筑的素面混凝土和预制混凝土板构成，以减少维护成本。混凝土板都经过精心设计，以与各个朝向的立面充分对应。

MPA Building

MPA大厦

Location/地点: Porto, Portugal/葡萄牙，波尔图
Architect/建筑师: Paulo Lousinha / Lousinha Arquitectos
Photos/摄影: Luís Ferreira Alves
Built area/建筑面积: 9,913m^2
Completion date/竣工时间: 2011
Key materials: Façade – precast concrete panel (Pré-Gaia), steel plates, thermal insulation composite system, bonded - ETIC
主要材料： 立面——预制混凝土板、钢板、复合隔热系统、ETIC外墙保温系统

Overview

The project is implanted on a ground with a total area of 1,364 m^2 and takes place in the old industrial zone of Porto, which is now being regenerated, by using the land for activities related to enterprises, replacing the previous industries. Making part of this process of regeneration, the building is implanted perpendicularly to Manuel Pinto de Azevedo (MPA) Street.

The southwest façade defines a new square opened to Manuel Pinto de Azevedo Street. It looks for dialogue with this public space, designing a front marked by the relation of the entrances and shops with this "square".

The building, of mixed use, shelters a school and offices. In order to allow autonomy of uses, two columns of vertical communications serve each of the programmes independently. On the ground floor, a "front" of commerce turns to a broad square in the northwest. In this front, two shops with public access by the exterior of the building – a cafeteria and a stationery centre – work associated with the school programme.

The school has a door on the southwest façade, near Manuel Pinto de Azevedo Street. This programme takes place on five floors: ground floor, first floor, second floor, third floor and half of the fourth floor. On the ground

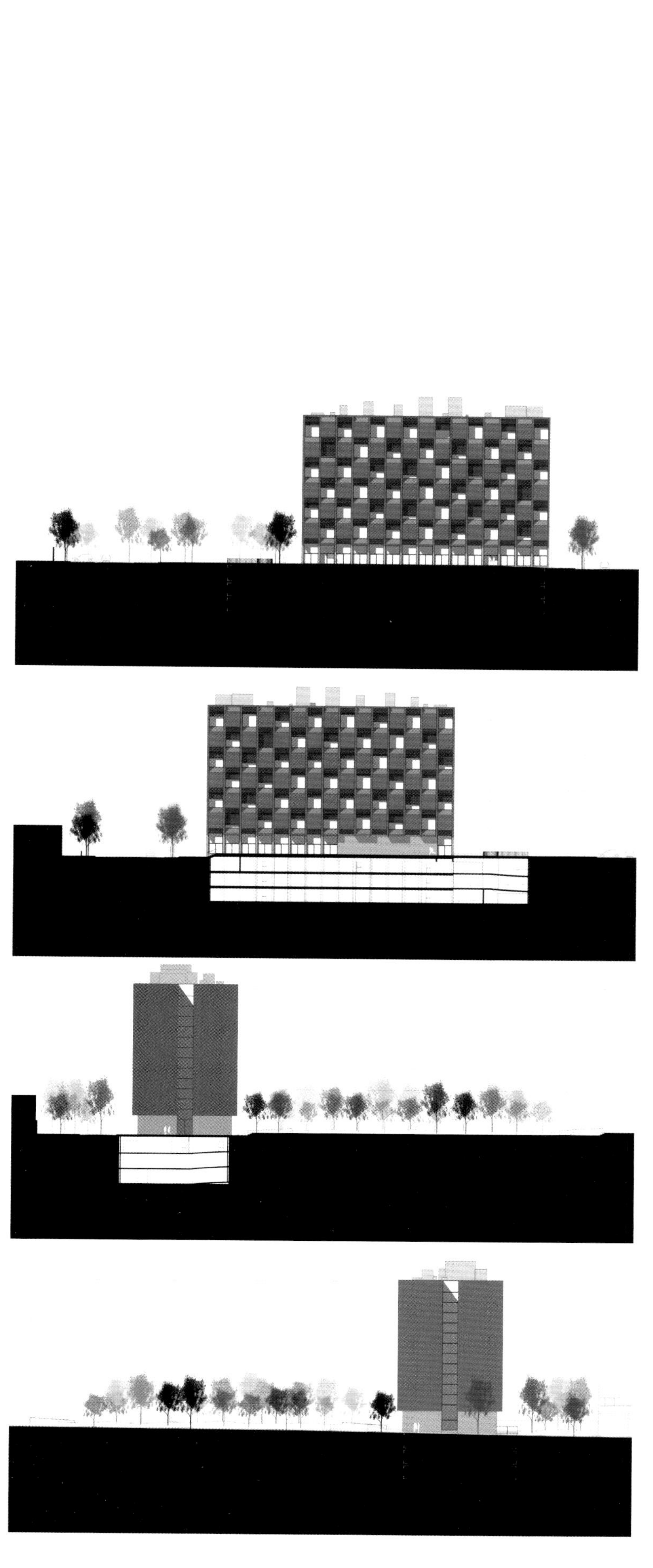

floor there is the reception and a foyer supporting the auditorium with 200 seats. The entrance of the school is through a space with triple ceiling height, marked by a staircase that besides the function of a strolling way assumes a sculpture character. The first floor is constituted by the school office, the library and six classrooms. The next two floors have seven classrooms each. The library – which is organised in two floors – occupies part of the second floor and the staff room is located on the third floor. On the fourth floor, there are two classrooms and the Head Office. The interior corridor, naturally illuminated, is broad enough to allow other uses besides distribution. Flexibility in the use of the classrooms was a main concern in the whole process of conceptualisation. Therefore is presented a solution that makes possible, either on the first floor or on the second or the third floors, create a unique space by joining four or five classrooms from the southeast wing. The northeast entrance will allow the distribution to the offices that occupy half of the fourth floor and the whole fifth and sixth floors.

Detail and Materials

Formally restrained, the built volume wants to show two moments: the contact with the ground, very clear (transparent), in glass, where all opaque plans are covered with steel plates; and the relation with the more opaque horizon line, where the alternation of the openings introduces tension in the drawing of the elevations. The superior floors are covered with precast concrete panels and overlaid with thin plaster applied on thermal insulation – thermal insulation composite system. By being protected by exterior awnings of the same colour of the plaster, the openings contribute, in these floors, to a dynamic relation of the façade, which therefore can have multiple combinations. These materials, steel and thin plaster of a dark colour, remind the industrial past of the zone. The reference to the industrial past of the place has always been present in the options concerning the design and the materialisation of the building.

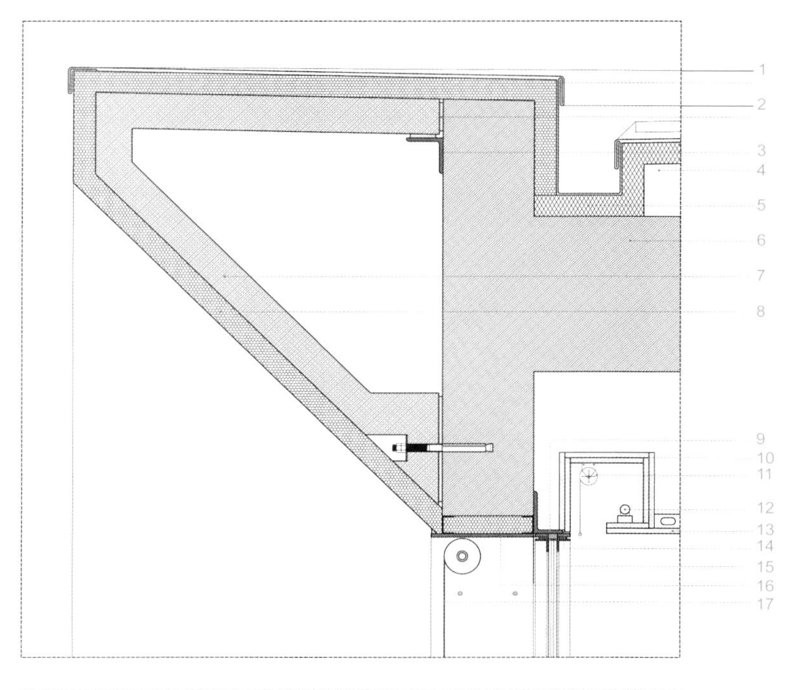

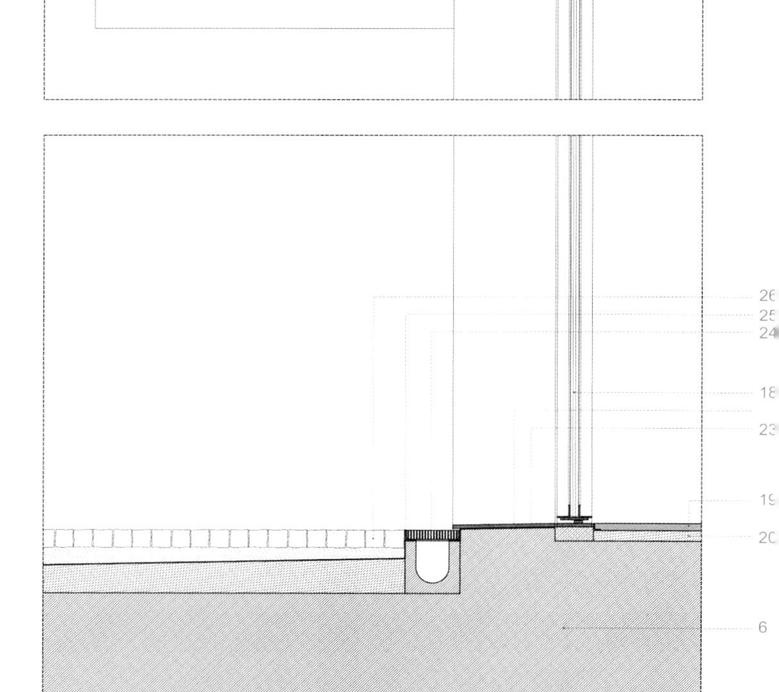

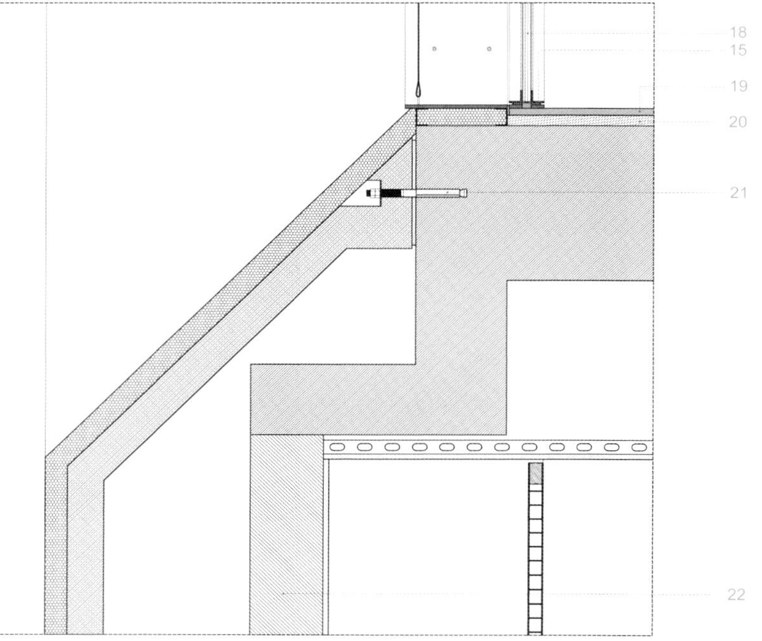

1. Sheet metal roof covering
2. Neopren strip
3. Steel equal angle section, 100x100x35mm
4. Fill layer with lightweight concrete
5. Insulation XPS, 60mm
6. Reinforced concrete
7. Prefabricated concrete panel
8. Thermal insulation composite system, 60mm
9. Steel unequal angle section, 120x80x10mm
10. Plasterboard, 15mm+15mm
11. Interior blind
12. Flurescent lamp
13. Plasterboard, 15mm
14. Steel flat section, 25x6mm (2x)
15. Steel equal angle section, 35x35x4mm
16. Steel plate, 270x10mm
17. Exterior awning
18. Double glazing, 8+6mm
19. Self-leveling mortar
20. Fill layer
21. Mechanical fixation
22. Thermal concrete block
23. Water run-off layer
24. Galvanised grating
25. Steel equal angle section, 30x30x6mm
26. Granite 50x50x50mm

1. 金属屋顶盖板
2. 氯丁橡胶条
3. 等边角钢型材100x100x35mm
4. 轻质混凝土填充层
5. XPS隔热层60mm
6. 钢筋混凝土
7. 预制混凝土板
8. 复合隔热系统60mm
9. 不等边角钢型材120x80x10mm
10. 石膏板15mm+15mm
11. 内置百叶窗
12. 荧光灯
13. 石膏板15mm
14. 扁钢型材25x6mm（2x）
15. 等边角钢型材35x35x4mm
16. 钢板270x10mm
17. 外遮阳
18. 双层玻璃8+6mm
19. 自流平砂浆
20. 填充层
21. 机械固定
22. 隔热混凝土砖
23. 排水层
24. 镀锌钢栅
25. 等边角钢型材30x30x6mm
26. 花岗岩50x50x50mm

项目概况

项目占地面积1,364平方米，前身是波尔图的一个老工业区，目前正通过改造逐渐利用商业活动取代工业活动。作为复兴计划的一部分，建筑垂直于MPA街。

项目南侧建造了一个面向MPA街的新广场，力求与公共空间建立起对话，使大厦入口和店铺与新广场之间产生密切的联系。

这座多功能建筑内同时设有学校和写字间。为了实现功能的独立，两个功能区分别使用不同的垂直交通通道。一楼的商业“前脸”与宽敞的广场相对。从公共入口可以进入两家店铺——餐厅和文具销售中心，二者都与学校有所关联。

学校在东南面设有入口，靠近MPA街。学校占据了5个楼层：一楼、二楼、三楼、四楼和五楼的一半。一楼是前台、门厅和可容纳200人的礼堂。学校的入口大厅有三层高，拥有一个极富雕塑感的大楼梯。二楼是学校办公室、图书室和6间教室。上面两层各有7间教室。另一间图书室占据了三楼的一部分空间，而教师办公室则位于四楼。五楼是两间教室和校长办公室。室内走廊采用自然采光，足够使用者双向通行。建筑师在设计过程中充分考虑了教室的灵活应用。因此，在二楼、三楼和四楼，4至5间教室被共同安排在建筑的东南侧。写字间占据了五楼的一半和整个六七楼层，可以通过西北入口进入。

细部与材料

建筑外观呈现出两个方面的诉求：一是通过玻璃和钢板实现了与地面的透明联系；二是通过各个楼层之间交替的开窗设置实现了与远处更为封闭的地平线之间的联系。上层楼梯由预制混凝土板覆盖，上方覆有薄石膏板保温层——复合隔热系统。窗口外有同色石膏遮阳板提供遮阳保护，从而实现了建筑外墙的多种组合配置。深色钢材和石膏薄板等材料令人回忆起过去的工业区。这种暗喻在项目的设计和材料选择中比比皆是。

Office Building and Logistic Centre

办公楼与物流中心

Location/地点: Nola, Italy/意大利，诺拉
Architect/建筑师: modostudio
Photos/摄影: Julien Lanoo
Site area/占地面积: 20,235 m^2
Built area/建筑面积: 13,760m^2
Completion date/竣工时间: 2011
Key materials: Façade – precast concrete panels
主要材料： 立面—— 预制混凝土板

Overview
The project is located in a strategic industrial area, well connected with the main highway which brings traffic from the north to the south part of Italy. The site area is highly visible from the highway, and the client requests were to create a very strong and recognisable façade.

The project, even in its extreme simplicity composition, aims to transmit the values of innovation, comfort, technology, relax and brand representation. Due to that, the façade represents the image of the building. It covers over 2,000 m2 of company offices on two storeys. The offices are located in the south part of the building along the short side of the building. All the offices are faced towards the Vesuvio Vulcan. The roof hosts a photovoltaic plant of 550KW powered.

The ground floor hosts the main entrance hall which is located in a bari-centric area. The main hall brings employees and clients to the other office departments: marketing, administration, design, product, retail, direction and the showroom.

Detail and Materials
Due to the fact that the structure and the main envelope were made of precast concrete, the architects designed the main façade with the idea of pushing at the extreme value the use of the concrete. Reinforced concrete panels with a rhomboidal pattern in different sizes characterise the main façade. These concrete panels are fixed to the main structural façade through a steel frame system. The panels are of 4 different sizes and the position and the rotation of them give to the façade a various image. The windows are realised opening a side of the panels which are connected with the main structural envelope through metal sheet plates.

The main façade is realised with these particular panels on the first floor, while on the ground floor the façade is a continuous glass curtain wall with strong low emission value able to give the best comfort inside the office.

All the other sides of the building are made by precast concrete panels with a vertical texture. The loading and unloading gates will be covered with a metal roof totally integrated in the precast concrete structure.

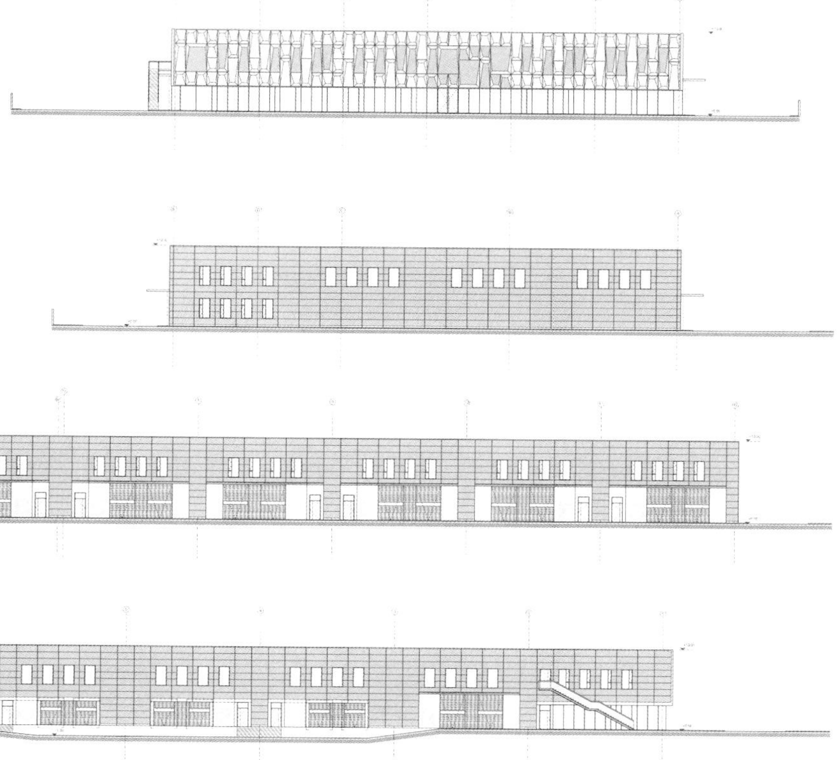

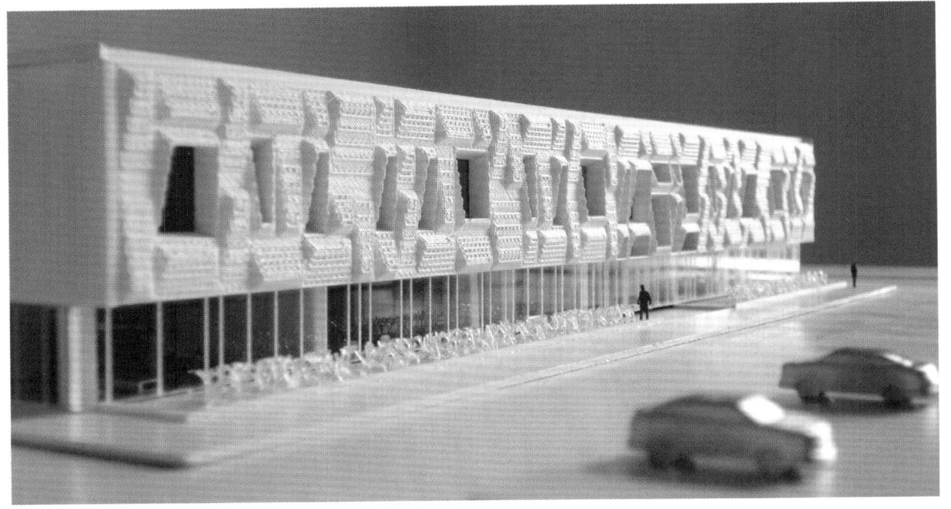

项目概况

项目位于一块战略工业区内，与意大利南北两向的主干线公路有着良好的连接，从高速公路上可清晰地看到项目场地，因此客户要求建筑师打造一个具有强烈视觉冲击和高辨识度的建筑立面。

项目采用了极简的结构，旨在传递创新、舒适、技术、轻松的价值理念和品牌形象。因此，建筑立面体现了建筑的形象。2,000多平方米的办公空间分布在两个楼层内。办公室设在建筑的南侧，沿着建筑的短边排列。所有办公室都朝向维苏威火山。建筑屋顶的光电伏板的功率可达550kW。

一楼正中是主入口大厅。大厅引领着员工和客户前往不同的办公部门：市场部、行政部、设计部、产品部、零售部、指导部和产品展示厅。

细部与材料

由于建筑结构和主外壳都由预制混凝土构成，建筑师在主立面的设计中充分发挥了混凝土材料的价值。主立面以不同规格的带有长菱形图案的钢筋混凝土板为特色。这些混凝土板通过钢架系统固定在主结构立面上。板材分为四种不同的尺寸，它们的错落排列和旋转角度赋予了立面多变的形象。混凝土板的一面被打开形成了窗户，它们通过金属薄板与主结构外壳相连接。

主立面的二楼是这些独特的混凝土板材，而一楼则是连续的玻璃幕墙。幕墙所装配的低辐射玻璃充分保证了内部办公区的舒适度。

建筑的其他几个侧面均由竖条纹预制混凝土板构成。承重和非承重门上方的金属屋顶完全与预制混凝土结构整合起来。

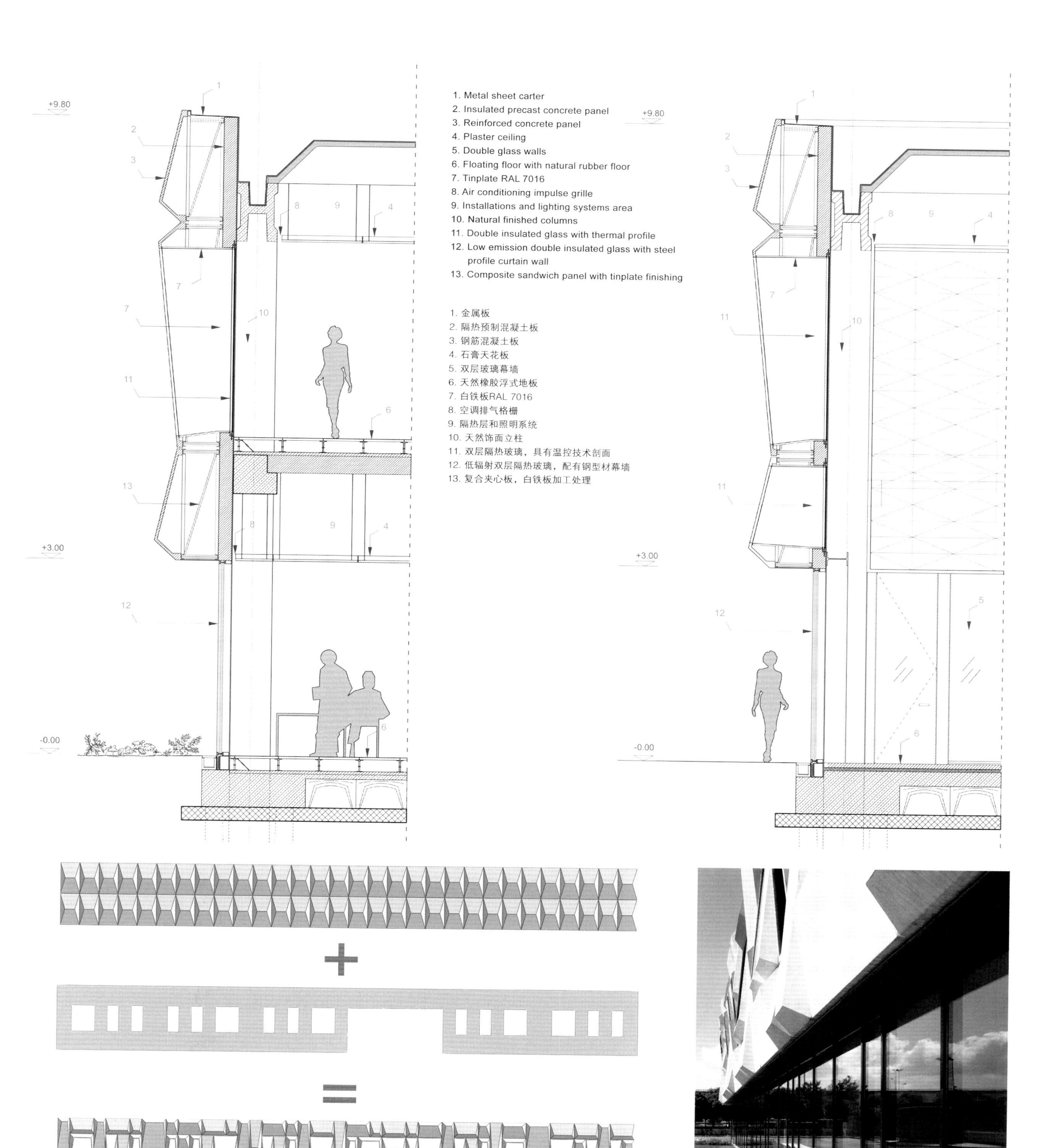
1. Metal sheet carter
2. Insulated precast concrete panel
3. Reinforced concrete panel
4. Plaster ceiling
5. Double glass walls
6. Floating floor with natural rubber floor
7. Tinplate RAL 7016
8. Air conditioning impulse grille
9. Installations and lighting systems area
10. Natural finished columns
11. Double insulated glass with thermal profile
12. Low emission double insulated glass with steel profile curtain wall
13. Composite sandwich panel with tinplate finishing
1. 金属板
2. 隔热预制混凝土板
3. 钢筋混凝土板
4. 石膏天花板
5. 双层玻璃幕墙
6. 天然橡胶浮式地板
7. 白铁板RAL 7016
8. 空调排气格栅
9. 隔热层和照明系统
10. 天然饰面立柱
11. 双层隔热玻璃，具有温控技术剖面
12. 低辐射双层隔热玻璃，配有钢型材幕墙
13. 复合夹心板，白铁板加工处理
+9.80
+3.00
-0.00
+9.80
+3.00
-0.00

Atrium Amras

安布拉斯中庭办公楼

Location/地点: Innsbruck, Austria/奥地利，因斯布鲁克
Architect/建筑师: Zechner & Zechner
Photos/摄影: Thilo Härdtlein, München
Built area/建筑面积: 9,450m^2
Completion date/竣工时间: 2013
Key materials: Façade – precast concrete components
主要材料: 立面——预制混凝土构件

Overview
The brief was to plan an office building with low energy consumption on the periphery of Innsbruck. The site operates as a gate to the city, due to its proximity to the Innsbruck Ost transport hub.

The building has an almost square ground plan, resulting from the layout of the development, and was designed as a type of atrium. The ground floor is raised by approximately 1m from the surrounding street level to avoid the underground parking coming into contact with groundwater. The entrance, which is connected to the neighbouring building that contains a furniture store, is also raised in this way to create a shared forecourt. The main entrance of the office building is therefore oriented towards this forecourt, and in addition the façade is rotated over two floors to mark the entrance and create a roof. Access to the pedestrian forecourt is via inclines, a wide staircase and a wheelchair ramp.

The word "atrium" inspires the building's name, and the atrium also provides important spaces within the structure to facilitate communication and interaction.

On the ground floor of the five-storey building there is commercial floor space, including a pharmacy. Office space with flexible configurations is located above, as well as studios and doctors' practices. Access to the units is via arcade-like corridors. The office spaces located around the atrium are organised to provide variable layouts of differently sized rental units. Alongside the interaction space

ATRIUM
amras
390296
atrium-amras.at
Gewerbe
fläche
212

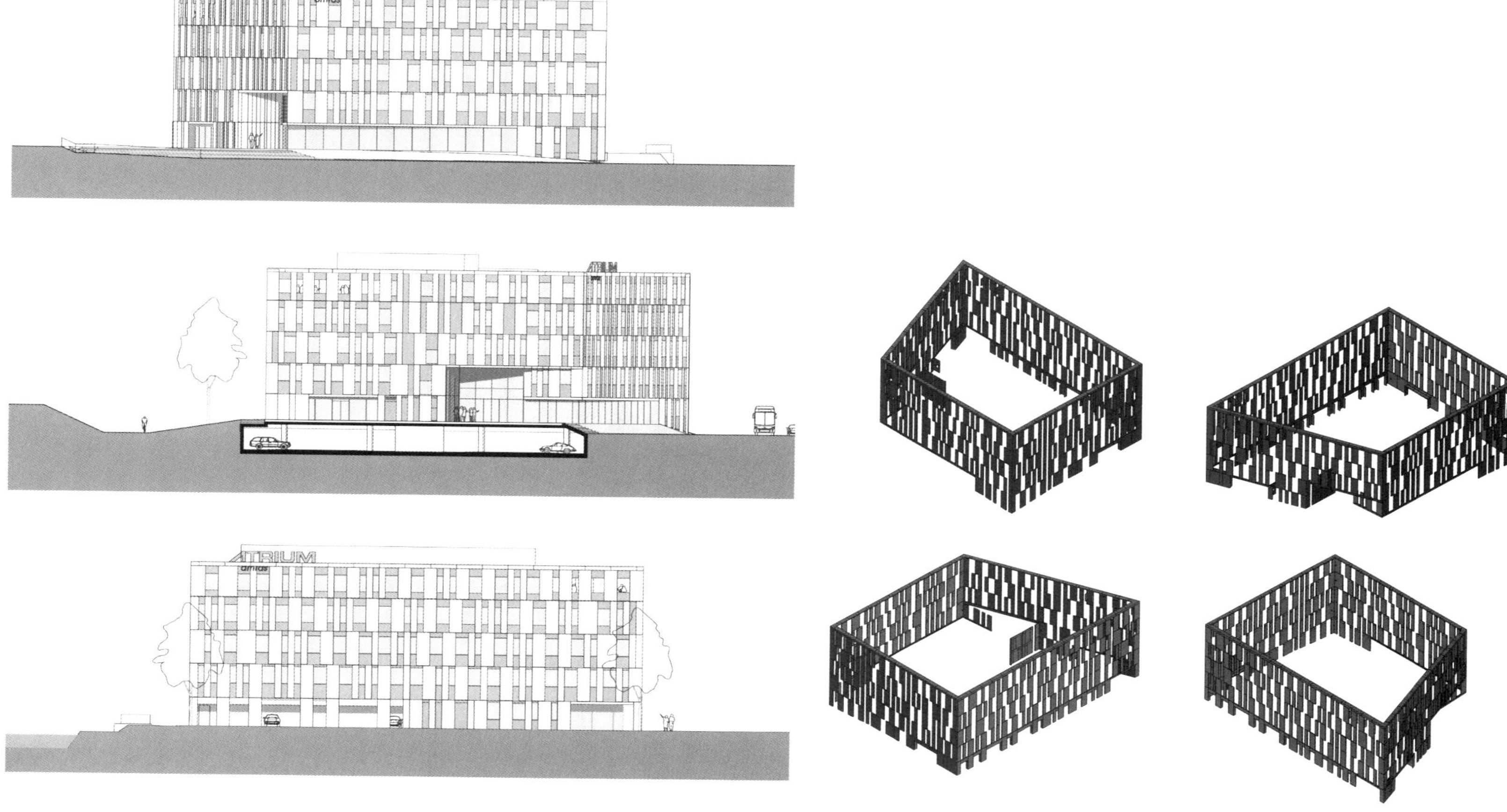
ATRIUM
amras

of the atrium, a roof terrace provides another semiprivate meeting point for people employed within the building. The large terrace on the top floor also emphasises the views that the site provides of the surrounding mountains and Ambras Castle to the south.

The optimum mix of transparent and closed exterior wall surfaces, high-quality exterior insulation and glazing that provide waste heat recovery give the building an energy efficiency class of A+, with HWBref of 13.6 kWh/m^{2}a. The office windows are a composite design that has optimum building physics performance. The structural properties allow a lean HVACR design.

With the neighbouring project, "Leiner/Mpreis", synergies were created, both in relation to the surrounding buildings and in shared transport systems. Alongside the jointly designed and used forecourt, shared entrances and exits to the underground parking reduce the space required for private vehicle access.

Detail and Materials

The façade is constructed from precast concrete components, each the height of an individual storey. These create a non-uniform grid composed of two differently sized window openings that alternate to form a varied pattern. The windows are a composite design with slim profiles and sun protection systems fitted in the middle for protection from the wind.

The slight offset of the façade elements at each storey creates an interplay of light and dark, marks the edges of each level and makes it possible to read the floors. The contrast of light and dark is intensified by differences in the roughness of the surfaces. The rough and matt sandstone-colour concrete surfaces are contrasted with smooth reflective glass elements. The exterior shell of the building is such a distinctive part of the structure that it becomes a corporate identity and an icon for the project.

The exterior walls are load bearing and are constructed from site-mixed concrete, to avoid the use of supports near the façade. The large parapets prevent fire spreading from floor to floor. Two cores reinforce the building frame, and house the staircases and shafts. The ceilings are unstressed reinforced concrete without joists. The upper ceiling is constructed using glued-laminated boards that function as box girders.

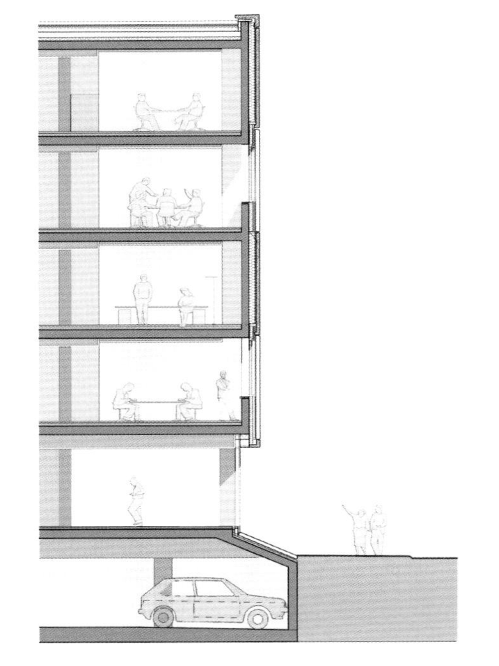

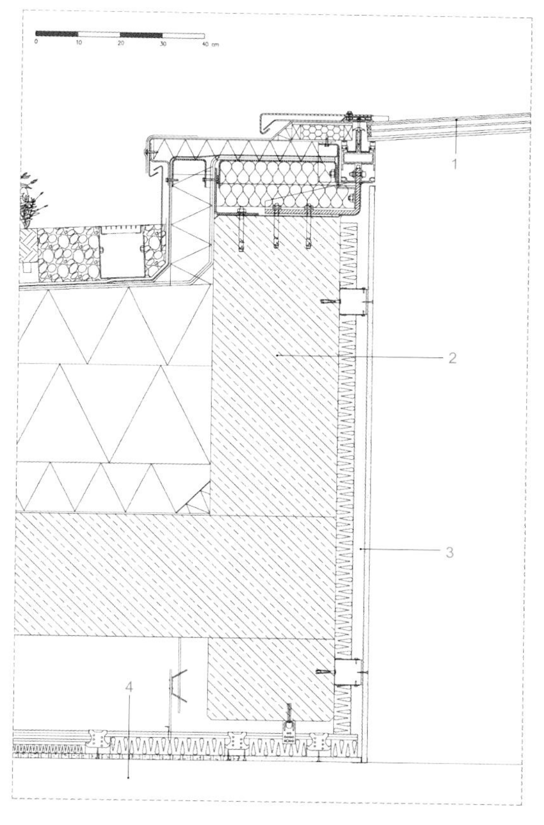

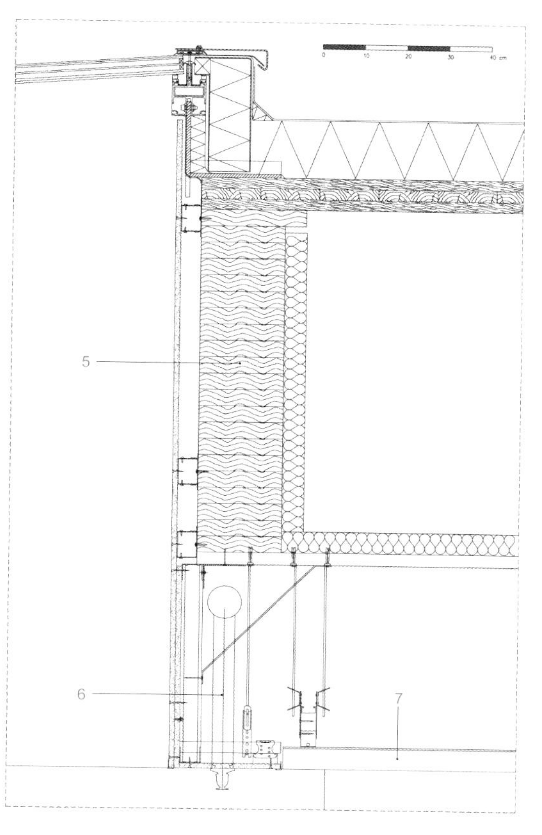

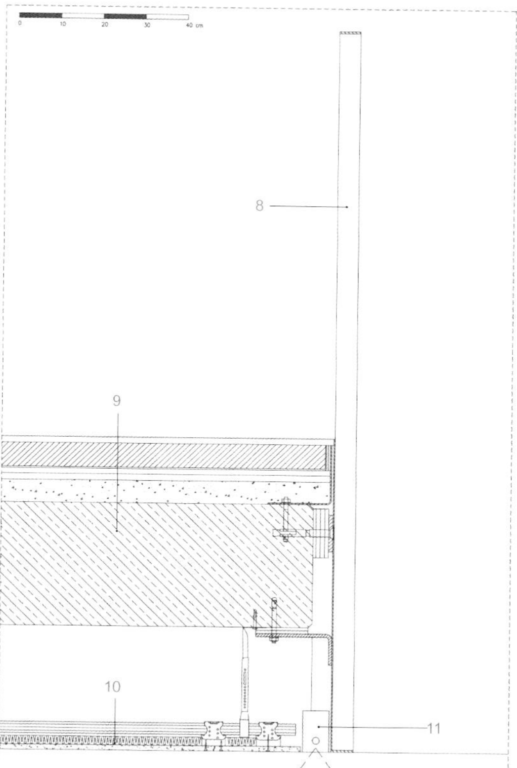

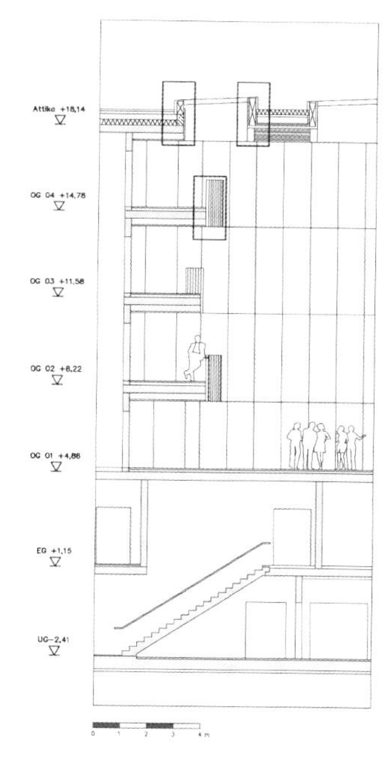

1. Glass roof: Post and rail façade; triple glazing
2. Reinforced concrete attic
3. Gypsum cardboard cladding
4. Suspended ceiling gypsum cardboard pedorated
5. Laminated wood beam
6. Sprinkler pipe
7. Lower ceiling aluminium lamellae
8. Railing flat steel 50/5
9. Reinforced concrete slab, composite stone floor
10. Suspended ceiling gypsum cardboard pertorated
11. Light band

1. 玻璃屋顶；立柱栏杆；三层玻璃
2. 钢筋混凝土阁楼
3. 石膏板包层
4. 石膏板吊顶
5. 胶合板梁
6. 洒水管
7. 下层天花板铝层
8. 扁钢栏杆50/5
9. 钢筋混凝土板，复合式地面
10. 石膏板吊顶
11. 灯带

项目概况

项目要求在因斯布鲁克的外围地区设计一座低能耗办公楼。由于靠近因斯布鲁克交通枢纽，项目场地可以看作是城市的门户。

建筑的平面规划基本呈正方形，以中庭为中心展开设计。建筑低层比周边街道高出近1米，避免了地下停车场能接触到地下水。建筑入口与旁边的家具商场相连，通过这种抬高的方式形成了一个共享前庭。办公楼的正门正对前庭，而建筑墙面向内旋转，形成了入口屋檐。人们可以通过宽阔的楼梯和无障碍专用坡道进入前庭。

“中庭”一词是建筑名称的灵感来源，同时，中庭也为建筑结构提供了重要的交流互动空间。

一楼是商业空间，内设一个药房。办公空间、工作室以及一些医生诊所被灵活地配置在上面的楼层。人们可通过拱廊似的楼梯进入以上单位。办公空间环绕着中庭展开，拥有多样化的尺寸和布局。除了中庭互动空间之外，屋顶平台也为人们提供了一个半私密会面场所。顶楼的大型平台还享有周边山脉和南侧安布拉斯城堡的美景。

外墙立面透明区与封闭区的最佳配置比、高品质外墙保温系统和玻璃装配提供了余热回收功能，使建筑的节能等级达到了A+，平均能耗为13.6 kWh/m^{2}a。办公楼的窗口采用复合设计，优化了建筑的物理性能。结构特性为高效的供暖系统设计提供了基础。

建筑与旁边的商场项目实现了协同效应，并且与周边的其他建筑联系起来，共享交通系统。处理连接设计和共享前庭之外，它们还共享地下停车场的出入口，从而减少了私人车辆通道的空间需求。

细部与材料

建筑立面由预制混凝土构件建造而成，混凝土构件的高度与单独楼层的高度相同。这种设计形成了不统一的网格图案，两种不同尺寸的窗口相互交替，形成了多变的图案。细长的窗口采用复合设计；为了防风，遮阳系统被安装在中间。

每层楼外墙构件的轻微位移形成了独特的光影交错效果，标志出各个楼层的边界。墙面粗糙度的差异进一步凸显了光影的对比：粗糙的亚光面沙色混凝土墙面与光滑的高反光玻璃形成了强烈对比。独特的建筑外壳已经成为了企业的象征，也成为了项目的标志。

作为承重墙，建筑外墙由现场混合的混凝土构造而成，避免了靠近立面使用支柱。宽大的护墙能防止火势在楼层间蔓延。两个内核结构加固了建筑框架，内设楼梯和电梯。天花板为无应力钢筋混凝土结构，无托梁。上层天花板采用胶合板构造，起到了箱型梁的作用。

Extension of Martin Luther School

马丁路德学院扩建

Location/地点: Marburg, Germany/德国，马尔堡
Architect/建筑师: Hess / Talhof / Kusmierz Architekten Und Stadtplaner Bda
Photos/摄影: Florian Holzherr
Built area/建筑面积: 2,070m^2
Completion date/竣工时间: 2010
Key materials: Façade – prefabricated concrete panel
主要材料: 立面——预制混凝土板

Overview

In 2005 the university city of Marburg held a competition for both the expansion of the Martin Luther School and the redesign of the neighbouring city hall completed in 1969. This combination seems unusual at first glance, yet seems reasonable in terms of urban space and, at a closer look, offers potential for synergy. For instance, when a congress takes place in the city hall, utilisation of the school's classrooms or cafeteria is possible. On the other hand, the school can use spaces of the city hall for its own events.

Due to the scope of building measures, the project is developed in phases. The first construction phase includes the school addition. The main use areas of the compact passive house are oriented towards the south, while auxiliary rooms are situated in the north. In between a central circulation hall receives daylight from above and connects all floors across the entire length of the building. The recess hall serves as cafeteria, meeting point, or event space depending on the time of the day and, via an open staircase, leads to the classrooms. Galleries and bridges permit a multitude of views.

Detail and Materials

A limited number of durable surface materials suitable for use in a school characterise the interiors, such as exposed concrete with a fine wood surface pattern, slatted acoustic panelling with paint finish, as well as natural rubber flooring. Colours emphasise the functional

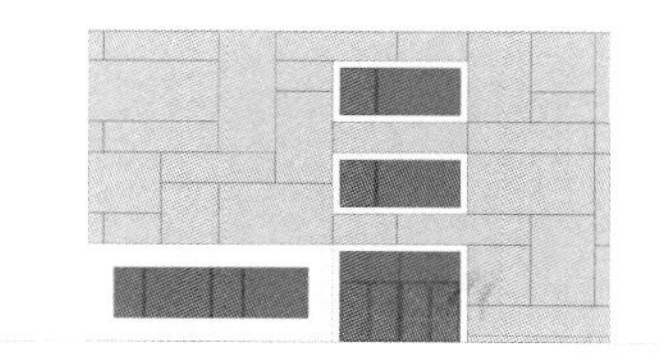

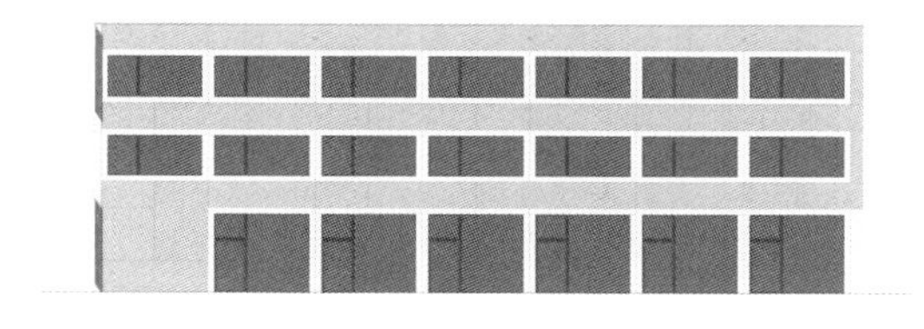

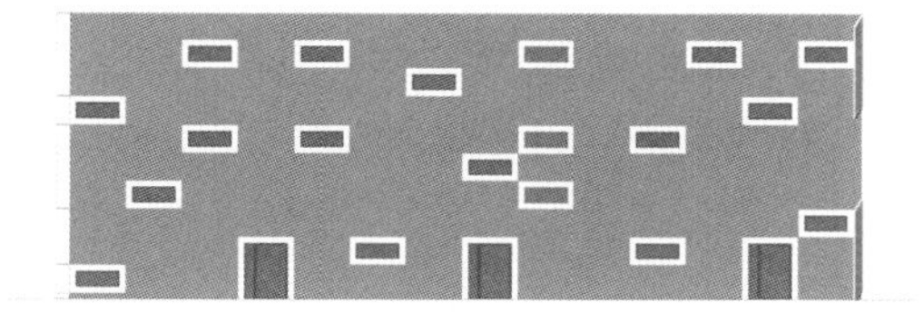

areas of the building complex. A powerful green colour denotes the school's circulation areas and recess hall. MDF panelling with natural finish provides the classrooms with a rather calm atmosphere.

The southern and eastern façades comprise the first part of the new city hall's building envelope. A prefabricated concrete panel curtain wall recalls the existing construction: its embossed surface displays the negative of the existing exposed aggregate concrete façade's positive. Plastic moulds served to transfer its structure to the new façade elements. They are based on an idealised exposed aggregate concrete element scaled to 120% of its original size. Variations were created by using a master element that is larger than the actual individual prefabricated elements of the façade. This enables creating castings of different parts of the master element.

The building envelope receives depth through prefabricated elements with greater thickness and even surfaces that serve as frames for window and door openings. Extendable fabric blinds also permit views towards the exterior when drawn guarantee sufficient sun protection along the graciously glazed southern façade. The temporary façades between existing and new feature an economic thermal insulation composite system with playfully applied small openings along the northern façade. In order to meet passive house standards, the entire building envelope is built as airtight as possible and is highly thermally insulated. Closed exterior surfaces don't exceed U-values of 0.15 W/m^2K.

项目概况

2005年，马尔堡大学城为马丁•路德学院的扩建和市政厅（建于1969年）的重建举办了一次设计竞赛。这二者的组合看起来不合常规，但是却能起到相互协助的作用。例如，当市政厅举办会议时，可以利用学校的教室或餐厅。另一方面，学校也可以利用市政厅举办自己的活动。

由于建筑规模较大，项目决定分阶段进行开发。第一阶段为学校的扩建。新建筑的主要功能空间朝向南面，辅助空间则设在北面。中央流通大厅能从上方接收日光，将所有楼层连接起来。根据时间的不同，休息大厅可用作餐厅、集会地点或活动空间，通过宽敞的走廊与教室相连。建筑内部穿插的走廊和廊桥保证了视野的多样性。

细部与材料

建筑内部采用了一系列适合学校使用的表面材料，例如带有精致木纹图案的素面混凝土、涂漆隔音板条、天然橡胶地面等。建筑内的功能区通过不同的色彩突出表现。通道空间和休息大厅以强烈的绿色色调为主。带有天然饰面的中密度纤维板为教室营造了平和的气氛。

建筑的东、南两个立面组成了新市政厅建筑外壳的第一部分。穿孔混凝土板幕墙与原有建筑结构相互呼应；浮雕表面的凹孔与原有的露石混凝土的凸出正好对应。塑料模具将自身的结构转换到新的立面构件上，以理想化的120%尺寸的露石混凝土构件为基础。建筑师通过使用一块尺寸大于实际预制构件的主构件实现了表面元素的多样化。各个小型构件都是主构件的不同部分。

建筑外壳通过加厚的预制构件以及窗框、门框等表面实现了厚重感。可伸展式布料百叶窗既能保证充足的遮阳保护，又能为室内带来良好的视野。新旧建筑之间的临时立面以经济的复合保温系统为特色，在北侧立面设有小型窗口。为了实现被动建筑标准，整个建筑外壳尽量保证密封性和高度隔热性。封闭的外墙表面的传热系数小于0.15 W/m^2K。

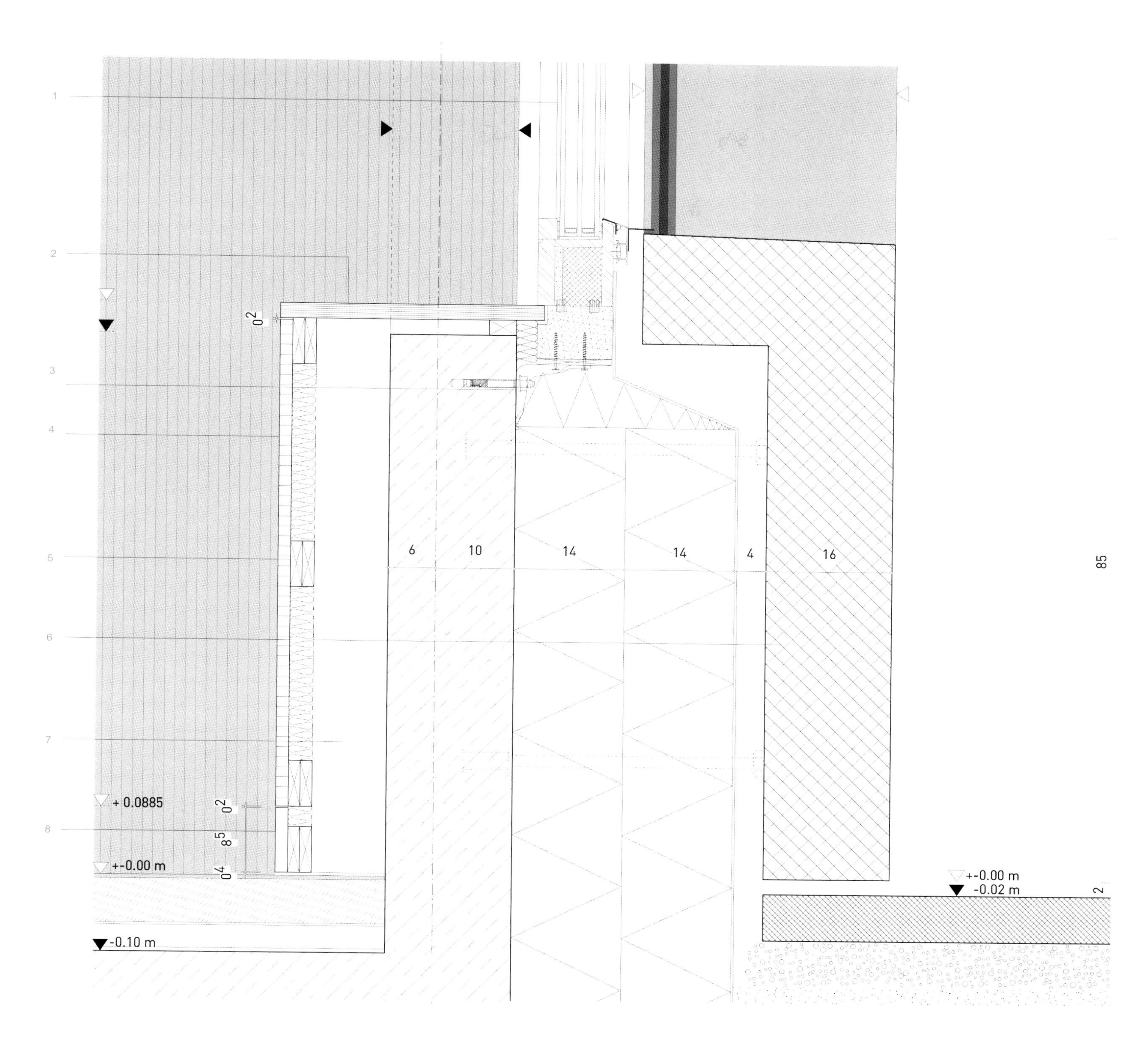

1. Post and beam façade wood-aluminum
2. Windowsill
3. Steamtight connection
4. Acoustic panel
5. Substructure
6. Precast reinforced concrete
7. Mounting bar
8. Skirting 16mm

1. 梁柱式立面，木材+铝
2. 窗台
3. 气密连接
4. 隔音板
5. 下层结构
6. 预制钢筋混凝土
7. 安装条
8. 壁脚板16mm

1. Precast reinforced concrete
2. Steamtight connection
3. Post and beam facade wood-aluminum

1. 预制钢筋混凝土
2. 气密连接
3. 梁柱式立面，木材+铝

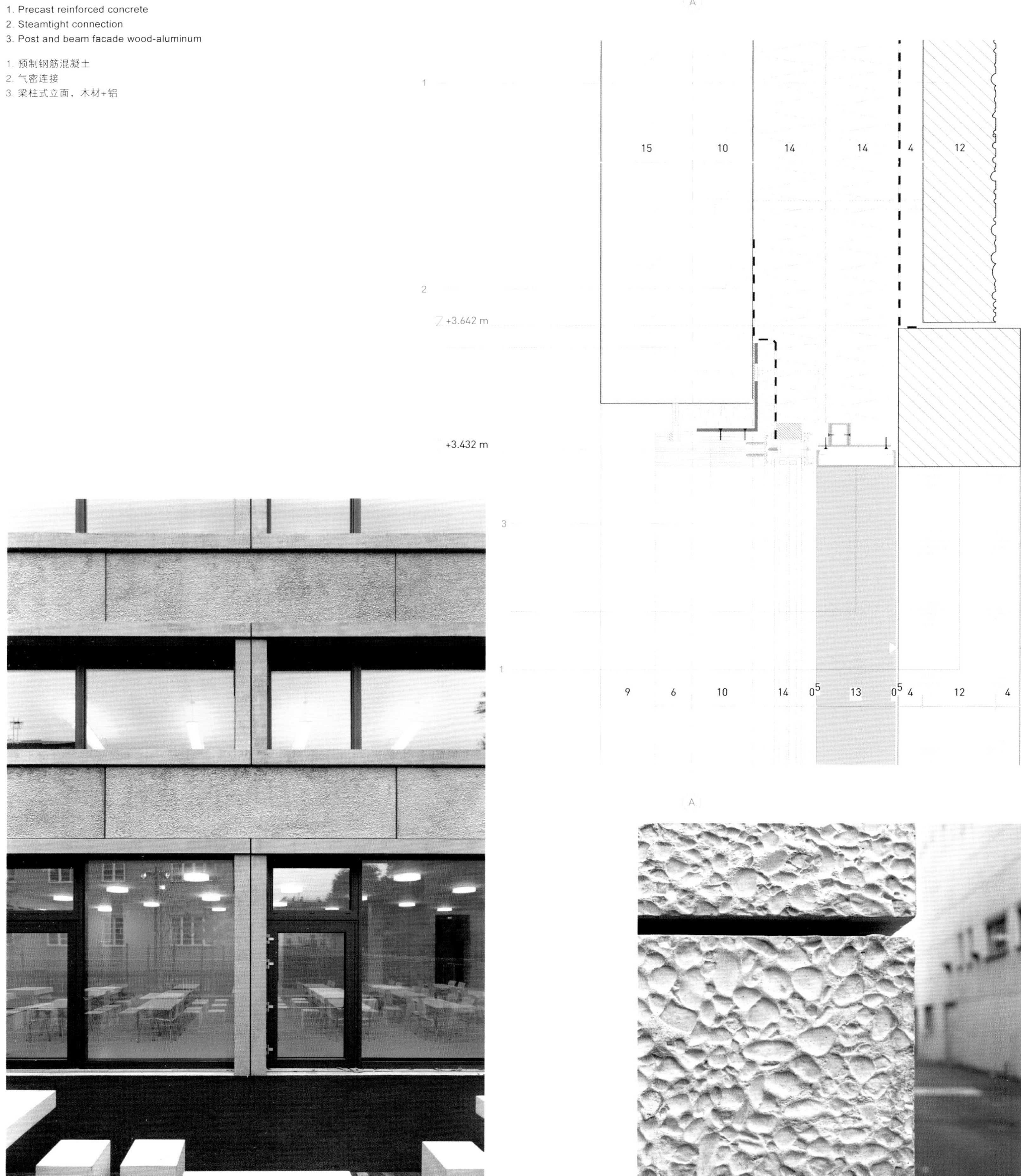

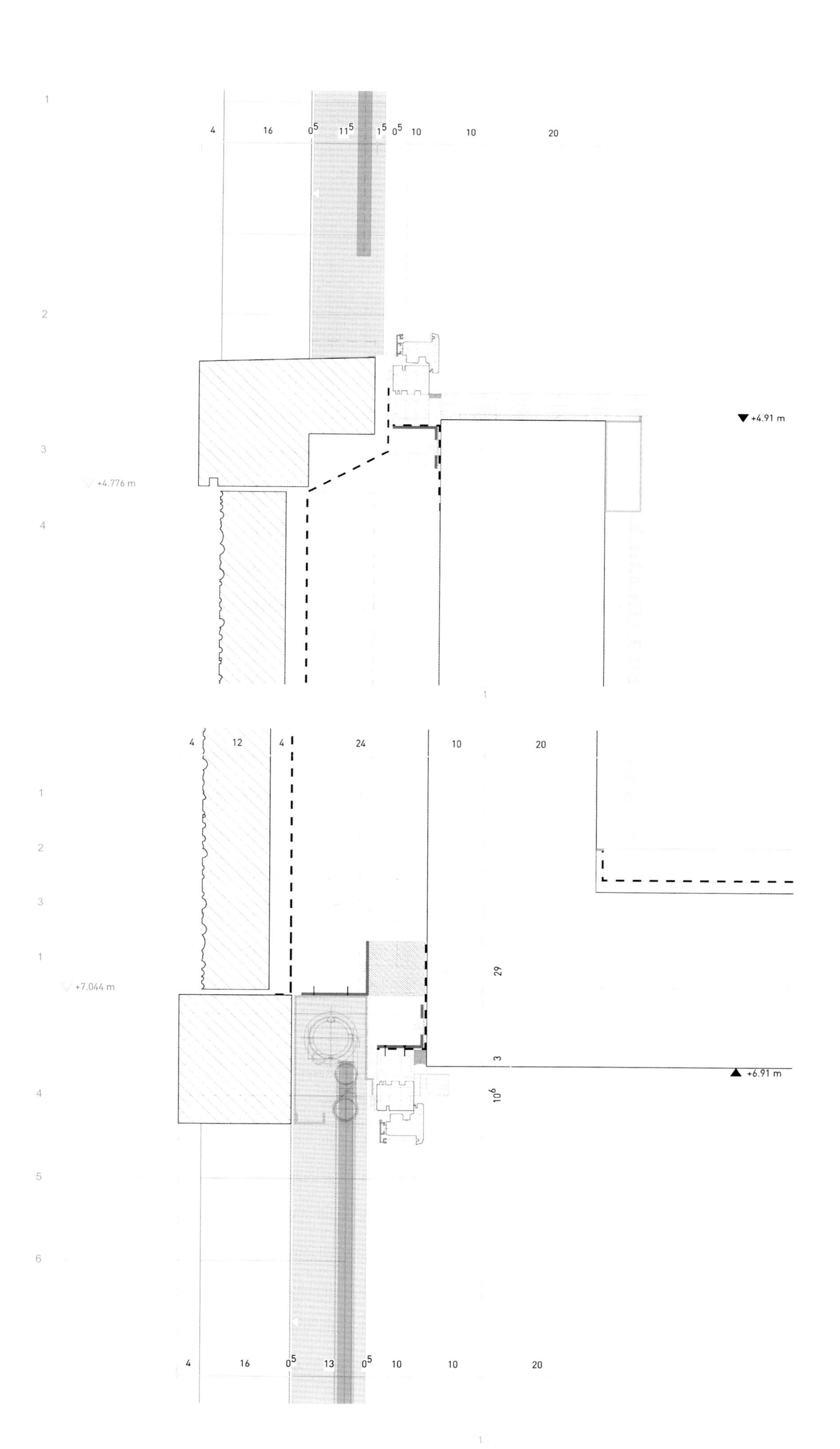

1. Passive house windows
2. Joint
3. Substructure
4. Precast reinforced concrete

1. 被动式住宅窗
2. 接缝
3. 下层结构
4. 预制钢筋混凝土

1. Precast reinforced concrete
2. Substructure
3. Conduit
4. Passive house windows
5. Joint
6. Textile sunscreen

1. 预制钢筋混凝土
2.下层结构
3. 导管
4. 被动式住宅窗
5. 接缝
6. 遮阳帘

Kindergarten Lotte

乐天幼儿园

Location/地点: Tartu, Estonia/爱沙尼亚，塔尔图
Architect/建筑师: Kavakava Architects
Photos/摄影: Kaido Haagen, Aivo Kallas, Lauri Kulpsoo, Kristo Nurmis
Site area/占地面积: 8,041m^2
Built area/建筑面积: 2,292m^2
Key materials: Façade – Precast concrete sandwich panel
主要材料：立面——预制混凝土夹芯板

Overview

The city of Tartu has the goal of implementing high-quality modern architecture in new public buildings. The new kindergarten, which is located in one of the most dilapidated areas of Tartu (the so-called Chinatown is a former Soviet military garrison), is a result of this policy.

The kindergarten's layout – a six-cornered, star-shaped floor plan forced into a square – arose from the desire to avoid long corridors and to create an orderly outer perimeter and street space for the building. Building is situated on one edge of the plot and leaves the southern side free as a play area.

External appearance of the low street-sided building has an inwardly character. Concrete side walls with small coloured glass openings and a bamboo fence of the height of the building itself make for some isolation of the kindergarten from the outside world.

The corridor winding around the central hall doubles as a play area for the children. The central hall is about a metre below ground level, creating the effect of a stage, and together with "play" corridor that surrounds it, it forms a single open space. The six petals contain kindergarten home rooms, creativity classrooms, kitchen, dining room and administrative offices. The star-shape allows lots of light to the interior. Intimate courtyards are formed between wings.

Continuous geometry throughout all design is used. Plan is formed by lines which intersect under 60 degrees with each other. Same way is developed triangle-pattern on façades and ceiling at the core of the building.

Detail and Materials

Cast-in-place concrete is used for the core. It has an impressive amount of thermal mass and need for cooling in spring and early sum-

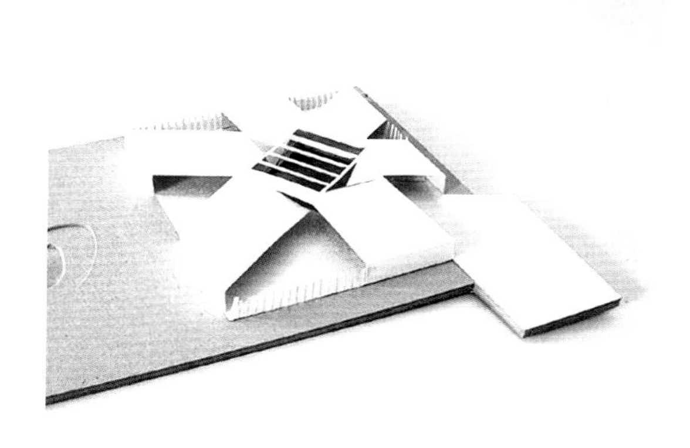

mer is reduced to minimal. Natural ventilation is combined with the mechanical one in this building. All rooms have plenty of natural light. Children day-rooms face south towards an open air play area. Prefabricated concrete is used for a walls with triangular windows – great accuracy was needed here because triangular glass panes were inserted directly (without frames) into the concrete wall.

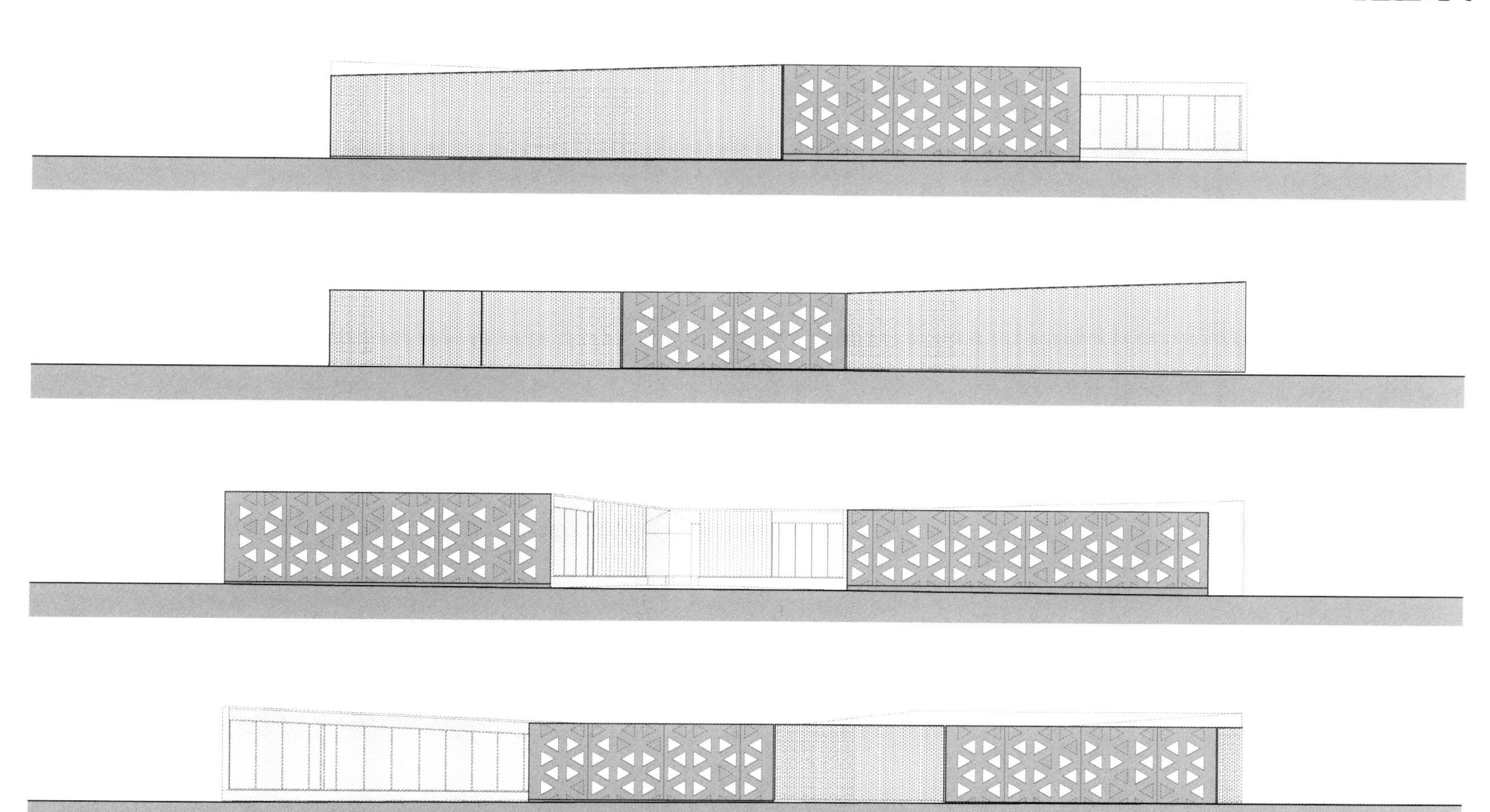

项目概况

爱沙尼亚的塔尔图市希望建造一批高品质现代公共建筑。新建的幼儿园位于塔尔图一处比较荒凉的地区，正是该政策的产物。

幼儿园采用六角星形的平面布局，外围空间呈正方形。这种设计避免了过长的走廊，形成了整齐有序的外围结构和临街空间。建筑位于场地的一边，将南侧留作游乐区。

建筑的外观具有明显的内向特征。混凝土侧墙上配有小型彩色玻璃开窗，而与建筑同高的竹栅栏使幼儿园与外面的世界隔绝开。

走廊环绕中央大厅展开，也可以兼作儿童的游戏区。中央大厅低于地面1米，形成了舞台的效果，它与游戏走廊共同形成了开放空间。六个花瓣结构分别设置着幼儿园的休息室、创意教室、厨房、餐厅和行政办公室。星星造型让室内充满了阳光，两个不同花瓣之间的空间形成了私密的庭院。

连续的几何造型贯穿了整个设计。整个布局都由夹角小于60度的交叉线条构成。外墙和建筑内部天花板上的三角图案同样采用了这种开发方式。

细部与材料

建筑用预制混凝土板构造内核结构。混凝土板具有极好的热质量，能保证春季和初夏的制冷需求。建筑采用了自然通风与机械通风相结合的通风方式。所有房间都有充足的自然采光。儿童看护房朝南，正对露天游乐区。由预制混凝土构成的外墙上设有三角形开窗，窗口构造要求十分精确，因为三角形玻璃板被直接嵌入到混凝土墙壁之中（没有采用任何框架设施）。

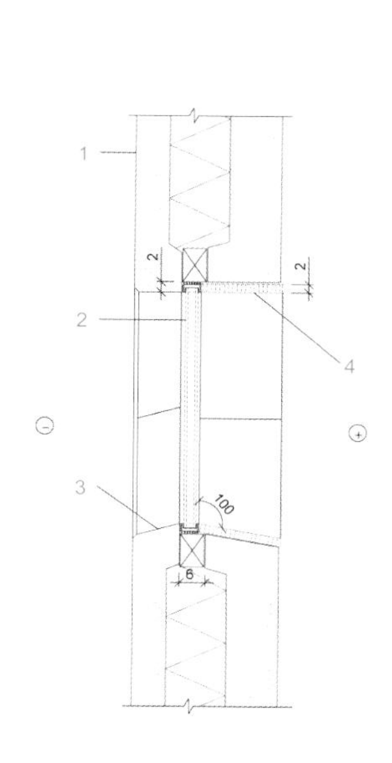

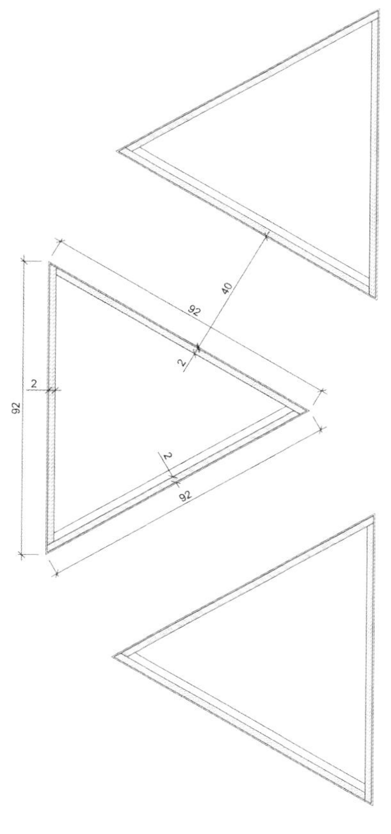

1. Precast concrete sandwich panel
   80mm concrete
   140mm thermal insulation
2. Double glazing w/coloured
3. Exposed concrete
4. 20mm plywood sheet

1. 预制混凝土夹心板
   80mm混凝土
   140mm隔热层
2. 彩色双层玻璃
3. 素面混凝土
4. 20mm胶合板

1. 混凝土板上嵌铸式三角窗
2. 预制混凝土夹心板：
– 混凝土80mm
– 隔热层
– 混凝土120mm
3. 三角窗
   彩色双层玻璃
4. 素面混凝土墙
5. 铺装石
– 石间圬工砂30mm
– 压实粗砂岩50mm
– 压实粗砂岩150mm
– 水平地面
6. 屋顶结构
– 屋顶薄膜，碎石覆盖20mm
– 隔热层30mm
（刚性矿物棉，带有气槽30mm）
– EPS隔热层，最小值200mm
– 隔汽层
– 瓦楞钢板153mm
– 吸音棉板50mm
– 空气腔20mm
– 穿孔吸音石膏天花板
7. 20mm胶合板（薄板）
8. 楼板结构
– 天然油毡
– 现浇筑钢筋混凝土板100mm
– 聚乙烯膜
– 沙子100mm
– 压实粗砂岩100mm
– 压实碎石100~200mm
– 水平地面

1. Cast-in triangular window shape in concrete panel
2. Precast concrete sandwich panel:
   - Concrete 80mm
   - Thermal insulation
   - Concrete 120mm
3. Triangular window double glazing w/ coloured glass
4. Exposed concrete wall
5. Cobblestones
   - Fine masonry sand between stones 30mm
   - Compressed gritstone 50mm
   - Compressed gritstone 150mm
   - Leveled ground
6. Roof construction
   - Roofing membrane covered w/ gravel 20mm
   - Thermal insulation 30mm (Rigid mineral wool w/air channels 30mm)
   - Thermal insulation EPS min 200mm
   - Vapour barrier
   - Corrugated steel sheet 153mm
   - Acoustic wool board 50mm
   - Air cavity 20mm
   - Perforated acoustic gypsum ceiling board
7. 20mm plywood sheet (veneer)
8. Floor construction
   - Natural linoleum
   - In situ reinforced concrete slab 100mm
   - Polyethylene membrane
   - Sand 100mm
   - Compressed gritstone 100mm
   - Compressed gravel 100-200mm
   - Leveled ground

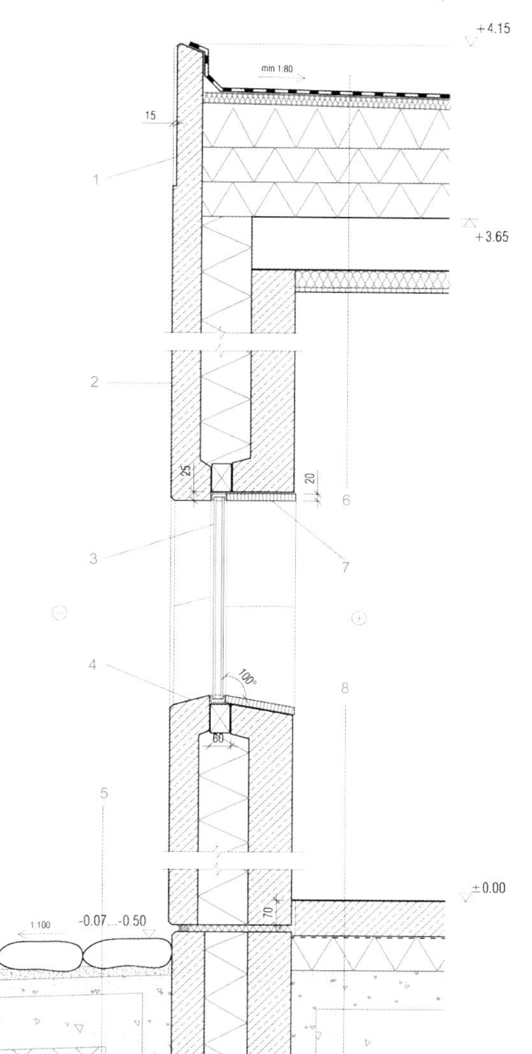

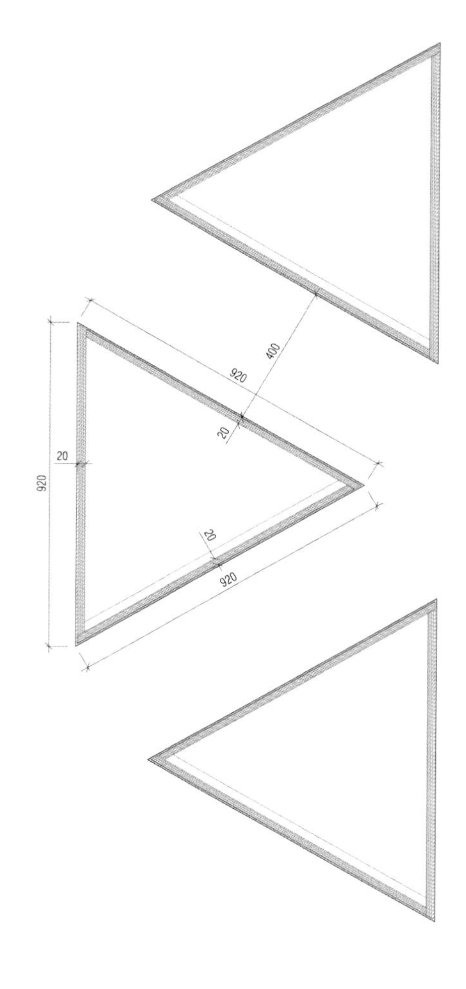

House of Music

音乐屋

Location/地点: Aalborg, Denmark/丹麦，奥尔堡
Architect/建筑师: Coop Himmelb(L)Au
Photos/摄影: Duccio Malagamba
Site area/占地面积: 20,257m²
Completion date/竣工时间: 2014
Key materials: Façade – precast concrete panels, perforated aluminium/aluminium panels, aluminium sandwich panels, PV Cell in glass triangulars, curtain wall and glass triangulars on steel shell structure
主要材料：立面—— 预制混凝土板、穿孔铝板、铝夹心板、PVC三角板、幕墙和三角玻璃

Overview

The "House of Music" in Aalborg, Denmark, was designed by the Viennese architectural studio Coop Himmelb(l)au as a combined school and concert hall: its open structure promotes the exchange between the audience and artists, and the students and teachers. "The idea behind the building can already be read from the outer shape. The school embraces the concert hall," explained Wolf D. Prix, design principal and CEO of Coop Himmelb(l)au.

U-shaped rehearsal and training rooms are arranged around the core of the ensemble, a concert hall for about 1,300 visitors. A generous foyer connects these spaces and opens out with a multi-storey window area onto an adjacent cultural space and a fjord. Under the foyer, three more rooms of various sizes complement the space: the intimate hall, the rhythmic hall, and the classic hall. Through multiple observation windows, students and visitors can look into the concert hall from the foyer and the practice rooms and experience the musical events, including concerts and rehearsals.

The flowing shapes and curves of the auditorium inside stand in contrast to the strict, cubic outer shape. The seats in the orchestra and curved balconies are arranged in such a way that offers the best possible acoustics and views of the stage. The highly complex acoustic concept was developed in collaboration with Tateo Nakajima at Arup. The design of

the amorphous plaster structures on the walls and the height-adjustable ceiling suspensions, based on the exact calculations of the specialist in acoustics, ensures for the optimal listening experience. The concert hall will be one of the quietest spaces for symphonic music in Europe, with a noise-level reduction of NR10 (GK10).

Instead of fans, the foyer uses the natural thermal buoyancy in the large vertical space for ventilation. Water-filled hypocaust pipes in the concrete floor slab are used for cooling in summer and heating in winter. The concrete walls around the concert hall act as an additional storage capacity for thermal energy. The fjord is also used for cost-free cooling.

The piping and air vents are equipped with highly efficient rotating heat exchangers. Very efficient ventilation systems with low air velocities are attached under the seats in the concert hall. Air is extracted through a ceiling grid above the lighting system so that any heat produced does not cause a rise in the temperature in the room.

The building is equipped with a building management programme that controls the equipment in the building and ensures that no system is active when there is no need for it. In this way, energy consumption is minimised.

Detail and Materials

The design of the precast concrete panels is based on an acoustic demand. They build the inner box in box wall and pattern with popped out and recessed amoebas ensures the optimal sound experience.

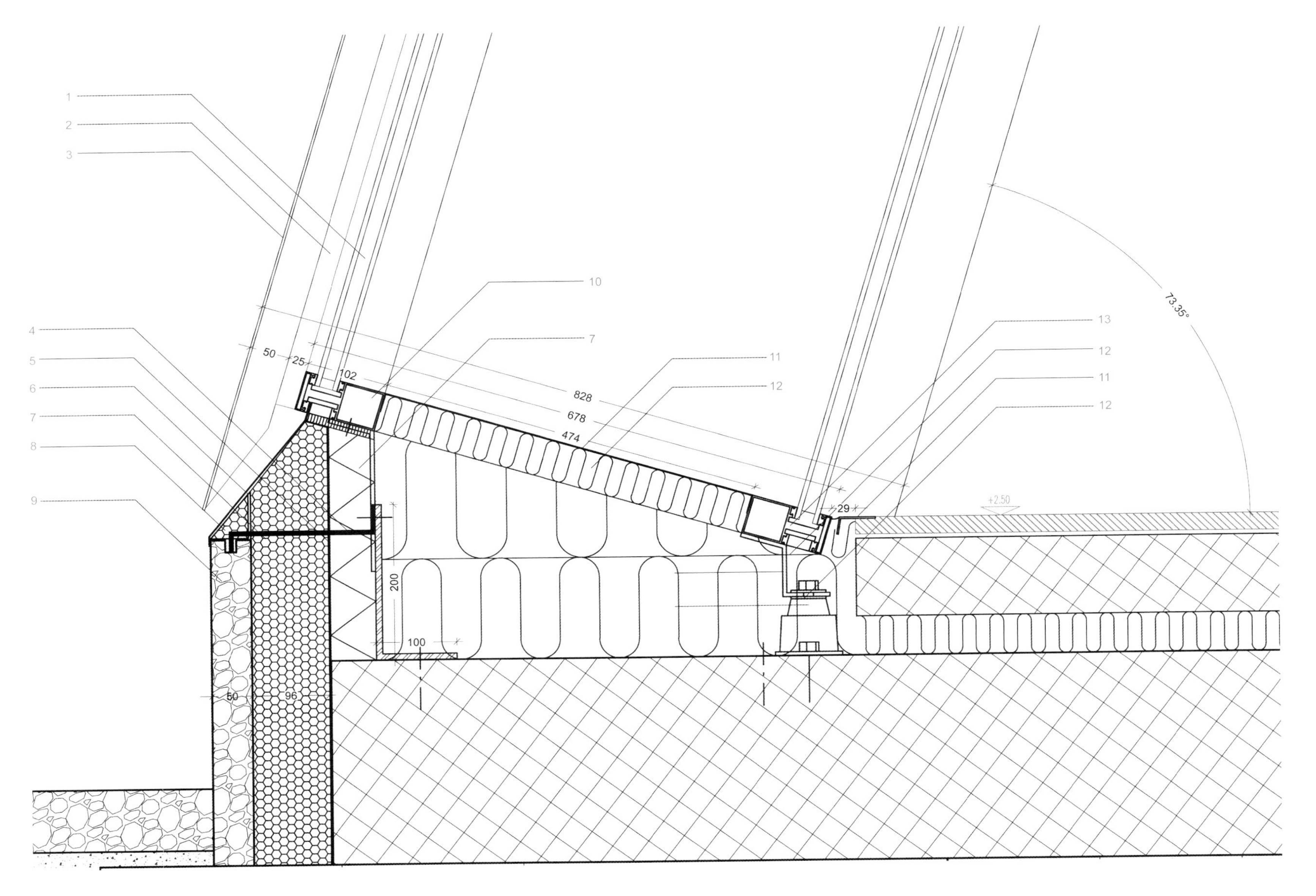

1. Exterior window
2. Hat profile
3. Perforated aluminium panel
4. Mechanical attachment
5. Bituminous felt
6. Flashing
7. Expanded polystyrene
8. Polystyrene triangle
9. Granite vertical
10. Exterior window frame
11. Aluminium profile
12. Mineral wool
13. Interior window frame

1. 外部窗
2. 顶盖
3. 穿孔铝板
4. 机械附件
5. 沥青毡
6. 防水板
7. 发泡聚苯乙烯
8. 聚苯乙烯三角
9. 花岗岩竖挡
10. 外部窗框
11. 铝型材
12. 矿棉
13. 内部窗框

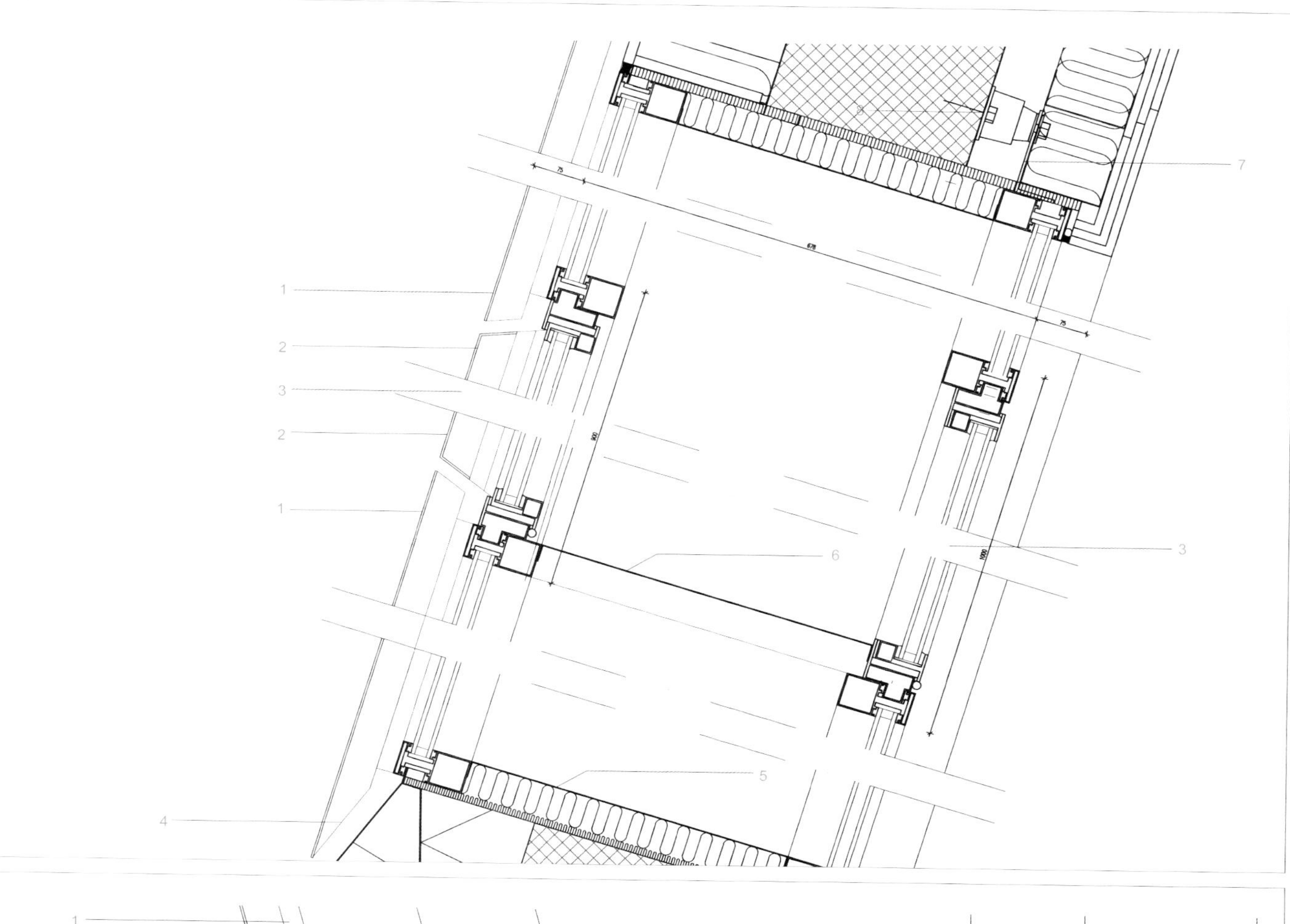

1. Fixed, perforated panel
2. Perforated panel mounted on operable window frame
3. Operable window (inward)
4. Customised profile at louver edge
5. Aluminium panel along window perimeter, surface and colour as the windows
6. "Step-safe" window shelf in rescue openings, surface and colour as the windows
7. Galvanised bracket
8. Mason BRB mount

1. 固定穿孔板
2. 穿孔板，安装在可开窗框上
3. 可开式窗（向内）
4. 百叶窗边缘定制型材
5. 窗围铝板，色彩与窗户相同
6. 安全窗架，色彩与窗户相同
7. 镀锌支架
8. Mason BRB安装架

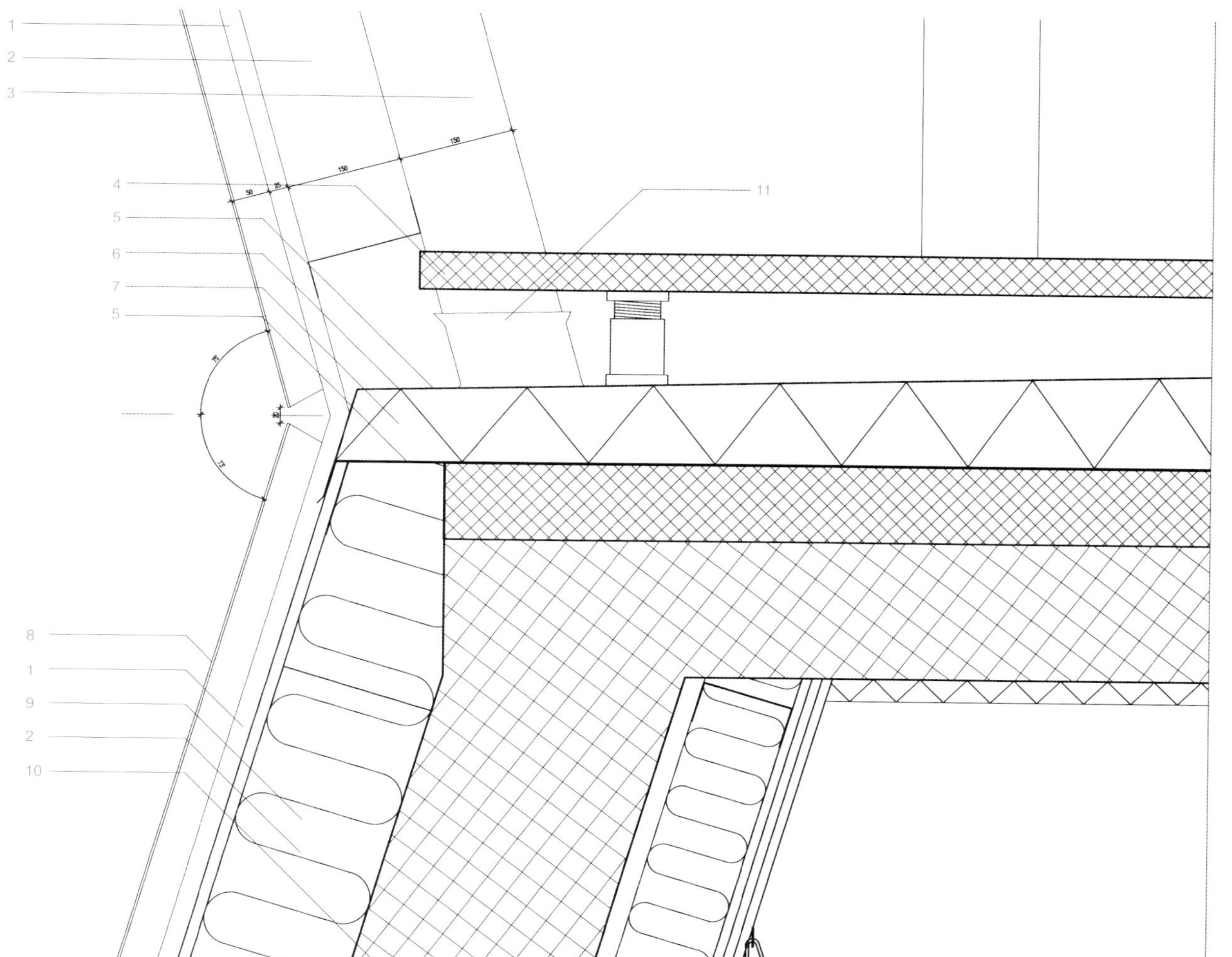

1. Hat profile
2. 150Mm light steel profile
3. 150X150 mm steel profile (lc2)
4. Concrete paving
5. Membrane
6. Aluminium flashing
7. Glass foam
8. Aluminium panel
9. 200Mm mineral wool
10. Concrete wall element
11. Sealed water proofing

1. 顶盖
2. 150mm轻钢型材
3. 150x150mm钢型材
4. 混凝土铺装
5. 膜结构
6. 铝防水板
7. 玻璃泡沫
8. 铝板
9. 200mm矿棉
10. 混凝土墙壁组件
11. 密封防水

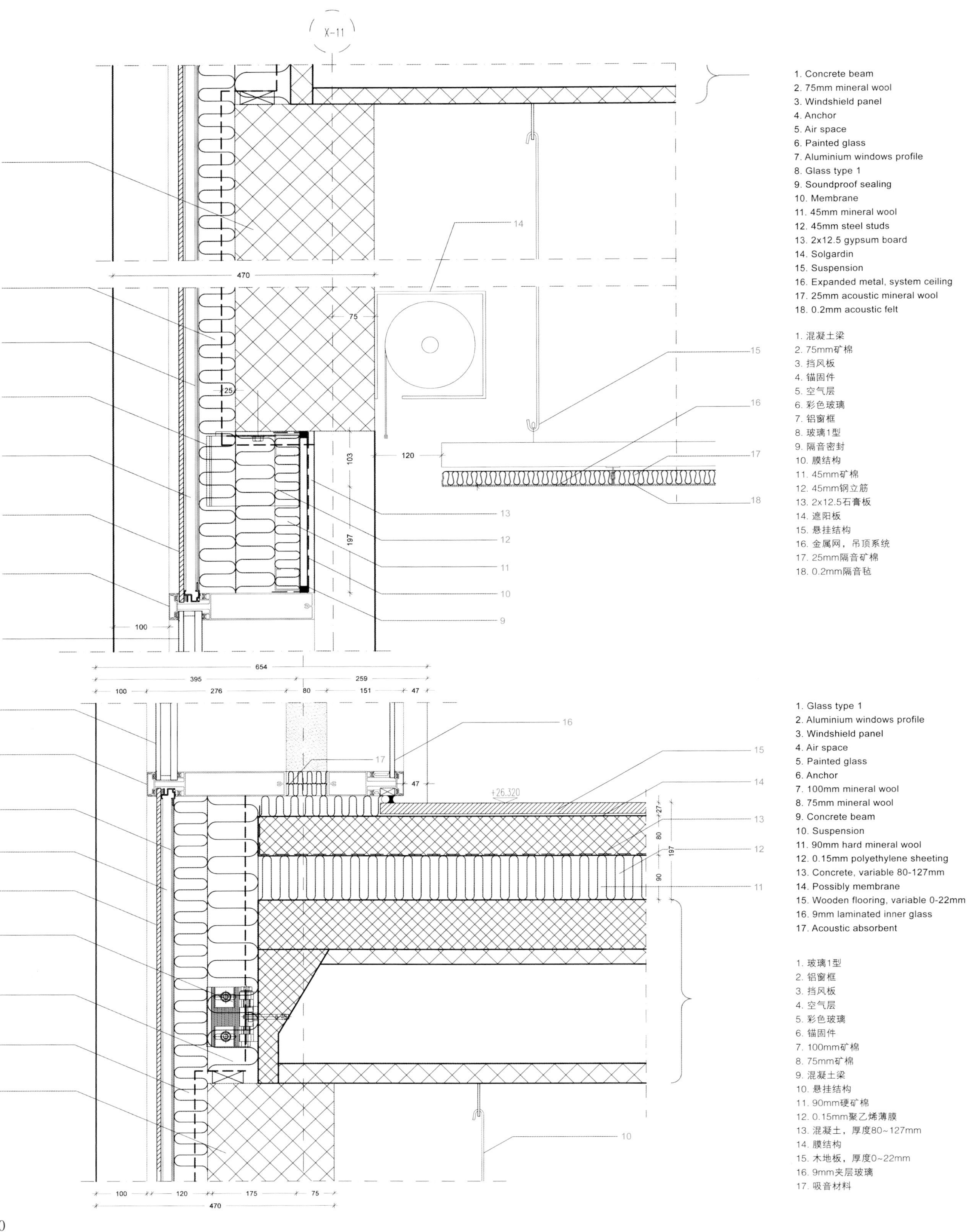
X-11
470
75
25
103
120
197
100
654
395
259
276
80
151
47
+26.320
27
90
175
1. Concrete beam
2. 75mm mineral wool
3. Windshield panel
4. Anchor
5. Air space
6. Painted glass
7. Aluminium windows profile
8. Glass type 1
9. Soundproof sealing
10. Membrane
11. 45mm mineral wool
12. 45mm steel studs
13. 2x12.5 gypsum board
14. Solgardin
15. Suspension
16. Expanded metal, system ceiling
17. 25mm acoustic mineral wool
18. 0.2mm acoustic felt
1. 混凝土梁
2. 75mm矿棉
3. 挡风板
4. 锚固件
5. 空气层
6. 彩色玻璃
7. 铝窗框
8. 玻璃1型
9. 隔音密封
10. 膜结构
11. 45mm矿棉
12. 45mm钢立筋
13. 2x12.5石膏板
14. 遮阳板
15. 悬挂结构
16. 金属网，吊顶系统
17. 25mm隔音矿棉
18. 0.2mm隔音毡
1. Glass type 1
2. Aluminium windows profile
3. Windshield panel
4. Air space
5. Painted glass
6. Anchor
7. 100mm mineral wool
8. 75mm mineral wool
9. Concrete beam
10. Suspension
11. 90mm hard mineral wool
12. 0.15mm polyethylene sheeting
13. Concrete, variable 80-127mm
14. Possibly membrane
15. Wooden flooring, variable 0-22mm
16. 9mm laminated inner glass
17. Acoustic absorbent
1. 玻璃1型
2. 铝窗框
3. 挡风板
4. 空气层
5. 彩色玻璃
6. 锚固件
7. 100mm矿棉
8. 75mm矿棉
9. 混凝土梁
10. 悬挂结构
11. 90mm硬矿棉
12. 0.15mm聚乙烯薄膜
13. 混凝土，厚度80~127mm
14. 膜结构
15. 木地板，厚度0~22mm
16. 9mm夹层玻璃
17. 吸音材料

项目概况

丹麦奥尔堡的“音乐屋”由威尼斯建筑工作室Coop Himmelb(l)au设计，是一所音乐学校和音乐厅的组合：它开放的结构能促进观众和艺术家、学生和老师之间的相互交流。Coop Himmelb(l)au 的设计总监沃夫•D•普利克斯称：“建筑的外形体现了背后的概念。学校将音乐厅包围起来。”

U形排练室围绕着可容纳约1,300名观众的中央音乐厅展开。宽敞的中庭将这些空间连接起来，通过落地窗区域眺望旁边的文化空间和峡湾。门厅下方是三个大小不同的房间：私人厅、韵律厅和古典厅。各种各样的景观窗让学生和观众可以透过门厅和排练室看到音乐厅内部，体验音乐会、排练等音乐活动。

音乐厅内部流畅的线条与建筑外观硬朗的造型形成了鲜明对比。管弦乐队和弧形包厢的座椅的布局为观众提供了最好的视听效果。高度复杂的音响设计由建筑师与英国奥雅纳工程顾问公司共同合作。墙面上无定形石膏结构和高度可调的天花板吊顶的设计都以音响专家的精确计算为基础，保证了最佳的听觉体验。音乐厅减噪等级为NR10（GK10），将成为欧洲交响乐最安静的场所之一。

门厅没有使用风扇，而是在大型垂直空间内利用热能浮力进行自然通风。混凝土楼板内的充水管道兼具夏季制冷和冬季供暖功能。环绕音乐厅的混凝土墙提供了附加的热能存储功能。峡湾的凉风同样能起到免费制冷的作用。空气从照明系统上方的天花板格栅中抽出，保证房间产生的热量不会引起室内升温。

建筑配有建筑管理系统，能控制建筑内的设备，保证系统不进行不必要的运转，从而实现节能。

细部与材料

预制混凝土板的设计以音效需求为基础。它们通过前后凹凸构成内部盒式墙壁和图案，保证了最佳的音响效果。

Swinburne University Advanced Technologies Centre

斯温伯格大学高新技术中心

Location/地点: Melbourne, Australia/澳大利亚，墨尔本
Architect/建筑师: H2o Architects
Photos/摄影: Trevor Mein
Built area/建筑面积: 19,000m²
Completion date/竣工时间: 2011
Key materials: Façade – precast concrete façade, recycled bricks, curtain wall & framed glazing
主要材料： 立面——预制混凝土立面、再生砖、幕墙、框架玻璃

Overview

The Swinburne University of Technology Advanced Technologies Centre (SUT ATC) is the most important recent development undertaken by the university and creates 19,000 square metres for state-of-the-art research and learning centres for Engineering for 2,500 students, staff and researchers.

The project has been created by H2o architects working with Swinburne University after H2o architects were selected from a design competition for the project in July 2007. The project was completed in January 2011.

The development is a single facility with four distinct elements, configured as twin separated ten level towers behind twin three level structures reinforcing the Burwood Road frontage. The entire site is also covered by a service intensive specialist research lower ground facility.

The SUT ATC is the first project to be awarded

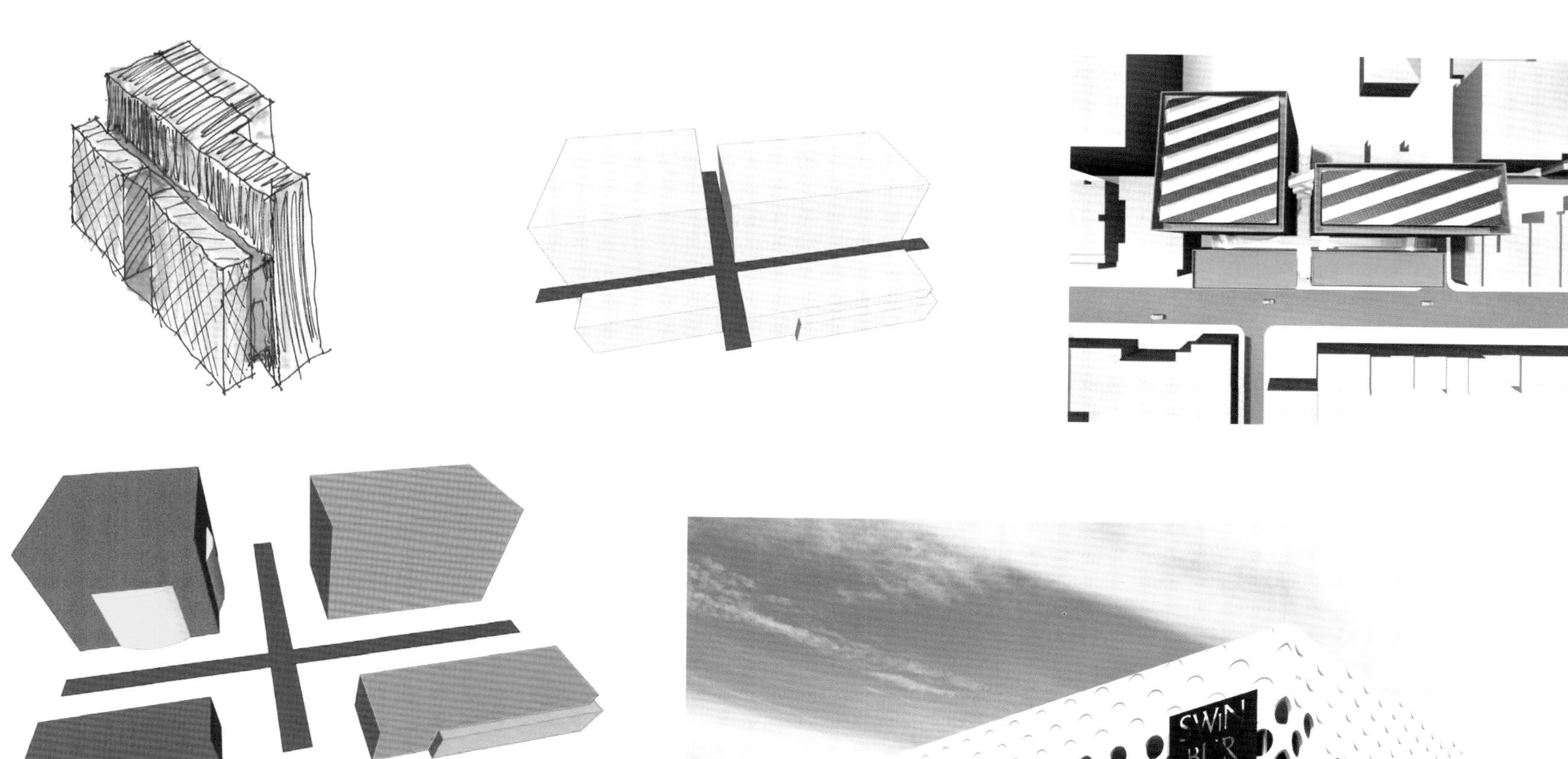

Five Stars by design, using the Green Star Education v.1 tool of Green Building Council of Australia.

Detail and Materials

The architects were seeking a universal approach to the façade tectonic to unify the four disparate building volumes. They had "locked in" precast concrete for structural, cost and thermal mass reasons. They initially drew a grid with openings of approximately 0.9m and curved corners, referencing the ubiquitous Beclawat rail car windows (a major railway line runs through the campus). This seemed to offer the opportunity of a multi faceted approach to vision, ventilation and servicing and by varying the density to orientation and views.

The architects sought a design solution that would be both visually engaging and simple to construct. An approach developed whereby a matrix of repeating circles (from 0.8 to 1 metre in diameter) serves as a dynamic screening device. They saw an analogy with skin pores which are variously coloured, breathe, allows moisture to penetrate but are also water proof and form a raised "scab" for protection.

The pattern translated into various forms, including circular glazing, framing views from within, ventilation through openable sashes (in coloured glass), mechanical louvres, completely open as a sun control device, as a raised "stud" in spandrel conditions or where windows were not required and as a graphic film to glazed workshop spaces at street level.

1. Silicone
2. Plasterboard revel on rolled furring channels
3. Rolled 50x25x3 channel perimeter
4. Remove vert. leg
5. EN 3609
6. Capral 50 series horizontal frame (EL 2074)
7. Flat filler to suit
8. Bulb seal
9. Olympic or similar structure glaze awning
10. Awning channel
11. Square glass panel silicone sealed to aluminum awning sash with S.S. stay to vertical window jamb
12. Compressible seal (PVC bulb seal)
13. 6mm glass as per schedule with structural silicone and double sided tape
14. Rolled 80x20x3 angle
15. Rolled 20x20x3 angle
16. Foam tape
17. Provide neoprene packing as required

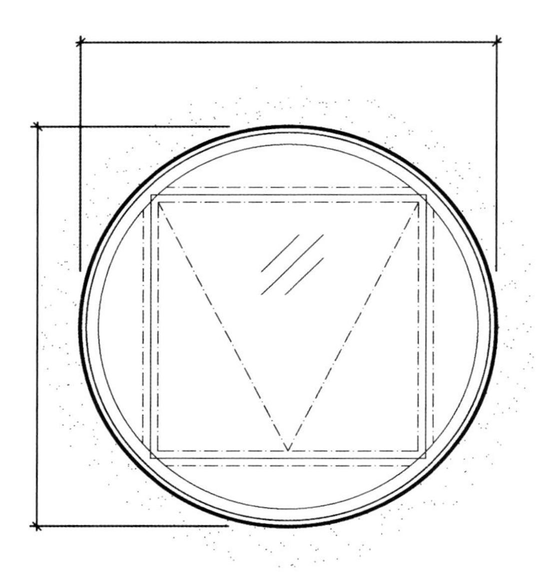

1:10

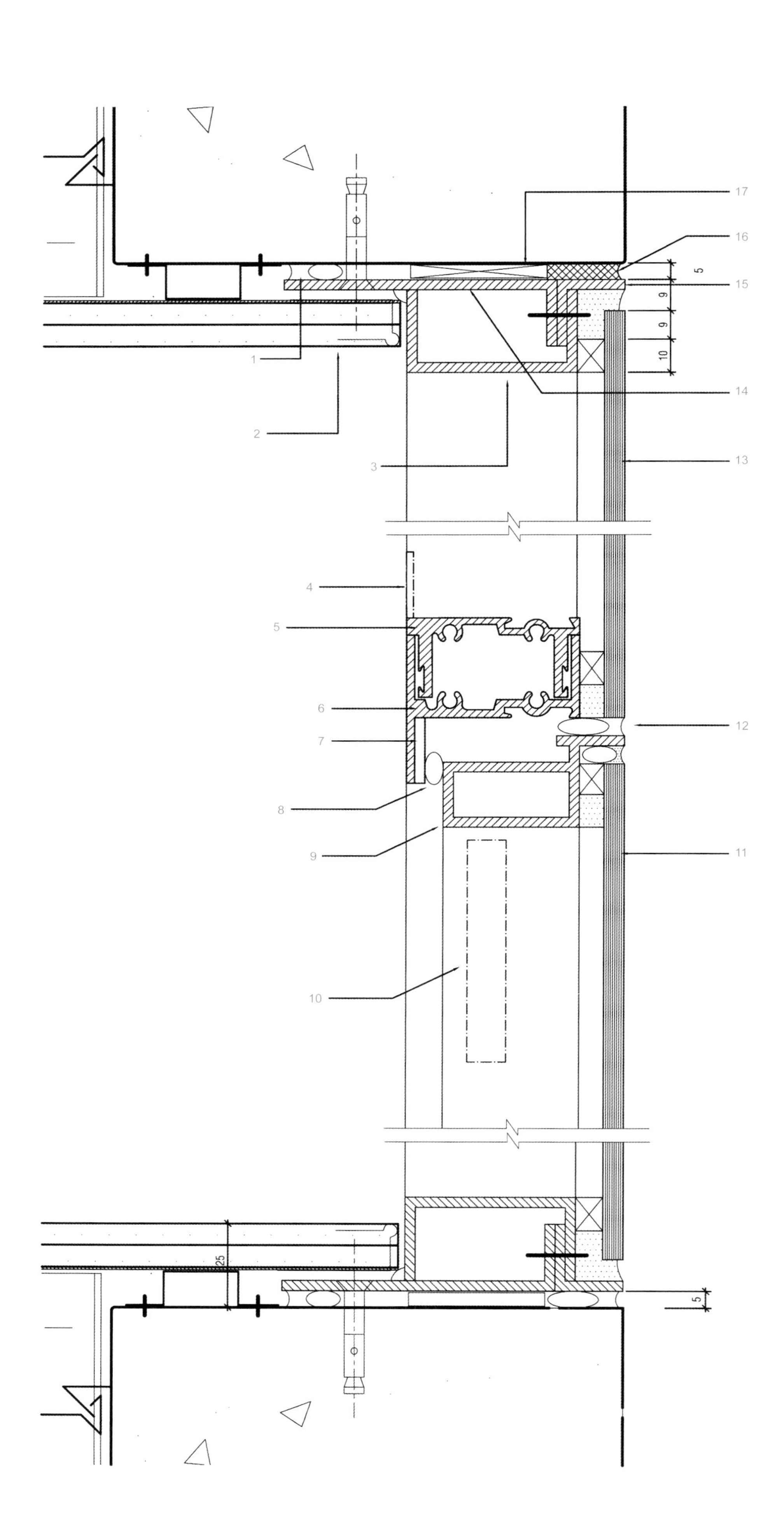

1. 硅胶
2. 石膏板，安装在轧制副龙骨上
3. 轧制龙骨边50x25x3
4. 移动垂直支柱
5. EN 3609
6. Capral 50系列水平框架（EL 2074）
7. 平填充
8. 球缘密封
9. Olympic或类似结构玻璃雨篷
10. 雨篷槽
11. 方形玻璃板，硅胶密封于铝制雨篷框架上，与窗框垂直
12. 可压缩密封（PVC球缘密封）
13. 6mm玻璃，结构硅胶和双面胶带密封
14. 轧制角钢80x20x3
15. 轧制角钢20x20x3
16. 泡棉胶带
17. 氯丁橡胶密封

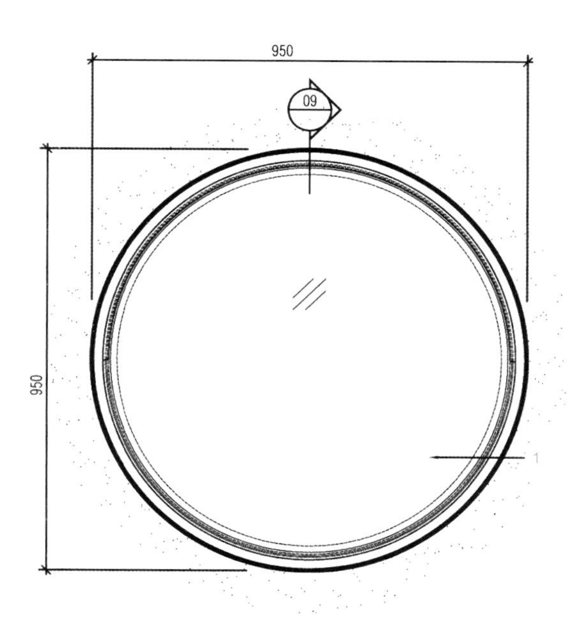

1. Laminated safety glass structural sealant set
2. 76 steel studs set diagonally to avoid window openings
3. Curved 333 furring channel around reveal opening, or alternative project short cuts of channel at right angles to provide fixing for P/BD reveal lining
4. Line of window reveal
5. Curved P/BD reveal lining (2x6.5 thick)
6. Curved steel stud (or c curved sheet steel angle flashing 0.8 BMT) cleated off furring channnels to provide backing support to P/BD reveal & wall lining around window edges
7. Rondo P10 corner bead curved to outline & trim edge of circular opening
8. 13 P/BD wall lining
9. R1.5 wall insulation
10. Rondo 129 furring channel attached to 237 clip fixed to wall (set diagonally)

1. 夹层安全玻璃，结构密封剂
2. 76钢立筋，对角装配，避开窗口
3. 窗侧弯形333副龙骨，或替代品，为P/BD窗口侧线提供固定安装
4. 窗侧线
5. 弯形P/BD窗侧线（2x6.5厚）
6. 弯形钢立筋（或C弯形角钢防水板0.8BMT），为P/BD窗侧和窗口墙衬提供支撑
7. Rondo P10弯形墙角护条，包围并装饰圆窗边缘
8. 13 P/BD墙衬
9. R1.5墙面隔热层
10. Rondo 129副龙骨，通过237夹固定在墙壁上（对角装配）

1:10

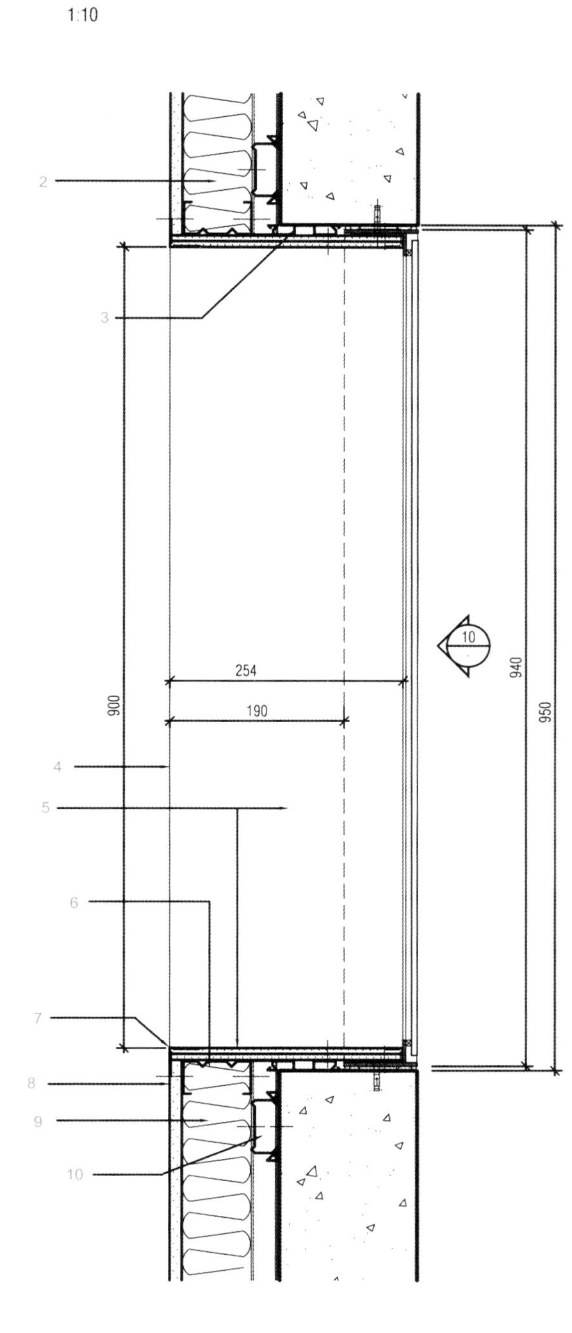

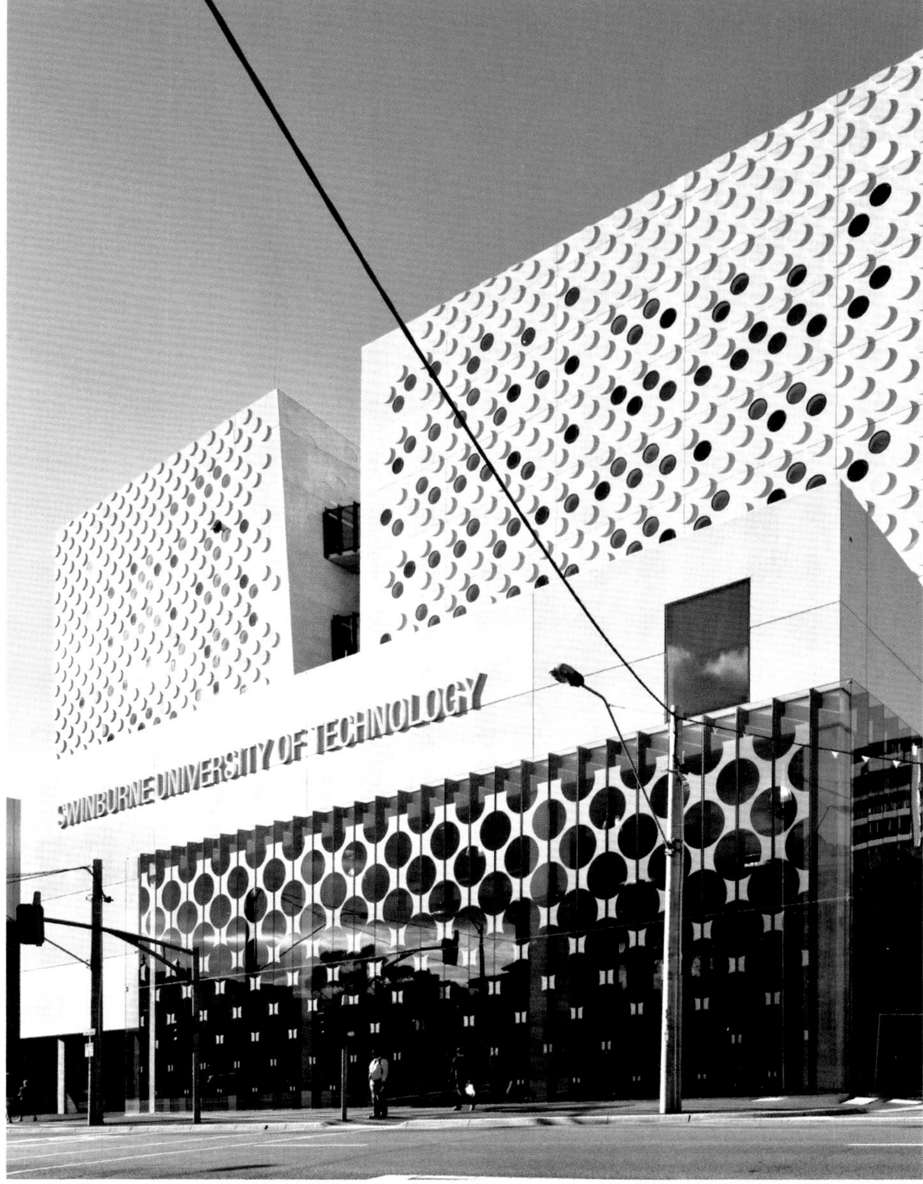

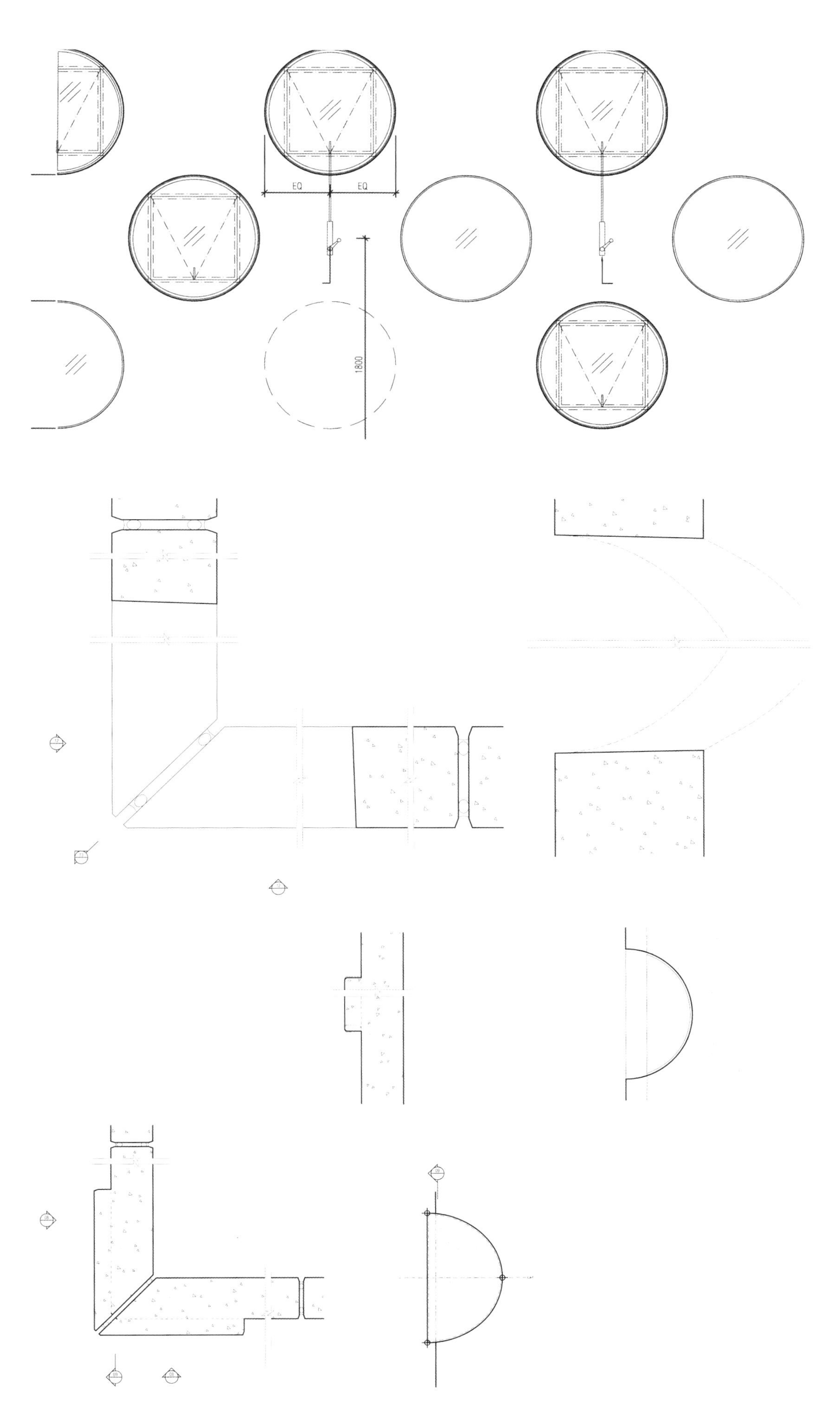

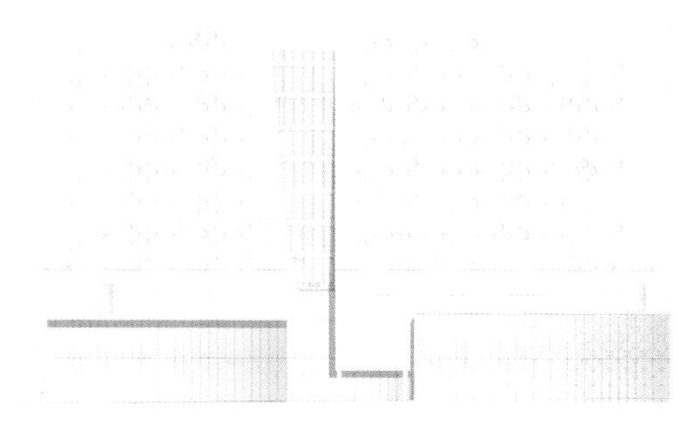

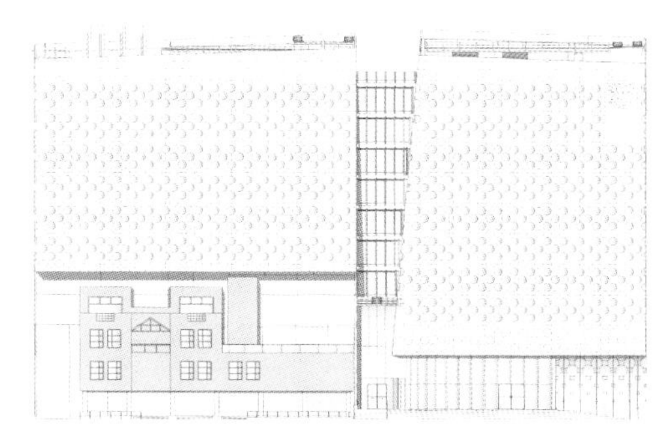

项目概况

斯温伯格大学高新技术中心是该大学近期最重要的开发项目，拥有19,000平方米的高新技术研究和学习中心，为2,500名师生和研究者提供服务。

H2o建筑事务所在2007年7月的一次设计竞赛中脱颖而出，获得了与文斯伯格大学合作开发项目的机会。项目于2011年1月正式完工。

开发项目拥有四个独特的空间结构，分别是10层高的双子塔楼和3层高的双子底层结构，后者正对布尔伍德路。整个场地上还覆盖着一个底层服务科研设施。

斯温伯格大学高新技术中心是澳大利亚绿色建筑委员会利用绿色星教育建筑评估v.1系统所评估的首个五星级绿色项目。

细部与材料

建筑师试图寻找一种通用的立面设计，将四个独立的建筑结构整合起来。出于结构特性、成本预算和热质量等因素的考虑，他们最终选择了预制混凝土作为主要材料。他们参考了Beclawat有轨电车的车窗设计了直径约为0.9米的圆窗，并将它们以网格的方式展开。这种设计为视野、通风、服务等功能提供了多重选择，可以根据需要调节窗户朝向和开口。

建筑师试图找到一种既吸引眼球又方便建造的设计方案。反复出现的圆形（直径在1.8米至1米之间）阵列正好可以作为富有动感的幕墙建造工具。这些建筑表皮上的开口被赋予不同的色彩，既保证了自然通风和水分进入，又起到防灰防护的作用。

圆形阵列被转化为各种形式，包括圆形玻璃窗、彩色玻璃通风开口、机械百叶窗等，可作为遮阳装置、楼层之间凸出的纽扣状装饰也可充当一楼临街工坊的图案幕墙。

University Campus in Tortosa

托尔托萨大学校园

Location/地点: Tortosa, Spain/西班牙，托尔托萨
Architect/建筑师: Josep Ferrando + Pere Joan Ravetllat + Carme Ribas
Photos/摄影: Pedro Pegenaute
Built area/建筑面积: 10,000m^2
Completion date/竣工时间: 2011
Key materials: Façade – lattice precast concrete panel
主要材料: 立面——格架预制混凝土板

Overview

The project is bound between both urban and natural landscape elements; the main avenue that leads out of the city and the river, the congress hall and the park. The proposal consists of one building that houses all four schools in order to obtain a larger mass of volume so that the building, in turn, becomes the gateway between the natural landscape and the town.

The brief avoids the traditional format of corridors and successive classrooms. There is a disproportion between the size of these corridors and their use. They are too narrow for any common activities or socialising, but too wide for circulation alone. In this format, the proportion of façade that the corridors occupy is also disproportionate considering that they have the same amount of façade as the classrooms.

These common areas are one of the most important spaces in a university building as a lot of social activity occurs outside the classroom. Therefore the architects have developed a plaza-type concept so that, like a wheel, the connecting point is the hub of the system. However, a circular format would fail as a gateway. The distance that a person would have cross along the routes below the building would always be the longest. The geometry of the hub is transformed from a circular to a cross-shaped mass. The centre remains a hub and gateway, and, thus, the distance walked to cross it is minimised.

The path connects the university campus and

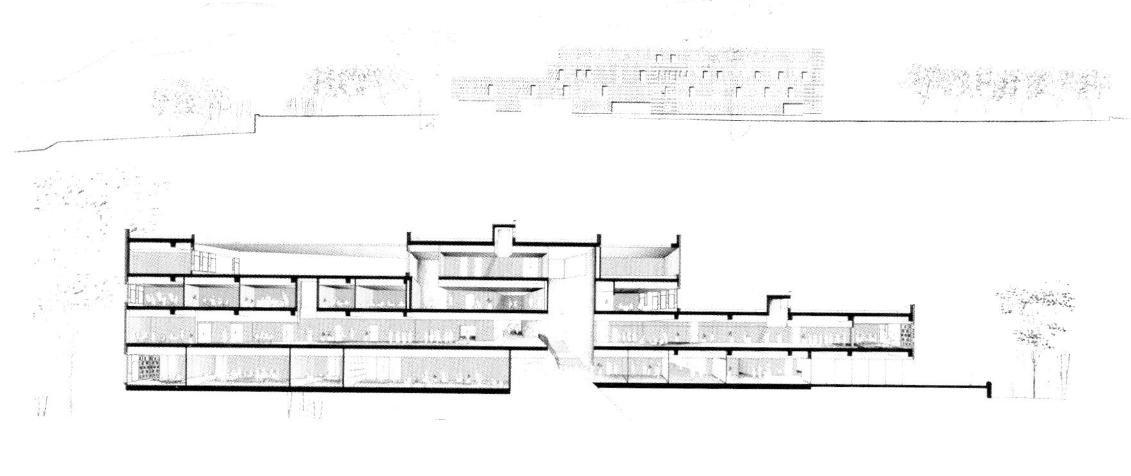

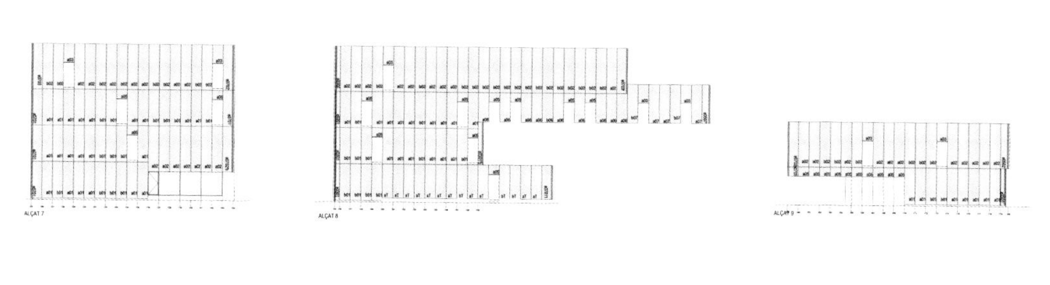

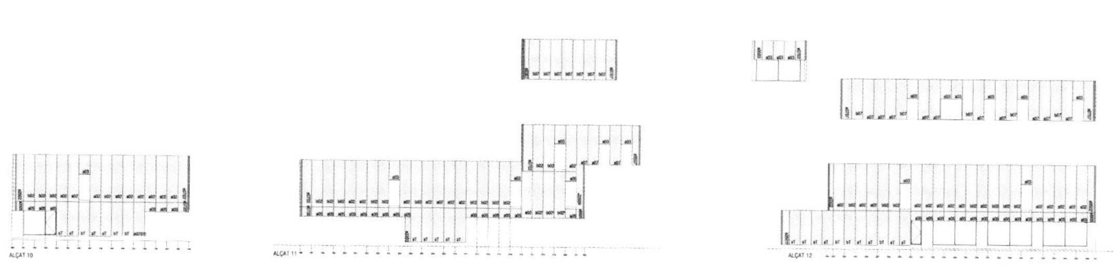

the congress hall, and they share part of their facilities. The congress hall has an auditorium, but no cafeteria-bar, and the university campus has a bar but no auditorium. A type of dialogue is set up, not only between the uses of the two buildings, but also in the materiality and geometry of the two volumes; the congress hall is a prism made of U-Glass while the university campus a more fragmented volume made of concrete.

This geometry echoes the historical buildings of the surroundings: the castle at the top of the hill and an old church. The project is related to both the natural and historical contexts. The lines of the geometry of the volume are parallel to those of the congress hall and the nearby housing blocks.

The building reflects the scales of the elements that surround it. At the back it recognises the height of the congress hall and at the front it is scaled down to meet the park and river. The rear is more vertical and hard relating to the city/urban elements. A longitudinal void signals the entrance.

The relationship of the building to its surroundings called for the rethinking of the typical façade of windows, and led to the development of a skin that resembles wicker. This also acknowledges the traditional trade of the area which is the manufacture of wicker baskets, and also relates the building to the surrounding fields of wicker. In the areas that allow light to enter, the architects simply stretched the wicker basket texture to make it less dense.

Detail and Materials

The entire façade is made of concrete cast in just two moulds. The rough side of the cast panel (the one without the wicker texture) is polished so that it resembles a fine terrazzo that relates to the interior paving of the building.

The rhythm of the façade panels is such that at the end of each room, the last panel is omitted. This allows for a clear view of the surroundings, and the far wall of each room becomes a light-reflecting panel.

项目概况

项目被夹在城市和自然景观元素之间；主通道将城市、河流、会议厅和公园连接起来。项目在一座建筑内设置了四个学院，形成了巨大的体量感，使建筑变成了自然景观与城市之间的门户。

设计方案没有采用传统形式的走廊和连续的教室。传统走廊的尺寸和功能并不匹配。狭窄的走廊不适合任何公共活动或社交，但是仅用作交通通道又太宽。在这种形式中，走廊所占外墙的比例同样也不太得当，它们所占的外墙面几乎与教室一样多。

公共区域是大学教学楼最主要的空间之一，许多社交活动都在教室之外进行。因此，建筑师开发了一种广场型的设计概念，像轮子一样，把连接点作为系统的枢纽核心。但是圆形布局不适合作为门户通道，人们穿越建筑的路线将是最长的。因此，枢纽的造型从圆形变成了十字形。这样一来，中心仍然是枢纽，而穿越距离却被缩到最短。

通道将大学校园与会议厅连接起来，使它们共享部分设施。会议厅有一个礼堂，而没有餐吧，而大学校园有餐吧没有礼堂。两座建筑不仅在功能上建立起对话，在物质与结构上也存在共同感：会议厅是由U形玻璃构成的棱柱结构，而大学楼则用混凝土打造更加分散的结构空间。

建筑造型与山顶的城堡和老教堂等周边历史建筑遥相呼应，不仅与自然环境相适应，并且与历史环境相联系。建筑结构的线条与会议厅以及附近的住宅楼的线条都是平行的。

建筑反映了周边建筑元素的比例。建筑后部与会议厅的高度相似，前面则按比例缩小至与公园与河流相匹配。建筑后部更加硬朗，倾向于城市元素。入口以纵向空间标志出来。

建筑与周边环境的关系让建筑师重新考虑了典型外墙窗口的设计，最终形成了类似于柳条编织品的建筑表皮。这种设计还暗示了该地区制作柳条篮的传统手工艺，并且使建筑与周边田地里的柳树联系起来。在允许光线进入的区域，建筑师简单地将柳条编织的纹理拉伸，形成更大的开口。

细部与材料

整个建筑立面都由预制混凝土构成，仅使用了两种模具。没有柳条编织纹理的一面所采用的预制板材经过了抛光，像精致的水磨石一样，与建筑室内的地面联系起来。

外墙板材的韵律感体现在各个房间一端最后的一块板是省略的。这种设计保证了各个房间都享有外界的视野，并且将房间的远墙变成了一块轻质反光板。

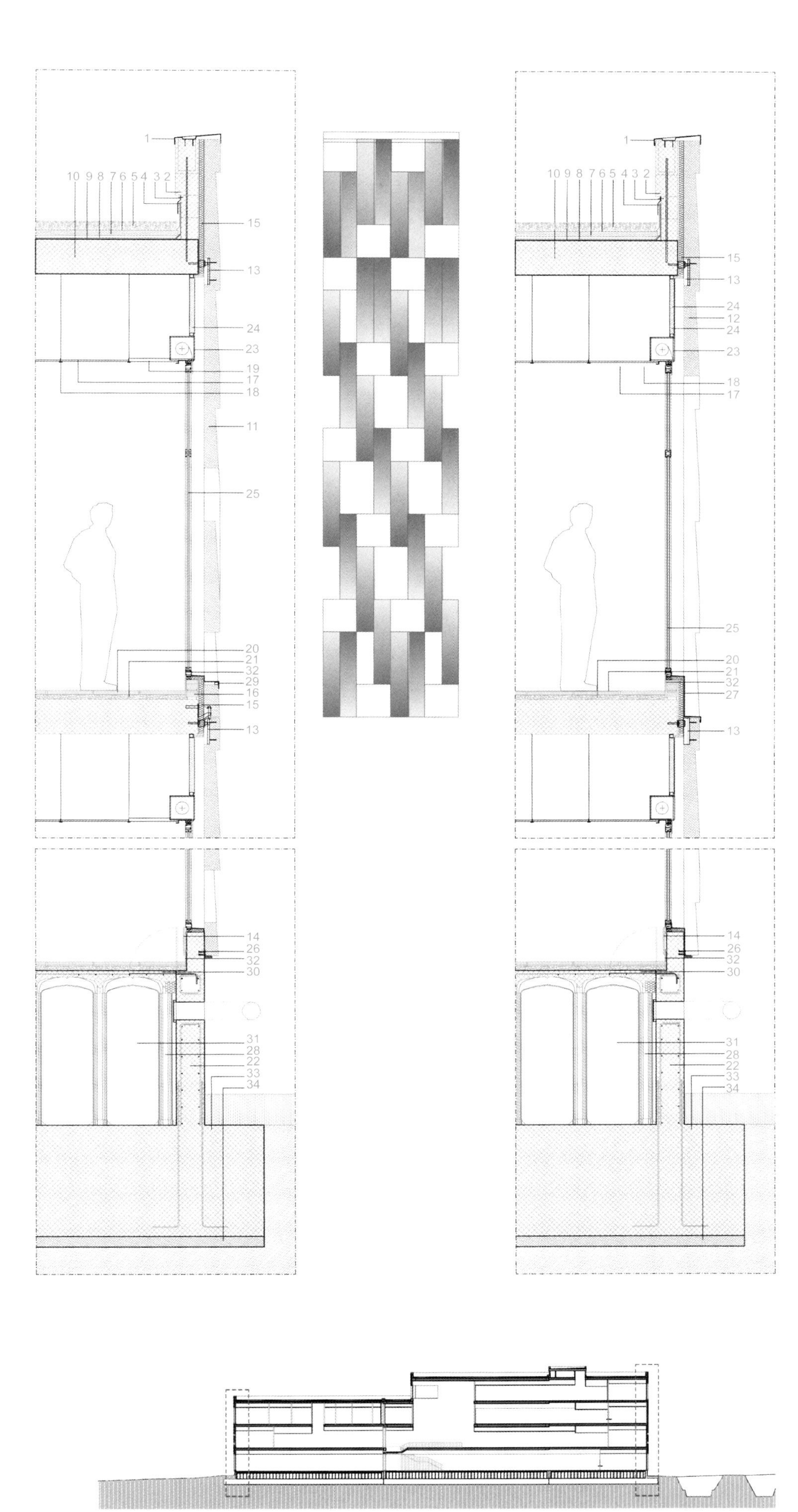

1. Anodised aluminum sheet
2. Concrete block parapet
3. Anodised aluminum plate bolted to the wall
4. TENSOFLEX sheet
5. Gravel 8.5cm
6. Filter layer FELTEMPER 150 P
7. Extruded polystyrene 5cm
8. Waterproof membrane
9. Separating layer of synthetic geotextile felt
10. Concrete slab
11. Lattice architectural concrete panel 4.20x1.20m
12. Opaque architectural concrete panel
13. Anchor between panel and slab
14. Black coloured socket DM
15. Thermal insulation, polyurethane
16. Perforated brick 10x19x9cm
17. Removable ceiling plates 60x60cm
18. Concealed aluminum joinery 15mm
19. Fixed wrap of ceiling
20. Terrazzo pavement 4cm
21. Mortar
22. Concrete perimeter wall
23. Shutter box MONOBLOCK, 20x20cm
24. Panel sheet + isolation + sheet
25. Window aluminum frame with thermal break
26. Lower anchor Profile L
27. Jamb of galvanised steel, 5mm
28. Elevator system, Caviti type
29. Piece of galvanised steel, 5mm
30. Suspended floor
31. Ventilated chamber
32. Black coloured socket Trusplas
33. Foundation slab. 90cm
34. Lean concrete

1. 阳极氧化铝板
2. 混凝土砌块护墙
3. 阳极氧化铝板，固定在墙上
4. TENSOFLEX板
5. 碎石层8.5cm
6. 疏水层FELTEMPER 150 P
7. 挤塑聚苯乙烯5cm
8. 防水膜
9. 合成土工布毛毡分隔层
10. 混凝土板
11. 格架混凝土板4.20x1.20m
12. 不透混凝土板
13. 板材与地板之间的固定件
14. 黑色插槽DM
15. 聚氨酯隔热层
16. 多孔砖10x19x9cm
17. 可拆除天花板60x60cm
18. 隐式铝合金框15mm
19. 固定天花板包层
20. 水磨石铺装4cm
21. 砂浆
22. 混凝土外墙
23. 卷帘式壁龛MONOBLOCK，20x20cm
24. 薄板+隔热层+薄板
25. 铝窗框，带断热效果
26. 下层锚固L型材
27. 镀锌钢侧壁5mm
28. 电梯系统Caviti型
29. 镀锌钢件5mm
30. 悬垂楼板
31. 通风间
32. 黑色插槽Trusplas
33. 基础底板90cm
34. 少灰混凝土

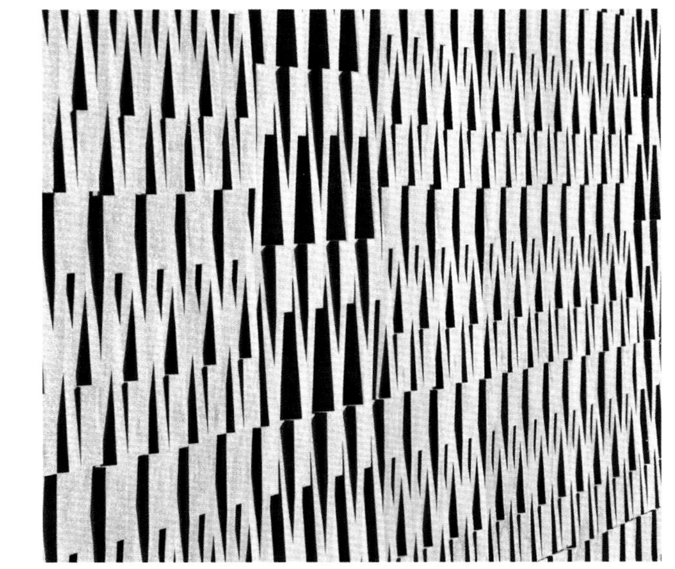

The Novium Museum

诺维耶姆博物馆

Location/地点: Chichester, West Sussex, UK/英国，奇切斯特
Architect/建筑师: Keith Williams Architects
Photos/摄影: David Grandorge
Site area/占地面积: 495m^2
Built area/建筑面积: 1,300m^2
Completion date/竣工时间: 2012
Key materials: Façade –precast concrete panels
主要材料：立面——预制混凝土板

Overview

The starting point for the architects was to understand both the Client brief for the Novium museum and the site context within Chichester's historic city plan which is characterised by a ring of partially intact enveloping Roman and Medieval walls. Within the walls the city is segmented into 4 roughly equal quadrants by a cruciform street pattern centred upon the medieval Cross.. The Novium Museum's site on Tower Street, sits in the north- west quadrant, running north from West Street the city's western arm, close by the medieval cathedral with its freestanding medieval bell tower.

Chichester's urban character has changed markedly once north of West Street. The immediate vicinity of the Novium's site was irrevocably altered during the 1960s and 1970s, when many of the historic buildings that once formed the street scene were swept away. Consequently, few buildings of evident architectural significance survive, beyond those listed on Tower Street adjacent and opposite the Museum site, and perhaps the circular 1960s library. The site itself was until this project, a brownfield car park beneath which lay the remains of the city's Roman baths. The Novium's site location, which is in line sight of the Chichester's Gothic Cathedral, marks the junction between the historic cityscape and the more fractured urban grain to the north.

The architects' design response to this diverse context attempted to establish a new urban grain by creating a unified block, centred on the Museum and an integral but later residential project, with a new set of buildings along the surrounding streets.

The unique aspect of the site itself centred on the presence of substantial archaeological remains of a series of Roman baths, which were discovered in the 1970s. The baths, part of Roman Chichester (Noviomagus Reginorum) date from the Flavian period (1st Century AD).

The new museum has been designed to span the remains of the baths (the hypocaust) which have been incorporated in situ within the main entrance hall and gallery as a permanent exhibit and an intrinsic part of a museum. The public galleries are stacked vertically on 3 levels and are linked by a processional stair, hovering above the remnant baths, culminating with views across the city to the cathedral from the Cathedral window, at the building's highest level. A slot between the stair flights brings a numinous light onto the baths and allows glimpses back over the baths from vantage points during the ascent through the building.

The city's set piece public and religious structures such as the market cross, cathedral and bell tower are all constructed in pale stone in a city that is otherwise largely brick or render, establishing a hierarchy of materiality which has informed the choice of materials of the new Museum. As a consequence the Novium has been clad in pale reconstructed stone on integral reinforced concrete structural panels, establishing its architectural and cultural connection with the tradition of Chichester's grander public structures, and providing an architectural accent amongst Tower Street's otherwise brick Georgian buildings. The proposed residential development (yet to be realised) is seen as fulfilling a supporting architectural role to the museum.

Both the Novium Museum and the residential scheme take a three storey parapet level consistent with the general scale of the Georgian buildings which define the east side of Tower Street. At its northern end, the Museum's main elevation incorporates a square form turret to introduce variety and accent to the street scene, and to signal the Museum from both West Street and the approach from the city walls to the north.

The main façade is carefully composed and incised to respond to the subtly undulating street scale of the adjacent listed Georgian buildings, whilst the square form turret provides accent and balance to the composition. The elevation is ordered by a proportional system including the Golden Section giving a precise geometry to its abstract expression.

The interior surfaces of the museum are made from high quality fair faced concrete. The architects spent much time researching and testing the concrete mix options before arriving at the final specification. Silver granite aggregates were used with silver grey granite fines, the surfaces of which were brushed to achieve a subtle sparkle to the concrete surfaces.

The new galleries have been designed to be flexible allowing a mix of permanent and temporary exhibitions, and provide education spaces, restoration, research and staff areas.

In summary, the new Museum is an elegant new insertion into an historic city fabric which resolves contemporary architectural expression within a mature historic city, as the first element within a wider urban project to reinstate the city fabric.

Detail and Materials

The building comprises 2,000 tonnes of reinforced concrete structure, envelope and finishes suspended over the archaeological remains of the ancient Roman baths. The footprint of the foundations was reduced to less than 3% of the overall plan and carefully threaded between the old walls onto high capacity bored piles.

A granular fill made a working platform above the delicate elements which were carefully mapped. The in-situ concrete superstructure was cast on formwork propped off this covering and enclosed with a pre-cast panel system. Subsequent excavation revealed the original forms below the new exhibition spaces.

New walls and floors are in very light coloured concrete with a glittering finish of mica particles. The aggregate is white granite from Derbyshire and the cement is brought up to a pale shade by mixing with ground blast-furnace slag. 65% concrete replacement helped reduce the building's carbon footprint and careful sampling was required to ensure that the mixes were workable to produce even results.

Exposed surfaces were treated to a soft sand-blasting of limestone grit to expose mica and quartz components of the granite. Very deep pours of concrete meant that the formwork had to be specially detailed and strengthened to prevent any grout loss or segregation from compromising texture or colour. Formwork joints were carefully coordinated with the architect's patterning of volumes and surfaces.

项目概况

建筑师的出发点是了解委托人对新博物馆的设计要求以及位于奇切斯特老城区内的项目场地环境。老城区以环形布局为特色，将古罗马围墙和中世纪城墙包围起来，城墙内被十字形街道划分为四个大致相等的区域。诺维耶姆博物馆位于钟楼街，属于西北城区，沿着西街向北延伸，距离中世纪教堂的独立钟楼很近。

奇切斯特市西街以北的城市风格经历了显著的变化。紧邻诺维耶姆博物馆的区域在20世纪60、70年代经历了巨变，许多构成街景的历史建筑被拆除了。因此，除了博物馆两旁和对面位于钟楼街上已列入保护的建筑之外，很少有重要的建筑得以保留下来。直至项目开发，这块场地都还是一片棕色地带，停车场下方隐藏着罗马浴场的遗址。

设计对如此复杂的环境做出了回应，力求通过打造一块统一的街区来重新塑造城市脉络。街区以博物馆为中心，未来将整合住宅项目，形成全新的建筑群。诺维耶姆博物馆标志着历史城区景观与北部城市景观的连接点。

项目场地的独特之处除了与奇切斯特的哥特式教堂列于同一直线之外，还在于它拥有罗马浴场的历史遗址。这些浴场属于罗马时代奇切斯特的一部分，可追溯到公元1世纪。新建的博物馆横跨浴场遗址之上，浴场与博物馆的入口大厅和展览厅融为一体，形成了永久性展览，是博物馆的重要组成部分。

新建的公共展览厅叠加在三层空间内，由楼梯相连接，盘旋于浴场遗址之上，从建筑的最高处可以穿过城市，欣赏到宏伟的教堂。楼梯段之间的开口为浴场带来了奇妙的光线，让人们可以在上楼梯的过程中从最佳角度观察浴场遗址。

市场十字架、教堂、钟楼等城市最主要的公共建筑和宗教建筑都采用白色石材构建，而其他建筑结构则大多为砖砌或粉刷，使建筑材料实现了一种等级感，这对新博物馆的建材选择十分重要。因此，博物馆的外墙选择以白色人造石材贴在钢筋混凝土结构板上，与奇切斯特宏伟的历史建筑实现了联系，同时也使自己从钟楼街上的乔治亚时期砖砌建筑群中脱颖而出。规划中的住宅项目将成为博物馆的辅助建筑结构。

博物馆和住宅楼都采用了三层护墙结构，与钟楼街东段的乔治亚时期建筑相一致。在博物馆北端，一个方形角楼结构突出了建筑的变化感，形成了独特的街景，使博物馆在西街和城墙北面看起来都十分显眼。

建筑的主立面经过精心构造，与略微前后起伏的历史建筑相互呼应，而方形角楼则起到了强调和平和的作用。建筑立面经过黄金分割等比例系统的规划，形成了精确而抽象的几何造型。

博物馆的内墙面采用高品质的琢面混凝土构成。建筑师花费大量时间对混凝土混合物配比进行了研究和测试，

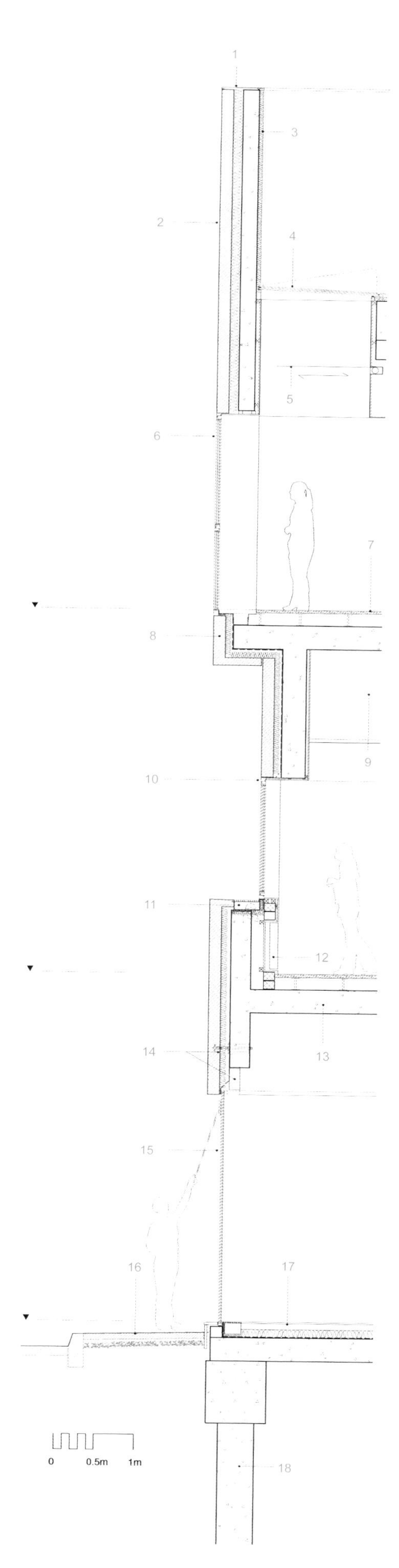

1. Pressed aluminium colour matched coping.
2. 150mm Honed finish precast concrete panels with PIR insulation prebonded to the back over a 50mm air gap to waterproofing.
3. Inner lining to r.c. upstand insulated and weather protected with bonded single ply membrane sheeting
4. Glazed rooflight in minimal frame automatic opening for ventilation
5. Interior blinds to mitigate strong daylight
6. Double glazed units in powder coated aluminium frames
7. Raised floors for flexible servicing with perimeter heating at window locations
8. Panels cast with integral reveal
9. Exposed r.c. deep rib beams in gallery spaces
10. Glazing positioned flush with cladding surface
11. Internalized drainage through channels on window ledge. No external pipework
12. Internal walls lined with plasterboard. Radiators recessed to be flush with wall surface
13. Main structure formed from fairfaced in-situ reinforced concrete
14. Stainless steel fabricated fixings and support brackets with adjustment possible in three directions bolted into "cast in" fixing points
15. Double glazed panels with glass sized to span floor to ceiling without additional wind load structure
16. External Portland stone paving paving
17. Internal floor finish Portland stone with underfloor heating system
18. Piled foundations with piling restricted to perimeter of site only

1. 加压铝，色彩与顶盖相配
2. 150mm塘磨表面预制混凝土板，配PIR预加隔热层+50mm空气层+防水层
3. 钢筋混凝土竖柱内衬，单层膜绝缘防护
4. 小框玻璃天窗，自动开合通风
5. 内部百叶窗，缓和强烈日照
6. 双层玻璃窗，粉末涂层铝框
7. 活动地板，方便窗边灵活的周边供暖
8. 板材，配整合窗侧
9. 清水钢筋混凝土肋梁
10. 玻璃板，与包层表面平齐
11. 通过窗台凹槽的内部排水，无外部排水管
12. 内墙内衬石膏板；嵌入式散热器与墙面平齐
13. 清水面现浇钢筋混凝土主结构
14. 不锈钢预制配件和支架，可三向调节，用螺栓固定在固定点上
15. 双层玻璃板，玻璃板跨度从地面到天花板，无附加风荷载结构
16. 外部波特兰石铺装
17. 内部地面波特兰石，配地板下供暖系统
18. 打桩地基

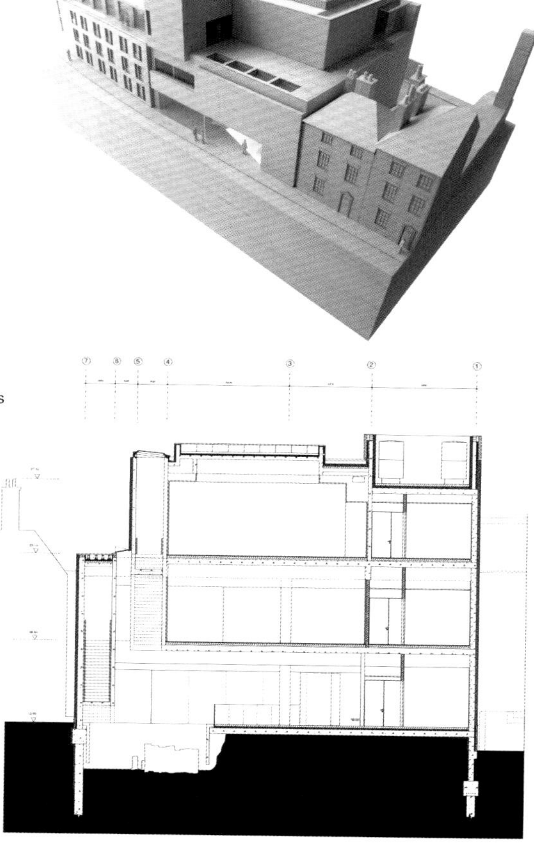

最终形成了完美的效果。银色花岗岩集料与银灰色花岗岩粉末共同使用，在混凝土表面形成了一种微妙的闪光效果。

展览厅的设计十分灵活，可以举办永久性展览和临时展览，同时还提供了教学、修复、研究空间以及办公区。总而言之，新博物馆为历史城市肌理注入了清新优雅的气息，解决了在历史城区内建造现代建筑的问题，是奇切斯特复兴城市脉络的第一步。

细部与材料

建筑由2,000吨钢筋混凝土结构、外壳和罩面构成，悬浮在古罗马浴场历史遗迹上。地基的占地被缩减至整体规划面积的3%以下，巧妙地穿梭在旧墙体和大容量灌注桩之间。

颗粒填充物在精巧的组件上方形成了工作平台。现场浇筑的混凝土上层结构在模架中形成，支撑着这层盖板，并且有预制混凝土板系统包覆起来。随后的挖掘展示了新展览空间下方的原始形态。

新建的墙壁和楼板采用极浅的混凝土，表面点缀着闪闪发光的云母颗粒。来自德比郡的白色花岗岩集料和水泥通过混合高炉矿渣而形成了灰白的色调。65%的混凝土置换帮助减少了建筑的碳排放量，建筑师通过精心采样保证了混合物的可行性。

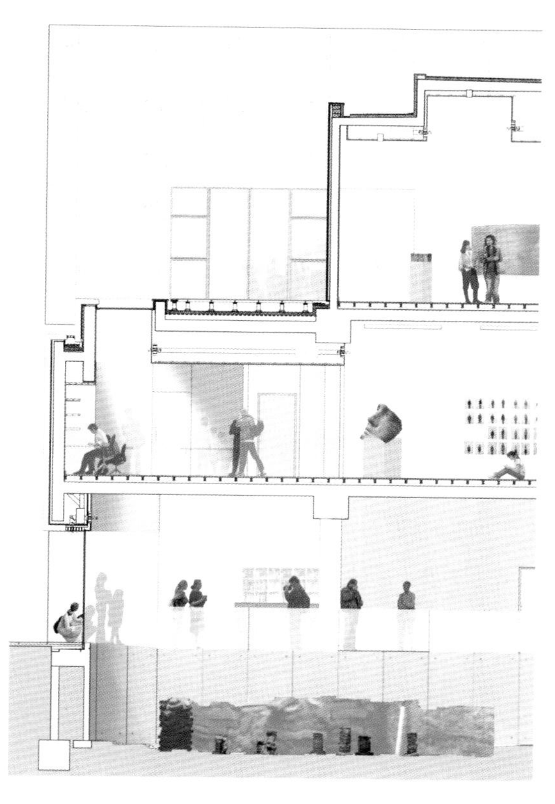

素面混凝土经过柔和的石灰岩砂砾喷砂，在表面显露出云母和花岗岩中的石英成分。厚重的混凝土要求模架必须十分精准牢固，以避免造成任何泥浆损失以及纹理或色彩的缺失。模架接缝与建筑师所设计的结构和界面十分协调统一。

Dyson Building: Department of Fine and Applied Arts

戴森楼：艺术与应用艺术系

Location/地点: London, UK/英国，伦敦
Architect/建筑师: Haworth Tompkins
Photos/摄影: Helene Binet, Philip Vile
Built area/建筑面积: 4,750m²
Completion date/竣工时间: 2012
Key materials: Façade – Polished precast concrete panel, double glazed curtain wall
主要材料：立面—— 抛光预制混凝土板，双层玻璃幕墙

Overview

The Dyson building, designed by award-winning architects Haworth Tompkins, is the most significant new development to the Royal College of Art since it moved to Kensington Gore in 1962 and forms the centrepiece of the RCA's new Battersea Campus.

The development sits alongside the RCA's existing Painting and Sculpture Programmes on the site. It connects with and provides a boost for "Creative Battersea", which currently boasts the headquarters of Vivienne Westwood, Foster + Partners and Will Alsop, amongst many others. As well as containing the Printmaking and Photography programmes, the building is also home to a 220-seat lecture theatre, a public gallery and the College's business incubator.

The building is conceived as a creative "factory" both in the industrial sense (as a place

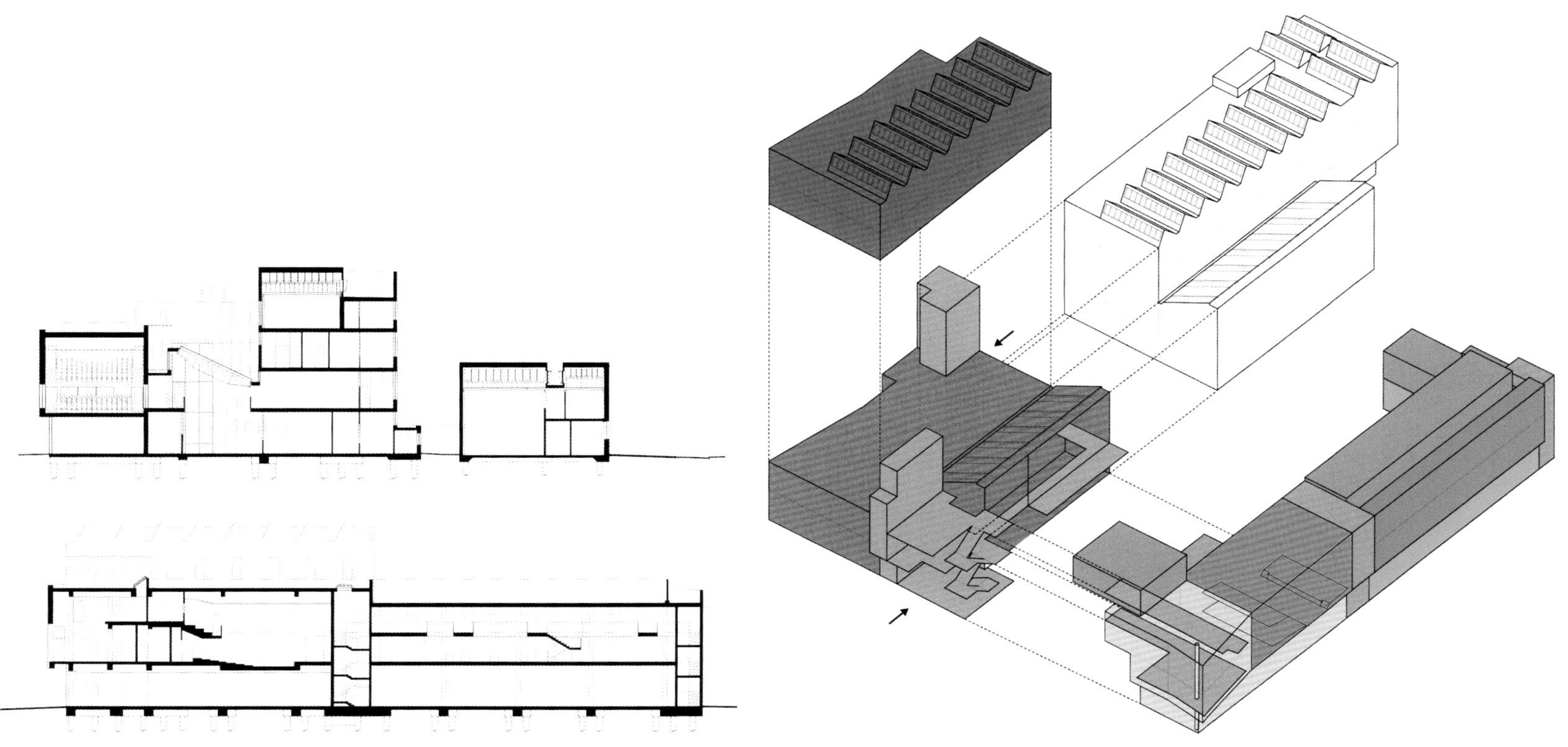

of industry), and through the reference to Andy Warhol's Factory as a place of art production. The building's aesthetic is functional and derives from the way it made with an in-situ concrete structure exposed throughout and used expressively to form a series of dramatic interlocking spaces.

A large top-lit "machine hall" links the two blocks and forms the heart of the building, designed to house the large presses used by printmakers. A more public zone of retail and business space is arranged along the street frontage, whilst the corner facing Battersea Bridge is cut away to give the RCA a public entrance, a gallery and foyer area for the lecture theatre.

A key characteristic of the RCA's success is the fluid relationship between programmes. The building has been designed to create "horizontal drift" between disciplines, and the creative processes take place in highly visible proximity to one another. The cross-fertilisation of ideas that is present and encouraged on the courses also is enhanced through the additional inclusion of InnovationRCA within the main building, blurring the boundary between the academic and the commercial. InnovationRCA provides business support and incubation services to help students and graduates protect and commercialise pioneering design-led technologies successfully.

项目概况

戴森楼由著名建筑师霍沃思·汤普金斯设计，是英国皇家艺术学院自从1962年迁到肯辛顿戈尔区以来最重要的新开发项目之一，它形成了皇家艺术学院巴特西校区的核心空间。

项目位于皇家艺术学院的油画与雕塑楼旁边，与“巴特西创意区”——目前拥有薇薇恩·韦斯特伍德、福斯特建筑事务所、威尔·艾尔索普等多家著名公司的总部。除了版画和摄影系学科项目之外，戴森楼还设有一个200座的演讲厅、一个公共画廊和一个学院产业孵化器。

建筑以“创意工厂”为概念进行设计，参考了安迪·沃霍尔艺术工厂的设计形式。建筑兼具美观性和功能性，由现场浇筑混凝土结构构成，形成了极富表现力的动感连锁空间。

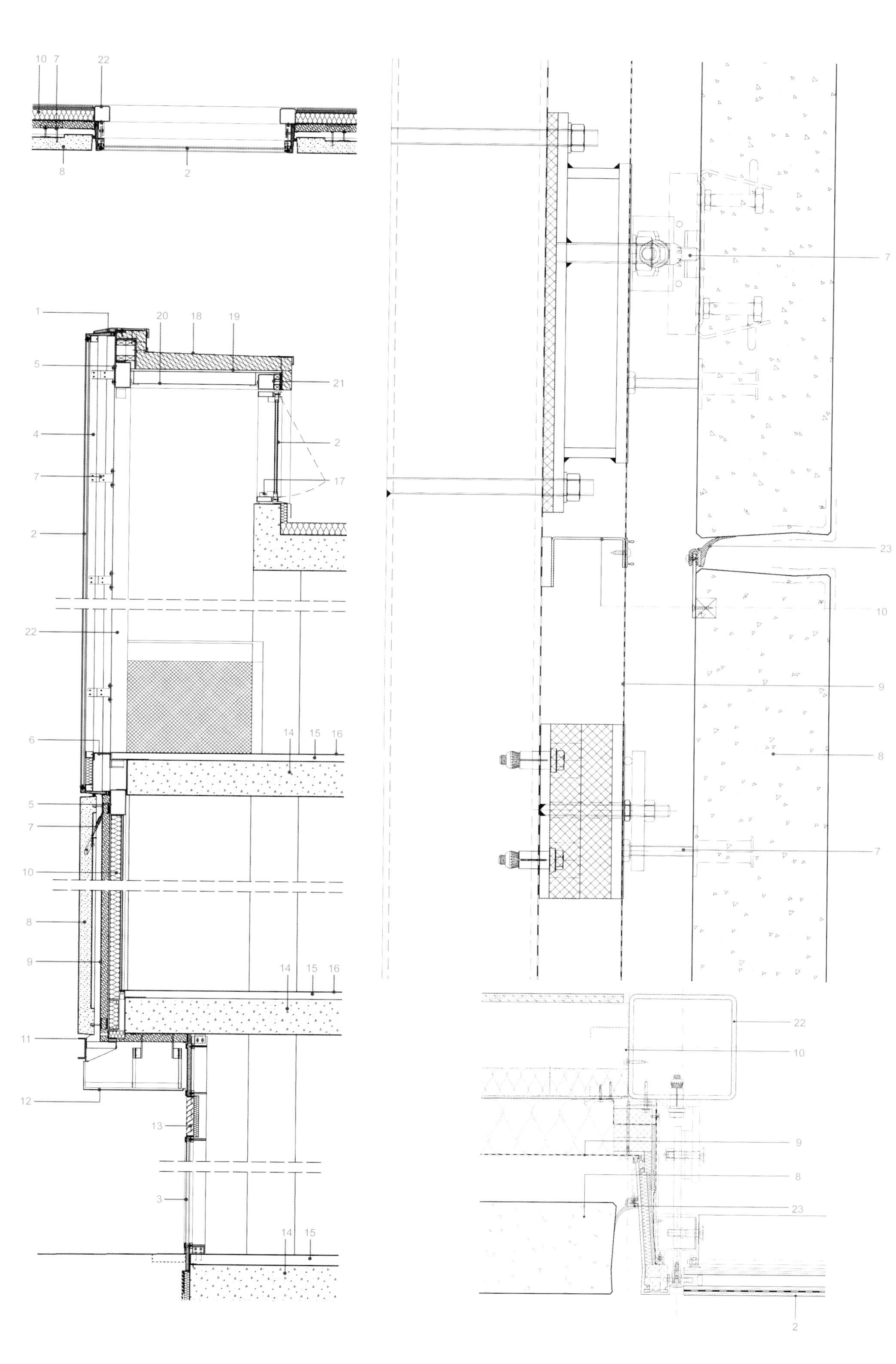

1. Aluminium capping
2. Double glazed curtain wall assembly with extruded aluminium framing
3. Double glazed curtain wall assembly with steel support profiles
4. Aluminium support framing
5. 250x150mm RHS beam
6. Painted steel floor plate
7. Hanger bracketry fixed to structural steelwork
8. Polished precast concrete panel
9. Breather membrane
10. Metal framed composite backing wall
11. Stainless steel channel
12. Aluminium soffit panel
13. Stainless steel louvres
14. 340mm reinforced concrete slab
15. 70mm screed topping
16. Linoleum sheet finish
17. Window opening actuator
18. Single ply roofing membrane
19. Metal deck
20. Structural liner tray
21. 150x150mm SHS beam
22. Steel frame
23. Flexible flipper gasket

1. 铝顶盖
2. 双层玻璃幕墙，装配挤制铝材框架
3. 双层玻璃幕墙，装配钢支架型材
4. 铝支架
5. 250x150mm RHS梁
6. 涂漆钢楼板
7. 悬吊支架，固定在结构钢件上
8. 抛光预制混凝土板
9. 透气膜
10. 金属框复合后墙
11. 不锈钢槽
12. 铝底板
13. 不锈钢遮阳
14. 340mm钢筋混凝土板
15. 70mm砂浆顶层
16. 油毡布饰面
17. 开窗控制器
18. 单层屋顶膜
19. 金属平台
20. 结构衬盘
21. 150x150mm SHS梁
22. 钢架
23. 可拆卸鳍状垫圈

一个巨大的天窗“厂房”将两座楼体，形成了建筑的核心，内设版画制作所需的大型压力机。建筑的临街面试更具公共特征的零售及商业空间，正对巴特西桥的一角成为了皇家艺术学院的入口，里面是画廊和演讲厅的门厅。

项目设计成功的主要特色在于各个空间设置之间流畅的联系。建筑在各个学科之间实现了“水平漂移”，所有创意流程都以明显的方式连接起来。学科之间创意的交汇与互动在皇家艺术学院创意中心的驱动下更能发挥作用。创意中心为皇家艺术学院的学生和毕业生提供支持服务，使得先锋设计技术成功地实现商业化。

Ibaiondo Civic Centre
伊拜昂多市民中心

Location/地点: Vitoria-Gasteiz, Spain/西班牙，维多利亚
Architect/建筑师: ACXT Architects
Photos/摄影: Josema Cutillas
Built area/建筑面积: 14,000m^2
Key materials: Façade – concrete prefabricated panel
主要材料：立面——预制混凝土板

Overview
Ibaiondo Civic Centre has a 14,000 sqm area and is located in Vitoria-Gasteiz (Spain). Sport, leisure and administrative services for neighbours at different parts of the city are joined together in these types of public buildings.

Once all interior functional, spatial and organisational requirements were defined, the project searched for an extroverted look to appeal the citizens, as to get the perception of the whole building to provide enough information of the public services to be provided there: theatre, leisure and sports swimming pool, solarium, café, indoor sports centre, library, workshops, council citizens help points, etc.

The project avoids forms of an elaborate façade composition, and shows itself as irregular and polyhydric, with a leisure personality. Because of such diversity at interior layouts, the exterior catches the citizen's eye, specially the polymer concrete façades, with a multidirectional groove to create an optical polychromatic illusion.

The building interior layout follows extensive and strict functionality criteria defined by the council technical team at competition phase. Sport services (swimming pool and indoor sports centre) are located to the north following a "Cartesian" geometry, due to their size and scale. So the rest of services are created to the south, with some sort of volumetric anarchy facing the residential area. Other uses are organised along a corridor separating and linking together different services. From this corridor, through glass enclosures, the visitor can recognise the different activities inside the building, as a suggestive "showroom".

Energy sustainability in the building is ratified by a high energy efficiency qualification, obtained by ensuring good thermal isolation and high equipment performances. Also an approximate 700 sqm area of solar thermal collectors provide energy to heat water for both swimming pool and building hot running water. This dedicated design generates an estimated C02 emissions saving of up to 1,900 Ton.

项目概况

位于西班牙的维多利亚市的伊拜昂多市民中心总面积14,000平方米，在公共空间内汇集了体育、休闲、行政服务等市民服务设施。

在确定了内部功能、空间和组织需求之后，项目开始寻求一个能够吸引市民的外观。建筑外观应当充分展示出建筑内部所提供的公共服务设施：剧场、休闲运动游泳池、日光浴室、咖啡厅、室内体育中心、图书馆、研讨工作室、市民服务问讯处等。

项目并没有选择精致的立面组合，而是将其呈现为不规则的多元结构，显得随意而轻松。由于内部功能布局的多样化，建筑外观十分引人注目，特别是聚合混凝土立面结构，多向的凹槽让建筑表面呈现出五光十色的视觉效果。

建筑内部布局严格遵守议会技术团队在竞赛设计阶段所提出的功能准则。体育设施（游泳池和室内体育中心）位于北侧，采用了笛卡尔坐标系方式进行布局。其他服务设施位于南侧，朝向住宅区，在空间设置上显得略为凌乱。其他功能区沿着走廊设置，将不同的服务区连接起来。访客可以通过这条走廊的玻璃窗辨识出建筑内部的不同活动区，整个走廊起到了“展览厅”的作用。

良好的隔热性能和设备效能使建筑实现了高度节能和可持续设计。总面积约700平方米的太阳能电池板能够为游泳池和内部用水提供加热。建筑的节能设计所节约的二氧化碳排放量可高达1,900吨。

KIROLA
TEATRO

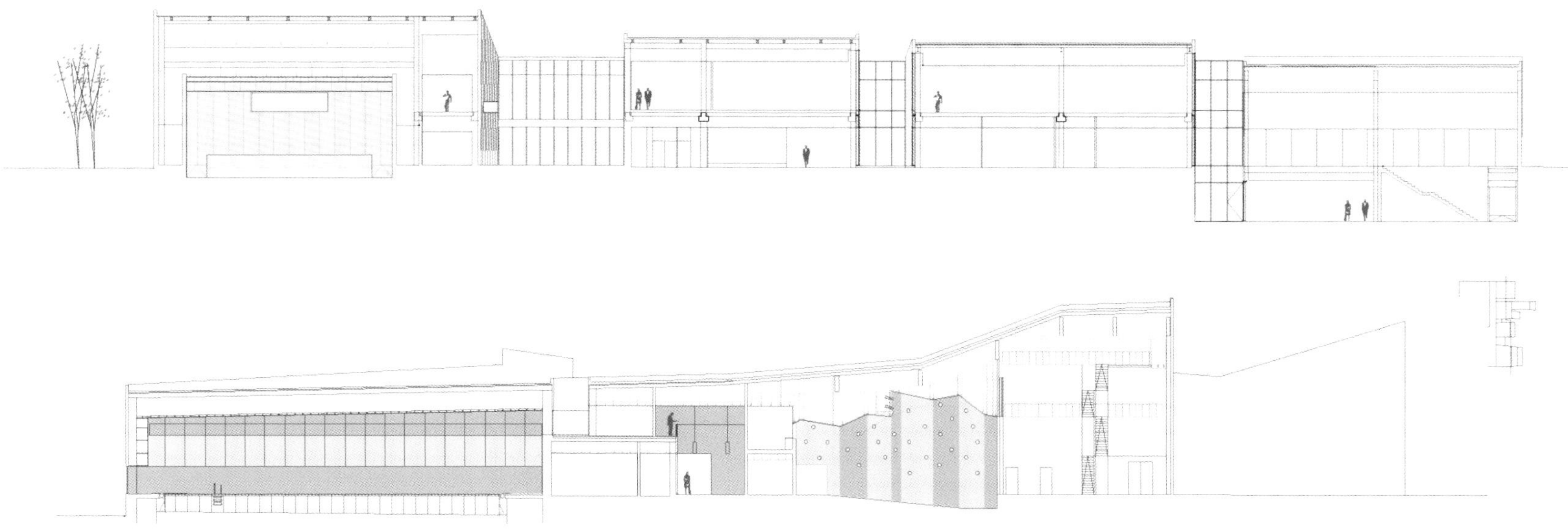

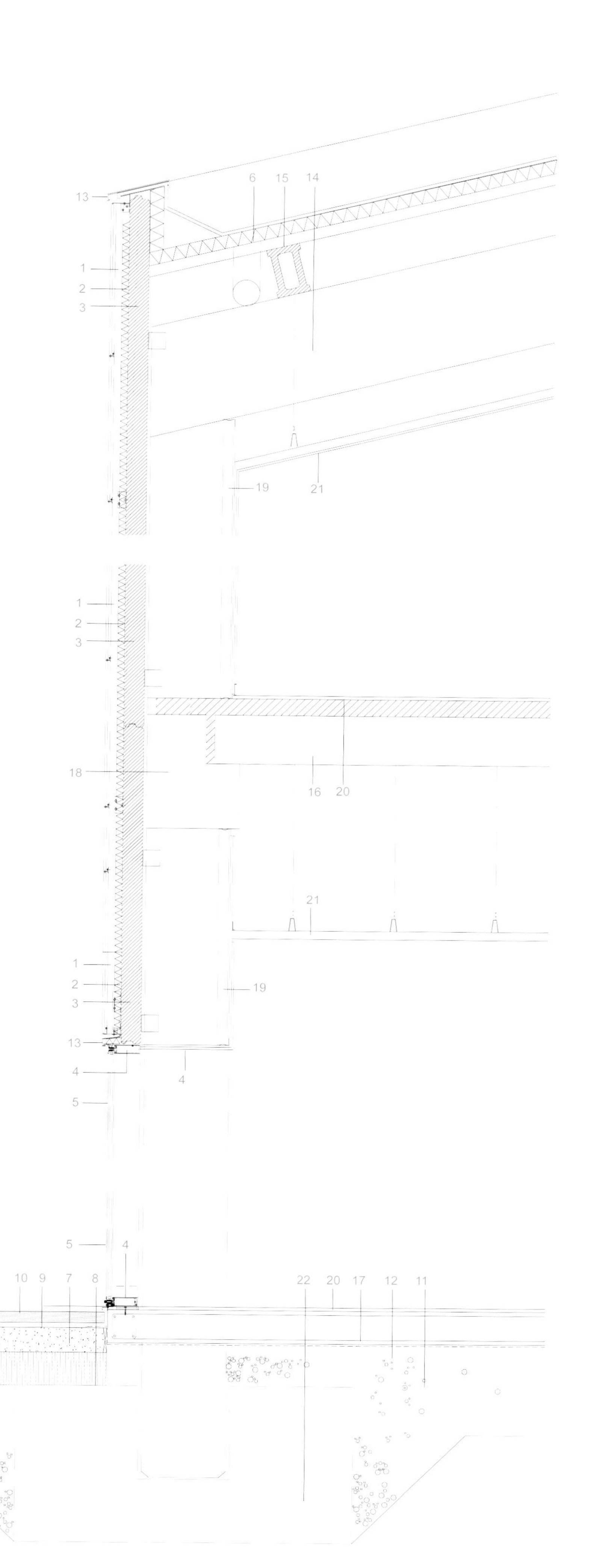

1. Cavity-wall façade of concrete polymer panels 14mm
   Aluminium profile 110mm
2. Polyurethane insulation 50mm
3. Concrete prefabricated panel 120mm
4. Glazed aluminium curtain wall
5. Double glass with low-E 70. 4+4.16.6
6. Roof decking system 150mm:
   Galvanised steel ribbed panel
   Rockwool panel insulation 80mm
   PVC laminate 3mm
7. Reinforced concrete bases
8. Fill of graded crushed aggregate
9. Mortar cement
10. Mass slab floor
11. Crushed rock
12. Polyethylene waterproof laminate
13. Flashing steel galvanised 1mm
14. Prefabricated concrete beam 20x60mm
15. Prefabricated concrete Tie-beam 300mm
16. Prefabricated concrete slab 300+50mm
17. Concrete base with steel fibres reinforced 15cm
18. Prefabricated T concrete beam
19. Concrete brick wall 15cm width
20. Industrial epoxy resin, over smoothed concrete base 5mm
21. False ceiling gypsum board
22. Concrete strip footing

1. 混凝土聚合板空气墙立面14mm
   铝型材110mm
2. 聚氨酯隔热层50mm
3. 预制混凝土板120mm
4. 铝框玻璃幕墙
5. 双层低辐射玻璃70. 4+4.16.6
6. 屋顶板系统150mm：
   镀锌钢棱纹板
   石棉板隔热80mm
   PVC层压板3mm
7. 钢筋混凝土底座
8. 碎骨料填充
9. 砂浆水泥
10. 大块楼板
11. 碎岩石
12. 聚乙烯防水层压板
13. 镀锌钢防水板1mm
14. 预制混凝土梁20x60mm
15. 预制混凝土连接梁300mm
16. 预制混凝土板300+50mm
17. 混凝土底座，钢纤维加固，15cm
18. 预制混凝土T形梁
19. 混凝土砖墙，15cm宽
20. 工业环氧树脂，涂于光面混凝土底座上，5mm
21. 石膏板假吊顶
22. 混凝土条形脚线

CaixaForum Zaragoza

萨拉戈萨文化中心

Location/地点: Zaragoza, Spain/西班牙，萨拉戈萨
Architect/建筑师: Estudio Carme Pinos
Photos/摄影: Ricardo Santonja, Estudio Carme Pinós
Site area/占地面积: 9,390m^2
Built area/建筑面积: 7,062m^2
Completion date/竣工时间: 2014
Key materials: Façade – prefabricated concrete panel and perforated aluminium sheet
主要材料： 立面——预制混凝土板和穿孔铝板

Overview

The architects start the project by posing two challenges: The first: to design a building that can "feel like a city" – both due to its uniqueness and to the public spaces it generates. The second: to design a building which connects with distant perspectives when walking around, providing at the same time introspection when inside its exhibition halls. In other words, a building which "feels like a city" and which makes us feel part of it when we inhabit it.

The architects solve these two challenges by raising the level of the halls. This allows freeing the ground floor, where they place the more open and transparent spaces: the lobby and the store. The aim is to create public spaces, make the park extend into the city by passing under the building – a space which is lit at night with drawings obtained by perforating the plate, which in addition hides the structure supporting the elevated halls.

Below the raised halls the architects place a semi-underground garden that serves as the exit to the auditorium and which can also be considered an anteroom or outdoors catering area. Thus, the auditorium – located underground and accessible through the lobby – can be considered as halfway underground and directly connected to the city thanks to the garden.

The two suspended halls face each other at different levels in a way that the visitor who exits one hall has a view of the city below the other hall. The architects believe decompres-

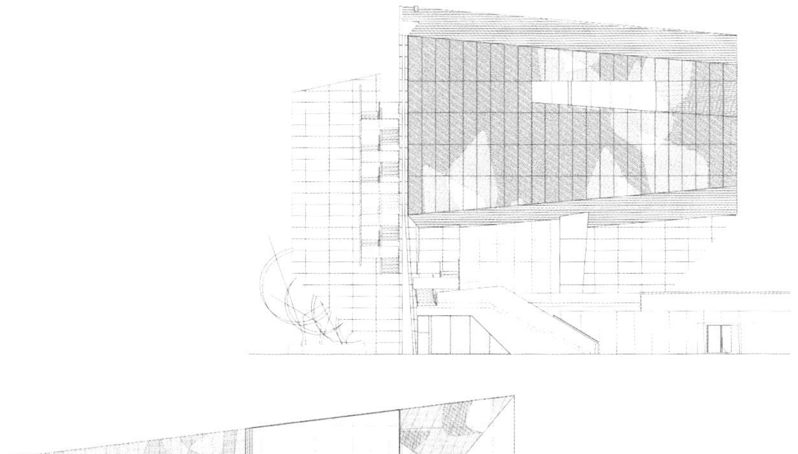

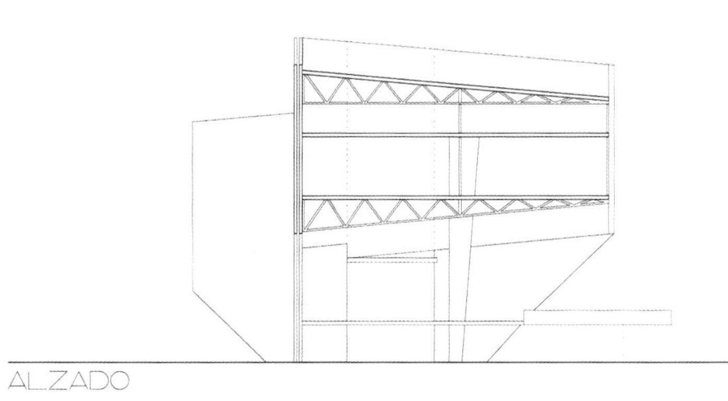
ALZADO

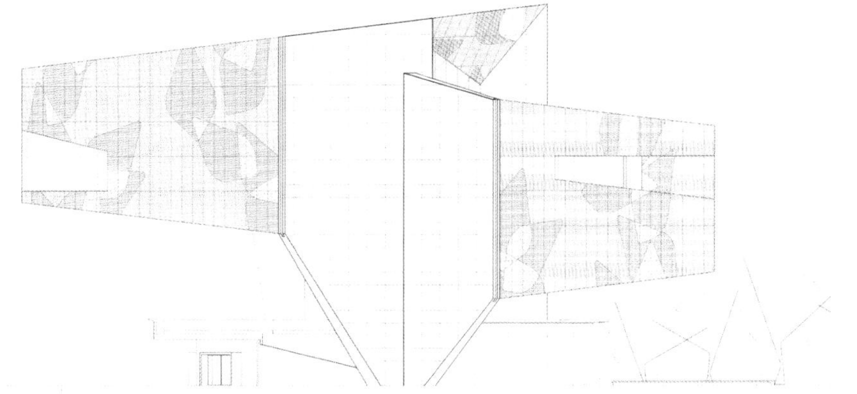

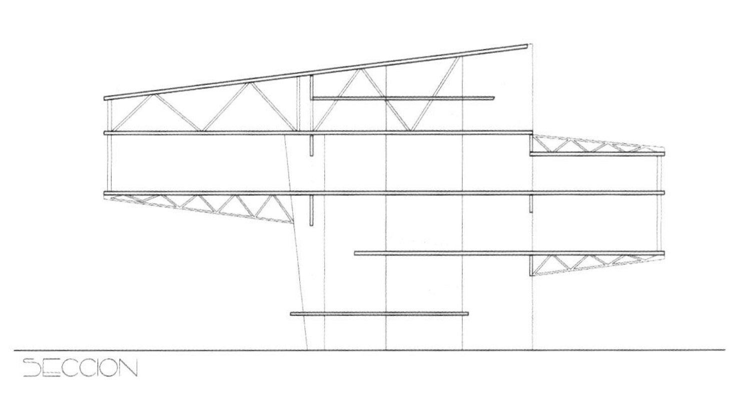
SECCION

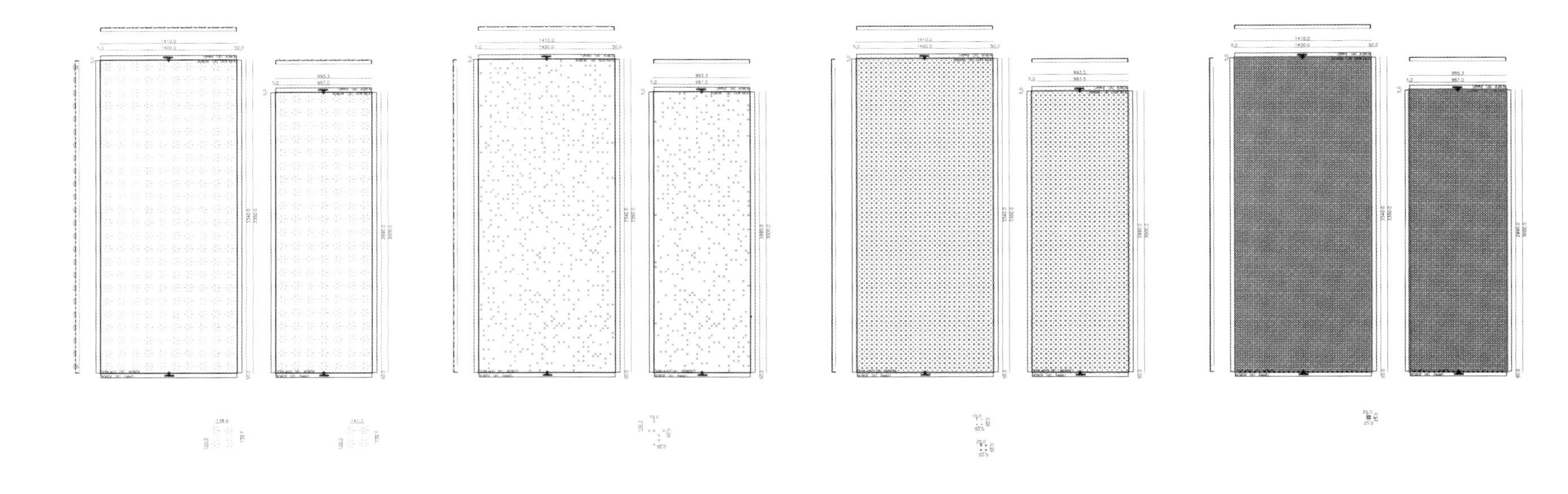

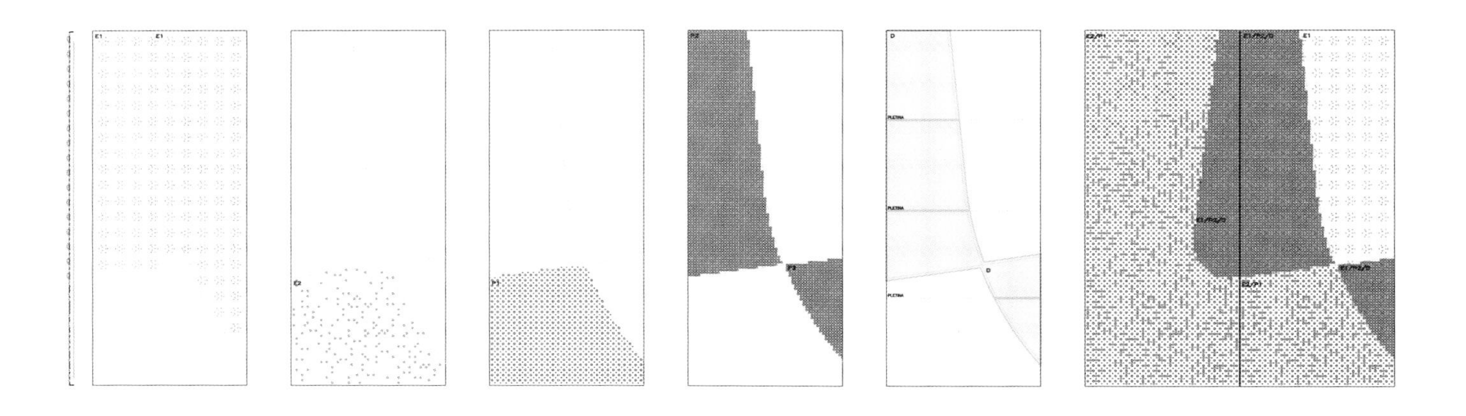

sion and relaxation areas are necessary between both halls – i.e., between exhibitions. With this aim, the two halls are connected by escalators, offering a journey which allows us to enjoy the distant views – completely different from the decontextualisation produced by elevators and which cannot offer the visitor any type of decompression.

On the upper part of the building and with views to the city are the coffee shop and the restaurant. Opposite to them and created by the different levels between halls, there is a terrace-bar which – keeping with the indoors restaurant – allows to enjoy fantastic views of the Ranillas meander and Expo Zaragoza.

Thanks to its unique and feasible structure, the building appears as a sculptural element amidst the park. The architects want the building to become a symbol of the progress of the technique and generosity of culture – a reflection of only the best things of our times.

项目概况

建筑师在项目设计中面临两个主要的挑战：一是要设计一座“感觉像一座城市”的建筑，独特并能提供公共空间。二是设计一座与远景相联系的建筑，为展览厅提供丰富的外部视野。换句话说，就是一座既“像城市一样”，又让我们能感到置身其中的建筑。

建筑师通过提升展览厅的高度来解决这两个问题。这种设计让一楼解放出来，配置了更开放通透的空间：大厅和商店。设计的目的是打造公共空间，让公园穿透建筑，拓展到城市中。建筑立面的穿孔图案在夜间被点亮，把支撑展览厅的结构基础隐藏起来。

建筑师在展览厅下方设置了一个半地下式花园，作为通往礼堂的入口。同时，花园也可以被用作前厅或户外就餐区。这样一来，位于地下的礼堂（也可由大厅进入）就能通过花园与城市直接联系起来。

两个悬浮的展览厅错落相对，游客们从一个厅出来能看到另一个厅。建筑师认为有必要在两个展览厅之间设置减压和休闲区。因此，两个展览厅之间由自动扶梯相连，人们可以在扶梯上欣赏远处的风景，这种感觉是厢式电梯所无法提供的。

在建筑的上半部分坐落着可以俯瞰城市风景的咖啡厅和餐厅。与它们相对的是由错层展览厅形成的平台休闲吧。人们可以在平台上欣赏莱尼拉斯散步路和萨拉戈萨世博会园区的美景。

独特而实用的结构造型让建筑看起来就像是公园里的一座雕塑。建筑师希望它能成为技术进步和文化发展的象征，只反映出我们这个时代最好的事物。

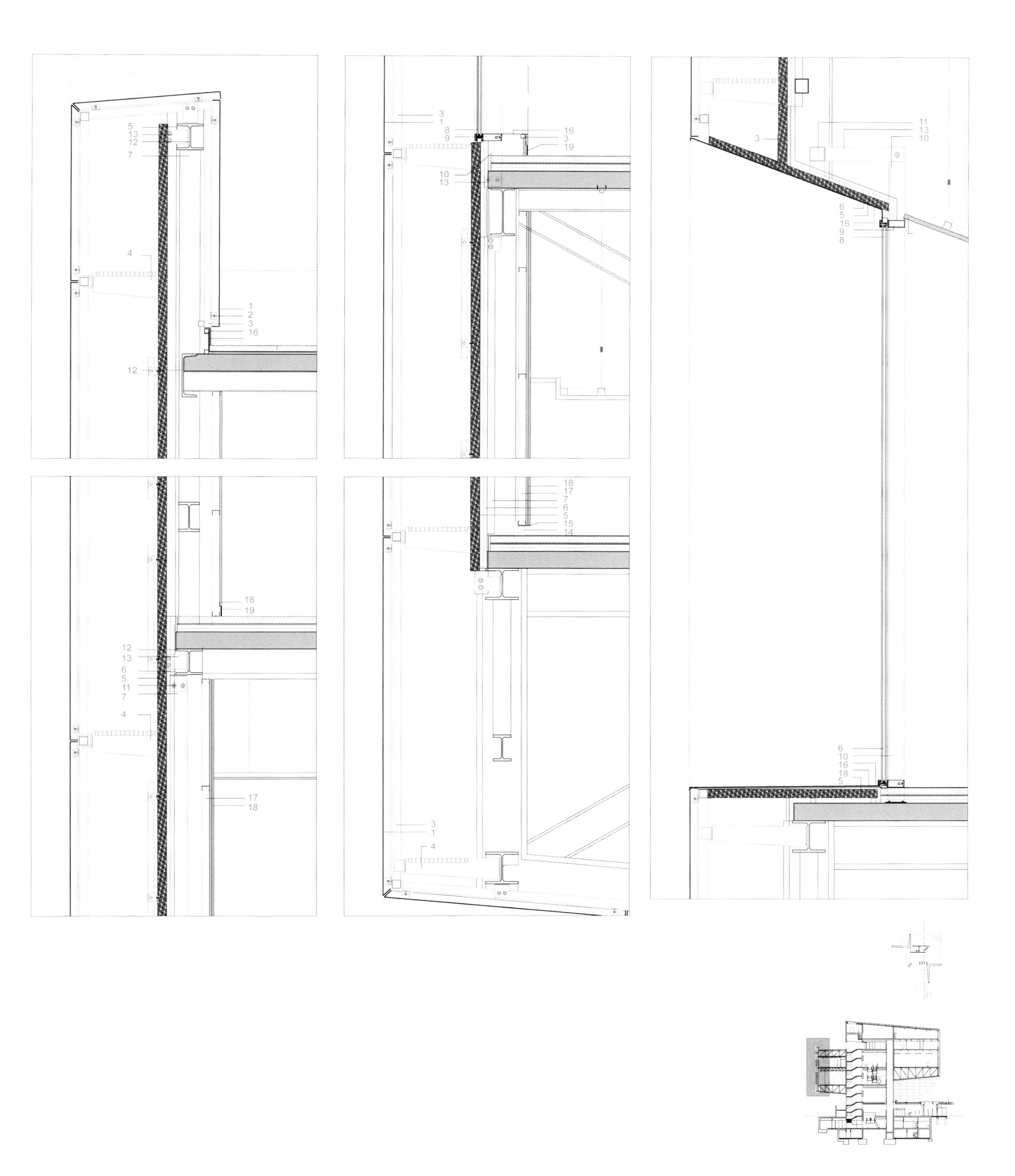

5
13
12
7
4
1
2
3
16
12
3
1
8
9
16
3
19
10
13
3
11
13
10
6
5
16
9
8
18
19
12
13
6
5
11
7
4
17
18
18
17
7
6
5
15
14
3
1
4
6
10
16
18
5

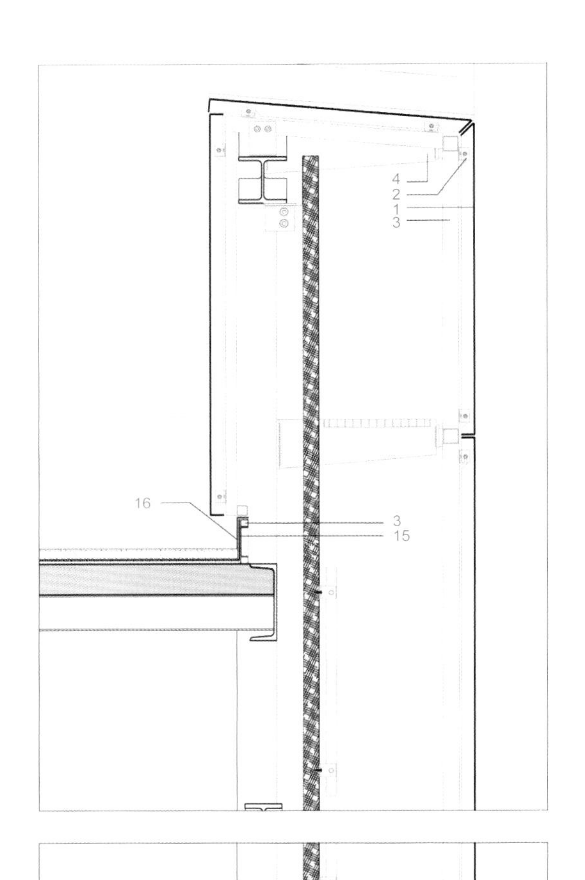

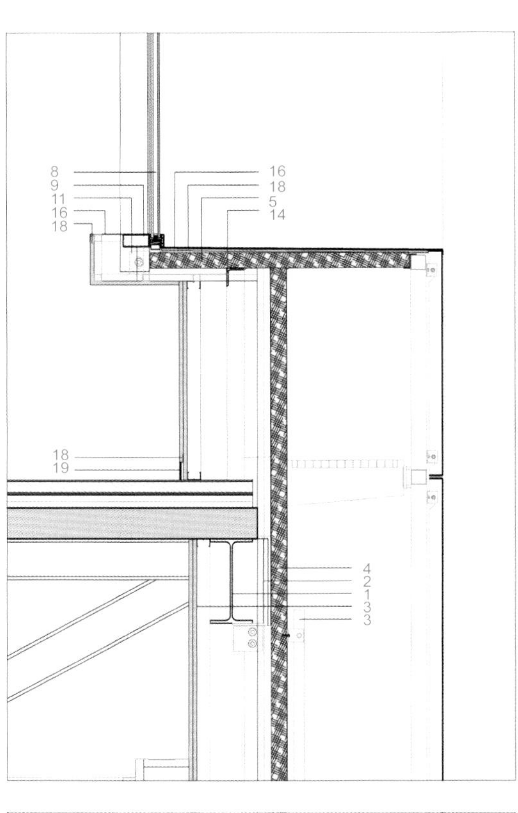

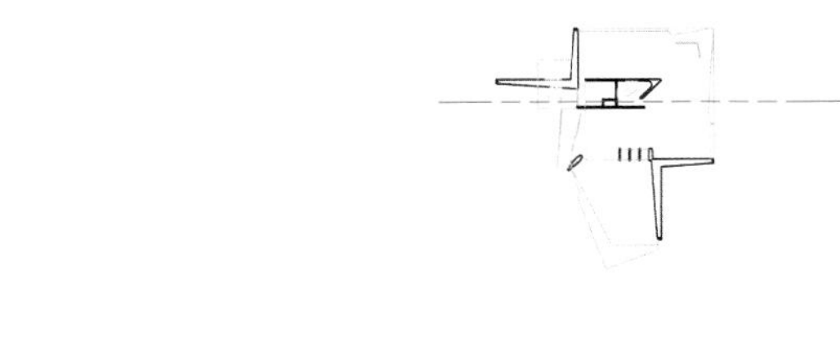

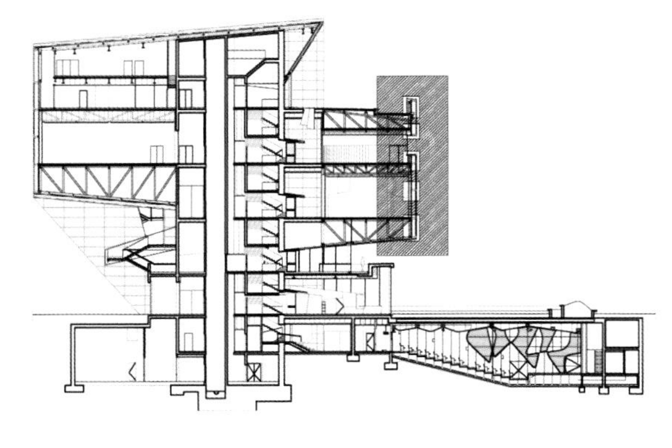

15
21
5
6
12
4
5
16
18
16
13
9
8

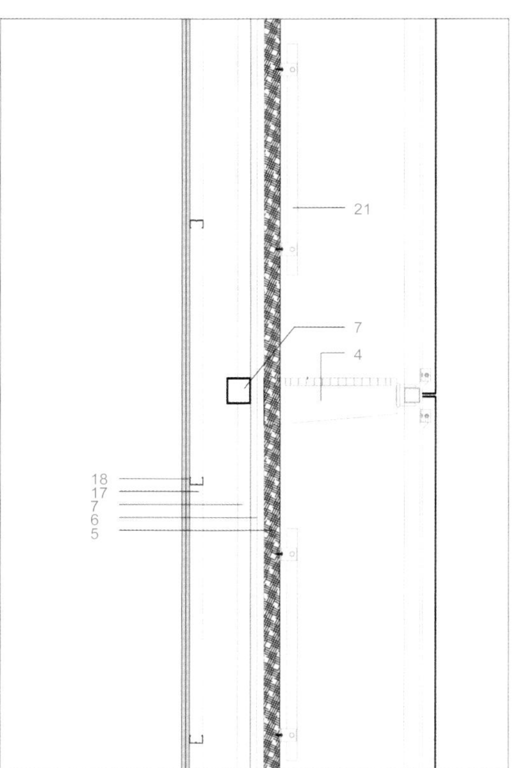

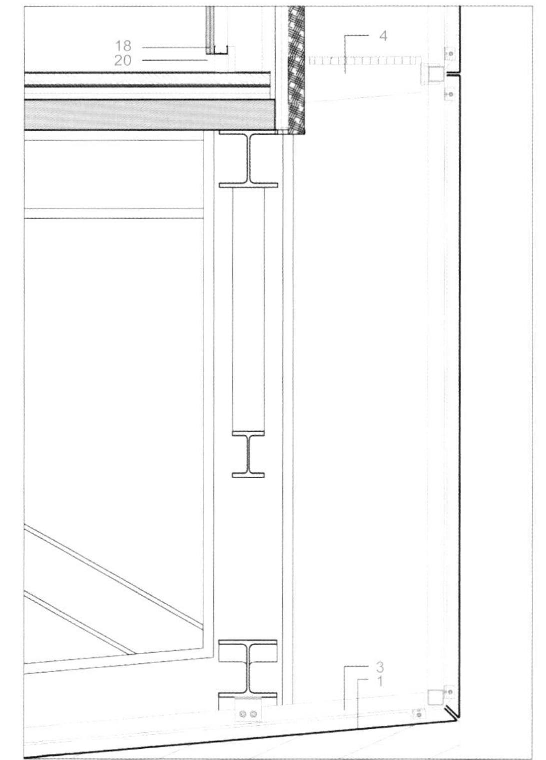

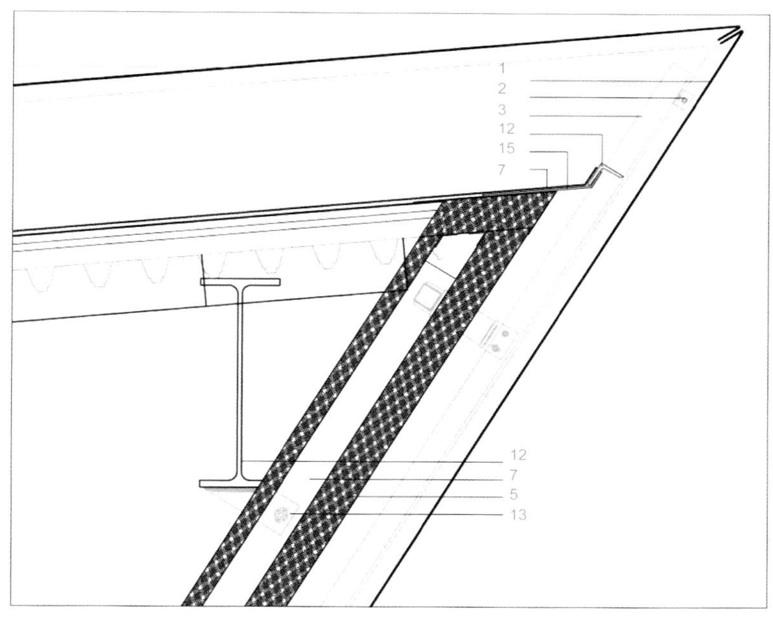

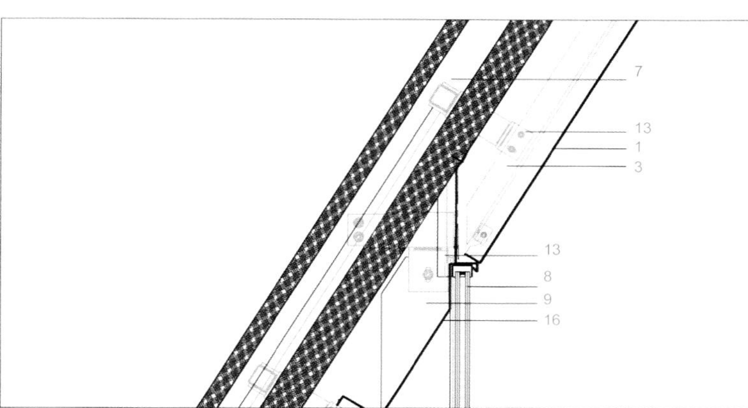

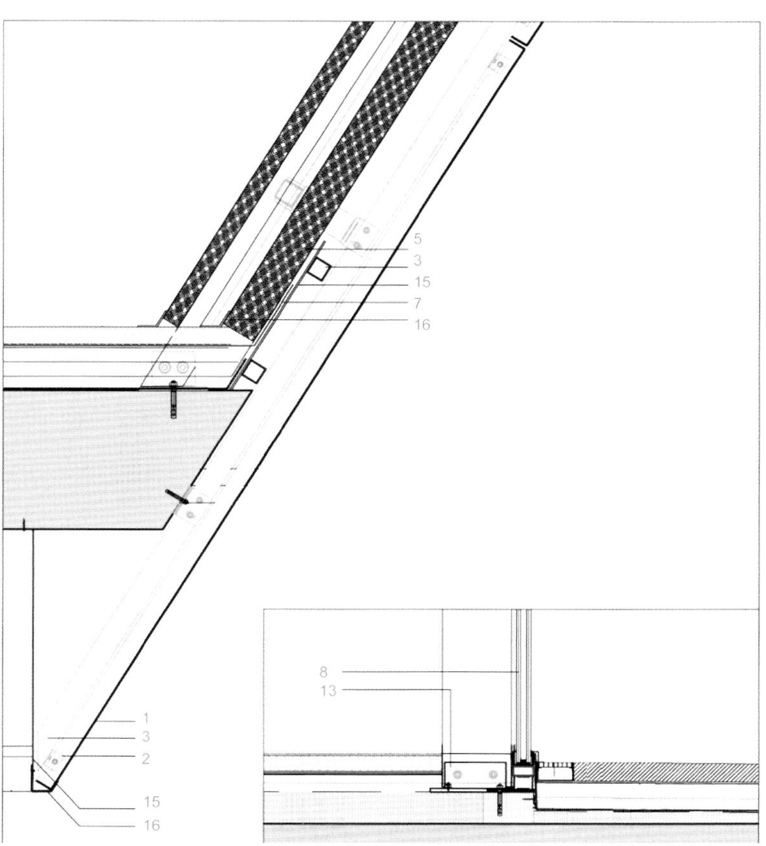

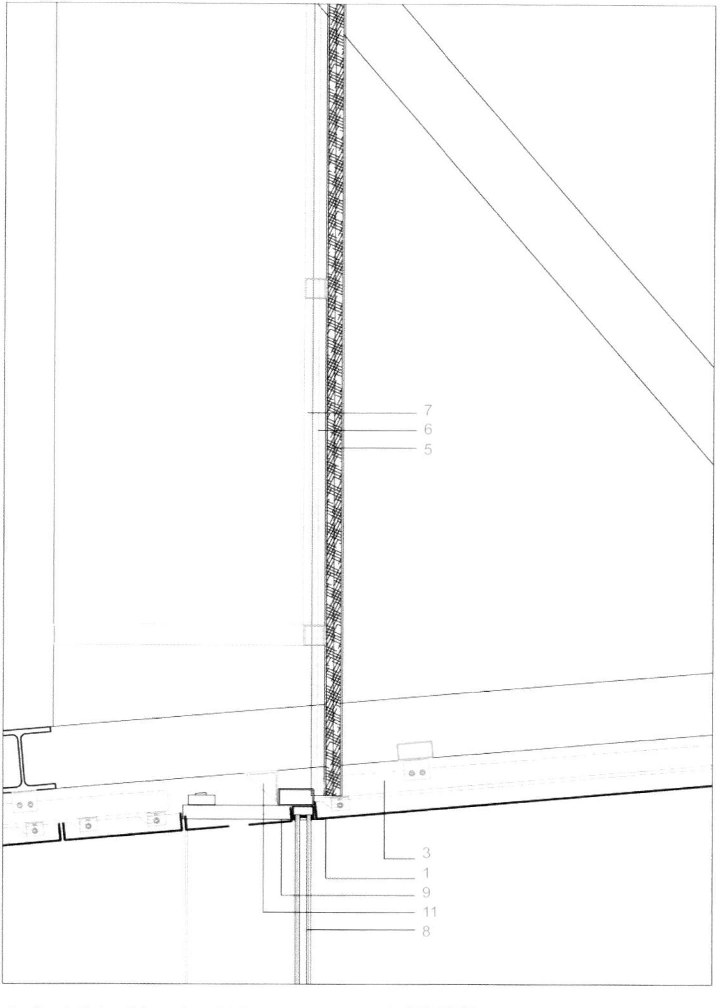

1. Perforated aluminium sheet E. 3mm
2. Fixing plate
3. Auxiliary steel tube structure
4. Structural gusset
5. Sandwich panel E. 60mm
6. Rockwool E. 50mm
7. Auxiliary structure, galvanised steel facade
8. Glass chamber. Curtain wall
9. Lacquered aluminum curtain wall
10. Laminated steel curtain wall
11. Fixing angular profile
12. Normalised profile
13. Fixing auxiliary structure
14. Normalised steel profile
15. Galvanised steel sheet E. 3mm
16. Painted aluminum plate E. 3mm
17. Plasterboard profiles
18. Plasterboard
19. Painted socket dm
20. Electrified socket
21. LED luminaire

1. 穿孔铝板E. 3mm
2. 固定板
3. 辅助钢管结构
4. 结构角板
5. 夹心板E. 60mm
6. 石棉E. 50mm
7. 辅助结构，镀锌钢立面
8. 玻璃腔；幕墙
9. 涂漆铝幕墙
10. 层压钢幕墙
11. 固定角材
12. 标准型材
13. 固定辅助结构
14. 标准钢型材
15. 镀锌钢板E. 3mm
16. 涂漆铝板E. 3mm
17. 石膏型材
18. 石膏板
19. 涂漆插槽
20. 电气插槽
21. LED灯

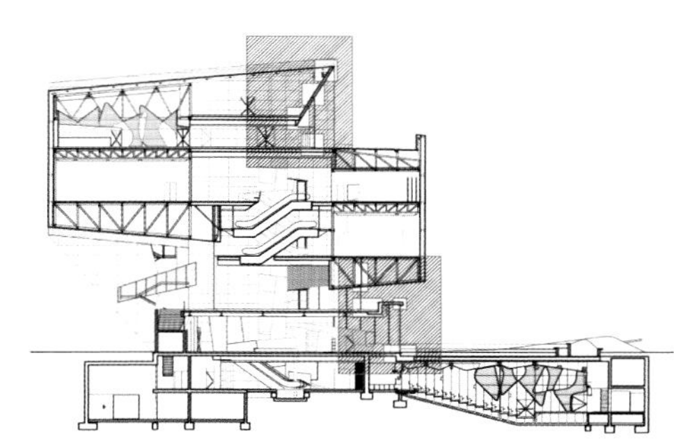

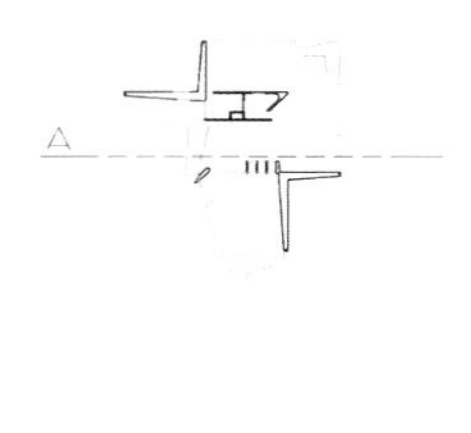

Espai Ridaura

瑞德拉公共中心

Location/地点: Santa Cristina d'Aro, Girona, Spain/西班牙，吉罗那
Architect/建筑师: Capella Garcia Arquitectura
Photos/摄影: Rafael Vargas
Built area/建筑面积: 1,215m²
Completion date/竣工时间: 2011
Key materials: Façade – precast concrete panels, colored and repellent treatment
主要材料： 立面—— 预制混凝土板，彩色和防水处理

Overview
This is a multi-purpose facility in Santa Cristina d'Aro, Girona, in the district of Baix Empordà. Santa Cristina d'Aro is situated close to the coast, in the upper reach of the Vall d'Aro, and the river Ridaura flows through it close to the town centre. The building stands on a triangular site of about 5,600 square metres, classed as being for public facilities. It occupies part of the site, while the remainder is left free and planted with trees as an area for waiting or strolling, so that the flows of people entering or leaving can be handled without difficulty. The site also has enough surface parking nearby to cope with the demand when there are performances.

During the design process, close attention was paid to the integration of the building into the landscape. The building is set against the backdrop of the leafy woodland that lines the banks of the Ridaura torrent, with the mountainside in the distance.

The municipality is growing as a result of the large amount of tourist activity during the summer months, which means that it already possesses considerable leisure and cultural facilities. Nonetheless, the facilities in Santa Cristina were very basic and quite a few years old, as well as being at some distance from the town centre. So much so that the concerts of the Santa Cristina International Chamber Music Festival took place in the parish church.

It therefore became essential to build a new facility that could accommodate the rich cultural and leisure life of the town in a venue that would meet the requisite standards of

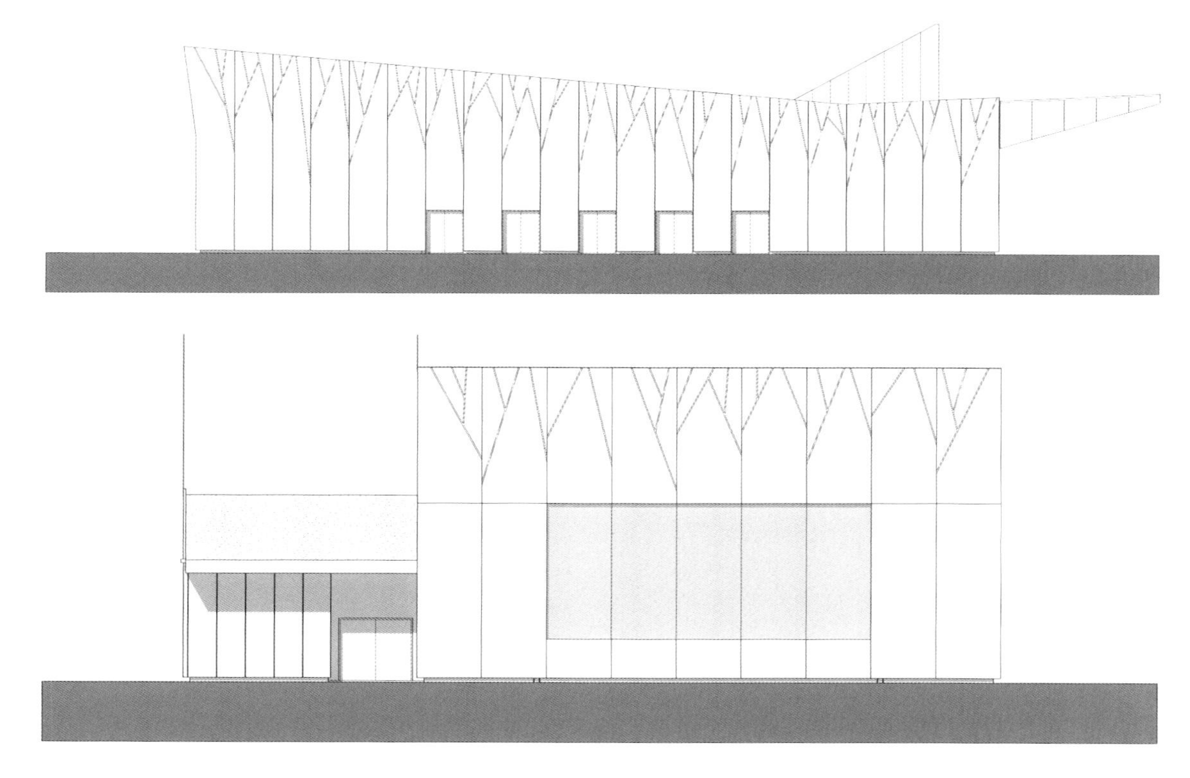

quality, with sufficient capacity, and equipped with the latest technology, making it possible to programme both professional performances and local gatherings.

The Espai Ridaura therefore provides a multipurpose space with room for 345 seated spectators (1,286 standing), suited to the fundamental functions of a small auditorium – stage performances, dances, small trade fairs, lectures, small conferences, etc. A system of fold-away seating enables the level central space to be cleared, making it possible even for vehicles to enter.

Detail and Materials

Because this is a public facility, it represents the community. Both the façade and the outer skin have to reveal this function, differentiating it from the architecture of private housing, as well as that of industry and commerce.

The main entrance is sheltered by a large projecting canopy which emphasises the entrance and adds symbolic character to the building, creating a large covered area that is protected from the sun and the rain.

The entire roof employs the environment-friendly "green roof" system, which as well as maintaining a rich and varied vegetation, provides excellent acoustic and thermal insulation for the interior, as well as requiring minimal maintenance. There are also thermal solar panels to generate hot water, and a low-consumption lighting system. Large skylights light the central passageway.

Both elements of the building employ a standardised system of prefabricated panels that are individualised by patches of earthy pigments in the same range of colours as those of the land on which the building stands, and which do not require subsequent maintenance.

Although the building is intended to be multifunctional, its use as an auditorium was regarded as primordial; for this reason excellent acoustic insulation has been provided. Furthermore, the materials used in the interior have been chosen to ensure the appropriate acoustic response with regard to reverberation time for music. In addition, all the necessary lighting, sound and stage equipment has been installed to meet the needs of the whole range of intended activities.

项目概况

这是一个位于西班牙吉罗那市圣克里斯蒂那达罗的多功能公共设施。建筑位于一块总面积约5,600平方米的三角形场地上，占据了部分场地，其他空间则种植着一些树木，方便人们在楼外等候或散步，保证了进出人流的流畅性。场地附近有充足的露天停车空间，能够满足演出时的停车需求。

建筑师在设计阶段十分注重建筑与景观的融合。建筑后方是河岸边茂密的树林，远处则以高山为背景。

随着夏季游客的数量剧增，圣克里斯蒂那原有的休闲文化设施已经不能满足城市的需求。一些设施十分基础，一些略显陈旧，还有一些则距离市中心较远。甚至连圣克里斯蒂那国际室内音乐节都不得不在教区教堂内举行。

因此，城市有必要新建一座能够满足丰富的文化和休闲生活需求的新设施。该设施必须满足高品质标准，具有充足的容量，配有先进的技术，能够举办专业演出并为本地居民的日常聚会提供空间。

瑞德拉公共中心拥有可容纳345个坐席（可站立1,286人）的多功能空间，能够举办小型演出、舞蹈表演、小型交易会、演讲、小型会议等活动。这些座椅系统能够将中央空间清空，即使汽车也能照常进入。

细部与材料

作为一座公共建筑，它代表着社区。因此，建筑立面和表皮都必须体现它的功能，使其与私人住宅、工业建筑以及商业建筑区分开。

建筑主入口处有一个向外伸出的屋顶，既突出了入口区域，又增加了建筑的象征意义，形成了一块能够遮风挡雨的巨大区域。

整个屋顶全部采用了环保的“绿色屋顶”系统，通过丰富的植被为室内提供了优秀的隔音隔热效果，同时也无需过多的养护。同时，项目还设有太阳能板来加热热水，采用了低能耗照明系统。巨大的天窗能照亮中央走廊。

建筑元件采用标准的预制板系统，上方添有一些土色染料补丁，其色彩与周边的土地色系一致，无需额外的养护。

虽然建筑具有多重功能，它的主要功能还是礼堂。因此，必须配备优秀的隔音效果。此外，考虑到音乐的混响时间，室内装饰材料保证了合适的声波效应。礼堂内安装了全套必要的照明、音响和舞台设备。

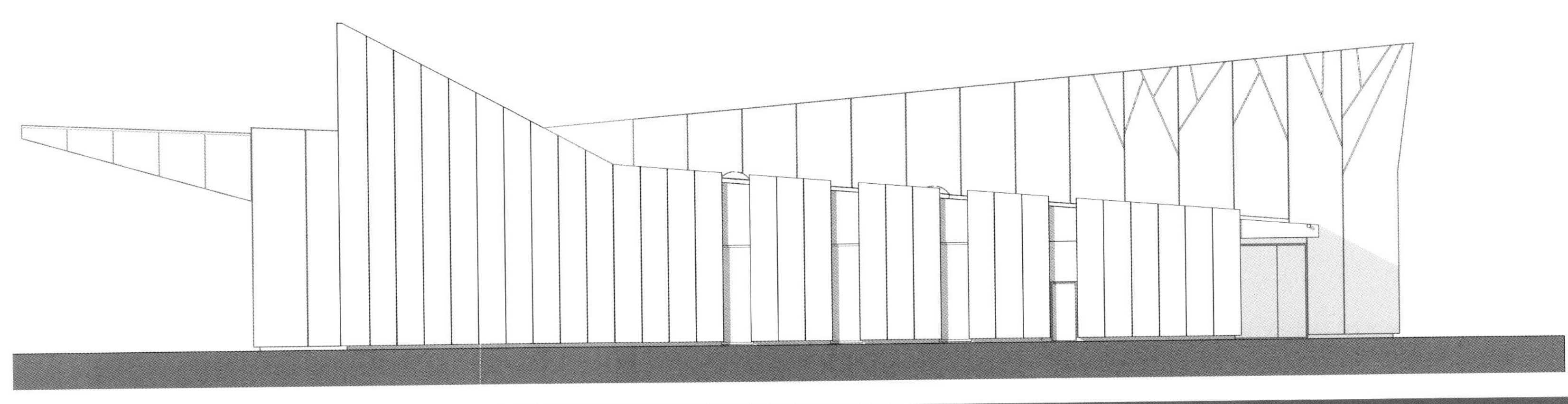

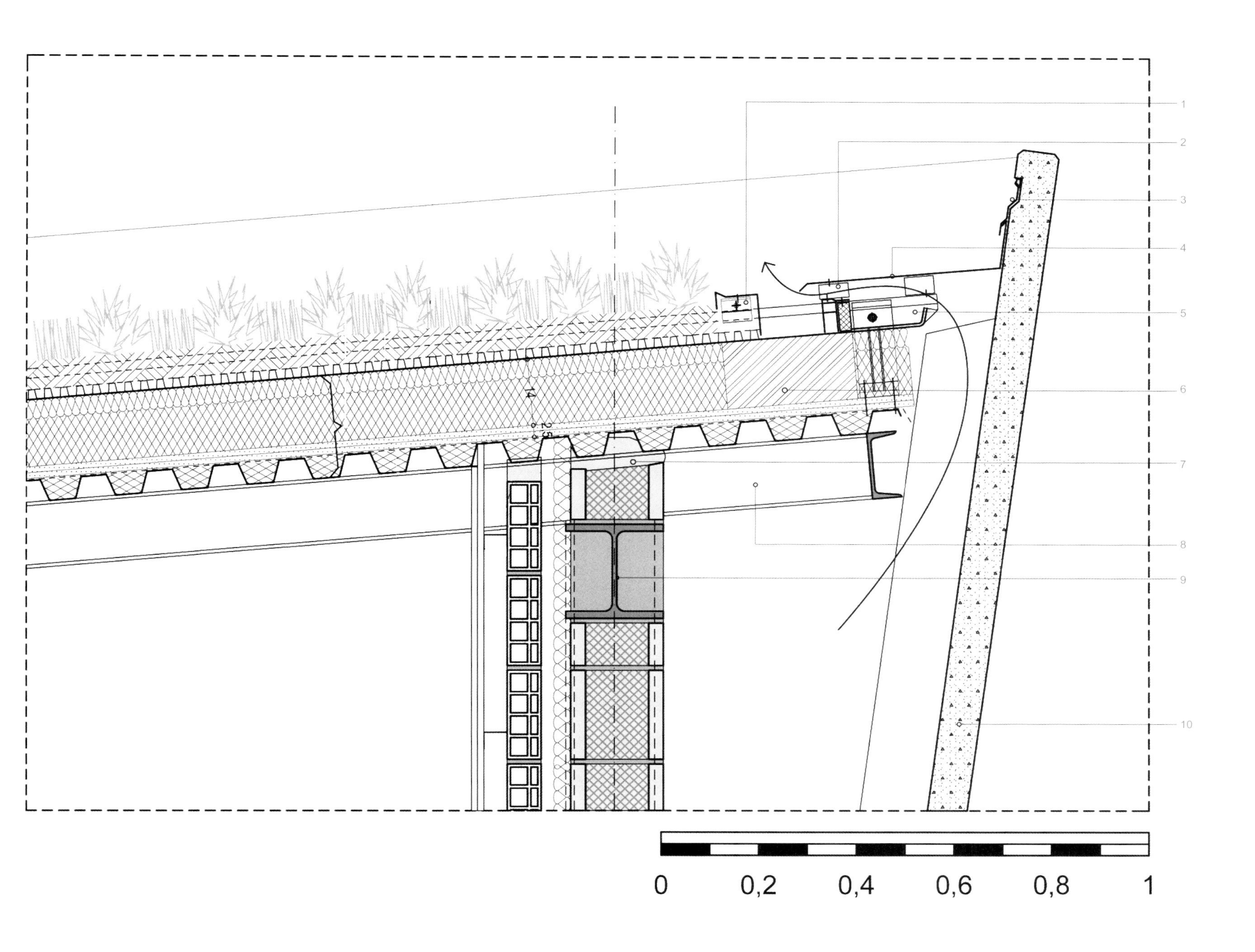

1. Aluminum clamp
2. Separador
3. Sealed aluminum finish
4. Aluminum sheet finish
5. Kalzip tray
6. Core insulation steel
7. Caulking foam of projected polyurethane, thickness min.: 3 cm
8. IPE belt 140 C/~ 150 cm
9. Fireproof metal truss
10. Prefabricated panels of reinforced concrete
    colored mass and repellent treatment
    fixed to the slabs

1. 铝制夹具
2. 隔断
3. 密封铝制表层
4. 铝板表层
5. Kalzip托盘
6. 钢制绝缘核心
7. 嵌缝泡沫聚氨酯，最小厚度3cm
8. IPE带140 C/~ 150 cm
9. 防火金属桁架
10. 钢筋混凝土预制板
    染色和防水处理
    固定到楼板上

Linx Hotel International Airport Galeao

加利昂国际机场林克斯酒店

Location/地点: Rio de Janeiro, Brasil/巴西，里约热内卢
Architect/建筑师: OSPA Arquitetura e Urbanismo
Photos/摄影: Marcelo Donadussi
Site area/占地面积: 4,800m²
Built area/建筑面积: 7,528.44m²
Completion date/竣工时间: 2013
Key materials: Façade – pre-fabricated concrete panels
主要材料： 立面——预制混凝土板

Overview

This is a project of an executive hotel, located near the Antonio Carlos Jobim International Airport (Galeão), into Infraero concession zone. The undertaking arises from World Cup and Olympic Games lodging demand in Rio de Janeiro before, during and after these events.

The hotel's main purpose is to serve transit passengers, business people and tourists which demand a quick connection to the airport. With 7,528.44sqm of floor area in a 4,800sqm lot, the building is composed of two blocks: one for the apartments, another for social and service areas.

According to airport zone height restrictions, the accommodation edifice is extended to the ground floor, optimising the available lodging units in 162 apartments. Besides the standard apartment, in both ends of every corridor two larger apartments are positioned, which can be united through the main circulation without losing each one's privacy.

The smaller volume is composed of two levels. In the front is located the porte-cochére, leading the guests to the main social areas, e.g. the lobby, the restaurant, the snack bar and the main conference room. In the first floor are found the gym, the bar, meeting rooms and a terrace, where is located a two-lane lap pool.

The building was fully designed with industrial elements, with precast concrete structure and prefabricated elements in order to optimize the construction runtime. The project combines energy efficiency, rational use of natural resources and automated operating systems.

Detail and Materials

From the beginning, one of the most important needs of the project was its fast construction.

Pre-fabricated elements were an obvious option, and they turned out to have an important role on the materialization of the concept of the hotel chain. In Brasil, the use of reinforced concrete was rapidly spread in the beginning of the last century and has played an important role on the search of the architectural identity of the country. Nowadays the material is still protagonist, and its application links the building with a vernacular Brasilian construction. With the definition of the use of pre-fabricated concrete panels for the façades, it was possible to choose a group of colours that would link the project not only with its country, but also with its natural surroundings.

项目概况

项目是一家紧邻加利昂国际机场的行政酒店，酒店隶属于机场特许管理区。该酒店的设置旨在满足里约热内卢在世界杯及奥运会期间的住宿需求。

酒店的主要目标是服务转机乘客、商务人士以及需要快速到达机场的游客。酒店总面积7,528.44平方米，占地面积4,800平方米，由两座楼组成：一座为客房楼，另一座是社交和服务空间。

根据机场区高度限制，客房楼尽量横向延伸，优化了可用的客房空间，共设置了162套客房。除了标准间之外，走廊两端各有两间较大的客房，这两间客房可以通过主通道联合起来，同时又不失各自的私密性。

较小的服务楼分为两层，楼前有一个门廊，引导着宾客进入主社交区，如酒店大堂、餐厅、小吃吧和泳池边的露台。

建筑完全采用工业元素建造设计，预制混凝土结构和预制构件大幅缩短了施工时间。项目结合了节能、理性使用自然资源和自动运营系统等高科技手段。

细部与材料

项目最主要的需求就是快速施工。预制构件显然是首选，它们在实现酒店设计概念的过程中扮演了重要的角色。在巴西，钢筋混凝土的使用从上世纪开始迅速扩张，混凝土在巴西建筑形象的塑造过程中起到了重要的作用。现在，混凝土仍然是主导者，它的应用将酒店与典型的巴西本土建筑联系起来。外墙的预制混凝土板被涂成一系列色彩，既保证了建筑的本土特征，又使其与周边的自然环境联系起来。

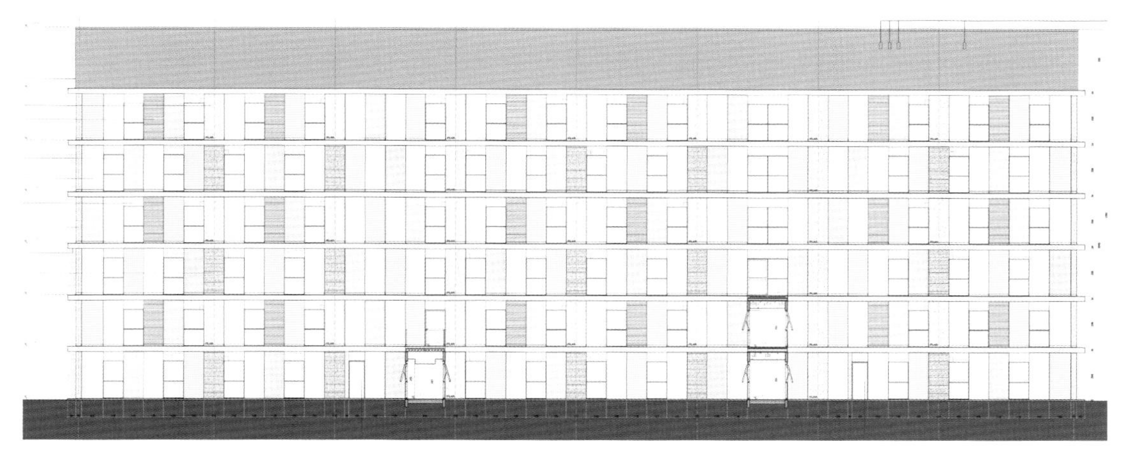

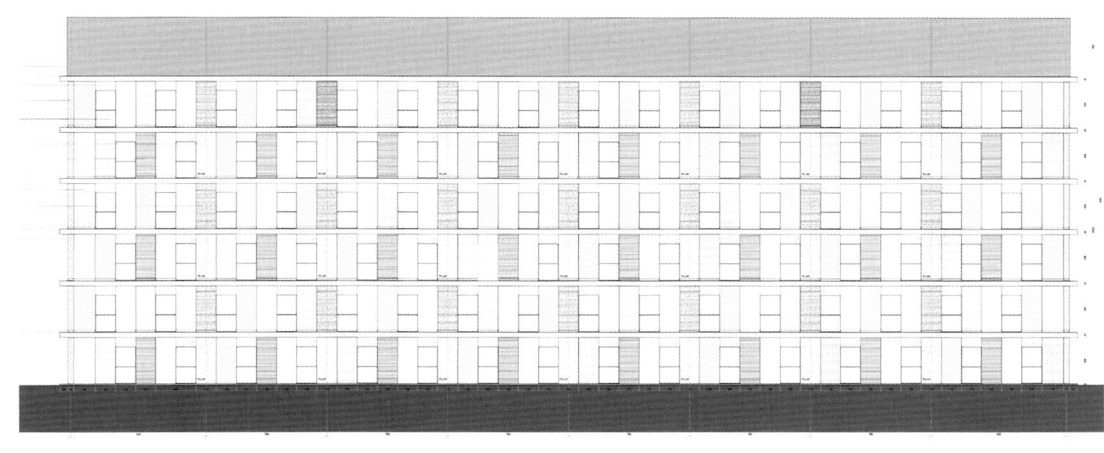

1. Steel capstone plate
   pre painted colour frosted black
2. Trapezoidal metallic shingle
   slant=5%
   with thermal-acoustics insulation
   pre painted
   colour bright grey or similar
3. Metallic beam
4. Steel gutter pre painted plate
5. Metallic profiles
6. Metallic structure
7. Shuttered gridiron
   composed by aluminium corners
   hidden internal structure
   anodised black aluminium
8. Waterproofing over all the pavement
   leveled and polished concrete floor
   for fibre tanks
9. Precast concrete beams
10. Masonry wall
    + waterproofing
    and mechanical protection
11. Eaves – capstone slant=4%
    Pre painted zinc plate
    folded, with dripping = #6.5mm
    colour grey – RAL 9006
12. Alveolar slab LP26
    precast concrete
13. Slab's capstone
14. Gaps
15. Beam 40x60cm
    precast concrete
16. Metallic insert
17. Gap between
    structure and panel = 50mm
18. Molding – light string
19. Plaster ceiling – PVA paint colour white
20. Prefabricated concrete panel
    125x298x10cm
21. Half plasterboard wall
    Satiny acrylic paint finishing
    colour white
22. Glass wool insulation
23. Ceramic floor
    60x60cm natural
24. Wooden mop board 5.0x1.2cm
    melamine colour frosted white
25. Pre fabricated concrete panel
    gap for capstone
26. Eaves – capstone slant=4%
    Pre painted zinc plate
    folded, with dripping = #6,5mm
    colour grey – RAL 9006
27. Precast concrete alveolar slab LP26
28. Slab's capstone
29. Metallic insert
30. Beam 40x60cm
    precast concrete
31. Molding – light string
32. Plaster ceiling – PVA paint colour white
33. Double structure
    simple-plate plasterboard
    until the slab
34. Glass wool insulation
35. PM01 – door
    melamine colour dark grey
    card lock
36. JA01 – glazing window
    anodized black aluminium
    window
    double-laminated glass colourless
    window's dimensions: 125x236cm
37. Roll-on blackout curtain colour black
    embed in the window's gap
38. Ceramic floor
    60x60cm natural
39. Stone sill
40. Stone floor
    slant = 4%
41. Gravel
42. Drain
43. Beam 20x50cm
44. Reinforced concrete subfloor
45. Black plastic canvas 200 weight
46. Compacted gravel
47. Compacted natural soil
48. Foundation stone
49. Concrete
50. Plaster ceiling – PVA paint colour white
51. Beam / Steel profile
52. Removable acoustics ceiling
    62.5x62.5cm

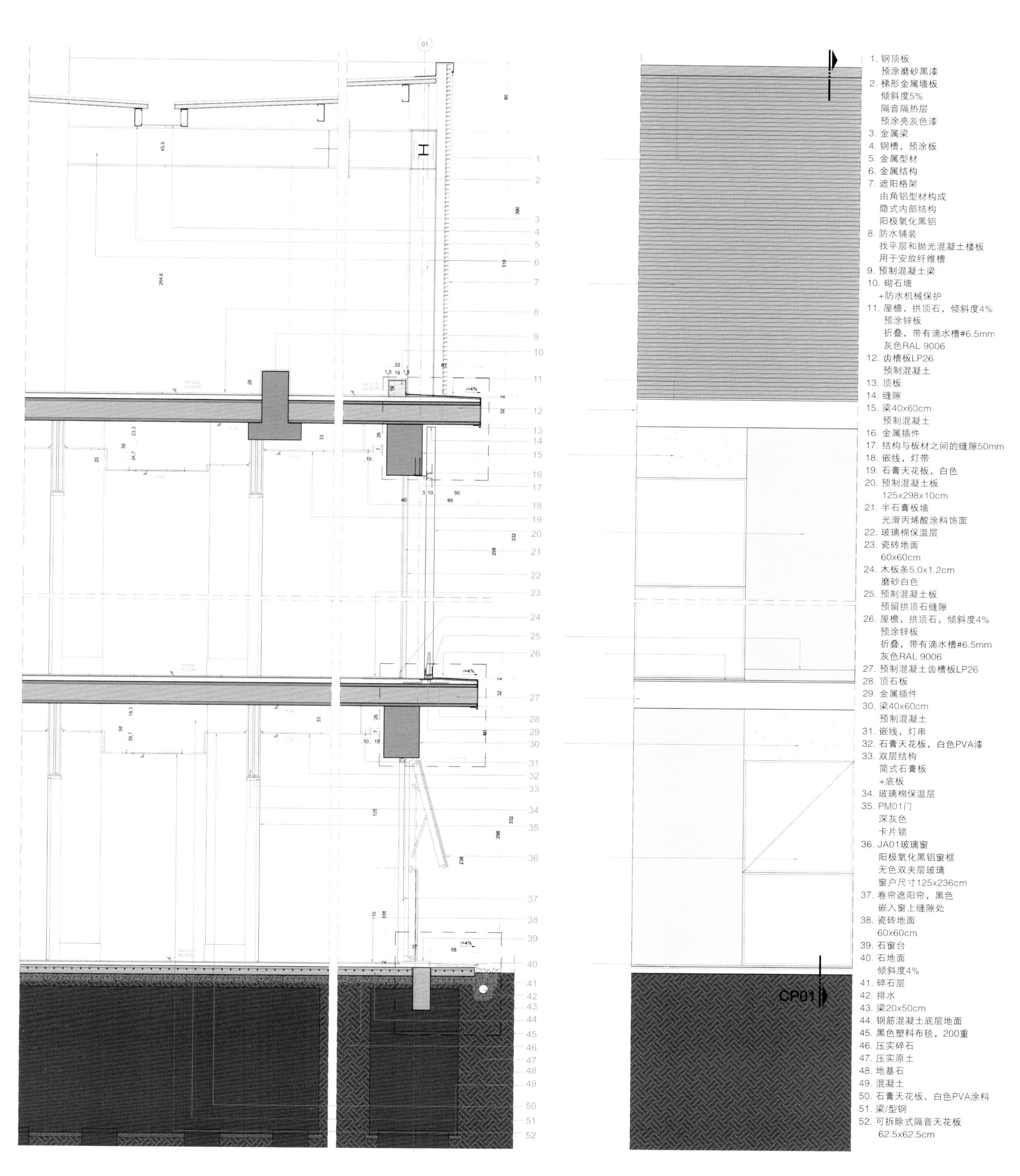
1. 钢顶板
预涂磨砂黑漆
2. 梯形金属墙板
倾斜度5%
隔音隔热层
预涂亮灰色漆
3. 金属梁
4. 钢槽，预涂板
5. 金属型材
6. 金属结构
7. 遮阳格架
由角铝型材构成
隐式内部结构
阳极氧化黑铝
8. 防水铺装
找平层和抛光混凝土楼板
用于安放纤维槽
9. 预制混凝土梁
10. 砌石墙
+防水机械保护
11. 屋檐，拱顶石，倾斜度4%
预涂锌板
折叠，带有滴水槽#6.5mm
灰色RAL 9006
12. 齿槽板LP26
预制混凝土
13. 顶板
14. 缝隙
15. 梁40x60cm
预制混凝土
16. 金属插件
17. 结构与板材之间的缝隙50mm
18. 嵌线，灯带
19. 石膏天花板，白色
20. 预制混凝土板
125x298x10cm
21. 半石膏板墙
光滑丙烯酸涂料饰面
22. 玻璃棉保温层
23. 瓷砖地面
60x60cm
24. 木板条5.0x1.2cm
磨砂白色
25. 预制混凝土板
预留拱顶石缝隙
26. 屋檐，拱顶石，倾斜度4%
预涂锌板
折叠，带有滴水槽#6.5mm
灰色RAL 9006
27. 预制混凝土齿槽板LP26
28. 顶石板
29. 金属插件
30. 梁40x60cm
预制混凝土
31. 嵌线，灯串
32. 石膏天花板，白色PVA漆
33. 双层结构
简式石膏板
+底板
34. 玻璃棉保温层
35. PM01门
深灰色
卡片锁
36. JA01玻璃窗
阳极氧化黑铝窗框
无色双夹层玻璃
窗户尺寸125x236cm
37. 卷帘遮阳帘，黑色
嵌入窗上缝隙处
38. 瓷砖地面
60x60cm
39. 石窗台
40. 石地面
倾斜度4%
41. 碎石层
42. 排水
43. 梁20x50cm
44. 钢筋混凝土底层地面
45. 黑色塑料布毯，200重
46. 压实碎石
47. 压实原土
48. 地基石
49. 混凝土
50. 石膏天花板，白色PVA涂料
51. 梁/型钢
52. 可拆除式隔音天花板
62.5x62.5cm
CP01

Housing and Urban Development Project in Manresa

曼雷萨住房与城市开发项目

Location/地点: Barcelona, Spain/西班牙，巴塞罗那
Architect/建筑师: Pich-Aguilera Architects
Photos/摄影: Pich-Aguilera Architects
Built area/建筑面积: 8,504m²
Completion date/竣工时间: 2013
Key materials: Façade – Industrialised concrete panels
主要材料：立面——工业化混凝土板

Overview

The proposal of the buildings and the open spaces has been structured from the consideration that the environment where the project is developed has got several points of view and confrontation with the existing city.

We can resume in three points this complexity:

• The relation with the "high" city which gives a leadership to the roofs and the different treatment of the pavements and the levels of the open spaces make up a "horizontal façade" which maintains the same strength and urban intention as the real façades. Green water-tank roofs are proposed with this objective.

• The relation with the St. Ignasi Street as a principal access in the future city; the project sets a typological organisation and a rhythm in the façade with a constant alignment for generating a set that marks a clear urban boundary.

• The relationship with Montserrat Street, where the buildings are planned to interpret the existing city with its irregularities and minimum parcel. The proposed arrangement tries to generate open spaces for pedestrians but at the same time it looks for spaces of relation between the different buildings. All the halls to the apartments are facing the Montserrat Street becoming a public space of connec-

P

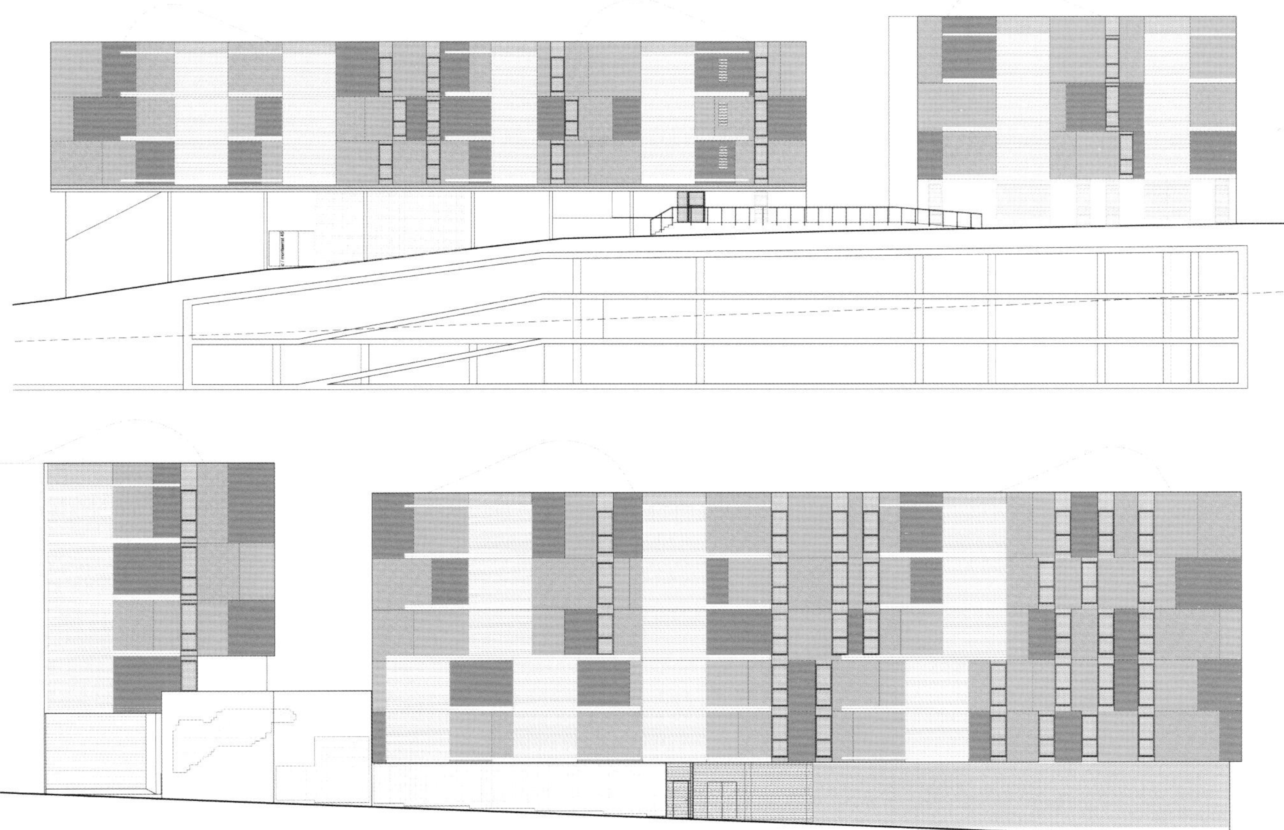

tion between the houses.

Detail and Materials

The façades of the project contemplated a cluster of holes and an array of blades, which generate a flow mattress or climate provide perfect cross ventilation of all types and control them from radiation.

Industrialised concrete panels allow: Higher quality façades and tightness; A great opportunity to colours and textures; Continuous insulation across the front; A high speed assembly and manufacturing, thereby saving resources and energy; Qualitatively improve safety in work.

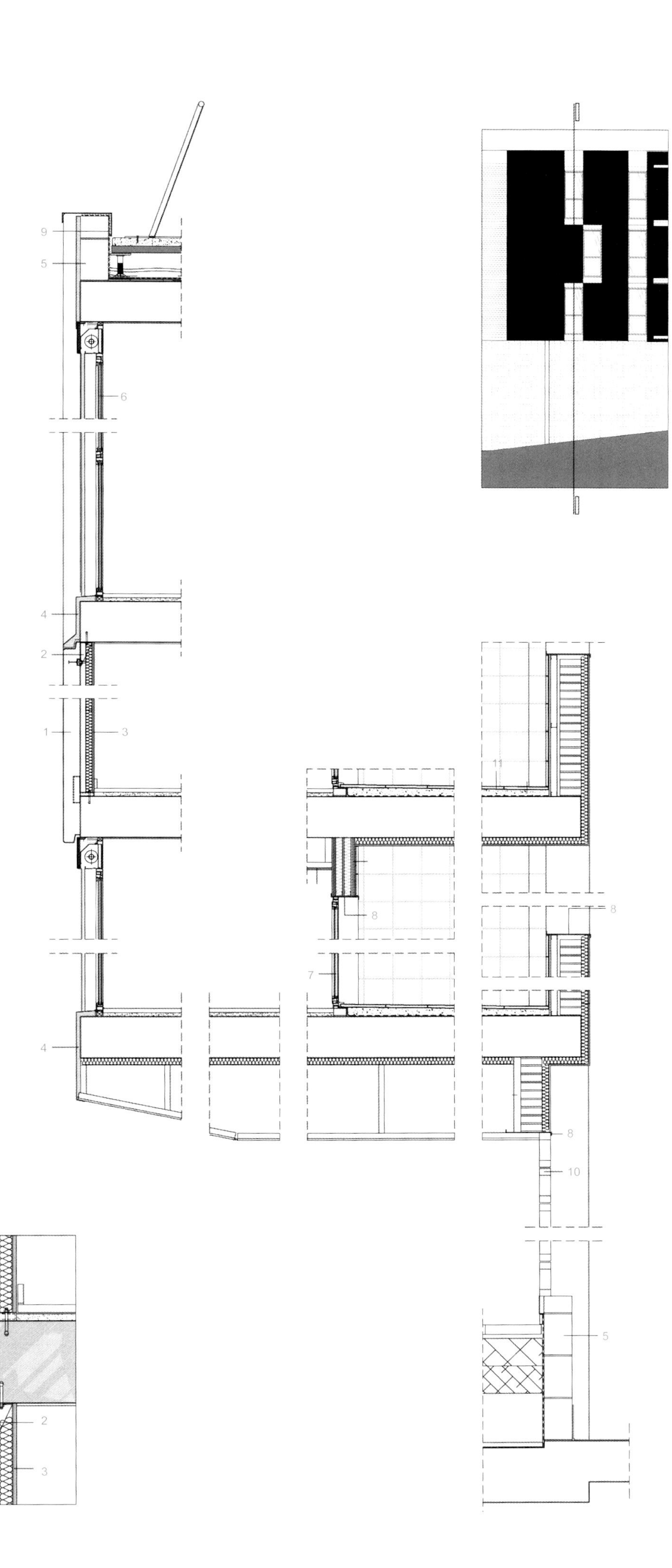

1. Precast concrete panel e=12cm
2. Anchor stabiliser, metallic guide, sliding bolt and metal angular
3. Plaster board, insulation, rockwool
4. Extruded polystyrene sandwich
5. Concrete block wall
6. Carpentry lacquered aluminum, with box blinds, monobloc type
7. Balcony clotheslines, lacquered aluminum
8. Metal threshold
9. Band connection
10. Ceramic piece, 25x25cm
11. Concrete slopes

1. 预制混凝土板e=12cm
2. 锚固件，滑动螺栓和金属角
3. 灰泥板，隔热层，石棉
4. 挤塑聚苯乙烯夹心板
5. 混凝土砌块墙
6. 涂漆铝窗框，配盒式百叶窗，单块型
7. 阳台晾衣架，涂漆铝
8. 金属门槛
9. 带状连接
10. 瓷砖25x25cm
11. 混凝土护坡

项目概况

建筑及开放空间的设计方案从几个方面充分考虑了开发所在的环境以及项目与已有城市结构之间的相互关系。

项目整体可以从三个方面进行考量：

• 与“城市高空”的关系。屋顶设计、地面铺装的处理、开放空间的层次共同组成了一个“水平立面”，它与真正的建筑立面在城市建设方面起到了相同的作用。建筑师通过绿色水池屋顶的设计实现了这一目标。

• 与圣伊格纳斯街的关系。圣伊格纳斯街是连接项目与城市的主要通道。项目通过典型的布局和富有动感的立面设计形成了清晰的城市界限。

• 与蒙特斯拉特街的关系。建筑计划通过不规则的造型和小块设计来重新诠释已有的城市景观。项目规划试图生成一些开放的步行空间，同时也试图在不同建筑之间寻找一种空间联系。所有通往公寓的大厅都朝向蒙特斯拉特街，在住宅之间形成了一个公共空间。

细部与材料

项目的立面由聚集的孔洞和一系列面板组成，它们形成了流动的空气垫，为建筑内部提供了完美的交叉通风，并且能保护室内不受阳光直射。

工业化混凝土板材的使用具有以下优点：更高品质的立面及密封性；更多的色彩和纹理选择；连续的隔热保温层；快速装配和生产，从而节约了资源和能源；大幅提升了施工的安全性。

Kindergarten Cerkvenjak – "Path of Learning"

柯尔克文雅克幼儿园——"学习之路"

Location/地点: Cerkvenjak, Slovenia/斯洛文尼亚，柯尔克文雅克
Architect/建筑师: Superform: Marjan Poboljšaj, Anton Žižek, Špela Gliha, Meta Žebre, Boris Janje
Photos/摄影: Luka Kaše
Site area/占地面积: 4,000m^2
Built area/建筑面积: 780m^2
Completion date/竣工时间: 2014
Key materials: Façade – Cement composite panel system (Swisspearl Carat and Eternit)) + Thin-layer structured plaster
主要材料：立面——水泥复合板系统（Swisspearl Carat and Eternit））+薄层结构石膏

Overview

The proposed kindergarten is in the village of Cerkvenjak, which is located in the centre of the Slovenske Gorice region, Slovenia. The kindergarten is inseparably connected with the natural surroundings of the trees and playground equipment. The concept of the kindergarten is similar to its local surrounding with the rhythmic string of volumes and roofs. Because of this, the kindergarten does not surpass the scale of an individual house and gives to the user – a child, a sense of home.

The proposed kindergarten design originated from an existing learning path of Klopotec, which runs nearby. The kindergarten is a new programme and function that upgrades the existing learning path; it becomes s a new branch of the Klopotec learning path. The result of using the principle of a learning path is a unique division and rhythm of the playrooms. The kindergarten offers rich experience to children, closer to the scale of the child.

Detail and Materials

Central strip of orange façade is embraced with grey roof and floor border. The façade is made of cement composite panels. Vertical raster gives it special design. Those divisions of façade panels create pattern with impression of flying birds. Therefore façade is pleasant and playful.

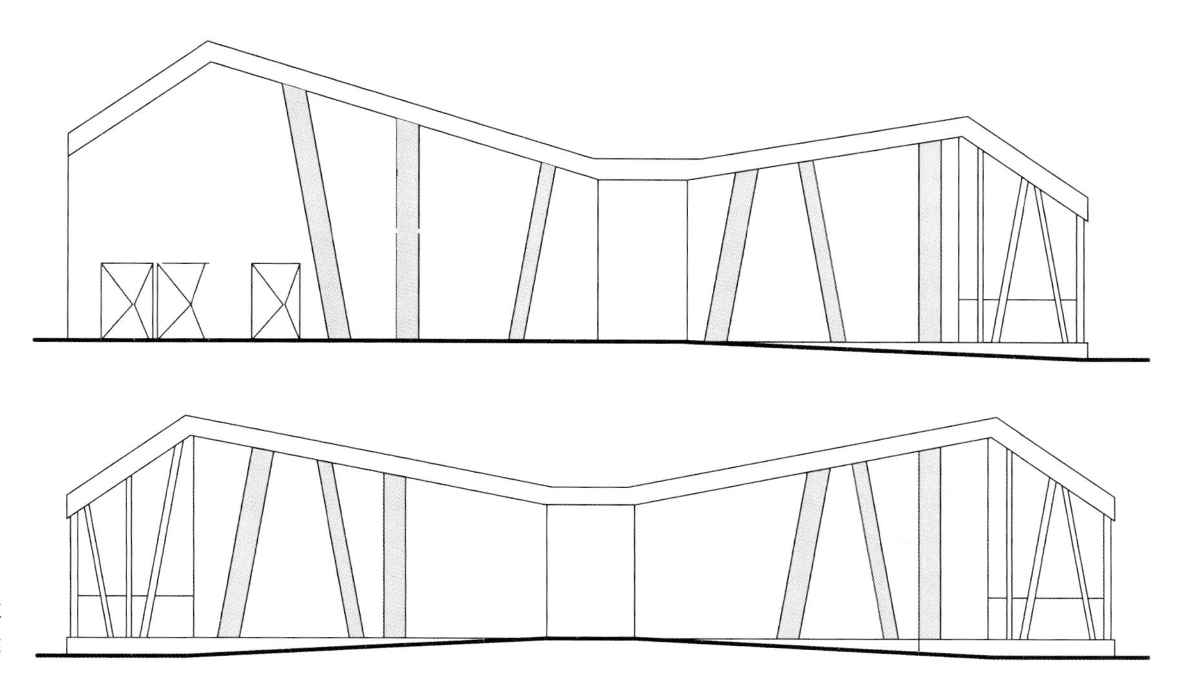

项目概况

这所幼儿园位于斯洛文尼亚斯洛文斯克格里斯地区中心的柯尔克文雅克村。幼儿园与自然景观和操场设施紧密地连接在一起，它的设计概念与当地连成串的建筑结构和屋顶设计十分相似。因此，幼儿园的规模并没有超越独立住宅，赋予了它的主要用户——儿童一种家的感觉。

幼儿园的设计灵感来源于附近的克洛普泰克学习路。幼儿园对学习路进行了升级和改造，成为了克洛普泰克学习路新的分支。学习路的设计概念对各个游戏室进行了独特的分区。幼儿园为儿童提供了丰富的体验，十分贴近他们的尺度。

细部与材料

橙色外墙的中央条带被灰色屋顶和地面边界所包围。外墙由水泥复合板构成，垂直栅格赋予了它独特的外观。外墙板的分割形成了类似于飞鸟的独特图案，显得生动而有趣。

1. Attachment of rafters, crafted each other
2. Threaded rod
3. Thermal insulation d=12cm, OSB panels
4. Metal box profile, vertically attached to the AB binding
5. Glass wall SS5
6. Vertical windows substructure anchored in AB panel
7. Waterproofing vertically to the aluminum window profile
8. Drain water in the void
9. Fibre cement panel, vertical metal profiles, galvanised
10. Border, aluminium sheet d=2mm
11. Metal shoe
12. Lining, fibre cement panels
13. Metal L profile, fixed in position, 2x2 screws drive rafter, tricks
14. Rod, aluminium sheet, RAL
15. Waterproofing foil, such as. Rhepanol, s/g=18/17(13) cm

1. 梁架附件，相互连接
2. 螺纹杆
3. 隔热层d=12cm，定向刨花板
4. 金属箱式型材，垂直安装在AB包边上
5. 玻璃墙SS5
6. 垂直玻璃下层结构，固定在AB板上
7. 防水层，垂直于铝窗框
8. 排水槽
9. 纤维水泥板，垂直金属型材，镀锌
10. 边框，铝板d=2mm
11. 金属套
12. 内衬，纤维水泥板
13. L形金属型材，2x2螺丝固定
14. 拉杆，金属板
15. 防水膜，Rhepanol，s/g=18/17(13) cm

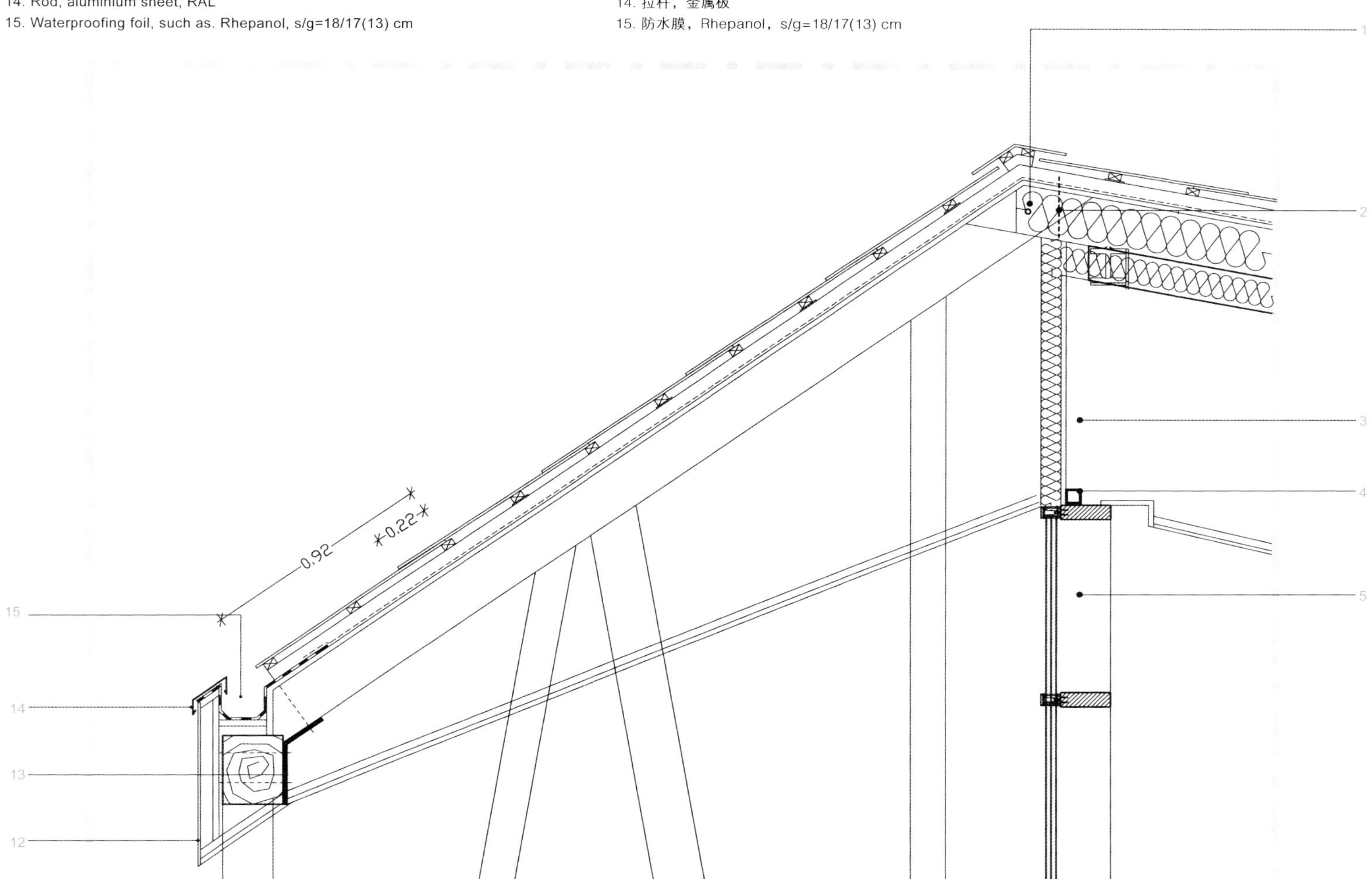

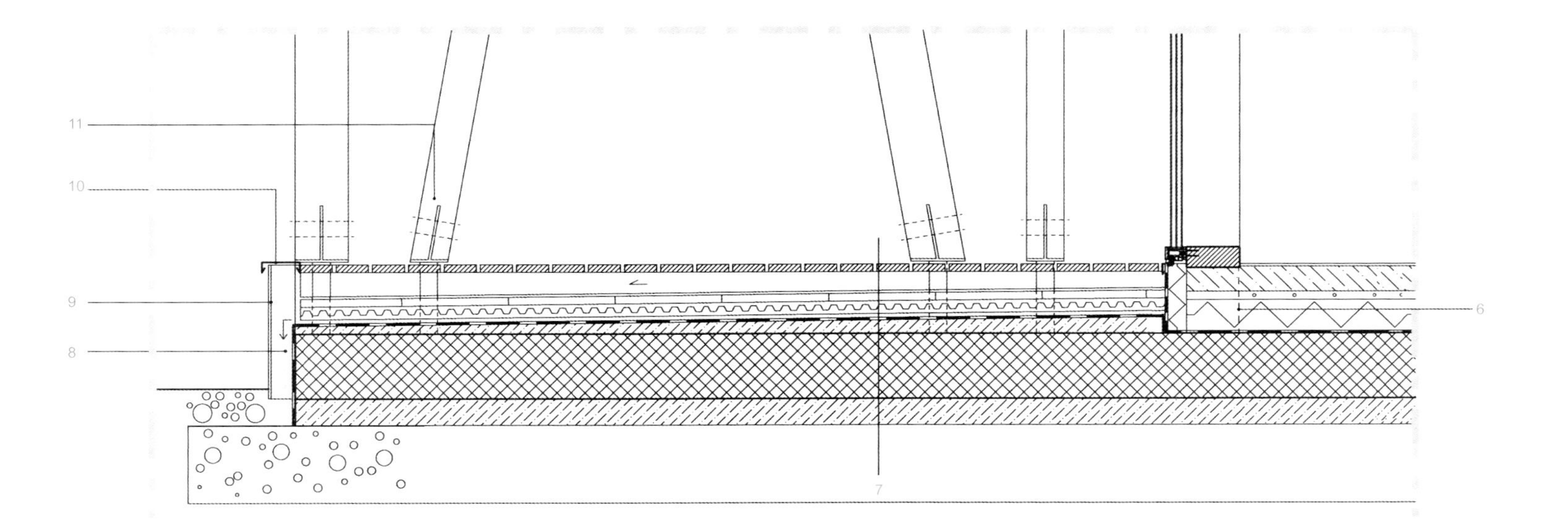

Alamillo Health Centre

阿拉米略健康中心

Location/地点: Seville, Spain/西班牙，塞维利亚
Architect/建筑师: Javier Terrados Cepeda, Fernando Suárez Corchete
Site area/占地面积: 1,139m^2
Built area/建筑面积: 3,610m^2
Key materials: Façade – precast reinforced concrete panels, stainless steel panels
主要材料: 立面——预制钢筋混凝土板、不锈钢板

Overview

The building site is located in the north edge of Grupo Residencial Bécquer block in Polígono Norte, Seville. The plot is triangular shaped in which the south-east façade belongs to Sánchez Pizjuán Avenue, the south-west façade belongs to Grupo Residencial Bécquer and the north façade, the longest one, belongs to a new street which connects Camino de Almez street with Sánchez Pizjuán Avenue.

Since the first analysis of the site, the required functional programme and the highest number of floors/levels allowed to this kind of building (3 in this case) was obvious that a very compact proposal should be the solution in response to all requests.

This is the reason why the proposed building occupies/completes all the perimeter under conditions of PGOU (Plan General de Ordenación Urbana de Sevilla). So it reproduces the triangular shaped of the plot and builds all functions along its perimeter which are available to users from the main lobby, that has three levels of height.

This simple functional diagram, that develops the building along the site edges from the main lobby, has resolved every level in order to specialise each one. The main entrance is located in the highest visible corner of the site, the one which is isolated/in the west side, at the intersection of Sánchez Pizjuán Avenue and the new street. In this point, it is created a wide porch that works as a prelude to the main

entrance. The main stairs /staircase with its elevators/lifts in the south-west façade, that is easy to see from the main entrance, resolve the main user path which is built around the perimeter of the main atrium. In the same façade, but in the opposite side, is hidden the secondary stairs/staircase that is protected in order to resolve the evacuation in case of fire.

With this organisation, the building has:
- In ground floor: the main lobby, reception, administration, pediatric clinic and rehabilitation area.
- In first floor: adults clinic and sanitary education area.
- In second floor: mental health area and private spaces for medical staff.

In addition, the building, which has a basement with 22 parking places, is approached by a ramp with trench shaped, which is built along the south-east façade, between the building and the edge of the plot.

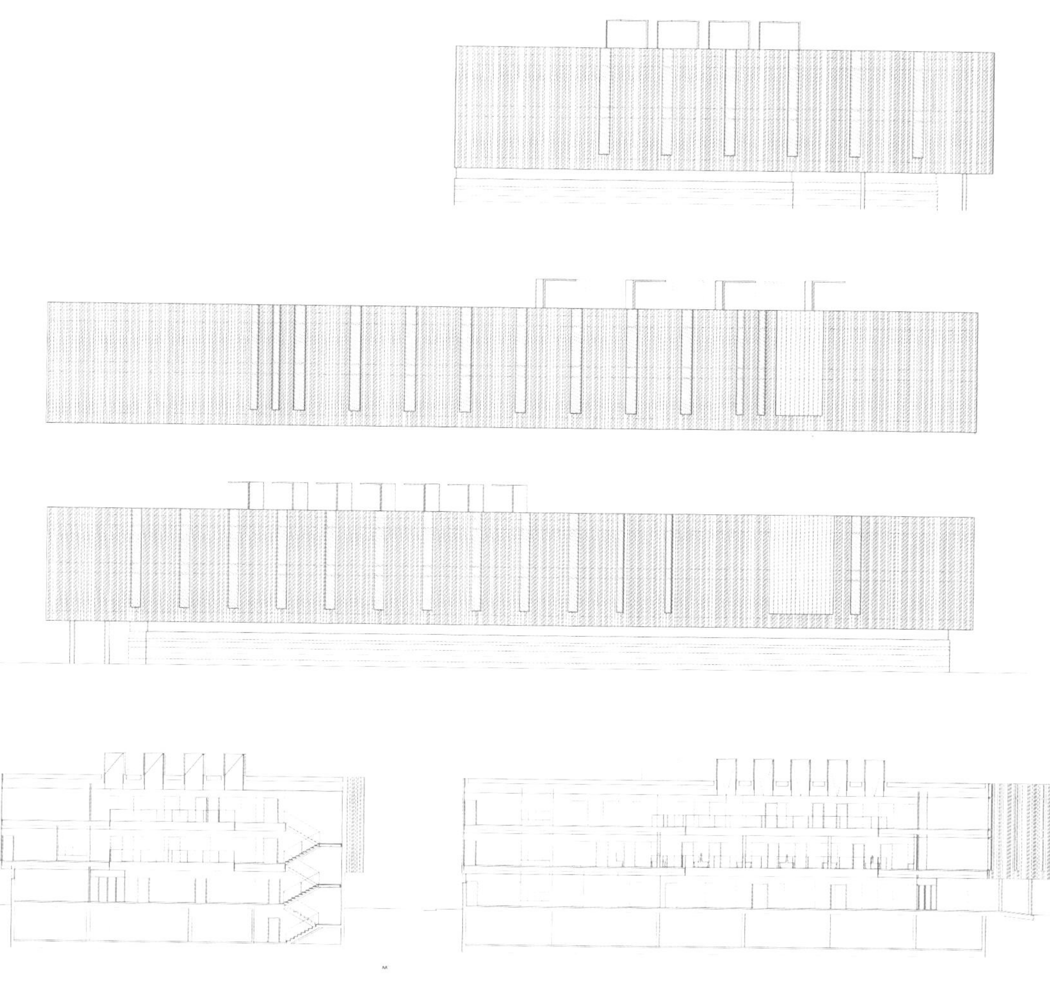

Detail and Materials

The envelope was designed as a tense continuous skin in which the joints were not perceptible. The solution consisted in using a series of 8-metre high precast reinforced concrete panels with a vertical corrugated texture that gave the whole a continuous appearance in which the vertical joints between panels could not be seen from the street. Coherently, the windows are arranged as vertical slots between panels.

Concrete panels are anchored to the floor slabs. Their most repeated measures are 8000 x 2700 x 15 mm. The rest of the envelope section comprises a 50 mm sprayed polyurethane insulation and a double 15 mm gypsum board panel as the inner layer.

The recessed plinth is clad with stainless steel panels in order to produce a sharp contrast with the corrugated concrete above and also to make an occasional urban graffiti easily removable. Those panels have a typical dimension of 2200 x 900 x 2 mm and are anchored on a 120 mm brick wall with an inner layer of polyurethane insulation and double gypsum board panels.

The "fourth façade" is the roof, which can be easily seen from the surrounding high residential blocks of that area. It is designed as an artificial lawn, furnished by the grid of protruding skylights.

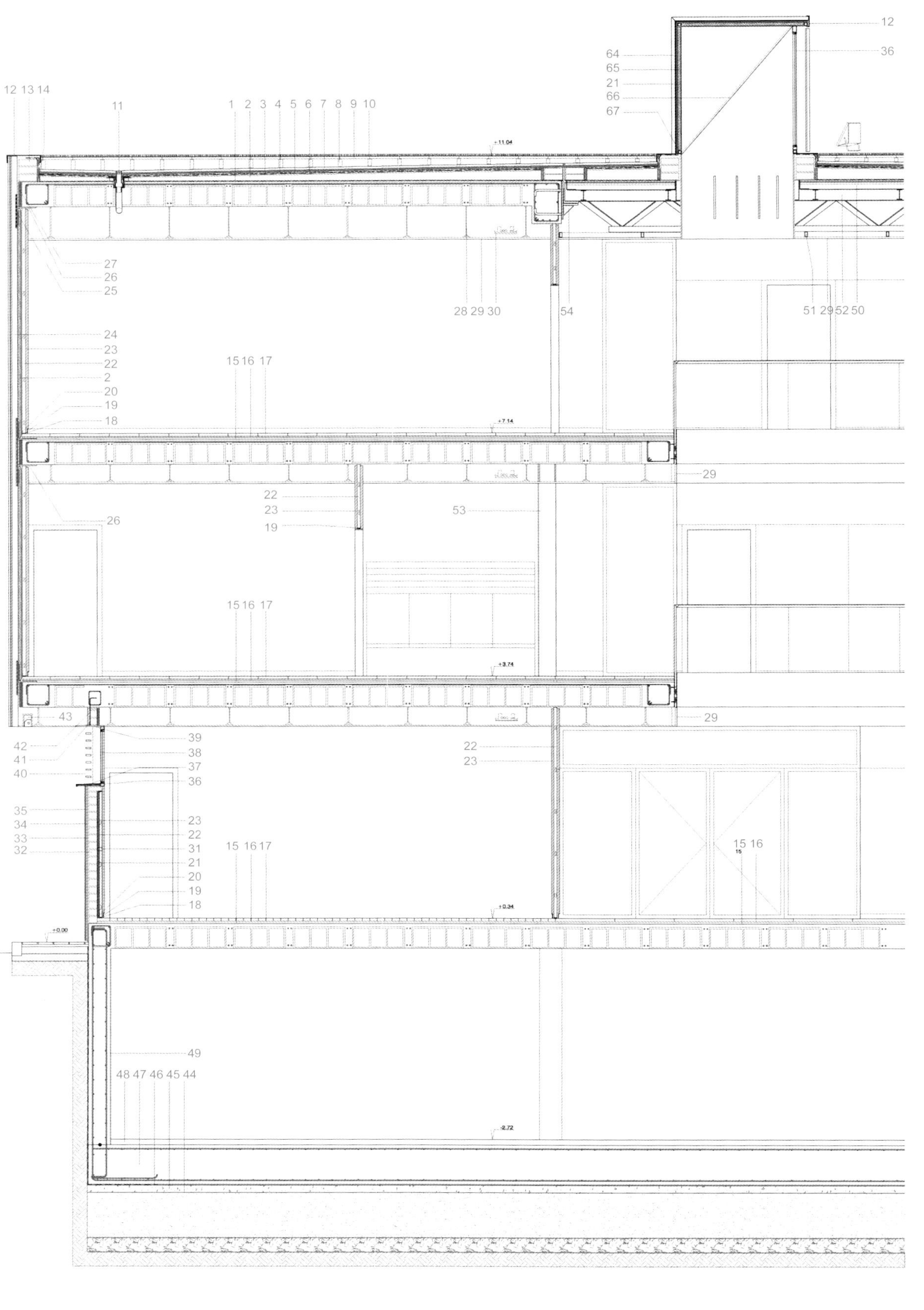

1. Vapour barrier with anti-rust paint
2. Cellular concrete for constructing a slope, min 2mm thick
3. Leveling layer of cement mortar
4. Modified bitumen waterproofing sheet with double polyethylene inner armor 4.8kg/m^2
5. Geotextile with 150gr/m^2 density
6. 30mm extruded polystyrene plates
7. 3mm protective mortar with steel mesh
8. Support of floating floor with prefabricated wedges
9. Floating floor in whitewashed Chinese tile
10. Synthetic grass carpet, weighted with quartz sand
11. EPDM prefabricated sump
12. 2mm folded aluminum finishing
13. 1ft perforated brick
14. 40mm expanded polystyrene element
15. 2cm sand bed over slab
16. 2cm M-40 grip mortar
17. Micro marble terrazzo paving, 50x50cm pieces
18. 2mm anodised aluminum skirting
19. 20mm thick rubber seal
20. 48x36mm galvanised steel sheet
21. 50mm polyurethane insulation
22. Stanchion in galvanised steel plate, 48x36mm
23. 15mm double plasterboard panel
24. Precast concrete, 15cm thick
25. Anchor plate built into precast panel
26. L200.20 profile for fixing prefabricated panels
27. Extruded polystyrene insulation, 3cm thick
28. Metal rods for fixing ceiling
29. Continuous laminated plasterboard false ceiling, 1.5cm thick
30. Perforated installations tray
31. 48mm bracing piece
32. Half brick wall
33. Waterproof mortar plastering, 1.5cm thick
34. Galvanised steel lintel, 8mm thick
35. Matte stainless steel sheet cladding panel
36. Stainless steel sill, 2mm thick
37. Sill coating in coloured phenolic board
38. Anodised aluminum metalwork with transparent safety glass, 3+3mm
39. Galvanised steel lintel, 8mm thick
40. Anodised aluminum rabbet, 2mm thick
41. 50.5 structure for fixing metal lintel
42. Round for fixing lintel
43. Lintel plasterboard with socket to accommodate light
44. Polyethylene film
45. Blinding concrete layer, 10cm thick
46. PVC sheet, 1.5mm thick
47. Reinforced concrete foundation slab, 55cm thick
48. Concrete slab finishing, surface treatment with silica arid corundum and quartz
49. Plaster, 2cm thick
50. 13cm composite slab on a corrugated galvanised steel backing sheet
51. 70.40.3 rectangular structure every 1m for ceiling fastening
52. 70cm laminated rolled truss
53. HEB-240 pillar, coated with 2mm anodised aluminum sheet
54. Tied beam, rolled section IPN-180

1. 隔汽层，涂有防锈漆
2. 多孔混凝土，用于构造斜坡，最薄处2mm
3. 水泥砂浆找平层
4. 改性沥青防水板，双层聚乙烯内甲4.8kg/m^2
5. 土工布，密度150gr/m^2
6. 30mm挤塑聚苯乙烯板
7. 3mm防护砂浆，配有钢网
8. 浮式地板支架，预制楔子
9. 粉饰中式地砖浮式地板
10. 人造草毯，石英砂加重
11. EPDM预制沟渠
12. 2mm折叠铝饰面
13. 1ft预制砖
14. 40mm发泡聚苯乙烯元件
15. 2cm沙层，位于楼板上方
16. 2cm M-40砂浆
17. 大理石水磨石铺装，50x50cm尺寸
18. 2mm氧化铝墙脚线
19. 20mm橡胶密封
20. 48x36mm镀锌钢板
21. 50mm聚氨酯隔热层
22. 镀锌钢板支柱48x36mm

项目概况

建筑场地位于塞维利亚市巴凯尔住宅区的北侧。地块呈三角形，东南边属于皮斯胡安大道，西南边属于巴凯尔住宅区，最长的北边则属于一条连接卡米诺阿尔梅斯街和皮斯胡安大道的新建街道。

对场地进行初次分析、此类建筑所必须的功能设置以及所规定的最高楼层数（本项目为3层）共同决定了项目将是一座十分紧凑的建筑结构。

这也是建筑之所以在满足《塞维利亚土地使用规划》的前提下，占据了场地的各个边缘的原因。因此，建筑复制了地块的三角形状，让所有功能区围绕着边界展开。建筑共分为三层，使用者由中央大厅进入。

为了实现各个楼层的特殊化，这个简单的功能规划实施逐层设计。主入口位于场地最显眼的一角，位于三角形的西边、新建街道和皮斯胡安大道的交叉口。门口设有宽敞的门廊迎接宾客。主楼梯和电梯位于西南边，从主入口可以一眼看到，解决了环绕中庭的使用者通道问题。在同一面的另一侧隐藏着附属楼梯，用作消防通道。

在这种组织结构下，建筑分为以下功能区：
一楼：主大厅、接待处、行政区、儿科门诊和康复区。
二楼：成人门诊和卫生教育区。
三楼：心理健康区和医务人员私人空间。

此外，建筑的地下室设有22个停车位，可经由建筑东南面、夹在建筑与地块边缘之间的壕沟式坡道进入。

细部与材料

建筑外壳被设计成紧凑的连续表皮，采用隐藏式接缝设计。建筑表皮由一系列8米高的预制钢筋混凝土板构成，板面上带有纵向瓦楞纹理。这种设计既实现了建筑外观的连续感，又通过纵向纹理将立缝隐藏起来。同样的，窗户也被设计成立式结构。

混凝土板与楼板固定在一起，最常用的尺寸是8000x2700x15毫米。建筑外壳的其他部分由50毫米的喷涂聚氨酯隔热层和两层15毫米的石膏板（作为内层结构）构成。

嵌入式底座被不锈钢板包覆起来，与上方的瓦楞混凝土板形成了强烈对比，同时也便于清除偶尔出现的城市涂鸦。这些板材的标准尺寸为2200x900x2毫米，固定在120毫米厚的砖墙上，墙面内层同样是喷涂聚氨酯隔热层和双层石膏板。

建筑的“第四个面”是屋顶，从周围的高层住宅楼都可以一眼看到。屋顶被设计成人造草坪，上面配有凸出的天窗。

23. 15mm双层石膏板
24. 预制混凝土，15cm厚
25. 锚定板，固定于预制板上
26. L200.20型材，用于固定预制板
27. 挤塑聚苯乙烯隔热层，3cm厚
28. 金属棒，用于固定天花板
29. 连续层压石膏板假吊顶，1.5cm厚
30. 预制安装托盘
31. 48mm支撑杆
32. 半砖墙
33. 防水砂浆抹面，1.5cm厚
34. 镀锌钢过梁，8mm厚
35. 亚光不锈钢板外墙板
36. 不锈钢窗台，2mm厚
37. 窗台，彩色酚醛树脂板覆面
38. 氧化铝窗框+透明安全玻璃3+3mm
39. 镀锌钢过梁，8mm厚
40. 氧化铝槽，2mm厚
41. 50.5结构，用于固定金属过梁
42. 圆口，用于固定过梁
43. 过梁石膏板，配有插口安装灯具
44. 聚乙烯薄膜
45. 基础垫层混凝土，10cm厚
46. PVC板，1.5mm厚
47. 钢筋混凝土地基板，55cm厚
48. 混凝土板饰面，表面由金刚砂和石英处理
49. 石膏，2cm厚
50. 13cm组合板，在瓦楞镀锌钢支撑板上方
51. 70.40.3直角结构，间隔1m，用于固定天花板
52. 70cm层压轧制桁架
53. HEB-240柱，2mm氧化铝薄板包层
54. 系梁，轧制型材IPN-180

PRATIC SPA Headquarters and Production Complex

PRATIC SPA总部及生产基地

Location/地点: Udine, Italy/意大利，乌迪内
Architect/建筑师: GEZA Gri e Zucchi Architetti Associati
Photos/摄影: Fernando Guerra / FG+SG
Site area/占地面积: 45,000m^2
Completion date/竣工时间: 2011
Key materials: Façade – concrete prefabricated panel and marble stone panel
主要材料： 立面——预制混凝土板和大理石板

Overview

The area, anchored in the industrial zone of the town, is also intimately tied to the surrounding agricultural landscape and is framed at north by the mountainscape. The project, carried out with the simplicity of the concrete prefabrication, joins the renewable energies and the natural light control to achieve a strong architectural identity, becoming a landmark for the passerby.

The design of the outside spaces takes advantage of the slight slope already existing on the lot to emphasise the lightness of the office core and sculptures the land with the materials obtained from the foundation groundworks. The parking lot is within a circular perimeter slightly buried in order to buffer its visual impact. The two main buildings straddle the garden, a semi-private area that opens towards west to the agricultural landscape. The nature of the space creates a human-nature relationship, uncommon in an industrial ambiance.

The use of clean energies has constrained all choices for the technical installations but in a discrete way, controlled by the architectural project. The production core's roof is composed of a series of photovoltaic panels, completely hidden from the sight. The total electric consumption is covered by energy obtained from renewable sources, generating a production environment which through the architectural project, becomes zero impact.¬

Detail and Materials

The production core is characterised by a prefabricated façade marked by pronounced vertical lines, the alternating glazed surfaces and solid panels of diverse widths, which are the total volume height. The differing dimensions of black marble stone and dark cement that make up the finish of the panels allow for a varied façade alive with the changing atmospheric conditions. The office core is protected on the south by a dark concrete beam of grand dimensions (5x80m) that highlights at a larger scale, the horizontality of the project. Its reflection on the glazed façade yields a great "floating shade" that serves a climatic function by mediating temperature within the work spaces.

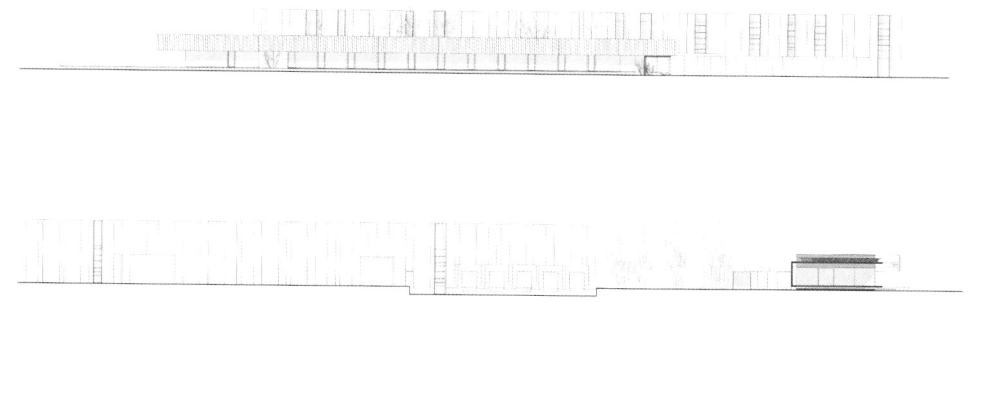

项目概况

项目位于工业区，同时也紧邻周边的农业景观，北面是远山的美景。项目通过预制混凝土构件实现了简洁的外观，并且结合新能源和自然光控技术呈现出强烈的建筑形象，为过往行人留下了深刻的印象。

外部空间的设计充分利用了斜坡地势，突出了办公楼的轻盈感，并且以地基结构重新塑造了地面。为了缓冲自身的视觉效果，停车场略微嵌入地下。两座主楼中间夹着一座花园；作为半开放区域，花园朝向西侧的农田开放。整个空间在人与自然之间建立起了亲和的联系，这在工业环境中是不多见的。

清洁能源的运用虽然约束了技术设备的选择，但是这对建筑师十分有利的。生产车间的屋顶由一系列光电伏板构成，它们完全隐藏在人们的视线之外。项目的所有电能消耗都由可再生资源所产生的能源覆盖。这样一来，通过建筑项目，生产环境实现了对环境的零影响。

细部与材料

生产车间以预制混凝土外墙为特色，不同宽度的落地玻璃面与实心墙面相互交错，形成了强烈的垂直线条。不同尺寸的黑色大理石板和深色水泥板利用变化的氛围让外墙变得丰富活泼。办公楼的南侧由一条巨大的深色混凝土梁（5x80米）所保护，突出了建筑的宏大感和横向趋势。它在玻璃墙面上的倒影形成了“浮动的阴影”，为办公空间提供了可随气候变化的遮阳效果。

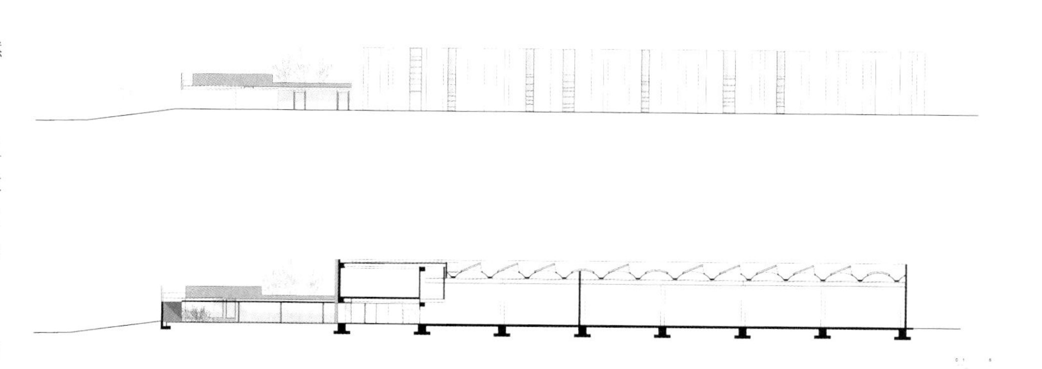

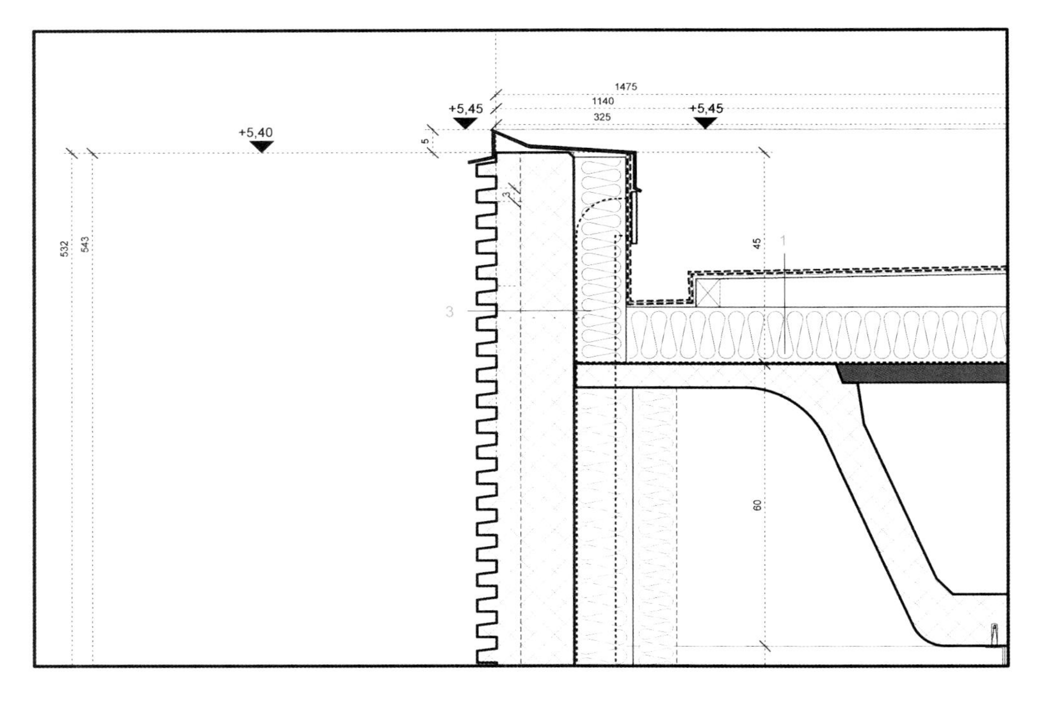

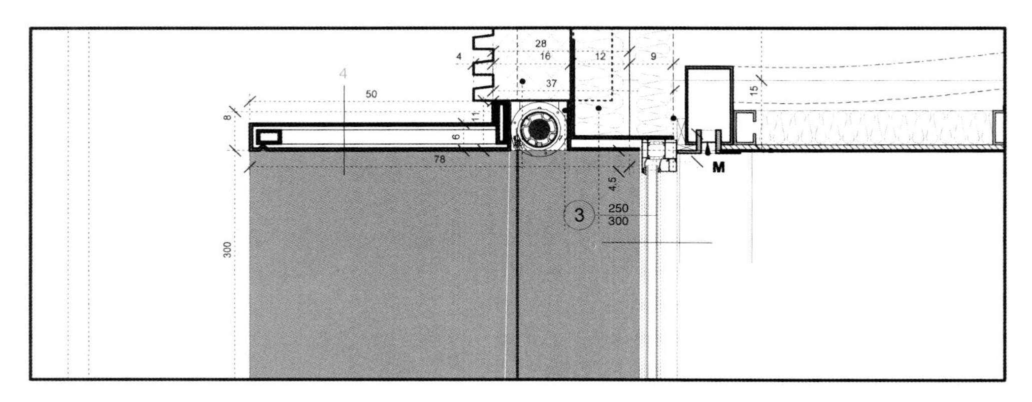

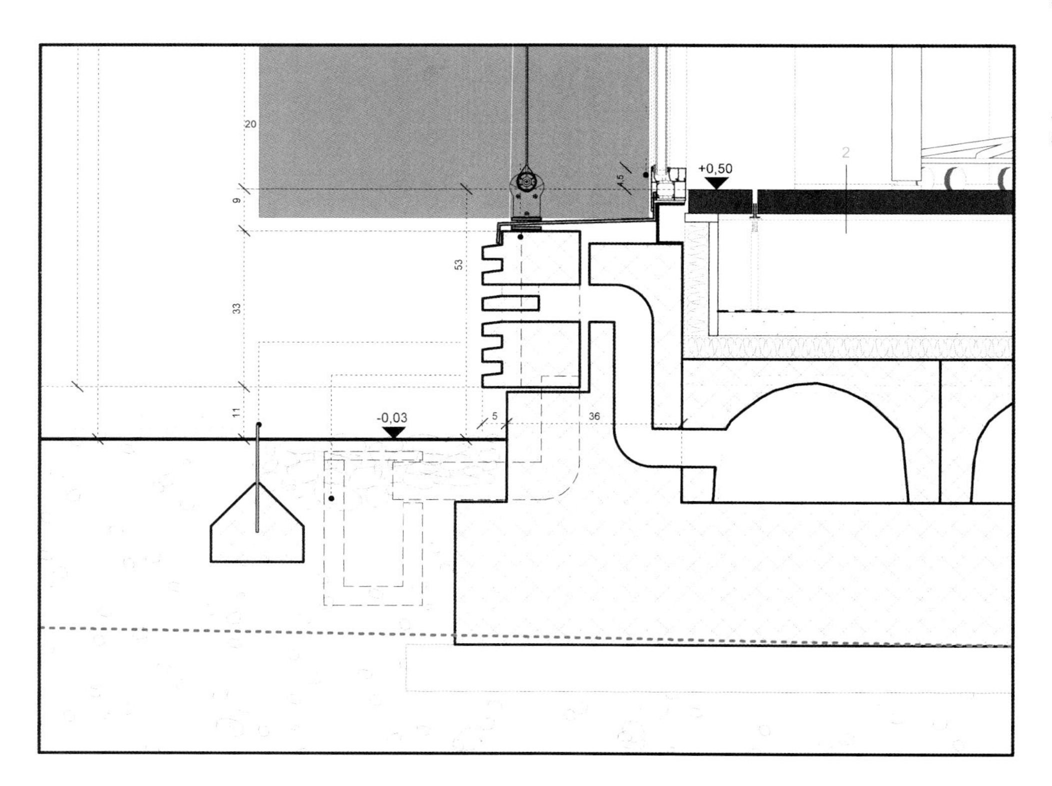

1. Roof:
   - Double waterpoofing
   - Inclination obtained with the insulation's shape
   - Rainwater collection on the north facade via rainpipes inside the prefabricated panels
   - Eps insulation panel (120mm) with polyethylene fim underneath
   - Concrete screed
   - Polyethylene film
   - Prefabricated concrete structure (60 cm)
   - Prefabricated concrete beam (60 cm)
   - Plasterboard suspended ceiling with double skimming and painting (100 mm)
2. Underground floor:
   - Porcelain tiles floating flooring (60 × 60 cm) laid on metal substructure joint with eps insulating strip (overall thickness 260 mm)
   - Self leveling screed (50 mm)
   - Thermal insualation (60 mm)
   - Ventilated foundation with upper layer (270+30mm)
   - Lightwieght concrete (100 mm)
3. Opaque vertical enclosure:
   - Reinforced concrete prefabricated panels (16 + 4 cm)
   - Lightweight concrete insualation (120 mm)
   - Rockwoll insulation (75 mm)
   - Internal plasterboard counter partition (18 mm + aluminium sheet) on metal substructure (75mm)
4. External black-out screen
5. Transparent vertical enclosure:
   - Aluminium window with thermal break
   - Double- glazing (6+ 0.76Pvb + 6 mm. 20 Mm argon. 6+0.76 Pvb + 6 mm; uw=<1.3=
   - Sides finished with steel sheeting

1. 屋顶:
   - 双重防水
   - 倾斜角度取决于绝缘层的形状
   - 北立面雨水收集，通过预制板内轨管道
   - eps 绝缘板 (120mm) 下面有聚乙烯薄膜
   - 混凝土刮板
   - 聚乙烯薄膜
   - 预制混凝土结构 (60 cm)
   - 预制混凝土梁 (60 cm)
   - 双层喷漆的石膏板吊顶(100 mm)
2. 地下室:
   - 瓷质砖浮动地板(60 × 60 cm) 与eps 绝缘带共同安装在金属架构上 (整体厚度260 mm)
   - 自流平砂浆 (50 mm)
   - 保温层 (60 mm)
   - 通风的地基与上层 (270+30mm)
   - 轻质混凝土 (100 mm)
3. 不透明的立式围墙:
   - 预制混凝土板 (16 + 4 cm)
   - 轻质混凝土绝缘层 (120 mm)
   - 岩棉绝缘层 (75 mm)
   - 内置的石膏板隔墙 (18 mm + 铝板) 在钢架构上 (75 mm)
4. 外部的遮光屏
5. 透明的立式围墙:
   - 隔热铝窗
   - 双层玻璃 (6+ 0,76 pvb + 6 mm, 20 mm氩, 6+0,76 pvb + 6 mm; Uw = < 1,3)
   - 双面安装钢板

Cemetery Road Housing

公墓路住宅

Location/地点: Sheffield, UK/英国，谢菲尔德
Architect/建筑师: Project Orange
Photos/摄影: Project Orange
Site area/占地面积: 3,600m^2
Built area/建筑面积: 2,800m^2
Completion date/竣工时间: 2010
Key materials: Façade – black stained cement board
主要材料：立面—— 黑色水泥板

Overview

The site is located along Cemetery Road, one mile southwest of Sheffield City centre, on the boundary of the established 19th century inner suburb of Nether Edge. It is a brownfield site, previously home to a redundant low quality two-storey commercial building and associated car park. Immediately adjacent lies the Sheffield General Cemetery, which has for many years been disused and is now classed as a conservation area and listed on the English Heritage Register of Historic Parks and Gardens. The site also borders a conservation area to the South.

Cemetery Road is Project Orange's second collaboration with the client further to completion of the award winning Sinclair Building, also in Sheffield. Project Orange were approached initially to carry out a feasibility study for a residential led redevelopment of the site, including also some commercial space, car parking and a mix of residential units including dwellings for sale, a large apartment for the client's sister and her family and a number of apartments to be retained by the client for rental as serviced apartments.

The design concept aimed to provide high quality, high density housing on the basis of a prototype atypical to the essentially suburban pattern of the immediate context – a secure courtyard/mews type configuration where the principle amenity space (both visual and practical) is provided by the central landscaped courtyard around which the new properties, both houses and apartments, are arranged and from which all are accessed. With the small number of units arranged in this way, the development is intended to promote a strong sense of community. Of the nine townhouses, there are two house types. Both are designed with dynamic cross sections and make a generous provision of balcony and roof terrace space. The internal arrangements are planned to maximise privacy between neighbours, with the principal habitable rooms of the eastern townhouses facing windows to staircases and

ancillary accommodation of the western townhouses across the courtyard. Likewise the section has been developed to create variety in internal ceiling heights with emphasis given to first floor reception rooms in the tradition of the piano nobile.

Circulation through the townhouses is organised on the principle of the architectural promenade. The temptation to locate stair flights one above the other in the interest of efficiency was resisted in favour of dramatic and meandering routes through each house whereby primary and ancillary spaces are linked together like beads on a chain. It is this arrangement of stairs, circulation, primary and secondary habitable volumes, roof terraces and balconies that generates the strong tectonic composition of the façades, a formal expression emphasised in the application of strongly contrasting cladding materials.

In contrast to the highly modeled and graphic internal elevations, the façades addressing the conservation areas are clad in traditional York stone, a material that has been widely used in and around Sheffield throughout the 19th and early 20th century. This stone forms an armature sympathetic to its context yet through which, in the detail of windows, the penthouse storey and recessed ground floor commercial accommodation, the language of the courtyard beyond is revealed.

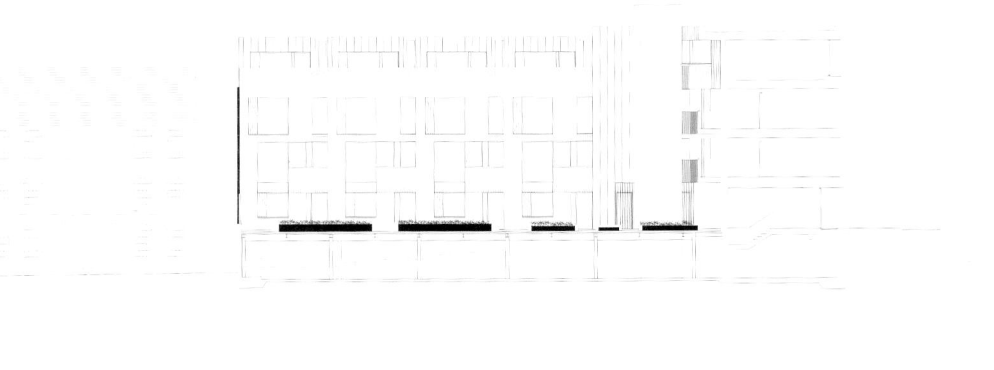

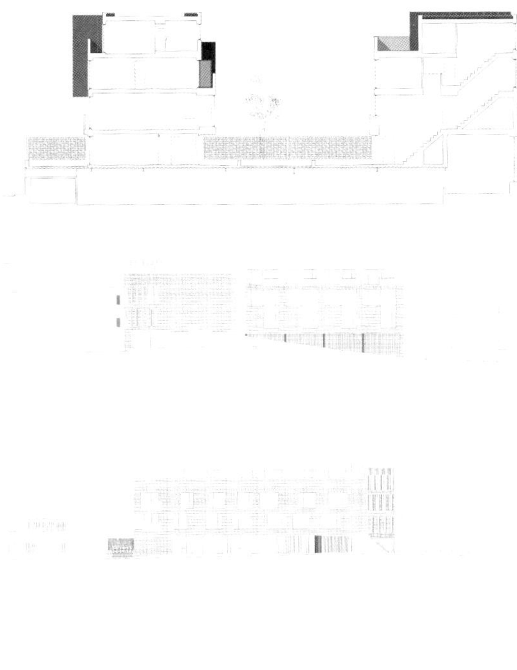

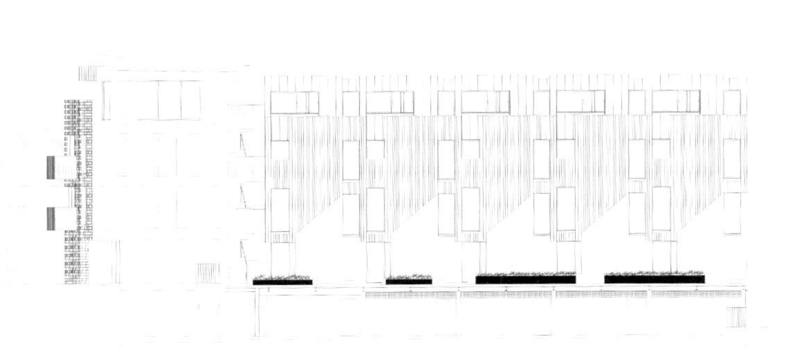

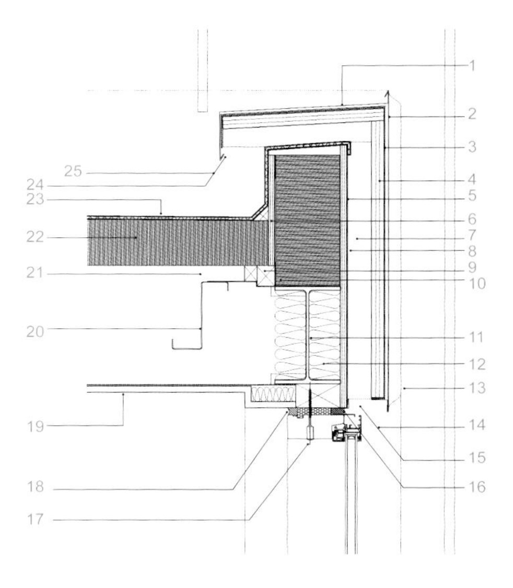

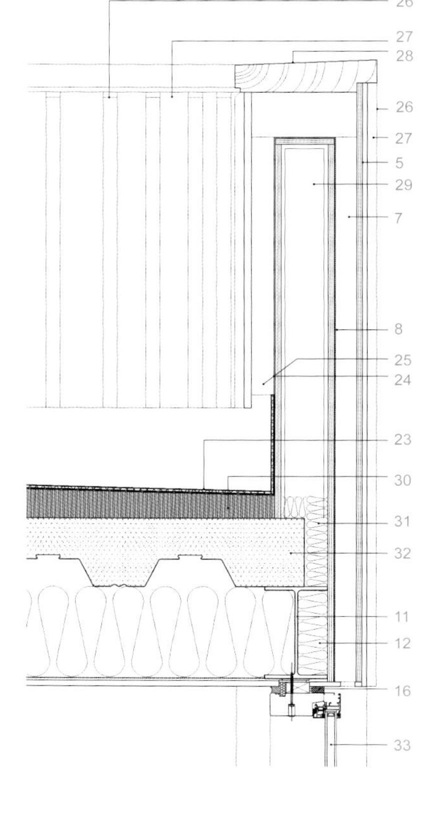

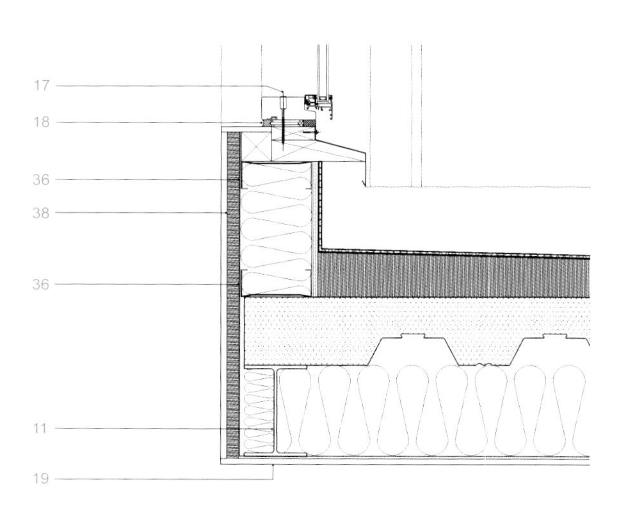

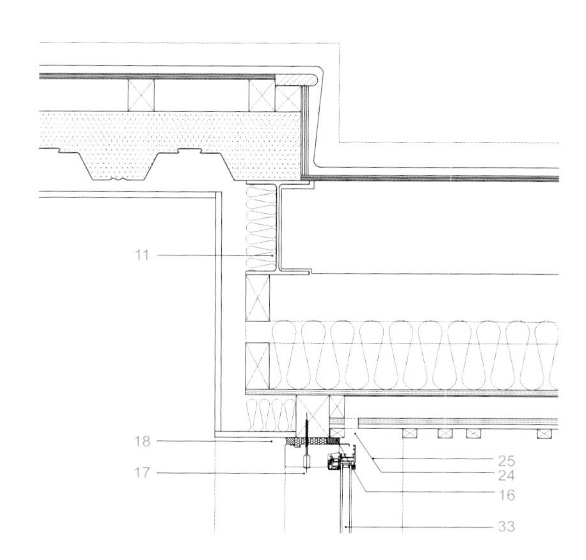

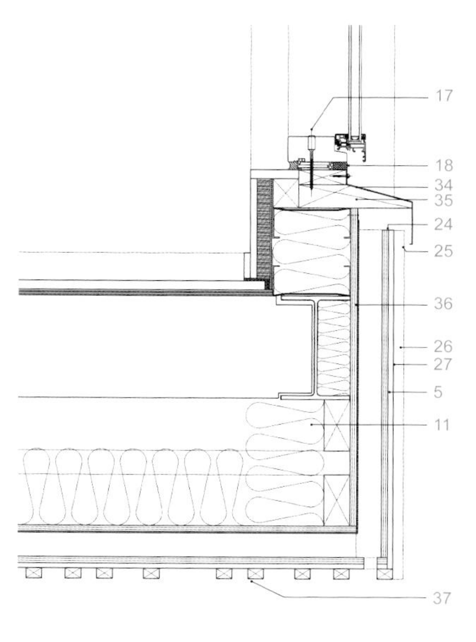

1. Zinc capping
2. VM Quartz-zinc pre-weathered zinc cladding
3. Tyvek Supro underlay
4. 25mm WBP ply
5. 12mm OSB ply
6. Timber fillet
7. 50mm battens to allow 50mm min ventilation gaps
8. Tyevek housewrap breather membrane
9. Timber stud bedded in butyl mastic
10. Galvanised steel angle flashing
11. Painted intumescent fire protection to all steel beams. Min 60 mins required to frame
12. Mineral wool packing
13. Zinc drip detail
14. Routed slot in WBP ply to allow 25mm min vent gap
15. Zinc insect mesh
16. Silicone seal by installer to window frame surround
17. Self drilling screw and cover cap by installer
18. PU expanding foam and silicone seal by installer
19. 12.5mm suspended plasterboard ceiling, taped and jointed with applied paint finish
20. Cold rolled roof "Z" purlins to engineer's specification and detail
21. 32mm profiled deck units
22. Tapered Kingspan insulation, mechanically fixed to metal deck
23. Resitrix EPDM membrane, continued up and over parapet
24. Insect mesh
25. 25mm min ventilation slot
26. 20×30mm black stained batten
27. 10mm black stained cement board
28. Black stained hardwood capping
29. Parapet post to engineers details and specification
30. Tapered Kingspan insulation
31. Mineral wool insulation
32. 150mm deep holorib concrete deck
33. Ideal Combi Futura window system
34. Anodised aluminum sill flashing to match window
35. Hardwood sill
36. 150mm metsec stud
37. Black stained batten and board build-up to soffit
38. 25mm K18 Kingspan insulated dryliner, taped and jointed with applied paint finish

1. 锌顶盖
2. VM石英锌预风化锌包层
3. Tyvek Supro衬垫
4. 25mm酚醛胶合板
5. 12mm定向宝华胶合板
6. 木嵌边
7. 50mm板条，最小通风缝尺寸50mm
8. Tyevek房屋包层透气膜
9. 木龙骨，嵌入丁基胶泥
10. 镀锌角钢防水板
11. 涂漆膨胀防火钢梁，最短防火时间60min
12. 矿物棉填充
13. 锌制滴水槽
14. 酚醛胶和板凹槽，最小通风缝尺寸25mm
15. 锌制防虫网
16. 硅胶密封，环绕窗框
17. 自钻螺丝和装饰盖
18. 聚氨酯发泡和硅胶密封
19. 12.5mm石膏板吊顶，涂料喷涂
20. 冷轧屋顶Z形檩条
21. 32mm压型平台装置
22. 锥形Kingspan装置，近些固定于金属平台上
23. Resitrix三元乙丙橡胶膜，延伸到护墙上
24. 防虫网
25. 25mm通风孔
26. 20x30mm黑色板条
27. 10mm黑色水泥板
28. 黑色硬木盖板
29. 护墙围栏
30. 锥形Kingspan隔热层
31. 矿物棉隔热层
32. 15mm holorib混凝土平台
33. Ideal Combi Futura窗户系统
34. 阳极氧化铝窗台，与窗口对齐
35. 硬木窗台
36. 150mm metsec螺柱
37. 黑色板条和底板
38. 25mm K18 Kingspan隔热干式套，涂料喷涂

Detail and Materials

The development is a steel framed construction with composite concrete and steel deck floors, all above an in-situ cast concrete basement. The stone walls to the front elevations are of cavity wall construction with partial fill insulation. The areas of rendered insulation are mechanically fixed back to a blockwork substrate, and deliver the highest performance insulation, giving the walls a U-value of 0.13 W/m^2. The areas of cladding are all fixed back to a Metsec stud substrate, with a cold air construction between the cladding and insulation. The windows are all by Ideal Combi, from their Futura system range, all thermally broken anodised aluminium and timber profiled frames. The internal courtyard is paved with black paving bricks, and the areas of decking to upper storey walkways and terraces are all Iroko hardwood.

项目概况

项目场地坐落在公墓路边，位于谢菲尔德市中心西南一英里（约1.6千米）处，在19世纪建成的耐德边界郊区的边缘。这属于一块棕色地带，前身曾是一座废弃的二层商业建筑和配套停车场。由于场地靠近谢菲尔德公墓，多年来一直被废弃，现在则被列为英国历史园林遗产保护项目的保护区之内。这块场地的南侧同样与一块保护区接壤。

公墓路住宅项目是Project Orange建筑事务所与该客户所合作的第二个项目。Project Orange首先对场地进行以住宅为主的开发进行了可行性分析，项目包括一些商业空间、停车场和多种规格的商业住宅单元，其中还包括客户妹妹家的大型公寓以及一些归客户所有的服务型出租公寓。

项目的目标是打造与所在环境相符的高品质、高密度的住宅。项目以中央庭院的模式展开：主要休闲便利设施都设在中央景观庭院中，四周则由别墅和公寓环绕。项目试图通过这种方式来强调一种社区感。九座联排别墅采用了两种住宅类型。这两种类型都选择了动态横截面设计，拥有宽敞的阳台和屋顶露台。住宅的内部配置优化了相邻住宅的隐私保护，东侧联排别墅的主居室朝向楼梯的窗户，而西侧联排别墅的辅助生活区则朝向庭院。同样，多样化的横截面设计形成了变化多样的天花板净高，突出了传统贵族住宅中的二楼会客室。

联排别墅的内部流线设计以建筑中的散步长廊为基本原则。建筑师并没有为了方便而将楼梯段直接叠加，而是在每座别墅里通过夸张的散步路线将主要空间和辅助空间连接起来，就像连成一串的珠子一样。这种楼梯、路线、主次居住空间、屋顶露台和阳台的布局实现了建筑立面强烈的构造感，具有强烈对比的包覆材料则进一步突出了这种造型。

与高度模式化和图形化的室内立面相比，建筑的外立面通过传统的约克郡石材向保护区表达了敬意。这种材料在19世纪和20世纪早期的谢菲尔德及周边地区有广泛的应用。石材与环境之间形成了联动感应，而窗口、顶楼以及一楼商业空间的细节设计则展示了项目与庭院之外空间的联系。

细部与材料

项目采用钢铁框架配负荷混凝土和钢板楼板，下方采用现浇筑混凝土地基。建筑正面的石墙是空心墙，仅采用部分填充隔热。隔热层通过机械固定在砌块基质的后面，从而实现最佳的隔热性能，墙面的传热系数仅为0.13 W/m^2。外墙包层固定在Metsec螺柱基板后方，在包层与隔热层之间由冷空气隔开。窗户全部采用Futura系列的Ideal Combi材料，配有断热阳极氧化铝框和木框。内部庭院铺设着黑色地砖，而木板路和平台区域则采用绿柄桑硬木进行铺装。

Police Station in Salt

萨尔特警察局

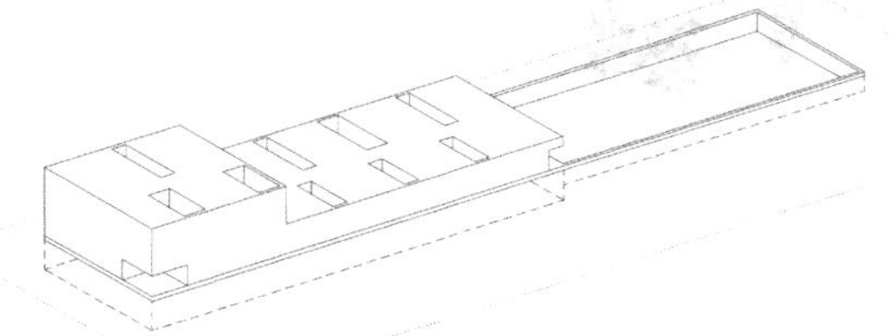

Location/地点: Salt, Spain/西班牙，萨尔特
Architect/建筑师: Josep Ferrando + Sergi Serrat
Photos/摄影: Adrià Goula
Site area/占地面积: 4,320m²
Built area/建筑面积: 1,650m²
Completion date/竣工时间: 2011
Key materials: Façade – precast concrete panel textured white
主要材料：立面—— 白色纹理预制混凝土板

Overview

The new police station is located on the outskirts of the neighbourhood in a zone reserved for council buildings. The proposal is placed in a pre-existing grove of magnificent beech trees that determine the main decisions for the project. The main building mass is concentrated at the north end of the site, in the unoccupied area, while the car park is organised around the grove. The remainder of the site forms a green cushion that separates the building from the road becoming a transition space for the visitors.

The building is separated from the street by no more than a wooded garden. There are no railings to bar the public. The police should be seen to be accessible and friendly. A compact, horizontal, abstract and introverted volume is proposed due to the requirements of high-security, use and flexibility of the programme. The public entrance is formed by raising the head at one of the ends of the building forming a large access porch. The staff entrance is placed at the opposite end of the main façade.

Once inside everything changes. A system of double height spaces and patios organise the private zones and flood the interior with light. The colour white becomes the main player. Its texture changes to differentiate the different walls: textured concrete for the exterior, polished stone on the floor, white brick and wood for the walls, perfect white for the ceiling...

The lightness of the interior is also reflected in the organisation of the building. The programme is ordered around strict hierarchical and functional criteria.

Detail and Materials

All panels are modulated so all the building can be completely solved with a single 3 metre wide piece. What's different is the length of the panel. All waterproofing solutions (upper and bottom) are precast with the panel so no extra elements are needed. A corrugated texture is chosen as panel finish so joins between elements are almost invisible so the volume can be read as a big massive white concrete cube.

项目概况

新建的警察局位于市郊专门为政府办公楼所预留的区域内，具体位置在一片茂密的桦树林里，这决定了项目的主要设计方向。主楼集中在场地的北端的一片空地，而停车场环绕树林展开。场地的其他部分形成了绿色缓冲带，将建筑与道路隔开，并且可作为来访者的过渡区。

建筑仅通过园林与街道隔开，并没有设置任何栅栏，这让警察局看起来亲和而友好。为了保证警察局对安全性、功能性和灵活性的高度需求，项目采用了紧凑抽象的内向型水平设计。建筑的一端向上提升，形成了一个巨大的门廊作为公共主入口。工作人员入口则设在建筑正面的另一端。

建筑内部截然不同，双高空间和天井将不同的办公空间组织起来，同时也为室内带来了充足的自然光线。白色是整个场景的主色调，它以不同的纹理来标志出不同的墙面：外墙是纹理混凝土，地面是抛光石板，内墙是白砖和木材，而天花板则是纯白色。

室内的轻盈感同样反映在建筑的组织结构上。警察局内的功能空间遵守了严格的层级和功能标准划分。

纵向的脊柱式空间布局围绕着中庭展开，在公共区域与私人区域之间建立起清晰的过渡。

细部与材料

所有混凝土板材都采用模块化设计，因此整座建筑立面都以3米宽的板材构造而成，不同点仅在于板材的长度。所有防水措施（包括上层和下层）都是预制的，无需使用额外的附件。外墙的瓦楞纹理作为一种装饰性元素，几乎将所有构件的接缝隐藏起来，让整个建筑结构看起来是一个统一的白色混凝土方块。

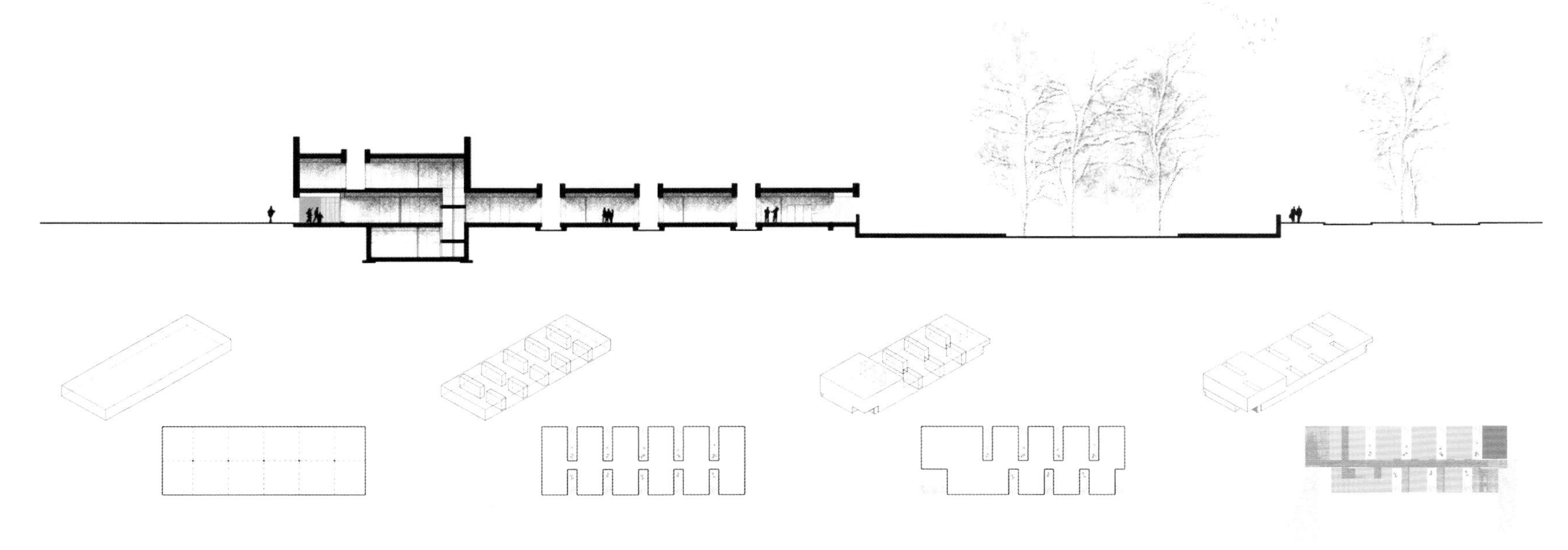

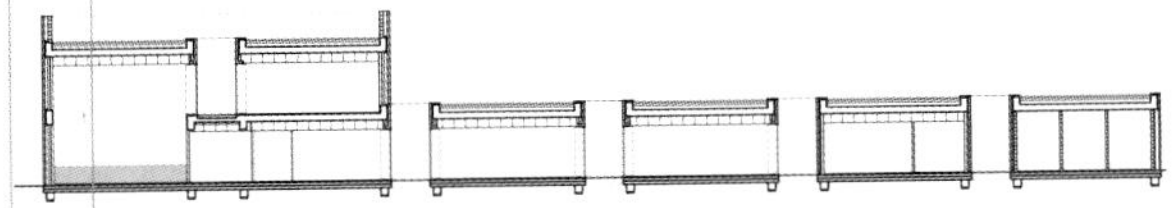

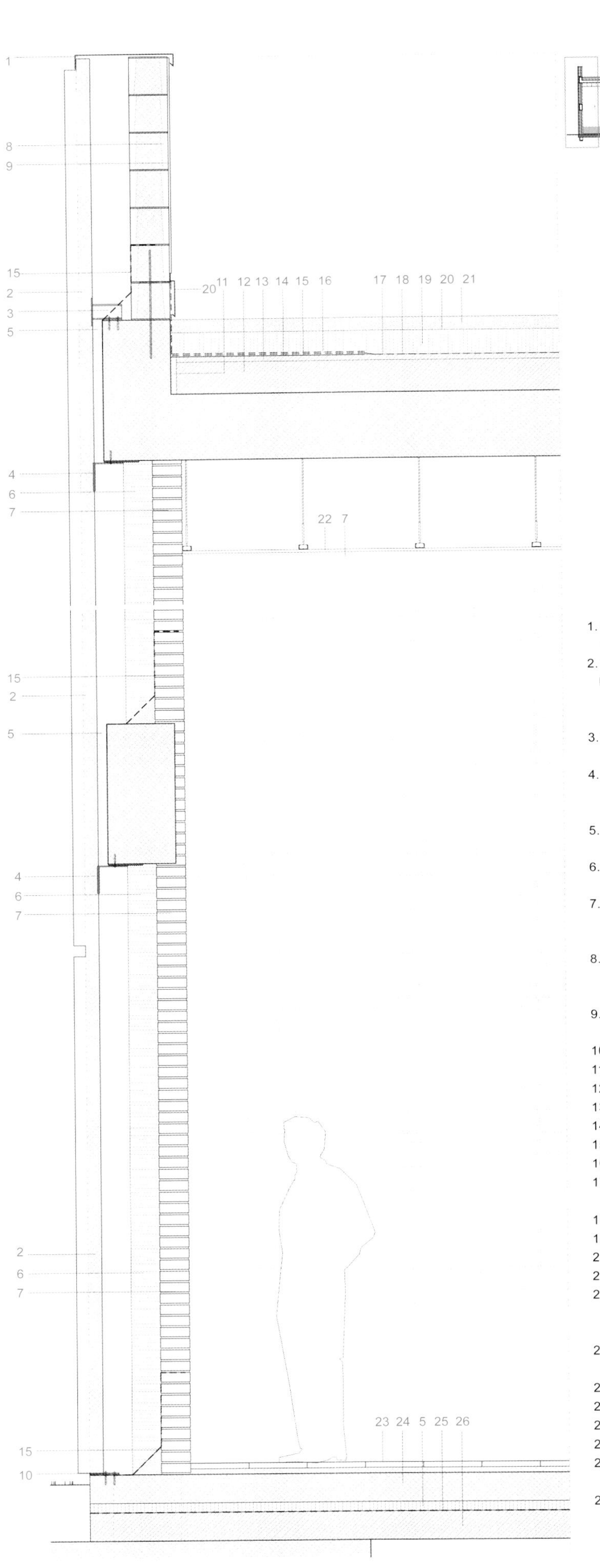

1. Coronation drip galvanised steel plate 0.7mm thick preformed and development 60cm affixed with mechanical fasteners
2. Precast concrete panel textured white. Cutting dimensions to project. Manufactured from white cement and limestone barren 15mm maximum diameter and armed with steel ferrules AMOUNT 500s in not less than 50kg/m^3. Formwork on one side of the rubber mold 1/148 Lanzarote Reckli house with a nominal thickness of 60+70mm (relief-solid)
3. Top fixing: Support. Assembly consisting of base plate 180x12x8 H-29 reinforced with stiffeners embedded in the precast concrete panel.Mechanically fixed to the slab
4. Fixing Lower: Rollover adjustable. The set consists of smooth Gripalon Rail 40/20 (L = 200) with connectors. Bolt with washer and lock nut simple toothed plate 60.8 (180 +100). Posted mechanically forged a taco Save Fixed M12x110 and Brake
5. 50x50x8 Insulation with EPS expanded polystyrene plate S. 50mm thick. 1.1 m^2K/W thermal resistance, smooth surface and sides of plain edge, placed with mechanical fasteners
6. Insulation with EPS polystyrene sheets expanded S. 150mm thick. 3.2m^2K/W thermal resistance, smooth surface and sides of plain edge, placed with mechanical fasteners
7. Ceramic enclosure to go inside view of production. Catalan and white format 29x14x5 sided view. Place a break together, with the arid white. Armed with prefabricated trusses inside hot galvanized steel or equivalent type and similarly MURFOR
8. Wall enclosure 20cm thick hollow block 400x200x200mm smooth, mortar, category I, according to the UNE-EN 771-3, standard cement to coat, placed with mortar mixed 1:0.5: 4 of Portland cement limestone and sand stone granite. Posted on wrought iron rods 0 10 and placed every 60cm length 120cm
9. Good view plaster vertically with cement mortar, trowelled finish and painted with plastic paint with smooth finish: two layers of sealant and finishing. Colour to be defined by D.F.
10. Regulating the slab. Profile hot galvanized metal L 140x140 mechanically attached to the structure.
11. Band Edge extruded polystyrene 2cm thick
12. Formation of dry cellular concrete without hangers. Density 300 kg/m^3, average 15cm thick, trowelled finish
13. Base Support: mortar layer = 3cm
14. Asphalt primer 0.5 kg/m^2
15. Edge Reinforcement Sill: Lamina waterproofing type ESTERDAN 40/GP
16. Corner Edge Reinforcement: Waterproofing sheet type ESTERDAN 30 P elastomeric
17. Double waterproofing sheet formed by sheets of asphalt modified bitumen LBM (SBS)-40 armor felt 100g/m^2 fiberglass, bonded between them
18. Pinch geotextile layer of 150g/m^2
19. Heavy protection: boulder layer 16 to 32mm in diameter, 10cm thick, placed un-bonded.
20. Pinch geotextile layer of 200g/m^2
21. Heavy protection: boulder layer 16 to 32mm in diameter, 10cm thick, placed un-bonded.
22. False continuous sound absorbing ceiling plasterboard panels with continuous linear circular perforation. Roofing with records 500x500. The same plate reattached with the second flat ceiling. Painted with Plastic Smooth finish: two layers of sealing and finishing. Color to be defined by D.F.
23. MATA terrazzo flooring type reference number 2000 2000 or equivalent and similar. Specimen size 30x30, placed discontinuously according to plans
24. Reinforced concrete floor HA-25/B/20/IIa 150mm thick with armed # 1 0 8c/20
25. Polyethylene sheet
26. Gravels pitching 150mm thick
27. Black limestone. Exploded 30x15 according to plans. Bush hammered finish.
28. Flashing galvanised steel plate 0.7mm thick and 20cm preformed development, placed with mechanical fasteners as detailed
29. Ceramic screed: a row of perforated brick 290x140x100mm.

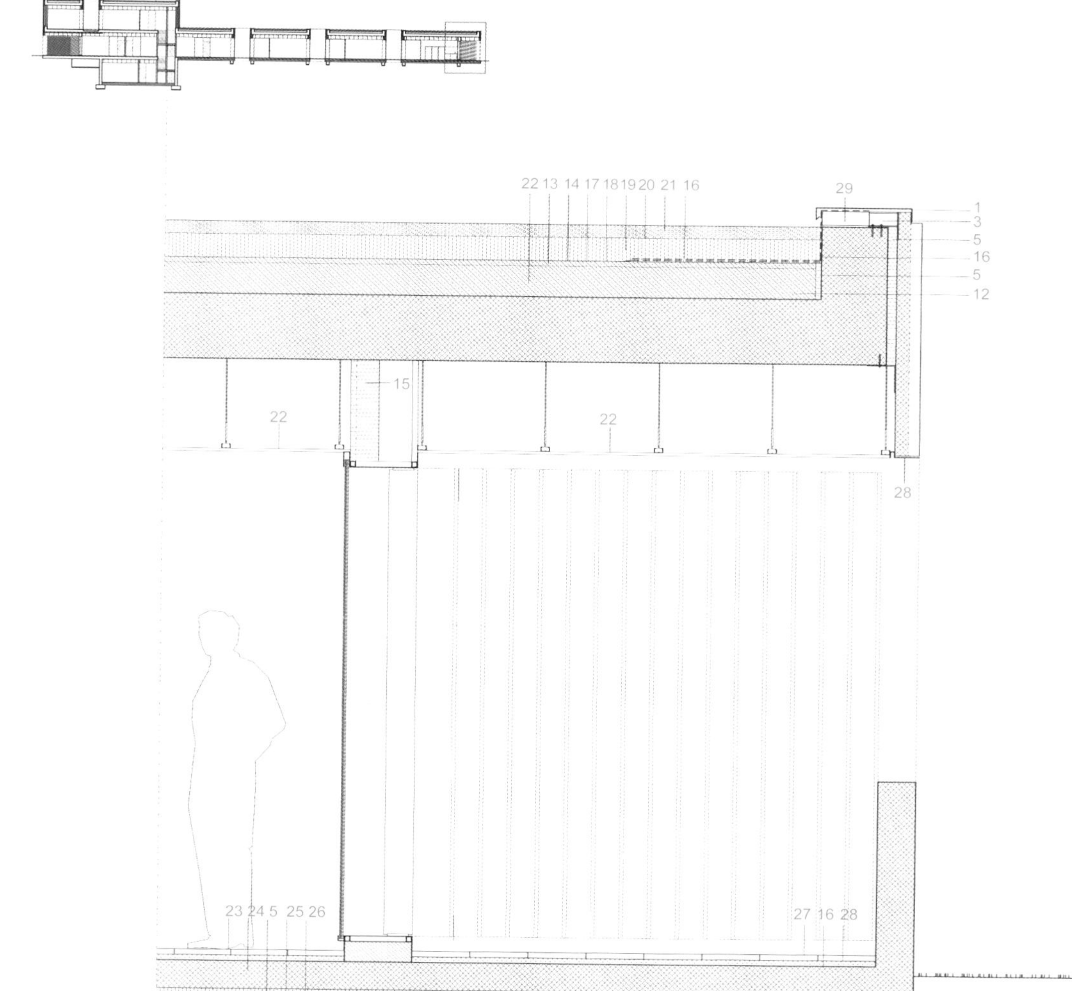

1. 预制镀锌钢板0.7mm厚，60cm固定在机械固定件上
2. 白色纹理预制混凝土板；切割成项目适用尺寸。由白色水泥和生石灰石（最大直径15mm）制作而成，钢圈AMOUNT 500s加固量不少于50kg/m³；模架位于橡胶模具1/148 Lanzarote Reckli一侧，标准厚度为60+70mm
3. 顶部固定：支架。底板组件装配，180x12x8 H-29（加强剂加固，嵌入预制混凝土板）；机械固定于底板上
4. 底部固定：可翻转调节。由光滑的Gripalon栏杆40/20（L=200）和连接件组成；螺栓包括垫圈和缩进螺母齿板60.8（180+100）；机械锻造固件Save Fixed M12x110和闸件
5. 50x50x8隔热层，EPS泡沫聚苯乙烯板S. 50mm厚；1.1 m²K/W热阻，光滑表面，平边，固定在机械固件上
6. EPS泡沫聚苯乙烯板S. 50mm厚；1.1 m²K/W热阻，光滑表面，平边，固定在机械固件上
7. 陶瓷墙围，Catalan白色版本29x14x5侧面；内置预制桁架加固，热镀锌钢或同等级的MURFOR
8. 墙围，20cm厚空心砖，400x200x200mm光滑砂浆，I级，标准水泥涂层，与波兰水泥砂浆和砂岩花岗岩混合比例为1:0.5:4；贴在锻铁杆010上，间距60cm，长120cm
9. 与水泥砂浆垂直的装饰石膏，抹光面，塑性涂料光滑饰面：两层密封剂及装饰面
10. 校正板材；热镀锌L形金属型材，140x140，机械固定于结构上
11. 边缘，挤塑聚苯乙烯2cm厚
12. 干型多孔混凝土结构，无吊架；密度300kg/m³，平均厚度15cm，抹光面
13. 底座支撑：砂浆层3cm
14. 沥青底漆0.5kg/m²
15. 边缘加固窗台：防水薄板，ESTERDAN 40/GP型
16. 角边加固：防水薄板ESTERDAN 30 P型人造橡胶
17. 双层防水板，由改性沥青LBM（SBS）薄层构成，100g/m²纤维玻璃加固
18. 收缩土工布层150g/m²
19. 重保护：卵石层16~32mm直径，10cm厚，未黏合状放置
20. 收缩土工布层200g/m²
21. 重保护：卵石层16~32mm直径，10cm厚，未黏合状放置
22. 连续隔音假吊顶石膏板，配有连续线形圆形穿孔，屋面500x500；附着第二层假吊顶。涂有塑性光滑涂料：双层密封和饰面
23. MATA水磨石地面，参考号2000 2000或类似；样本尺寸30x30，不连续铺装
24. 钢筋混凝土板HA-25/B/20/IIa，150mm厚，# 1 0 8c/20加固
25. 聚乙烯板
26. 碎石沥青涂层，150mm厚
27. 黑石灰石，30x15，凿制饰面
28. 镀锌钢板防水板，0.7mm厚，20cm成形制作，机械固定
29. 陶瓷砂浆：一排多孔砖290x140x100mm

Andalusian Institute of Biotechnology

安达卢西亚生物技术研究院

Location/地点: Seville, Spain/西班牙，塞维利亚
Architect/建筑师: Sol89 : María González, Juanjo López de la Cruz, Salvador Méndez and Francisco González
Photos/摄影: Jesús Granada
Built area/建筑面积: 6,398.74m²
Key materials: Façade – precast concrete panels
主要材料：立面——预制混凝土板

Overview

Located in one of the lots of the former International Exposition 1992 in Seville, a tertiary landscape saturated with high-flown singularities, a heritage from the exhibition's pavilions, the project includes areas for certain medical specialties, the surface of which is proportional to the capital invested in each of them. From this proportionality, the architects proposed a section shaped as a bar chart, setting up three linked voids forming a large space that develops the Centre's circulation and activities.

A compact piece raises, rough and opaque to the outside world and dug in it, using the void and porosity as the fundamental work subject. The architecture becomes what is in the middle, between what is built, embodying the Debussy's quote: "Music is the space between the notes". The void is formalised through the combination of different floor plans with a similar porosity but distributed in different ways, creating patios, terraces and voids. The circulations run parallel to the void, one side for the patients and the other for the medical staff, so that both meet only on the trays that conforms each specialty and that floats over the interior void.

Detail and Materials

The structure is solved using a bridge type reinforced concrete beams of light 14 meters and 1 meter ridge that form the large central vacuum. For siding the architects use a series of concrete panels that meet the outer layer of the enclosure, maximum dimensions 12x1.10 m and 12 cm thick. These panels are

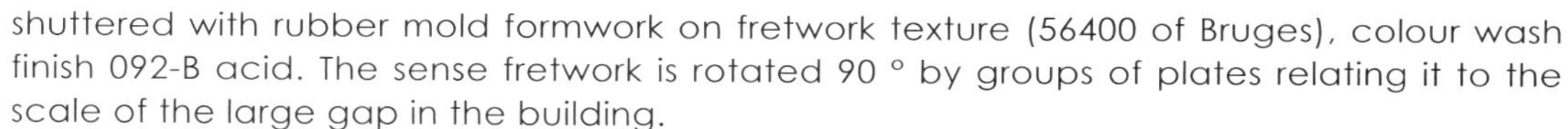

shuttered with rubber mold formwork on fretwork texture (56400 of Bruges), colour wash finish 092-B acid. The sense fretwork is rotated 90 ° by groups of plates relating it to the scale of the large gap in the building.

The panels are anchored to the concrete pillars and beams, and the concrete wall in the case of the east elevation, with anchorage plates with dimensions 150x150x10 with 2 anchors Hilti HST-M12 and 2 plates of variable length according to details. The lower kick plate is solved by a folded galvanised sheet 3 mm thick. At the upper edge this plate also protects the waterproofing layer overlaps introduced into the closure panel through a notch formed therein at the time of concrete in the workshop. For the meeting between panels the architects use seal plates "compriband" reinforced with sealing mastic high strength polyurethane the same colour as the plate is used. The seals are made in the lower portion of the fretwork so that they are hidden.

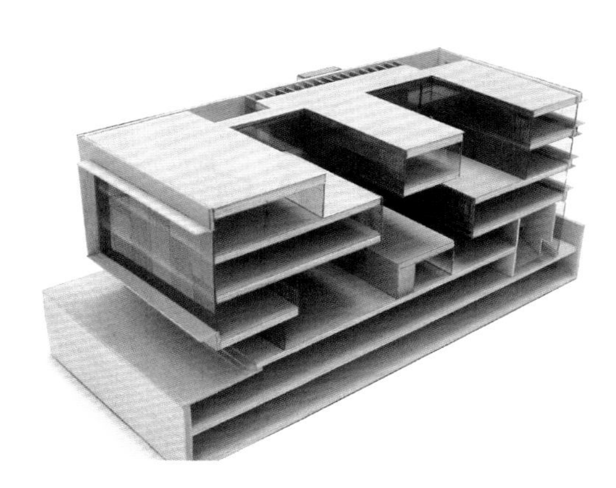

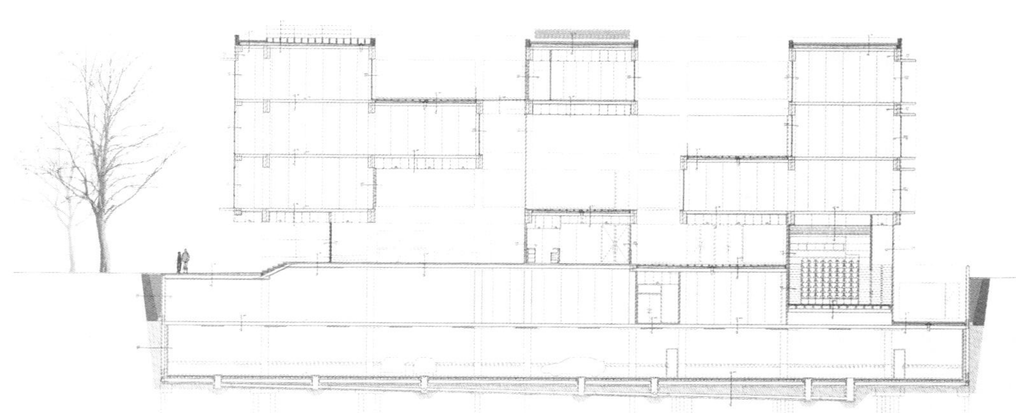

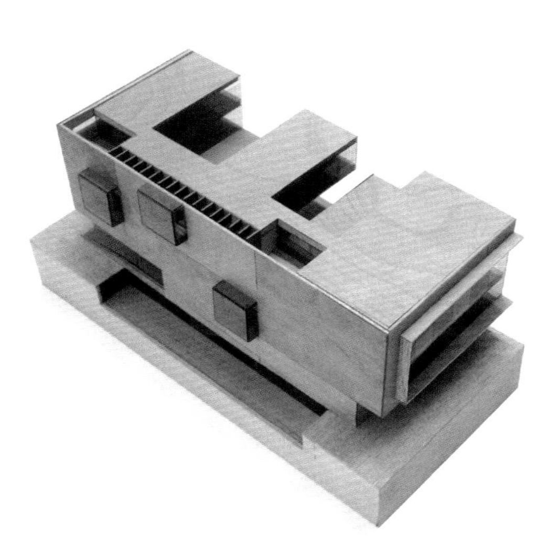

项目概况

项目场地是1992塞维利亚世博会的园区所在地之一。该地块曾遍布展览场馆，拥有奇特丰富的景观。项目包含若干个医学专业区域，各个区域的面积取决于该学科所获得投资的份额。因此，建筑师将建筑设计为一个长条造型，以三个相互连接的建筑结构构成一个大型空间。

项目显得紧凑、粗糙而封闭，以中空空间作为基础结构。正如德彪西的名言"音乐是音符之间的空间"一样，建筑也是建筑结构之间的空间。不同的楼面布局以相似的结构相互组合起来，形成了天井、露台和中空空间。建筑的交通流线与空间平行，一侧供患者使用，另一侧供医务人员使用，二者只能在相应的学科平台上相遇。

细部与材料

建筑利用由14x1米的桥型钢筋混凝土梁实现了大型中空结构。建筑师利用一系列混凝土板作为护墙板，使其与外墙包层贴合起来，混凝土板的最大尺寸为12x1.10米，12厘米厚。这些板材以带有浮雕纹理的彩色涂层橡胶模具建模。根据不同的室内结构，这些纹理分组成块，在水平和垂直的方向上构成了多样化的建筑墙面。

混凝土板固定在混凝土梁柱和东面的混凝土墙上，通过尺寸为150x150x10的锚固板固定起来。下层刮板采用3毫米厚的镀锌折叠板，在折叠板的上缘设有防水层。在各个板材的接合处，建筑师采用由高强度聚氨酯密封加固的同色密封板进行密封。密封处设在浮雕纹理的下半部，因此可以隐藏起来。

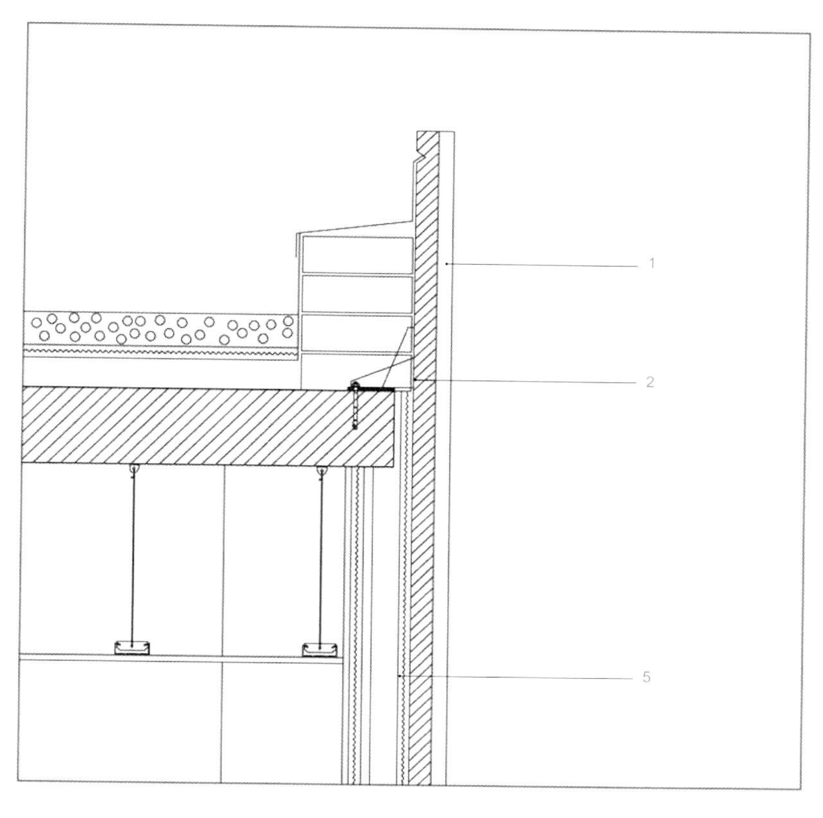

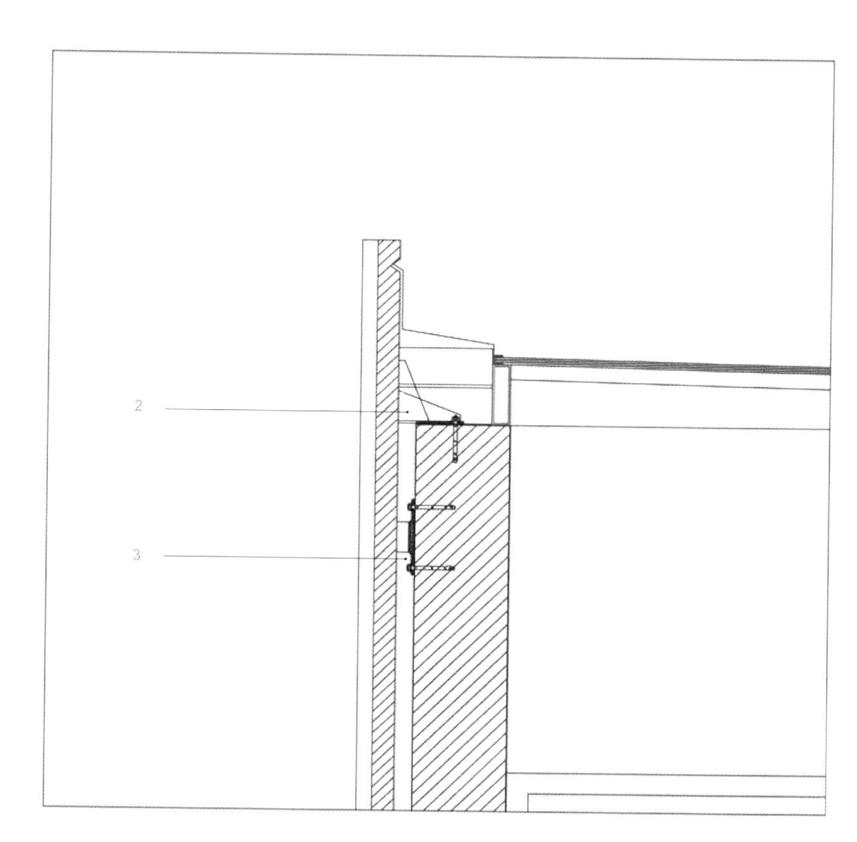

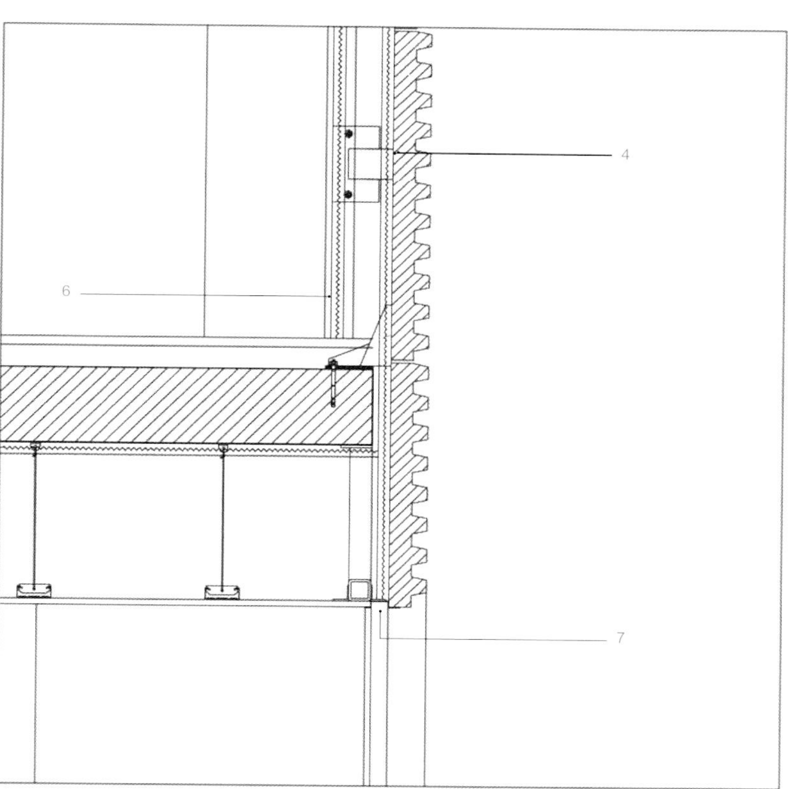

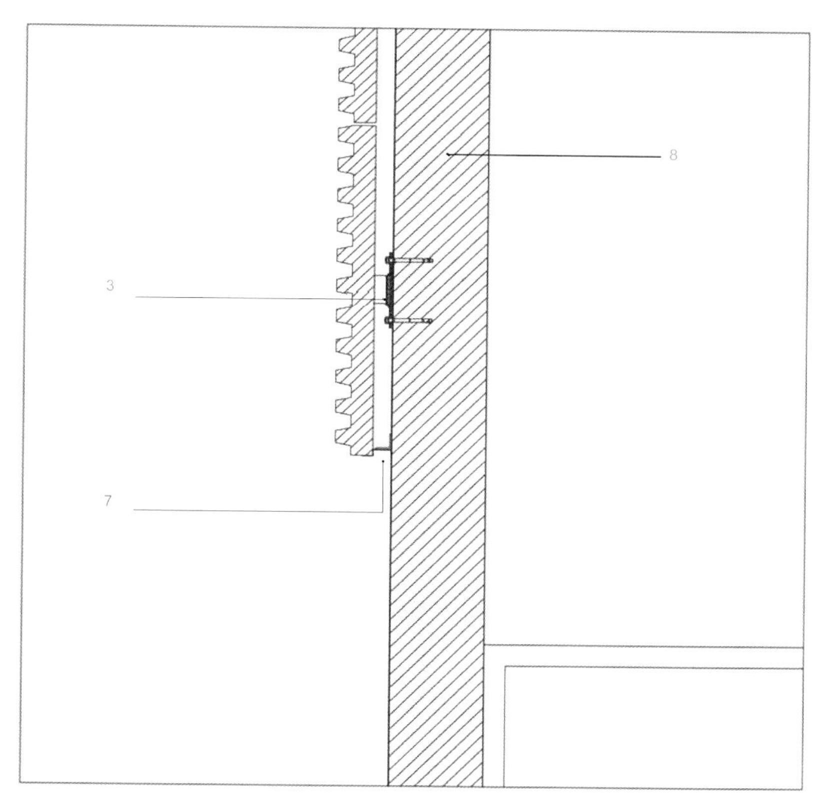

1. Prefabricated concrete pannel made with loose stones of limestone, Max. dimensions: 12x1.1x0.12 m, shuttering with rubber formwork (56400 de Brügge), acid finish surface colour 092-B
2. Steel sheet for mooring (150x150x10) with Hilti HST-M12 fixed to the top of the last slab
3. Anchor steel sheet 150x150x10 with Hilti HST-M12 fixed to the concrete wall
4. Anchor steel sheet 150x150x10 with Hilti HST-M12 fixed to the slab edges and pillars
5. Projected polyurethane foam (5cm) over the concrete pannel's back
6. Dry wall of plasterboards with acustic isolation
7. Lower finish piece between concrete pannels and ceilings with folded galvanised steel sheet (3mm)
8. Concrete wall

1. 预制混凝土板，由石灰石粉制成；最大尺寸12x1.1x0.12m，以橡胶模具建模，酸蚀表面色彩092-B
2. 固定钢板（150x150x10），用Hilti HST-M12固定在上一块板的顶部
3. 锚钢板150x150x10，用Hilti HST-M12固定在混凝土墙上
4. 锚钢板150x150x10，用Hilti HST-M12固定在板材边缘和柱子上
5. 挤塑聚氨酯泡沫（5cm），覆在混凝土板后面
6. 石膏板干墙，带有隔音层
7. 混凝土板和天花板之间的低层装饰组件，配有折叠镀锌钢板（3mm）
8. 混凝土墙

Nursery in La Pañoleta

帕诺勒塔幼儿园

Location/地点: Camas, Spain/西班牙，卡马斯
Architect/建筑师: Antonio Blanco Montero
Photos/摄影: Fernando Alda
Site area/占地面积: 1,000m^2
Built area/建筑面积: 593.95m^2
Completion date/竣工时间: 2012
Key materials: Façade – precast concrete panels printed with ribbed texture formliner colored green + double lock standing seam green zinc coating + coloured glass
主要材料：立面——预制混凝土板，印有绿色棱纹线条+绿色双锁立缝锌板层+彩色玻璃

Overview

The decision of building a nursery in this area tries to solve the historical lack of such services. The location of the building in a development area, next to a marginal zone, becomes an important point of the project, in addition to the morphology of the elongated plot.

The social environment of the building is characterised by an extremely contrasted situation, where social housing and high cost housing living on both sides of the street. This has caused a certain hermetic attitude of the high-cost housing, caused by excessive extroversion of social housing tenants, who make the street an extension of the house. Furthermore, the City Council has planned a concentration of school equipment in order to make easier the relations between these.

This has generated a certain kind of tensions in the design phase of the project, which has induced, in some way, an introverted character of the building, but the basic approach is to not turn away from that reality. The administrative area and professors are set at San Juan de Aznalfarache Street, next to the aforementioned marginal zone, orienting classroom to south and east, on the opposite side of the plot.

This configuration, together with the rectangular shape, very elongated, causes the appearance of a series of courtyards that articulates the interior of the building. Over this organisational scheme stands a continuous deck, which rises controlled in certain points to generate some skylights that illuminate the interior of the distribution areas and give indirect lighting to classrooms.

There are two fundamental aspects that link the building with the public promotion that generates it: 1. the search for a low maintenance and energy consumption. Highly durable materials such as zinc, concrete and HPC, and controls energy consumption and support renewable energy are used, and 2. the possibility of using of the building out of school time as a meeting facility of potential neighbourhood associations, by configuring of control and access area, along with the multi-purpose room and kitchen-dining room, separating these rooms from the rest of the centre.

Detail and Materials

To attend the need of short time work execution, it was decided to adopt pre-fabricated large format building systems, precast concrete panels façade, plaster-board partitions, zinc roof, which together with the steel structure and composite floor slab, has reduced execution times enormously.

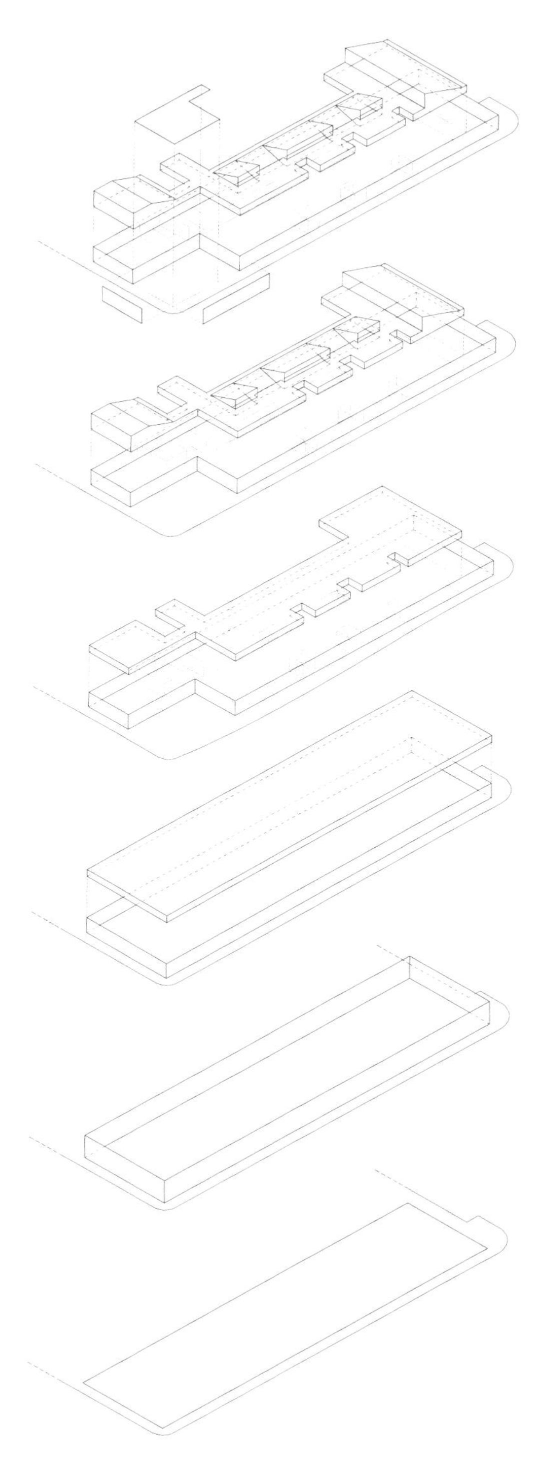

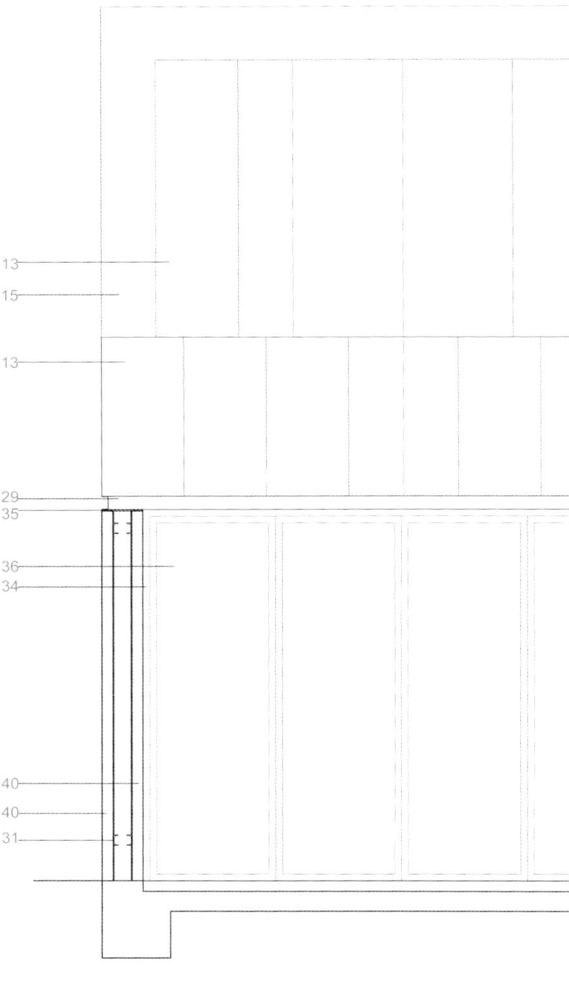

项目概况

这所幼儿园的建造旨在解决历史遗留下来的此类服务的缺失。建筑位于一个发展中区域，紧邻边缘地带，这一地理位置和建筑场地的细长地形是项目设计的关键所在。

建筑所在社会环境具有极大的差异，社会福利住宅和高端住宅分列街道两边。由于社会福利住宅的住户不断向外扩张，将街道变成了住宅的延伸，高端住宅的住户显得更加封闭。因此，市议会决定打造一个集中的学校设施来缓和二者的关系。

环境因素为项目设计带来了一定的压力，多少导致了建筑的内向型设计，但是设计的基本方案仍然是直面现实环境的。行政区和教师办公室都设在靠圣胡安街一面，紧邻前面提到的边缘地带；教室朝向东南，在地块的另一侧。

这种布局与场地细长的长方形造型共同决定了将室内空间连接起来的庭院的外观。在这种组织布局上是一个连续的平台，平台在某些点向上凸出形成天窗，照亮了室内流通区域，也为教室带来了间接照明。

建筑与公共空间的联系体现在两个方面。首先，建筑力求实现低维护需求和低能源消耗。项目选用了锌、混凝土、高性能混凝土等耐久性材料，对能耗实施控制，并且使用了辅助可再生能源。其次，建筑在教学时间外可以用作社区活动空间。建筑的出入区域以及多功能厅和餐厅都远离项目核心，方便供外界使用。

细部与材料

为了满足短期施工的要求，项目决定采用预制模块化建筑系统、预制混凝土板墙面、石膏板隔断、锌屋顶等结构，它们与钢结构和组合楼板一起，大幅缩短了施工时间。

1. S 275 J0H steel rectangular hollow section 80.40.3. for holding doors frames
2. S 275 J0H steel angle section LD 50.30.2 for holding façade boarding
3. Steel beam IPE
4. Thermal insulation polystyrene panel 60mm thick
5. Water-repellent DM board for holding VMZ Quartz-Zinc sheet
6. VMZ DELTA polyethylene layer
7. VMZ Quartz-Zinc lichen green double lock standing seam roofing
8. Galvanised steel deck and permanent formwork
9. Composite floor slab
10. S 275 J0H galvanised steel Z section 100.50.3. for holding deck boarding
11. Wood batten.
12. Galvanised steel channel section U 30.30.2. for holding glazing
13. SGG STADIP colour glass
14. Pinewood frame
15. VMZ Quartz-Zinc lichen green double lock standing seam façade
16. S 275 J0H galvanised steel angle section LD 100.60.3. for holding glazing
17. Light-weight concrete for slopes
18. Water proof layer
19. Plasterboard 13mm thick
20. Protection gravel layer 5cm thick
21. Brick masonry for parapets
22. Tile coating 20x10cm white
23. Terrazzo tiles pavement 40x40x3cm, light green
24. Quartz-Zinc lichen green folded sheet

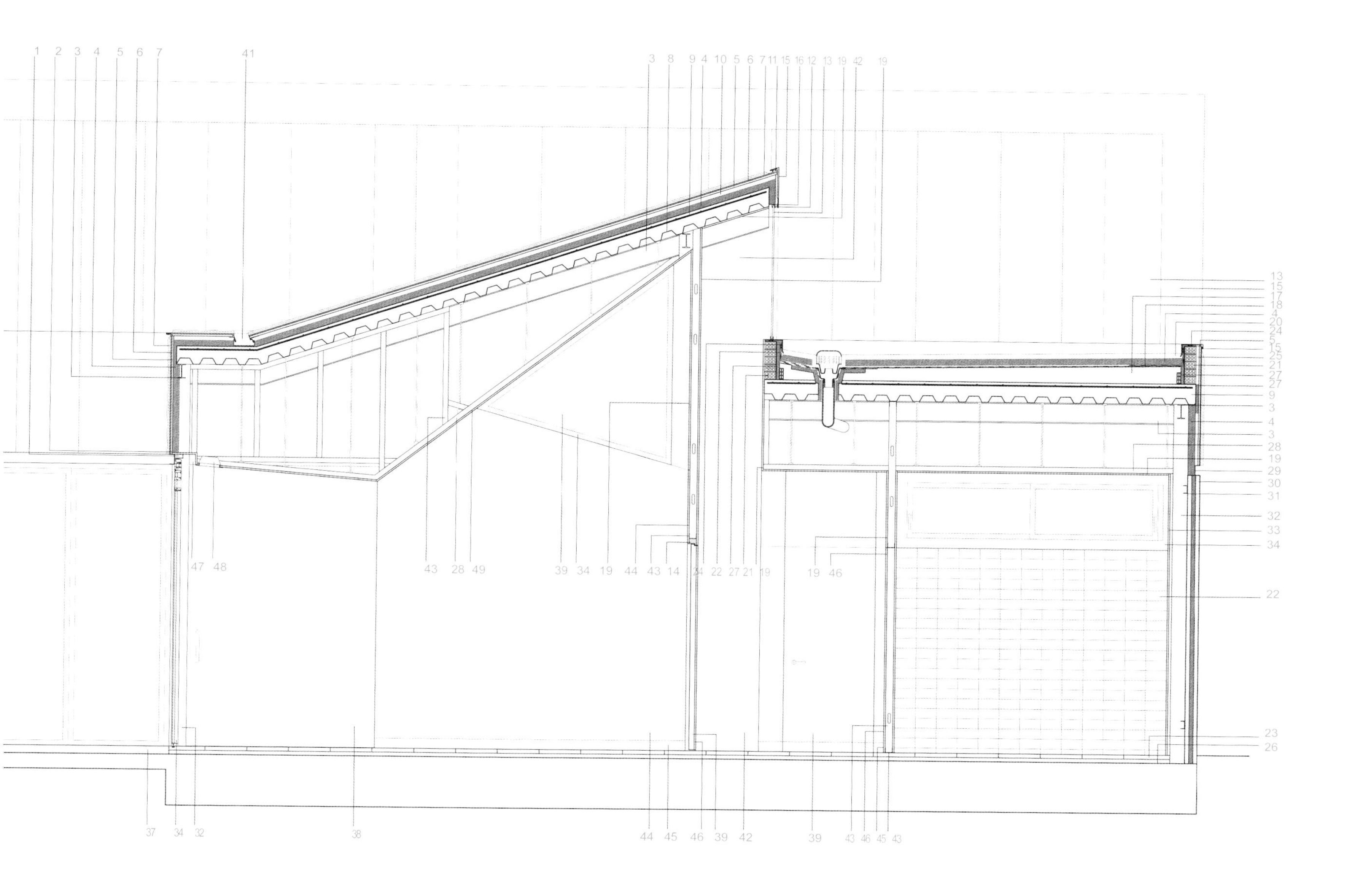

25. Wood batten 30x30mm
26. Sand layer 3cm thick
27. Thermal insulation polystyrene panel 30mm thick
28. Aluminium channel section for plasterboard system
29. Galvanised folded steel sheet gasket
30. Reinforced concrete panel printed with ribbed texture formliner, colored green
31. Steel channel section UPN80 for holding concrete panels
32. Steel beam HEB
33. Boarding for tile coating
34. Coloured aluminium frame RAL 1026
35. Galvanised folded steel sheet for covering concrete panels
36. Insulating glass SGG PLANISTAR 6mm+ air space12mm+SGG SECURIT 4+4mm
37. Rubber Flooring tiles 60x60x7cm
38. Sliding door
39. DM laminated HPL, colour RAL 9016 door, smooth surface
40. Reinforced concrete panel printed with ribbed texture formliner, coloured green
41. Zinc gutter
42. Ice finished tempered glass
43. Aluminium channel section for plasterboard system
44. HPL laminated board colour RAL 9016, granulated surface
45. Anodised aluminum folded sheet for skirting, 10cm high
46. HPL laminated board colour RAL 9016, smooth surface
47. Lateral plasterboard cavities in ceiling for hiding shade system
48. Recessed modular fluorescent lighting system, asymmetric flow
49. Acoustic plaster ceiling 13mm thick

1. S 275 J0H矩形空性型钢80.40.3，用于固定门框
2. S 275 J0H角钢型材LD 50.30.2，用于固定墙面板
3. 钢梁IPE
4. 隔热聚苯乙烯板，60mm厚
5. 防水DM板，用于固定VMZ石英锌板
6. VMZ DELTA聚乙烯层
7. VMZ石英锌青苔铝双锁立缝屋顶
8. 镀锌钢板层和永久性模板
9. 组合楼板
10. S 275 J0H镀锌Z形型钢100.20.3，用于固定平台板
11. 木板条
12. 镀锌U形钢槽，用于固定玻璃装配
13. SGG STADIP彩色玻璃
14. 松木框
15. VMZ石英锌青苔铝双锁立缝墙面
16. S 275 J0H镀锌角钢型材LD 100.60.3，用于固定玻璃装配
17. 轻质混凝土斜坡
18. 防水层
19. 石膏板，13mm厚
20. 防护碎石层，5cm厚
21. 砖砌护墙
22. 瓷砖覆层，20x10cm，白色
23. 水磨石地砖铺装40x40x3cm，浅绿色
24. 石英锌青苔绿折叠板
25. 木板条30x30mm
26. 沙土层，3cm厚
27. 隔热聚苯乙烯板，30mm厚
28. 铝槽型材，用于安装石膏板系统
29. 镀锌折叠钢垫片
30. 钢筋混凝土板，印有棱纹纹理，绿色
31. 槽钢型材UPN80，用于固定混凝土板
32. 钢梁HEB
33. 瓷砖板
34. 彩色铝框RAL 1026
35. 镀锌折叠钢板，用于覆盖混凝土板
36. 夹层玻璃，SGG PLANISTAR 6mm+空气层12mm+ SGG SECURIT 4+4mm
37. 橡胶地砖60x60x7cm
38. 拉门
39. DM层压HPL门板，RAL 9016色，光滑表面
40. 钢筋混凝土板，印有棱纹纹理，绿色
41. 锌槽
42. 冰面钢化玻璃
43. 铝槽型材，用于固定石膏板系统
44. HPL层压板，RAL 9016色，光滑表面
45. 阳极氧化折叠铝板墙脚线，10cm高
46. HPL层压板，RAL 9016色，光滑表面
47. 天花板上横向石膏板孔，用于隐藏遮阳系统
48. 嵌入式模块化照明系统，不对称流
49. 隔音天花板，13mm厚

RATP Bus Centre

RATP公交中心

Location/地点: Paris, France/法国，巴黎
Architect/建筑师: ECDM Architectes
Photos/摄影: Benoît Fougeirol, Philippe Ruault
Built area/建筑面积: 2,200m²
Key materials: Façade – Ductal® concrete prefabricated panel and colored glass
主要材料：立面——Ductal®预制混凝土板和彩色玻璃

Overview

The RATP wished to realise a new building intended for the bus drivers and the administrative staff of the existing centre. Located in Thiais, it controls all the bus lines of the south and the east of Paris city. It's an industrial context, characterised by a succession of boxes, at best basic. Characterised by a dense square plan (35m x 35m) developed on 2 levels, it presents itself as a dense building, inert, deaf, and enigmatic as "the hull of a Russian submarine in the waters of Murmansk".

Detail and Materials

The designers reduced the site to a bus park, a vast monolithic concrete slab, uniformed territory composed of one sole material. The building starts with the deformation of the ground, the distortion of the existing concrete slab, and continues it with an apparently similar material Ductal®, still concrete but a dazzling sheet of concrete, which responds to very sophisticated demands: informality of the structure, constant evolution of the plans, dematerialisation, precision, density, homogeneity of aspect according to the mould designed. It ensures a continuity of the ground from the road, to the skin of the façades, the suspended ceilings and the terrace rooftop without any rupture. Combined with the structural qualities of the material, the building has neither a beginning nor an end. It is a continuity of a surface of which we can, depending on what you aim at, not control the limits. It is an inflection of a slab which spreads on the whole site. The building appears like a monolith with rounded

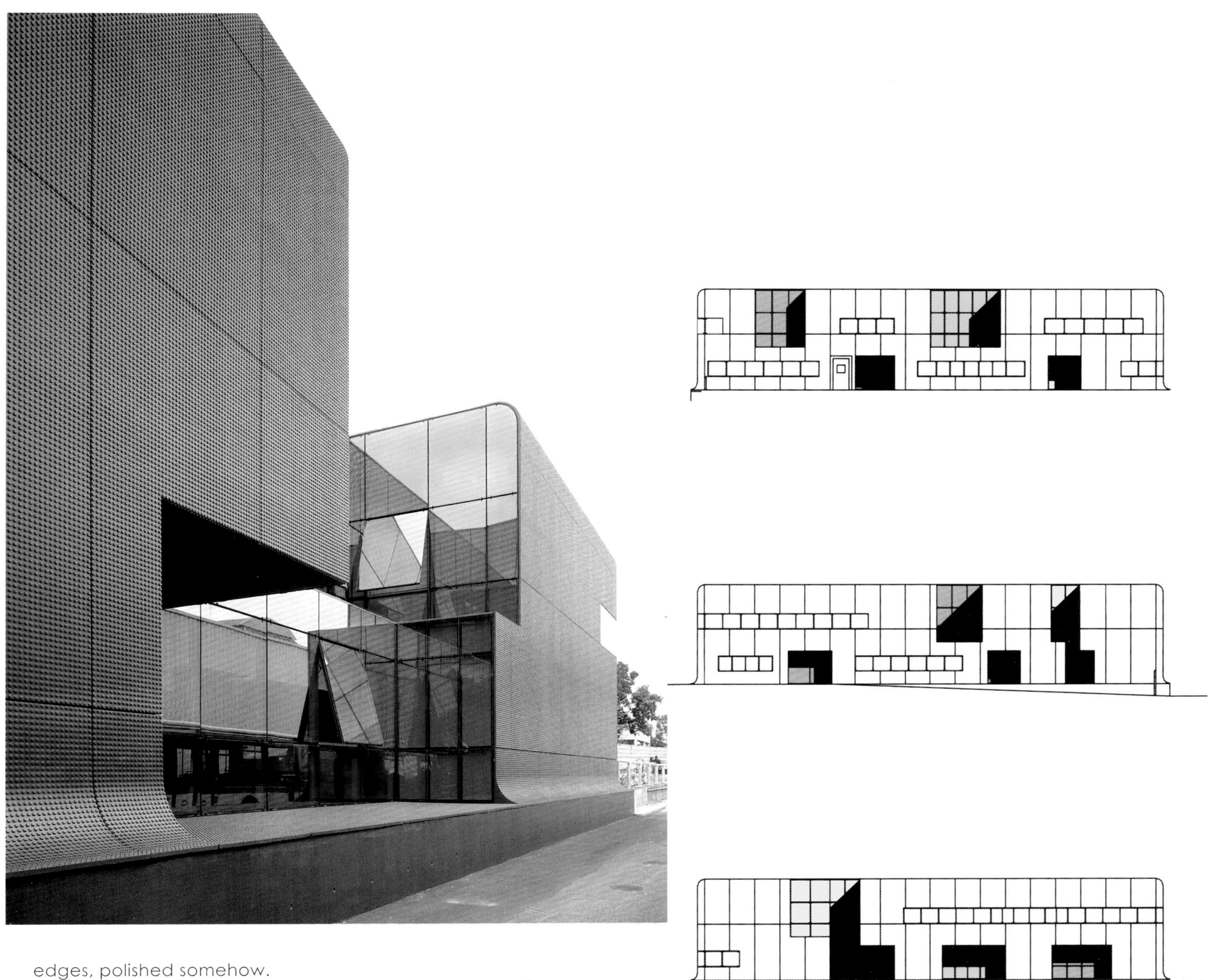

edges, polished somehow.

The texture games are facilitated by the flexibility of the material and its ability to be moulded with precision. The quality of finish and rigor of Ductal ® contribute to ensuring this concrete skin, accurate, continuous, perfect connections. The skin: 3cm thick displays a single texture of dots in relief like a game of "LEGO".

The "windows" are cut with a cutter blade: surgical incisions generating volumes in negative which reveal coloured mirrors under the thin crust of concrete. Treated with silver mirror dots, chromatics of the glass products are inspired by the tinted curtain wall frontages of the office buildings which border the main road.

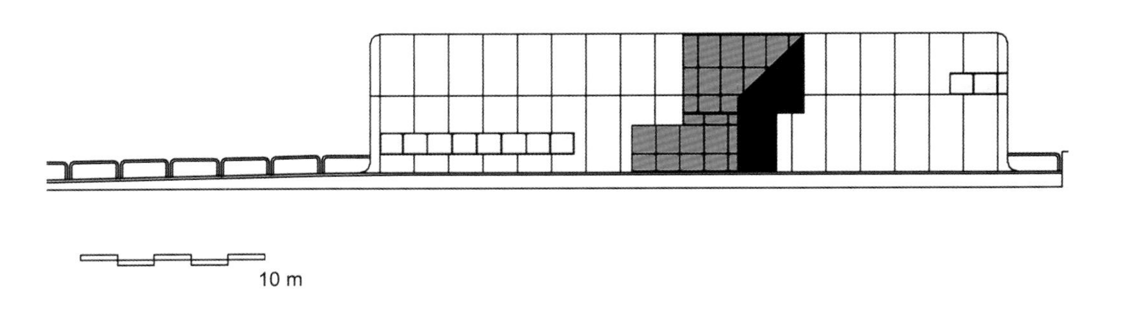

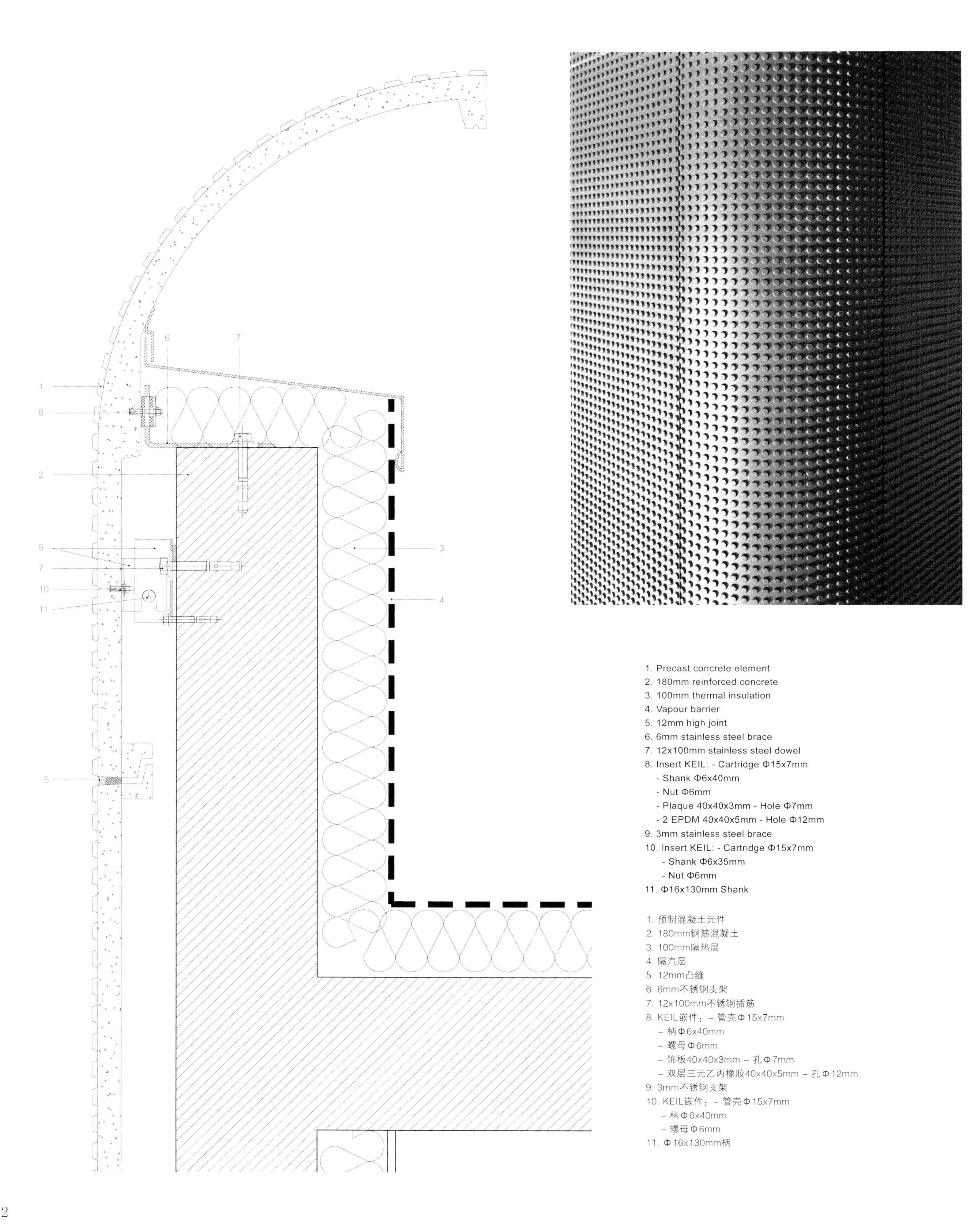

1. Precast concrete element
2. 180mm reinforced concrete
3. 100mm thermal insulation
4. Vapour barrier
5. 12mm high joint
6. 6mm stainless steel brace
7. 12x100mm stainless steel dowel
8. Insert KEIL: - Cartridge Φ15x7mm
   - Shank Φ6x40mm
   - Nut Φ6mm
   - Plaque 40x40x3mm - Hole Φ7mm
   - 2 EPDM 40x40x5mm - Hole Φ12mm
9. 3mm stainless steel brace
10. Insert KEIL: - Cartridge Φ15x7mm
   - Shank Φ6x35mm
   - Nut Φ6mm
11. Φ16x130mm Shank

1. 预制混凝土元件
2. 180mm钢筋混凝土
3. 100mm隔热层
4. 隔汽层
5. 12mm凸缝
6. 6mm不锈钢支架
7. 12x100mm不锈钢插筋
8. KEIL嵌件：– 管壳Φ15x7mm
   – 柄Φ6x40mm
   – 螺母Φ6mm
   – 饰板40x40x3mm – 孔Φ7mm
   – 双层三元乙丙橡胶40x40x5mm – 孔Φ12mm
9. 3mm不锈钢支架
10. KEIL嵌件：– 管壳Φ15x7mm
   – 柄Φ6x40mm
   – 螺母Φ6mm
11. Φ16x130mm柄

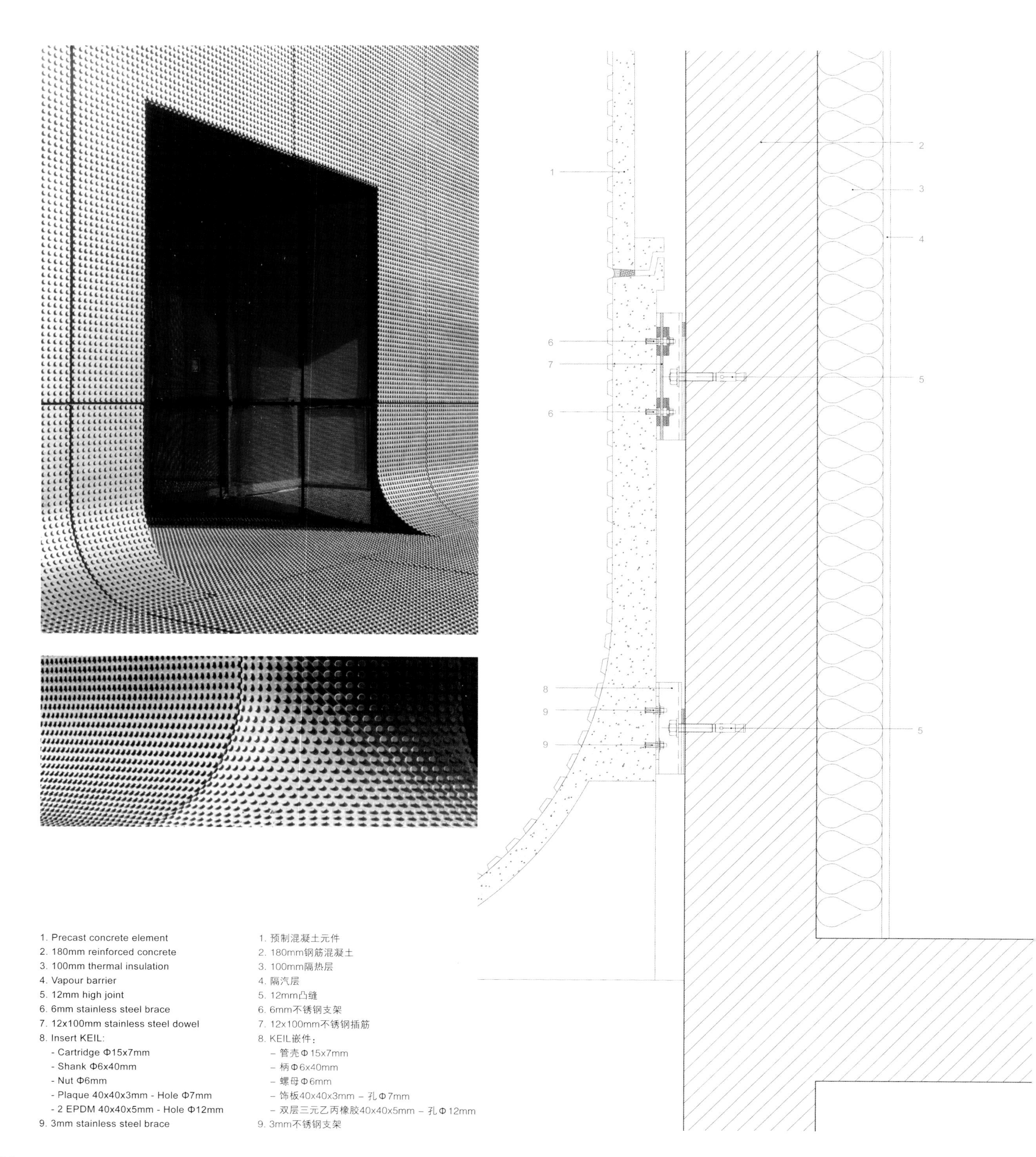

1. Precast concrete element
2. 180mm reinforced concrete
3. 100mm thermal insulation
4. Vapour barrier
5. 12mm high joint
6. 6mm stainless steel brace
7. 12x100mm stainless steel dowel
8. Insert KEIL:
   - Cartridge Φ15x7mm
   - Shank Φ6x40mm
   - Nut Φ6mm
   - Plaque 40x40x3mm - Hole Φ7mm
   - 2 EPDM 40x40x5mm - Hole Φ12mm
9. 3mm stainless steel brace

1. 预制混凝土元件
2. 180mm钢筋混凝土
3. 100mm隔热层
4. 隔汽层
5. 12mm凸缝
6. 6mm不锈钢支架
7. 12x100mm不锈钢插筋
8. KEIL嵌件：
   – 管壳Φ15x7mm
   – 柄Φ6x40mm
   – 螺母Φ6mm
   – 饰板40x40x3mm – 孔Φ7mm
   – 双层三元乙丙橡胶40x40x5mm – 孔Φ12mm
9. 3mm不锈钢支架

1. Bar wood
2. 19mm thermal insulation
3. Glued together outside glass
4. 6mm stainless steel brace
5. Insert KEIL: - Cartridge Φ15x7mm
   - Shank Φ6x40mm
   - Nut Φ6mm
   - Plaque 40x40x3mm - Hole Φ7mm
   - 2 EPDM 40x40x5mm - Hole Φ12mm
6. 12x100mm stainless steel dowel

1. 木条
2. 19mm隔热层
3. 胶合外层玻璃
4. 6mm不锈钢支架
5. KEIL嵌件：- 管壳Φ15x7mm
   - 柄Φ6x40mm
   - 螺母Φ6mm
   - 饰板40x40x3mm - 孔Φ7mm
   - 双层三元乙丙橡胶40x40x5mm
   - 孔Φ12mm
6. 12x100mm不锈钢插筋

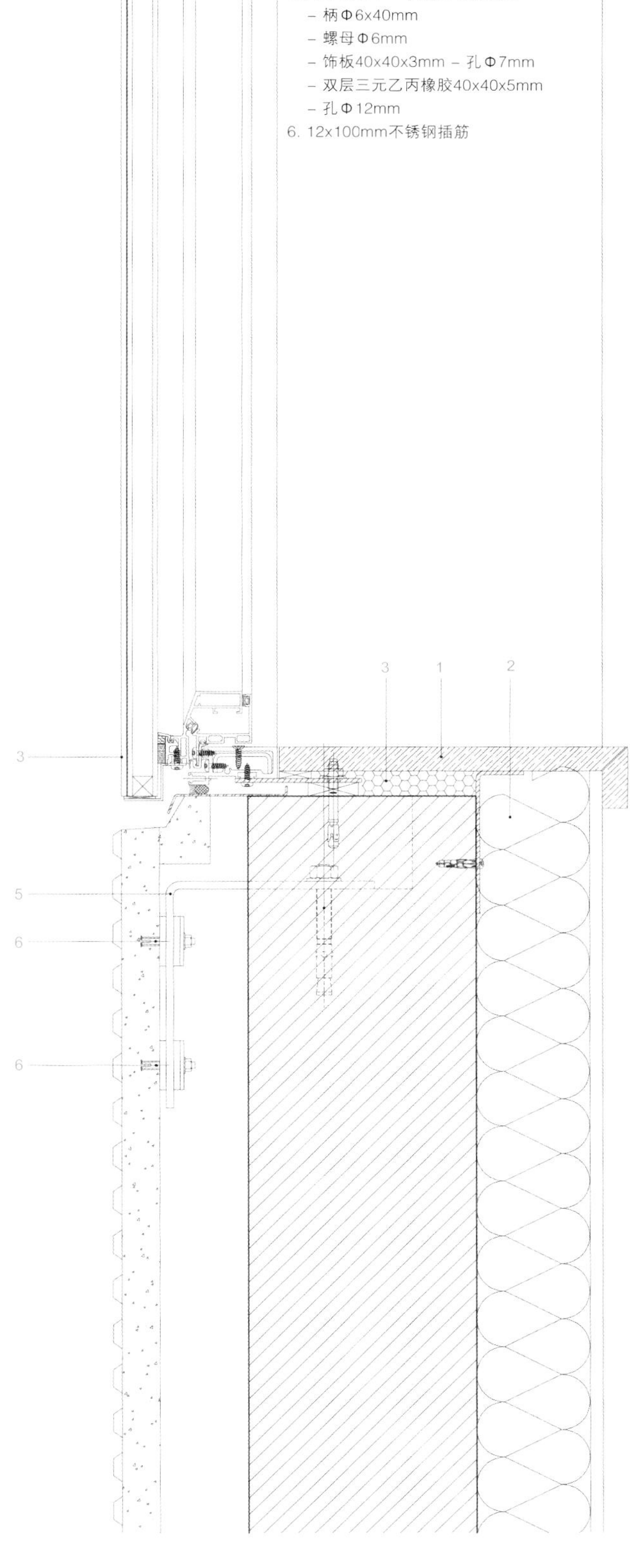

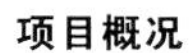

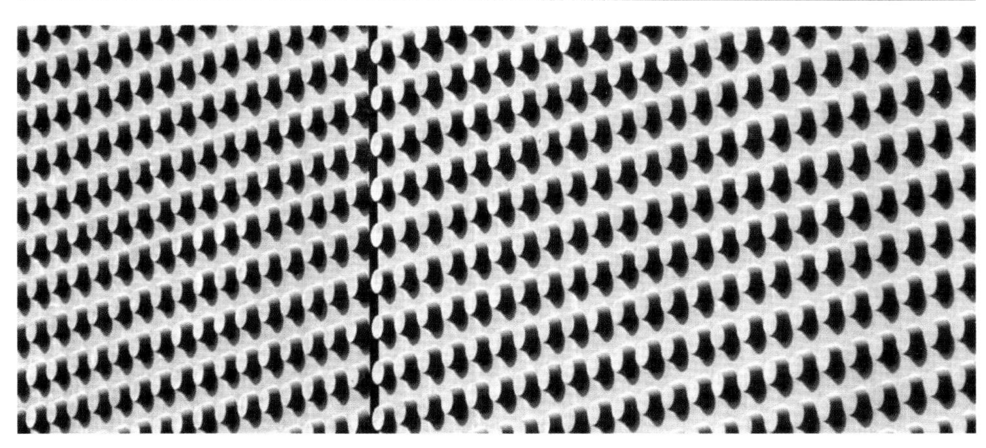

项目概况

巴黎大众运输公司（RATP）希望在原有中心的基础上为公交司机和行政办公人员新建一座办公楼。这座位于蒂艾的公交中心将控制巴黎城市东南部的所有公交线路。建筑位于工业环境内，以连续的盒式体块为特色。建筑采用紧凑的方形规划（35米x35米），分为两层空间，呈现出内向、神秘的姿态，就像是“摩尔莫斯克（俄罗斯港口城市）水域中的一艘俄罗斯潜艇”。

细部与材料

建筑师将项目简化为一个公交停车场，一块整体浇灌混凝土板，形成了由单一材料构成的空间领域。建筑从地面混凝土板的变形中逐渐升起，通过外观相似的Ductal®材料进行建造。这种材料是一种新型混凝土薄板，根据所设计模具的规制，它能够满足十分复杂的建筑需求：非常规结构、平面规划的连续演变、虚拟化、精确度、密度、均质性等。它能保证从地面到建筑外墙、吊顶、乃至露台屋顶的连续性，不会有任何断裂接缝。材料的结构特性让建筑“无始无终、无边无际”。这是一种没有限制的连续界面。确切来说，整个项目就是一块延伸于场地之上的板材。建筑看起来就像是一个进行了抛光的圆角巨石。

材料的灵活性以及精确的制模特性实现了建筑外墙巧妙的纹理设计。Ductal®材料的精准品质保证了这层混凝土表皮精确、连续、完美连接。3厘米厚的表皮上遍布小圆点，看起来就像乐高积木上的凸起一样。

窗户利用切割机切割出来，就像手术切口一样让彩色玻璃从混凝土外壳中显露出来。银色镜面圆点的加入让玻璃制品的色彩更加丰富，这一设计灵感来自于附近办公楼的彩色玻璃幕墙。

Fine Arts Museum

巴达霍斯美术馆

Location/地点: Badajoz, Spain/西班牙，巴达霍斯
Architect/建筑师: Estudio Arquitectura Hago (Antonio Álvarez-Cienfuegos Rubio & Emilio Delgado Martos)
Photos/摄影: Fernando Alda
Site Area/占地面积: 1,435.57m^2
Built area/建筑面积: 3,298.39m^2
Completion date/竣工时间: 2014
Key materials: Façade – perforated prestressed white concrete panel (2200x3000mm), steel frame and double glazing
主要材料：立面——白色预应力穿孔混凝土板（2200x3000mm）、钢架、双层玻璃

Overview

The backbone of the architectural strategy for the extension project of the Fine Arts Museum in Badajoz is meant to regain an Identity: a new built environment (Architecture) that interacts with the urban context (City) through its cultural content (Museum). The architects would like to make "the most beautiful building" to house the huge collection of Extremadura's great artists, like Picasso, Goya, Dalí, Zurbarán, Naranjo or Covarsí.

The architects project a complex whose starting point is the expansion of the existing museum, located in a listed building in the centre of Badajoz. The complex includes two new buildings that are connected through a courtyard and are opened to two different streets of the city. Due to the difficult circumstances that come together in this place (archaeological remains, party walls and the rehabilitation of the listed building) a powerful and coherent architectural response is required.

Two blocks are designed for different requirements. One building houses a permanent collection and it is attached to the original construction, adapting the new levels to the existing ones. The other block can be used for temporary exhibitions and it houses the headquarters offices in the fourth floor. They are connected by a courtyard, allowing to use the complex from different ways. In summer is possible to organise jazz crowded concerts.

The efficient and silent result allows the building to be integrated into the city without resorting to a compositional exercise, overflow-

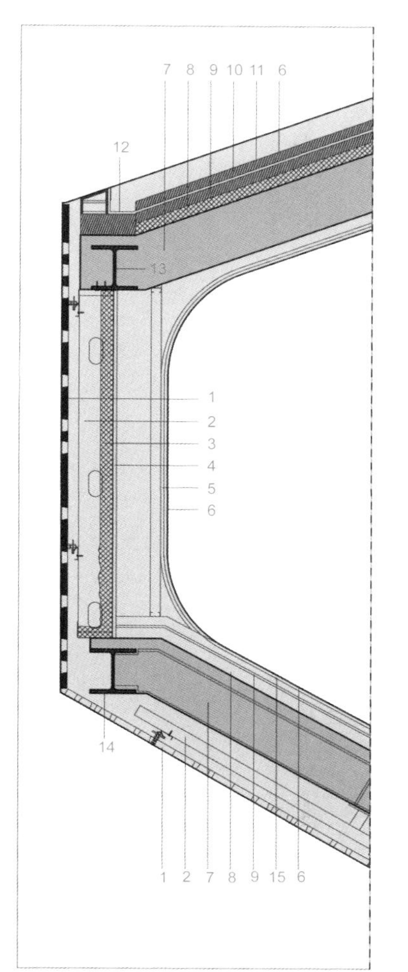

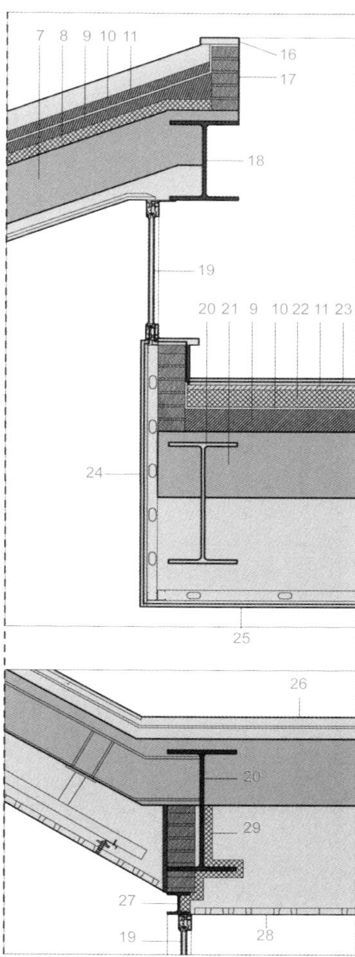

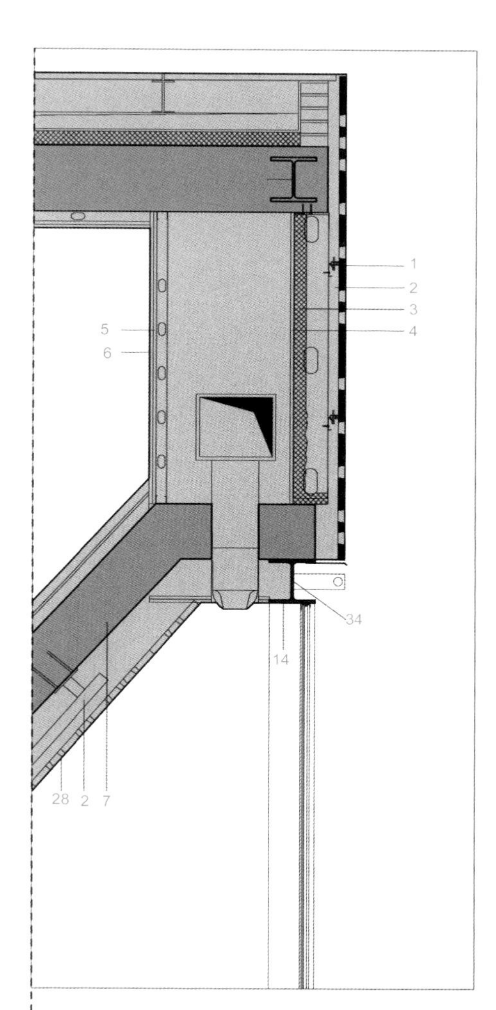

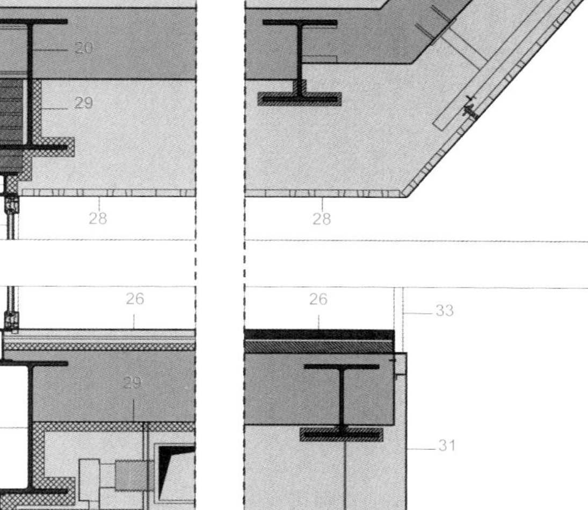

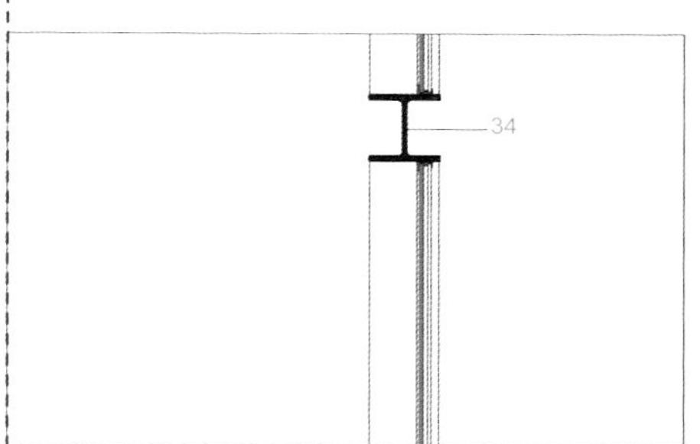

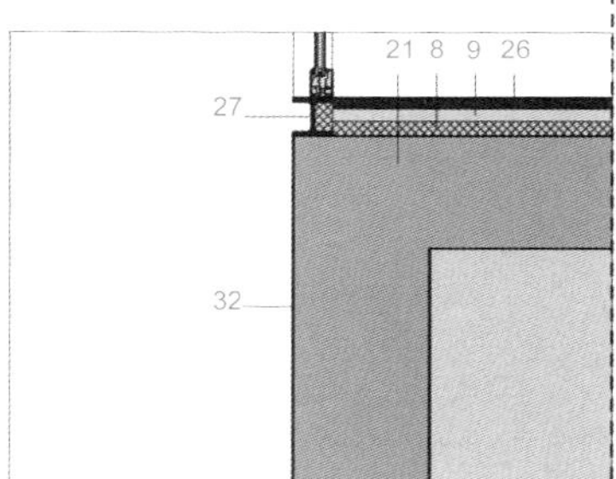

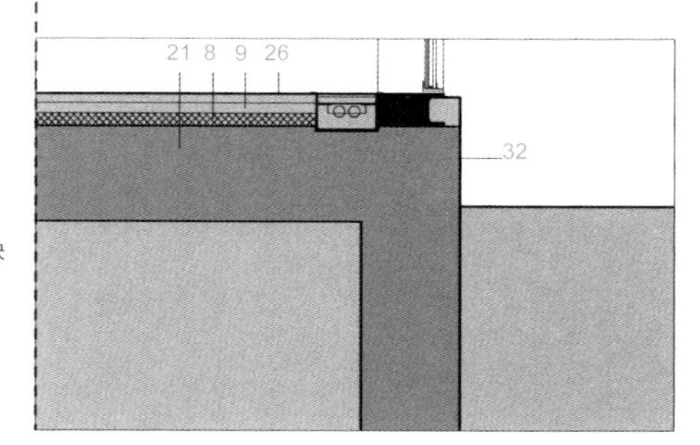

1. Perforated prestressed white concrete panel (2200x3000mm) horizontal composition
2. Galvanised steel substructure
3. Insulation (50mm)
4. Waterproof mortar panel
5. Plasterboard cladding (2x15mm)
6. White epoxy paint (1mm)
7. Reinforced concrete slab (250mm), metal joists and concrete blocks
8. Rigid polystyrene insulation (60mm)
9. Mortar levelling layer
10. Waterproof asphalt sheets
11. Coat of mortar (50mm)
12. Galvanised steel gutter
13. Upper truss chord
14. Bottom truss chord
15. Plasterboard cladding (6mm)
16. Coping stone
17. Low brick plastered
18. Steel beam (HEB-360)
19. Aluminium frame (thermal break) and double glazing
20. Steel beam (HEB-550)
21. Reinforced concrete slab (300mm), metal joists and concrete blocks
22. Rigid polystyrene insulation (80mm)
23. Cement tile (600x600mm)
24. Plasterboard cladding finished white RAL9010 (15mm)
25. Micro-perforated plasterboard cladding ceiling finished white RAL9010 (15mm)
26. Polished terrazzo floor (tiles 1200x1200mm)
27. End steel profile (HEB-100)
28. Perforated DM panel ceiling (30mm) finished white RAL 9010
29. Projected thermal insulation (60mm)
30. Plasterboard ceiling (15mm) finished white RAL 9010
31. Microperforated galvanised steel ceiling (tiles 600x1200mm)
32. Concrete: deactivated treatment
33. Steel handrail finished white RAL 9010

1. 白色预应力穿孔混凝土板（2200x3000mm），水平组合
2. 镀锌钢下层结构
3. 隔热层（50mm）
4. 防水砂浆板
5. 石膏板覆层（2x15mm）
6. 白色环氧漆（1mm）
7. 钢筋混凝土板（250mm）、金属托梁和混凝土砌块
8. 刚性聚苯乙烯隔热层（60mm）
9. 砂浆找平层
10. 防水沥青板
11. 砂浆层（50mm）
12. 镀锌钢槽
13. 上桁弦
14. 下桁弦
15. 石膏板覆层（6mm）
16. 顶石
17. 低层石灰砖
18. 钢梁（HEB-360）
19. 铝框（断热）、双层玻璃
20. 钢梁（HEB-550）
21. 钢筋混凝土板（300mm）、金属托梁和混凝土砌块
22. 刚性聚苯乙烯隔热层（80mm）
23. 水泥砖（600x600mm）
24. 石膏板覆层，白色RAL9010（15mm）
25. 微孔石膏板覆层吊顶，白色RAL9010（15mm）
26. 抛光水磨石地面（尺寸1200x1200mm）
27. 端钢型材
28. 穿孔DM板吊顶（30mm），白色RAL9010
29. 凸出的隔热层（60mm）
30. 石膏板吊顶（15mm），白色RAL9010
31. 微孔镀锌钢吊顶（尺寸600x1200mm）
32. 混凝土：去活化处理
33. 钢栏杆，白色RAL9010

ing the physical limits of the buildings and turning the proposed spaces into real stars of the urban environment. This project represents an encounter with beauty where the memory and its power remain in time; convert the whole museum into a new social, urban and artistic reality in the centre of Badajoz.

Detail and Materials

A typological scheme organised by an inverted "L", responds to the functional, spatial, structural and technical aspects. This scheme is fostered (inside and outside) through a cladding, a perforated prestressed concrete panel, whose scale and pattern seem to print a message in the skin of the building.

Large panels of glazing with white frames are set off-centre into bevelled metal-clad surrounds to form the entrance points to each of the buildings. Behind a double-height lobby faced with glass, a perforated metal platform with slim white railings forms the first floor of the permanent exhibition space. On the third level, the floor slopes dramatically upwards to account for the bevelled façade. Strips of glazing cut through the glossy white floors offer views into the galleries below and above. Levels in the new parts of the building are connected by a narrow white staircase, which is filled with natural light thanks to circular skylights overhead. On the upper floors of the temporary exhibition space, interior windows with deep black frames look from the new space into a terracotta-toned stairwell with ornate metalwork.

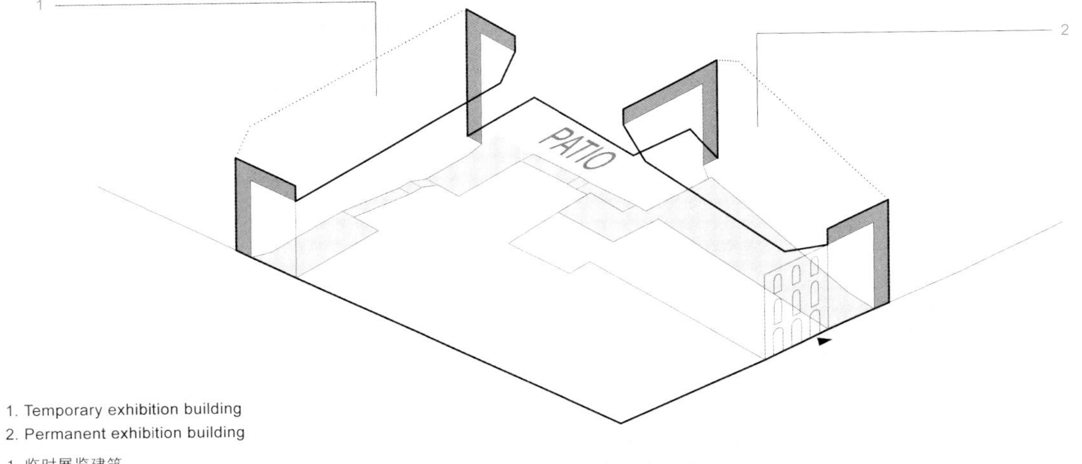

1. Temporary exhibition building
2. Permanent exhibition building

1. 临时展览建筑
2. 永久展览建筑

项目概况

巴达霍斯美术馆的扩建项目有意要重塑美术馆的形象：一个能与城市环境（城市）通过文化内涵（美术馆）产生互动的全新建成环境（建筑）。建筑师希望用“最美丽的建筑”来承放毕加索、戈雅、苏巴朗、纳兰霍、科瓦尔斯等埃斯特雷马杜拉地区最优秀艺术家的作品。

建筑师的项目规划以对原有美术馆(位于巴达霍斯市中心的一座历史建筑内)的扩建为出发点。项目包含两座通过庭院相连的新建筑，二者通过两条不同的街道与城市相连。复杂的项目场地环境条件（考古遗址、残垣断壁、历史建筑的修复）要求建筑师提出一个强大而连贯的设计方案。

两座楼的功能各不相同。一座楼内放置着永久性收藏品，与原始结构相连，新建的楼层与原有的楼层是统一的。另一座楼可用于临时展览，并且在四楼设置着美术馆的总部办公室。二者通过庭院相连，实现了功能上的连接。在夏天，庭院里可以举办爵士音乐会。

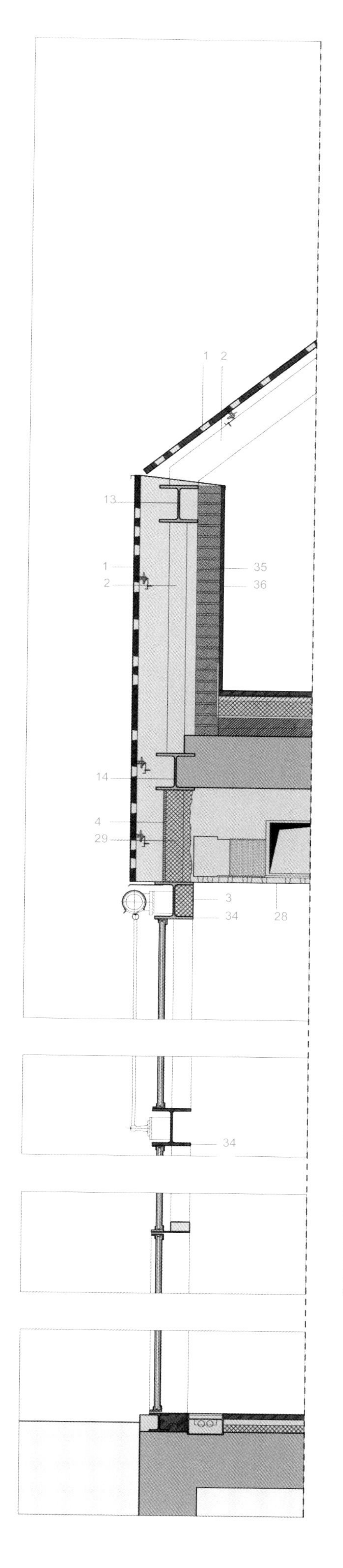

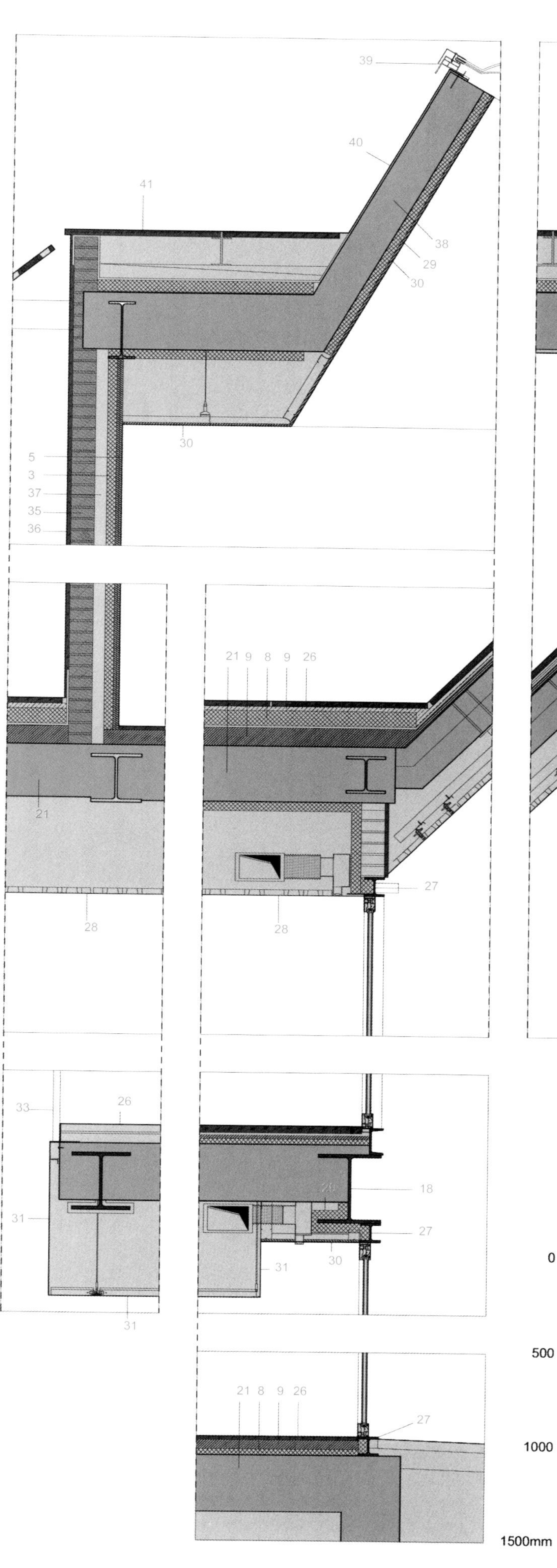

高效而低调的设计让建筑融入了城市，超越了建筑的物理极限，将项目空间变成了城市环境中的明星。项目是与美的邂逅，适时地留住了记忆和力量，将整个美术馆变成了巴达霍斯市全新的社交、城市和文化艺术中心。

细部结构与材料

反向的L形结构设计反映了功能、空间、结构、技术等方面的需求。建筑内外包覆着预应力穿孔混凝土板，这些板材的比例和图案就像在建筑表皮上印下了讯息一样。

以白色框架支撑的大块玻璃板被金属框切开，形成了建筑的入口。双高大厅后方的穿孔金属平台被白色栏杆所环绕，形成了永久性展览空间。三楼的地面向上夸张地倾斜，形成了倾斜的立面。玻璃条将光亮的白色地面切开，让人们可以看到上下两层的展览厅。新建的部分通过狭窄的白色楼体相连，头顶的天窗带来了充足的自然采光。在临时展览区的上层空间，透过配有黑色框架的室内窗可以看到一个陶土色调的楼梯间，里面配有华丽的金属装饰。

0
500
1000
1500mm

34. Steel beam (HEB-200)
35. 1FT wall perforated brick
36. Cement mortar (15mm) painted finish (white RAL 9010)
37. Ventilated air chamber (60mm)
38. Truncated cone concrete slab (200mm)
39. Transparent polycarbonate clamshell skyligtht (diameter: 500mm)
40. Waterproof mortar finished white
41. Floating cement tiles (600x600mm)

34. 钢梁（HEB-200）
35. 1FT墙面穿孔砖
36. 水泥砂浆（15mm），白色漆面RAL9010
37. 通风气腔（60mm）
38. 截锥混凝土板（200mm）
39. 透明聚碳酸酯蛤壳天窗（直径：500mm）
40. 防水砂浆，白色饰面
41. 浮式水泥砖（600x600mm）

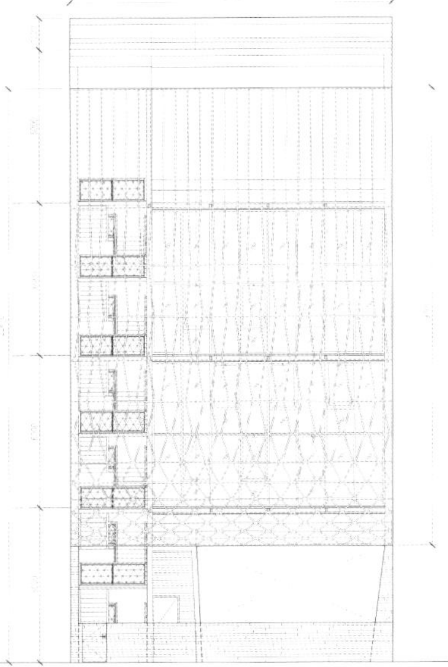

A Simple Factory Building

简单的工厂楼

Location/地点: Singapore/新加坡
Architect/建筑师: Pencil Office
Photos/摄影: Erik G.L'Heureux
Built area/建筑面积: 998m²
Completion date/竣工时间: 2012
Key materials: Façade – prefabricated lightweight concrete EIFS
主要材料：立面——预制轻质外墙保温系统

Overview

A Simple Factory Building addresses two contradicting demands: the mitigation of tropical solar radiation, and the openness, views, and transparency sought by the clients in a basic industrial typology. Completed in 2012, the 10,742-square-foot (998-square-metre) building is located in an industrial area of Singapore. It utilises a sophisticated 4-foot-deep (1.2-metre-deep) veil fabricated in lightweight EIFS and a bronze full-height window-wall envelope to reconcile this architectural conflict.

Wrapping continuously as a loop around the front elevation, car porch ceiling, rear elevation, and roof, the veil shields the building from the harsh tropical sunlight while calibrating views to the exterior. It also amplifies natural illumination, directs natural ventilation, and conceals mechanical equipment. It calibrates the performance of the building as a climatic engine.

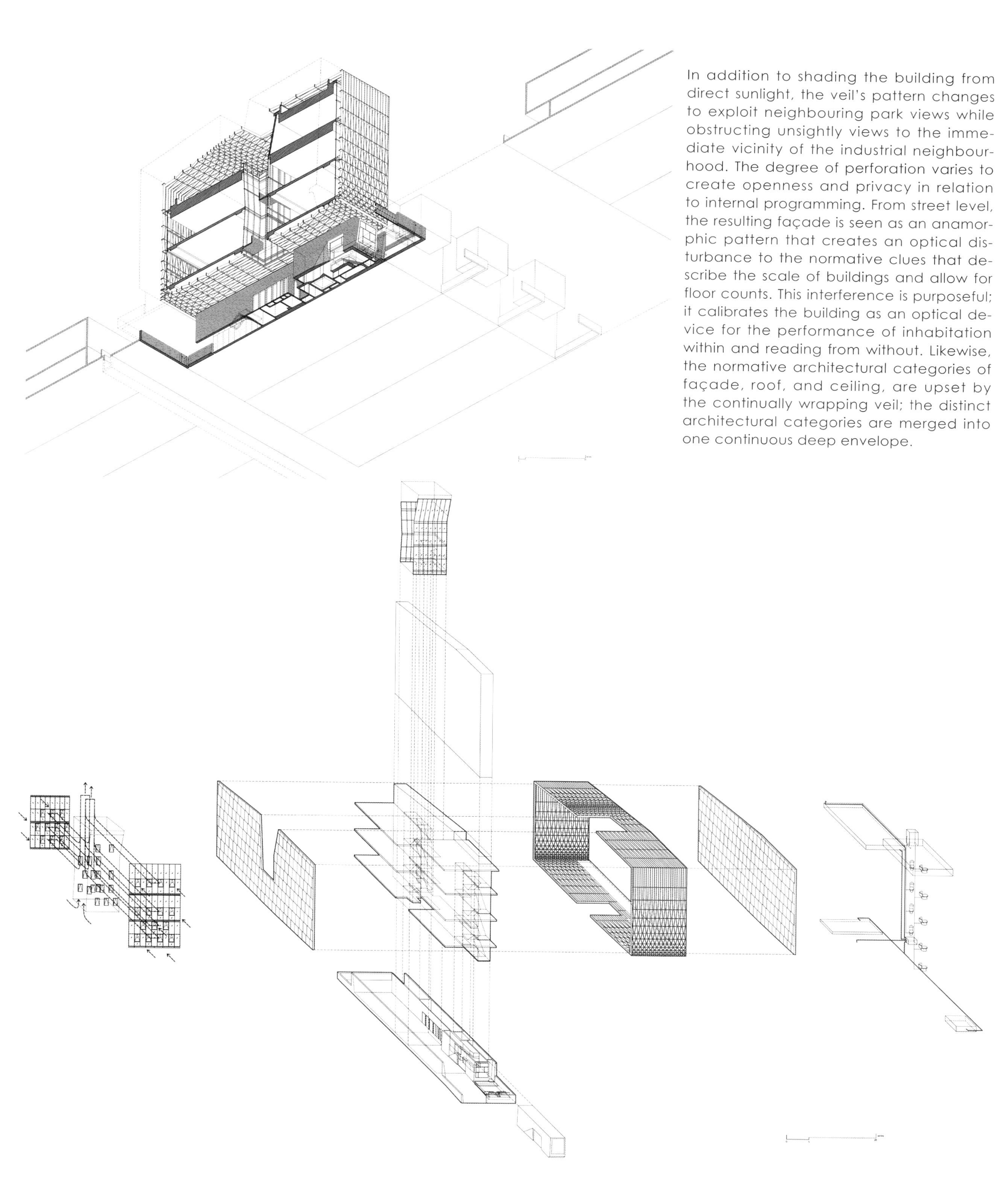

In addition to shading the building from direct sunlight, the veil's pattern changes to exploit neighbouring park views while obstructing unsightly views to the immediate vicinity of the industrial neighbourhood. The degree of perforation varies to create openness and privacy in relation to internal programming. From street level, the resulting façade is seen as an anamorphic pattern that creates an optical disturbance to the normative clues that describe the scale of buildings and allow for floor counts. This interference is purposeful; it calibrates the building as an optical device for the performance of inhabitation within and reading from without. Likewise, the normative architectural categories of façade, roof, and ceiling, are upset by the continually wrapping veil; the distinct architectural categories are merged into one continuous deep envelope.

GOLDEN SINGA ENGINEERING PTE LTD
32 TAGORE LANE

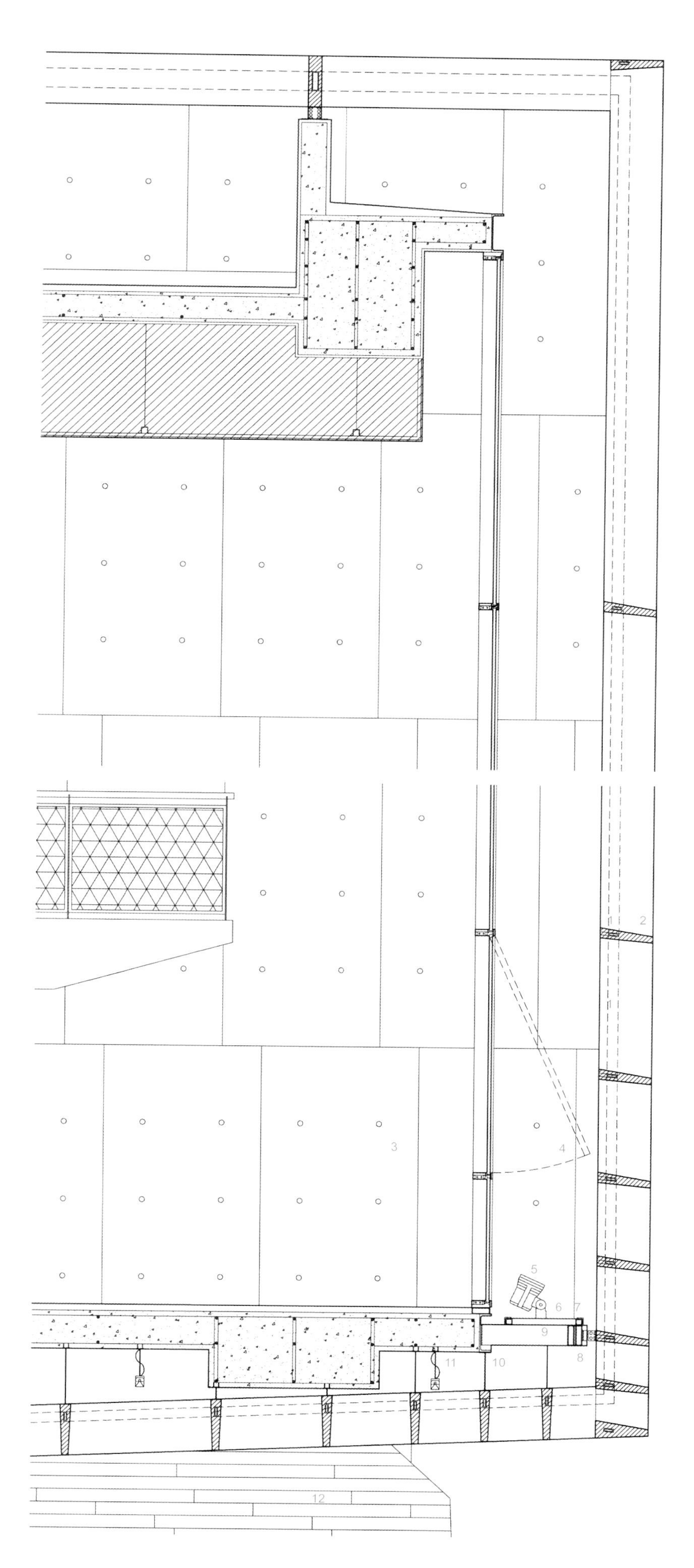

1. Galvanised metal hollow section frame
2. EIFS CNC cut panel
3. Fair face concrete pattern wall
4. Top hung aluminum window. bronze laminate glazing
5. LED light fixture
6. 50x50mm galvanised metal hollow section
7. 50x50mm metal angle
8. Internal metal sub frame
9. 8x80xl60mm galvanised rectangular metal tube
10. 300x90 mm galvanised metal parallel flange
11. Light fixture
12. Board form concrete

1. 镀锌金属空心型材框架
2. 外墙保温系统数控切割板
3. 光面混凝土图案墙
4. 上悬铝窗层压玻璃
5. LED灯具
6. 50x50mm镀锌金属空心型材
7. 50x50mm金属角材
8. 内层次级金属框架
9. 8x80x160mm镀锌直角金属管
10. 300x90mm镀锌金属平行凸缘
11. 灯具
12. 混凝土板

项目概况

这座简单的工厂楼满足了两个相互矛盾的需求：既缓和了热带地区灼热的太阳辐射，又实现了开阔的视野的通透性。项目于2012年竣工，总面积998平方米，位于新加坡的一个工业园区内。它利用一层纵深为1.2米的预制轻质外墙保温系统和青铜落地玻璃幕墙解决了以上两个建筑学上矛盾。

这层外墙保温层将建筑的正面、停车门廊、后面和屋顶环绕起来，使建筑不受刺眼的热带阳光照射，同时也保证了建筑室内的视野。它还有助于自然采光、自然通风，并且将机械设备隐藏了起来。它像一台气候引擎一样，调节了建筑的性能。

除了保护建筑不受阳光直射之外，保温层的图案还随着不同的视野而变化，为建筑提供了优美的公园景观，同时也阻挡了周边不美观的工业建筑。外墙的开口随着室内空间设置的开放性和隐私性而变化。从街道标高来看，建筑立面呈现为变形的图案，形成了一种独特的视觉干扰。这种干扰是有目的的，它将建筑变成了内部活动的视觉工具，利用建筑外观表现了内部活动。同样的，连续的保温层颠覆了规范的建筑立面、屋顶和天花板，各种独立的建筑元件融合成一个连续而厚重的外壳。

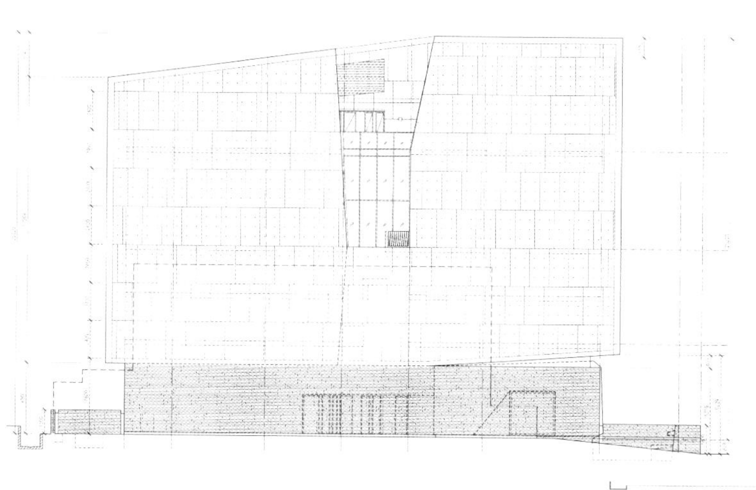

72 Screens

72幕墙楼

Location/地点: Jaipur, India/印度，斋普尔
Architect/建筑师: Sanjay Puri Architects
Site area/占地面积: 1,075m^2
Key materials: Façade – GRC concrete perforated panels
主要材料： 立面——玻璃纤维增强混凝土穿孔板

Overview

Enveloped in abstractly folded planes of perforated screens, this 6 level office building creates a sculptural presence. Located in the city of Jaipur in India which has a desert climate with average temperatures ranging from 30℃ to 50℃ through most of the year, the building is designed in response to the excessive heat imbibing traditional elements.

On a small plot of 1,075 sqm, the building envelope is restricted on all sides in plan as well as governed by height restrictions. The resultant floor area on each level after leaving

mandatory open spaces is only 326 sqm.

The entire service care and toilets are located on its southern side thus effectively reducing the heat gain substantially in a location where the sun is in the Southern hemisphere throughout the year.

In addition a glass reinforced concrete screen that takes its inspiration from the old traditional "jali" screens of the architectural heritage of the region sheaths the building on all sides further reducing the heat gain and rendering the building very energy efficient.

Designed as a corporate office headquarters, the ground level houses a reception and conference rooms above a car parking basement with office areas at the above 5 levels. The concrete screens around the building are supported by a steel framework with projections that vary from 0.9 to 1.5m. This creates an external periphery space for plants at each level that will act as further insulation from the external heat creating cooler office spaces within.

The office building thus overcomes the restrictions of its small plot creating office spaces that are very energy efficient to combat the excessive summers of its location while imbibing tradition in an abstract manifestation to create a sculptural quality for its inhabitants.

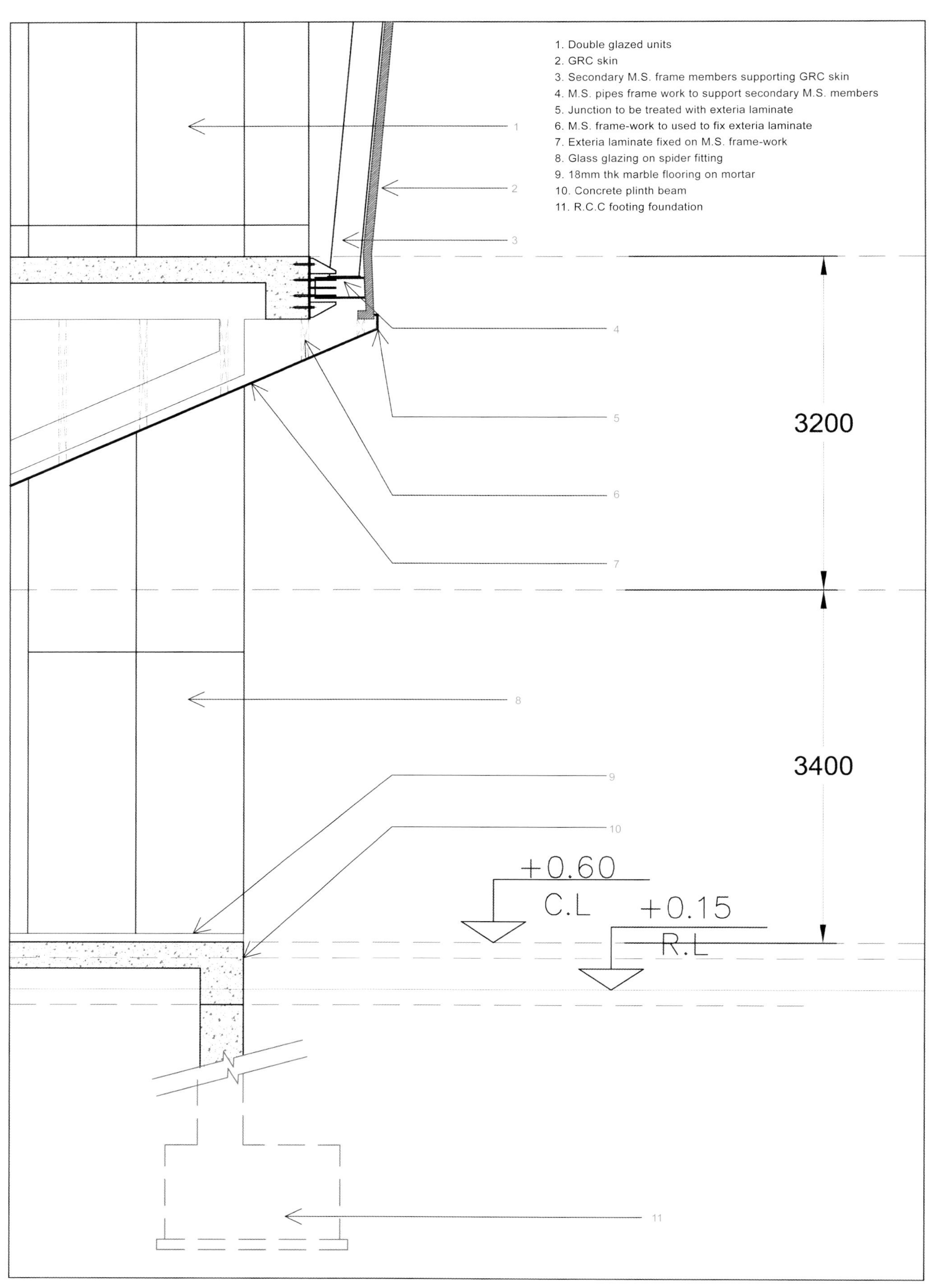

1. 双层玻璃单元
2. GRC（玻璃纤维增强水泥）表皮
3. 支撑GRC表皮的次级金属结构支架
4. 支撑次级金属结构元件的金属结构管架
5. 安装层压板的接合处
6. 金属结构支架，用于安装层压板
7. 层压板，固定于金属结构支架上
8. 玻璃装配，爪架安装
9. 18mm厚大理石地面，下面是砂浆层
10. 混凝土基座梁
11. 钢筋混凝土底脚地基

项目概况

这座六层高的办公楼完全包裹在抽象的穿孔板折叠平面之中，形成了富有雕塑感的独特外观。建筑位于印度斋普尔，该地属于沙漠气候，全年平均温度都在30~50℃之间，因此建筑设计采用了能充分吸热的传统元素。

项目的占地面积仅有1,075平方米，建筑的外形受到了来自四面以及高度的限制。排除强制开放空间之后，各个楼层之间楼面面积仅有326平方米。

所有服务区和洗手间都位于建筑南面，有效地大幅减少了南面的热增量（该地区太阳全年都位于南半球）。

此外，建筑师还从当地传统的“哥哩砖墙”中获得了灵感，制作了玻璃增强混凝土幕墙。这层幕墙将建筑四面包围起来，进一步减少了建筑的热增量，提升了建筑的节能性能。

作为一个公司办公总部，建筑一楼设置着前台和会议室，上面五层是办公区，地下设有停车场。环绕建筑的混凝土幕墙采用凸出的钢架（长度在0.9米至1.5米之间）支撑。这一设计形成了一层外围空间，每层楼都能栽种一些植物，从而进一步使办公区与外界的热源隔开。

办公楼克服了小场地的限制，打造了能应对炎热气候的节能办公空间，同时还以抽象的方式融入了传统设计，为建筑提供了独特的雕塑特质。

Automotive Industry Exporters Union Technical and Industrial Vocational High School

汽车工业出口公司技术与工业职业高中

Location/地点: Bursa, Turkey/土耳其，布尔萨
Architect/建筑师: Metin Kılıç, Dürrin Süer
Photos/摄影: Cemal Emden
Built area/建筑面积: 32,000 m²
Completion date/竣工时间: 2010
Key materials: Façade – glass fibre reinforced cement slab and tempered glass
主要材料: 立面——玻璃纤维加固水泥板、钢化玻璃

Overview

With its capacity of 720 students and various facilities including the dormitory, lodging places, sports hall and conference hall, the intense functional layout of the education building is designed as consisting of discrete parts with different elevations within an integrated whole.

The layout of discrete blocks in the site plan is connected with a social artery to which facilities of different strata are articulated in the third dimension, i.e., via a transparent gallery and a series of successive courtyards. Offering enriched means of social living as if resembling an urban square, this gallery is supported by the sports hall and the conference hall on the ground floor and spaces like the conference hall and the student canteen as well as workshops and classrooms on the entrance floor. The gallery acting as a social artery avails for a visual dialogue while it visually connects the gallery with outer courtyards. When walking along the urban streets, one can possibly have connections not only with the inner gallery, but also the outer courtyard. In order to support such diverse means of communication provided within the building, effort is also spent for creation of a dynamic environment with use of vivid colours.

The visual richness of the building is further reinforced with its fragmented building form as well. The existing green pattern – olive grove is purposely intertwined with the building mass in the design. While some parts of the various terraces provided at different elevations are converted into open areas for students, the focus at other parts has been to ensure continuity of the green.

The idea to cover the terraces of education

units with vegetation in harmony with the land slope has not only provided for continuity of the green, but also helped in placement of accommodation facilities (dormitory and lodging) separately from social and cultural facilities, at a location facing this green pattern created by the olive grove. The education building could therefore meet the idea of differentiating between the daily and private life of its users. The terraces designed to take place at different elevations of the building have been allocated to outdoor playtime area on the classrooms floor and were used as roof gardens on the resting hall floor of the dormitory. Landscape design intends to give the roof garden a rural atmosphere.

Detail and Materials

The notion of the building as a technical school has paved the way for the approach to reflect the mechanical systems upon its spatial formation. With the intention to design the project in a simple and clear-cut framework in structural terms, the reinforced concrete blocks are tied with steel beams and transparent planes covered with glass veneer. In order to minimise the diversity of building materials, the coating layer of ceilings, floors and walls has been removed. This has encountered the demand to accelerate the process of construction as well.

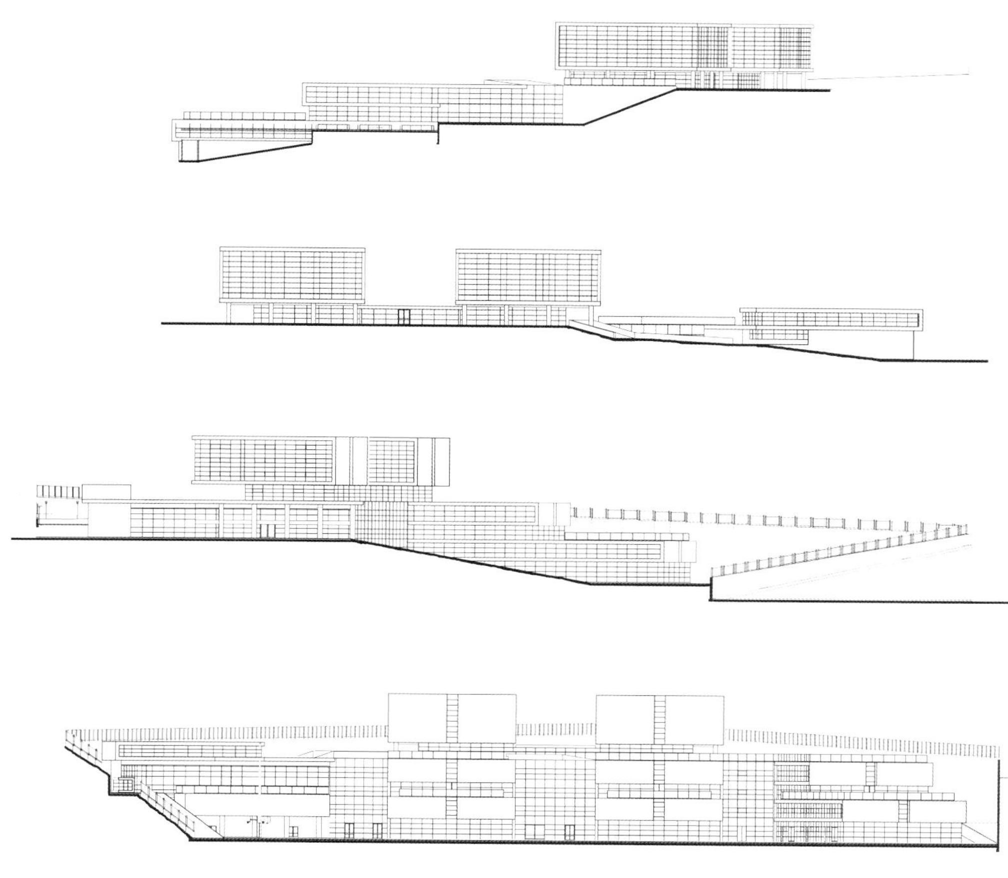

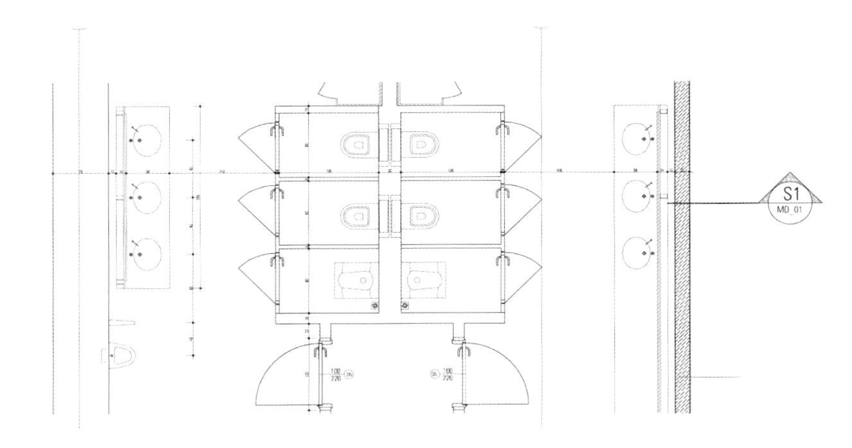

Despite the ambition against having any coating or casing over all interior and exterior concrete surfaces during the phase of schematic design, the decision has been taken during the application phase to coat the exterior surfaces with fiber-reinforced prefabricated concrete pieces due to insufficiency of trust in achieving any concrete construction of high-quality. It has been C40 reinforced concrete material poured. In uncoated concrete surfaces, circular columns are moulded in paper form (use and throw type). In addition to rapid continuation of the application process, this has also protected the building material during the phase of construction.

Finish plasters are applied upon partition walls. Specific details have been used to separate the finishes from concrete surfaces at points where plaster finishes meet the concrete. While the architectural difference between the load bearing and partition walls are emphasised further, the concrete surfaces could be protected especially during painting of plaster finish surfaces as well.

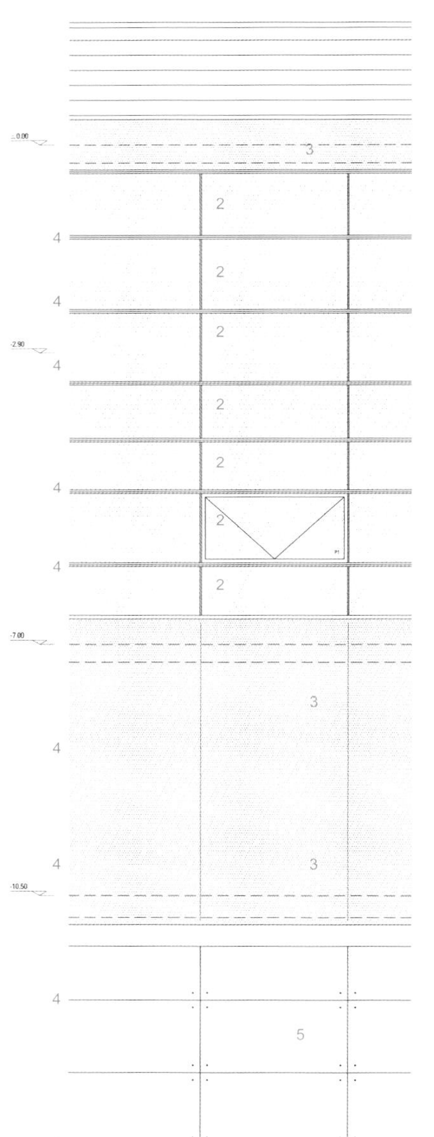

Regarding the pavement of floors, concrete epoxy hardeners are applied upon smoothed rendering. In classrooms, the canteen and the cafeteria/dining hall, suspended timber ceilings have been used for precautionary measures on acoustics.

1. Tempered glass (covered with green membrane) (6mm)
2. Tempered glass (6mm)
3. Glass fibre reinforced cement slab
4. Aluminium façade shape
5. Exposed concrete wall

1. 钢化玻璃（覆盖绿色薄膜）（6mm）
2. 钢化玻璃（6mm）
3. 玻璃纤维加固水泥板
4. 铝制外墙板
5. 素面混凝土墙

项目概况

建筑为720名学生提供了教学和相关设施，包括宿舍、寄宿区、体育馆和会议厅等，紧凑的功能布局被分散成若干个不同的立面，最终连接整合成一个整体。

独立楼体的布局通过一条社交干道连接起来，不同层次的设施通过第三维相互连接，如透明的长廊或一系列连续的庭院等。长廊像城市一样提供了丰富多彩的社交生活，由体育馆、会议厅、学生食堂、工作室、教室等共同支撑。长廊起到了社交干道的作用，并且将室内外有机地连接起来。从城市街道上，人们不仅可以进入内部的长廊，还能前往外面的庭院。为了辅助如此多种多样的空间设计，建筑师在空间中应用了丰富的色彩。

分裂的造型进一步增强了建筑丰富的视觉效果。橄榄树林的绿色被有意地融入了建筑设计之中。高低错路的平台为学生提供了丰富的露天空间，而其他部分的设计则保证了绿色主题的统一感。

在教学平台上种植植物，使其与周边山坡的景观和谐融合，这一概念不仅实现了绿色的延续，还有助于将住宿设施与社交文化设施区分开，让住宿设施正对橄榄树林。这样一来，教学楼就将教学和生活完美地隔开。教室外的平台被用作露天休闲娱乐区，而宿舍外的平台则成为了屋顶花园。景观设计为屋顶花园带来了浓厚的乡村气息。

细部与材料

作为一所技术学院，建筑通过空间设计反映了自身的机械系统。建筑师的目标是打造一个简单清晰的结构框架，钢筋混凝土体块通过钢梁和由玻璃薄片覆盖的透明平面连接起来。为了控制建筑材料的多样性，天花板、地板和墙板都没有使用任何包层。这种做法还有效缩短了施工的时间。

虽然建筑师在方案设计初期决定不在室内外的任何混凝土表面上添加包层或外壳，但在具体实施中，他们还是决定在外墙上添加一层玻璃纤维加固预制混凝土板，以

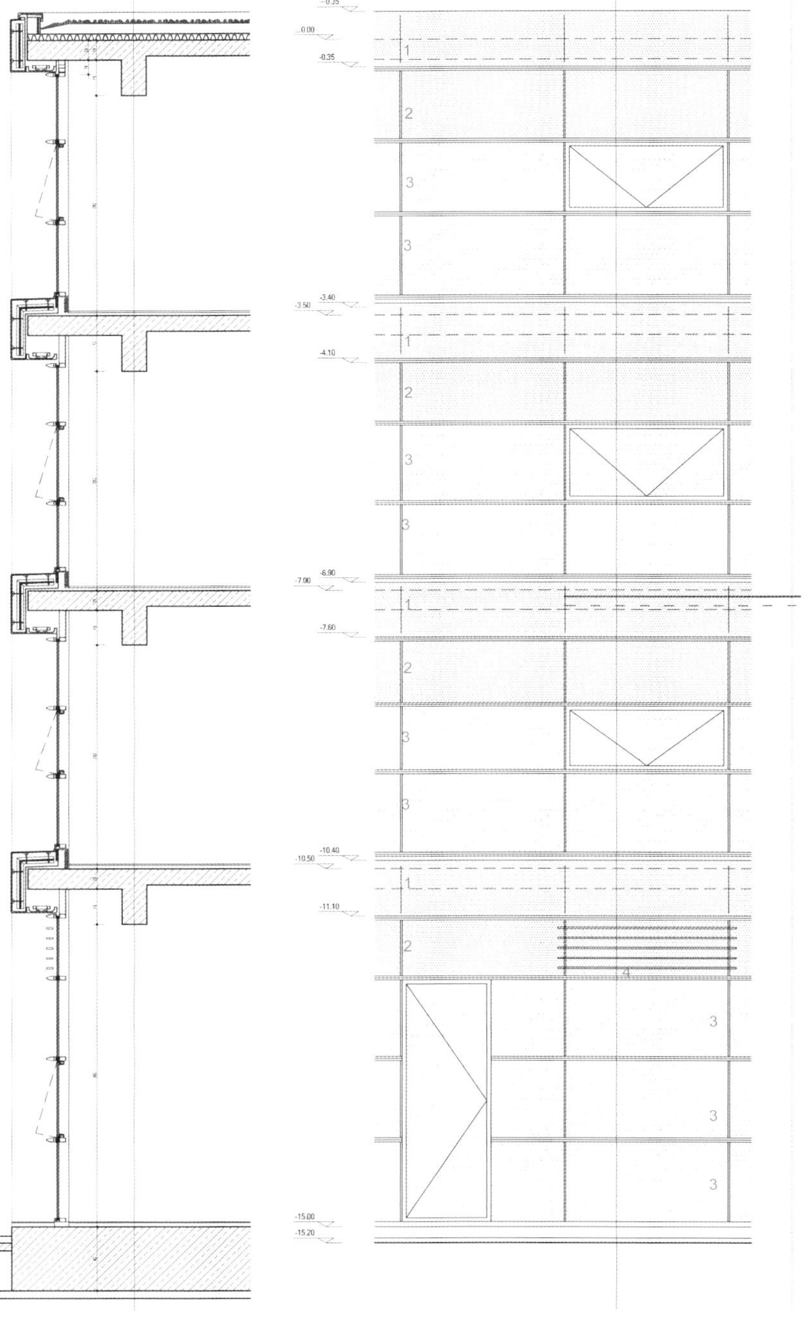

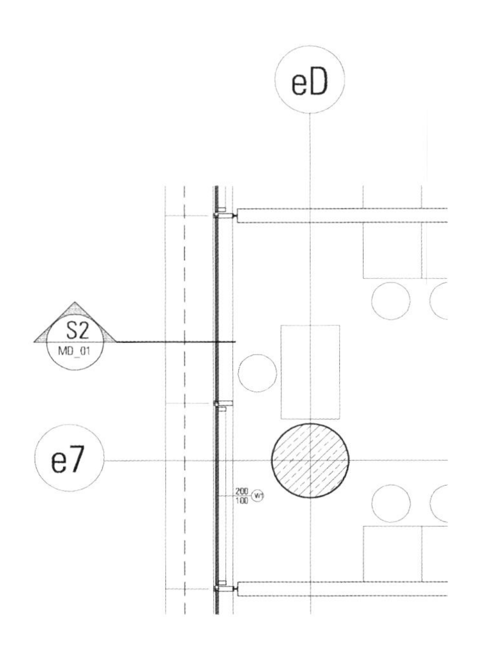

1. Glass fibre reinforced cement slab
2. Tempered glass (is covered with green membrane) (6mm)
3. Tempered glass (6mm)
4. Aluminium grid

1. 玻璃纤维加固水泥板
2. 钢化玻璃（覆盖绿色薄膜）（6mm）
3. 钢化玻璃（6mm）
4. 铝格栅

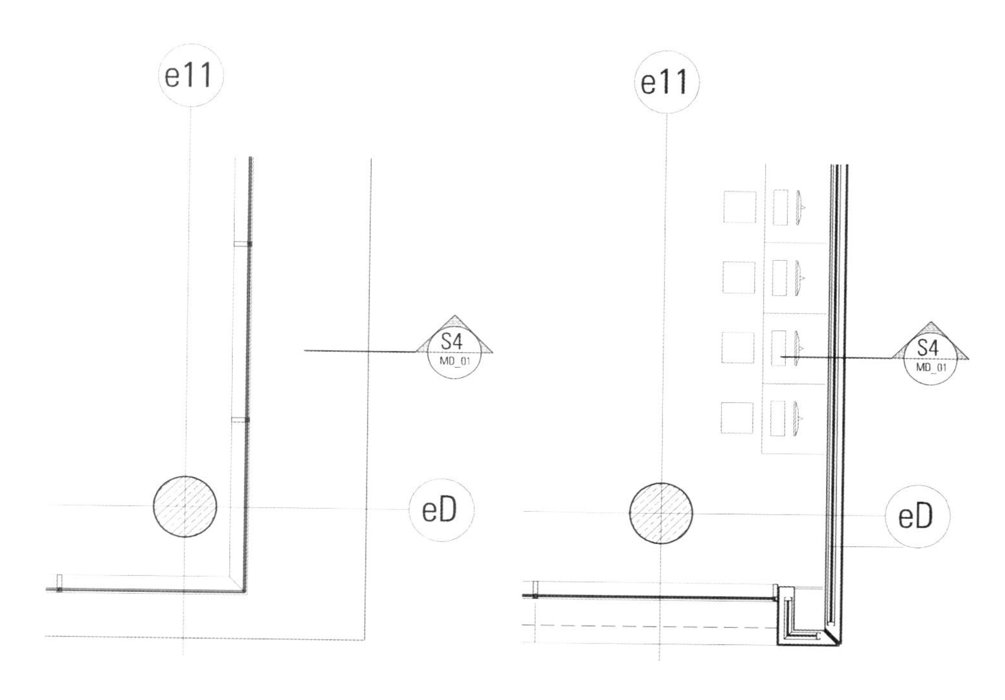

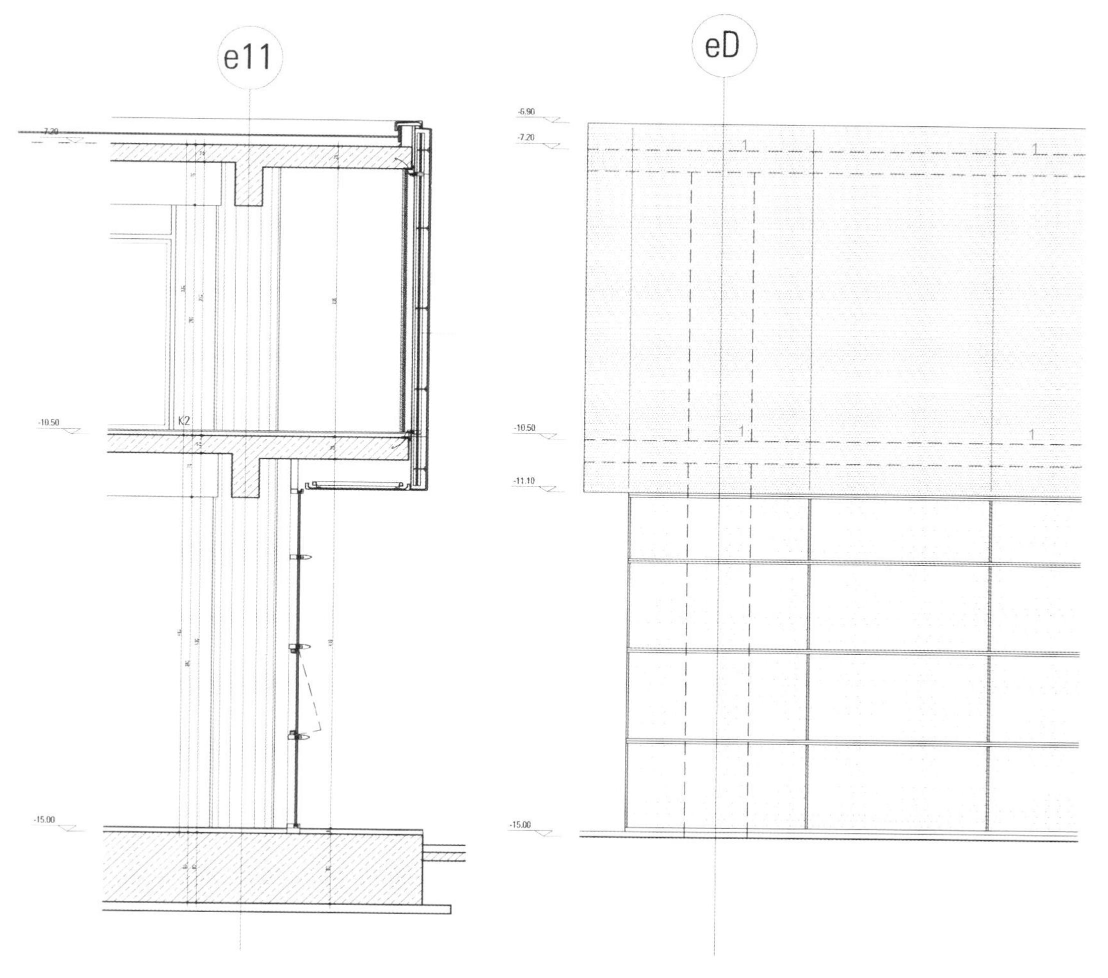

实现更高品质的混凝土结构效果。项目采用了C40钢筋混凝土材料。在不加涂层的混凝土表面，圆柱采用一次性纸膜塑造。除了能快速完成应用流程之外，这还能在施工阶段保护建筑材料。

修饰石膏被应用在隔断墙上。在石膏饰面与混凝土的接合处，建筑师利用特殊的细节将它们隔开。这一设计进一步突出了承重墙和隔断墙之间的建筑差异，并且还能在涂抹石膏饰面时实现对混凝土表面的保护。

在楼面铺装方面，混凝土环氧固化剂被应用在光滑打底上方。在教室、食堂和餐厅里，木板吊顶的应用可以实现有效的隔音。

New Theatre in Almonte

阿尔蒙特新剧院

Location/地点: Huelva, Spain/西班牙，维尔瓦
Architect/建筑师: Juan Pedro Donaire Barbero
Photos/摄影: Fernando Alda /Kavi Sanchez
Built area/建筑面积: 3,265.70m^2
Key materials: Façade – prefabricated GRC concrete
主要材料： 立面——预制玻璃纤维加固混凝土

Overview

The building is located on the site of an old winery. It has the challenge of integrate the existing old buildings, declared as cultural interest, and being part of a cultural complex of a total of three buildings and a common space. This space turns into the main place of the town and an important meeting area.

It is opportunity to work on light, material and space. The path chosen to work on these concepts, is the contrast. Contrast between outside and inside, between old and new, including a monumental scale and human scale. And the journey as the thread that sews and explains the intervention. A large area covered with large proportions and controlled height works as a high threshold. A monumental scale lobby welcomes the visitor showing the scale of a public building.

It is situated in the "Culture City" in Almonte, south of Spain, next to the Rehabiltation of the Warehouse for use as a public library, and School of Arts (Music, Dance and Art) and enclosed within a public square. The intervention rigorously respected and maintained the different morphologies of each of the next buildings and this itself constitutes a strong characteristic of the project.

Detail and Materials

The materials are very important in the building, a opportunity to have a large experience working in innovated architecture with traditional and new materials. The concrete was used as the main material to contrast with the wood and this established the main line of work. The traditional Warehouse image contrast with the contemporary theatre image, as a focal point for the public space. The brick contrast with the white concrete.

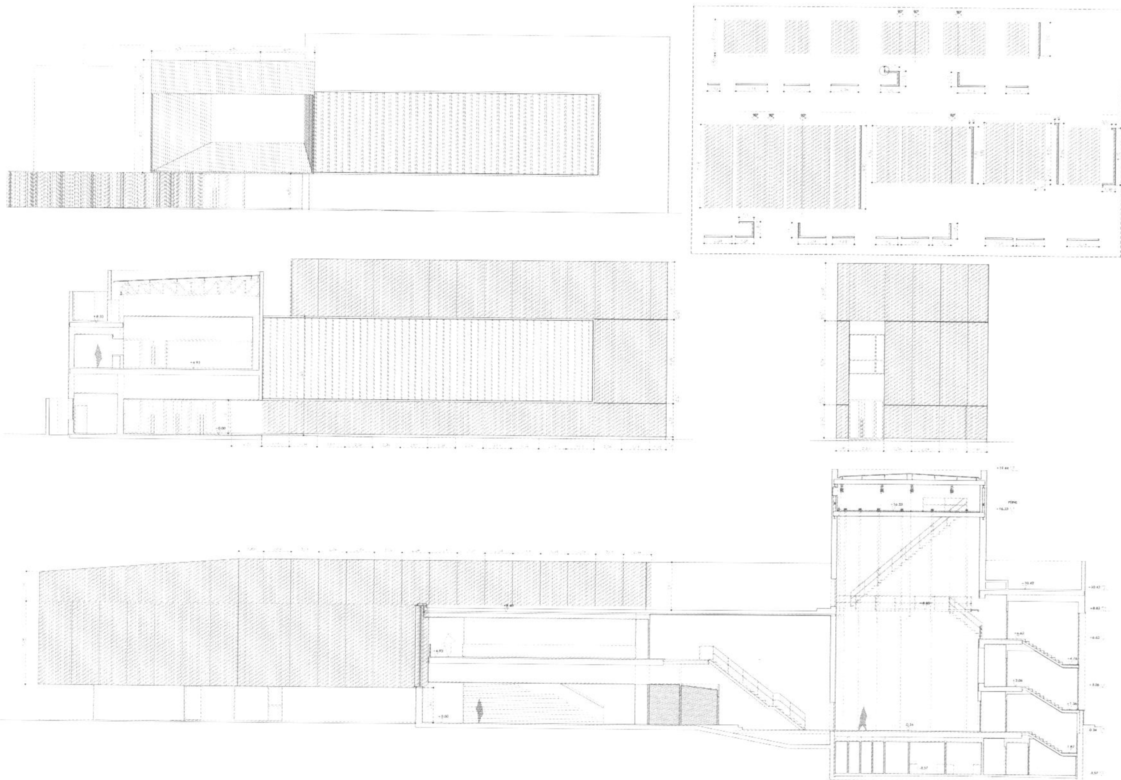

Teatro

项目概况

建筑位于一座旧酒庄的原址上。项目所面临的挑战是如何整合已有的旧建筑，使其以文化元素成为建筑群的一部分，打造由三座建筑构成的建筑群和公共空间。这一空间将成为城市的主要景点和重要的集会场所。

设计让建筑师将光、材料和空间整合在一起。最终，建筑师选择了对比这种形式。内与外、新与旧的对比包含整体尺度和人文尺度两个方面。设计的建筑旅程像线一样将它们贯穿起来。一个悬空于地面的巨大结构是项目的主入口，宽敞宏大的大厅欢迎着访客前来参观。

项目坐落在西班牙南部的“文化之城”——阿尔蒙特，紧邻由仓库改建而成的公共图书馆、艺术学校（音乐、舞蹈和艺术）和公共广场。项目十分注重建筑之间的联系，同时也保持了建筑形态的多样性，表现出强烈的特色。

细部与材料

材料对建筑来说至关重要。项目在创新建筑中融合了传统材料和新材料。作为主要材料，混凝土与木材形成了对比，奠定了设计的主要基调。传统的仓库形象与现代剧院形象相互映衬，构成了公共空间的焦点。土红色的砖块与白色混凝土形成了鲜明对比。

1. Prefabricated GRC concrete
2. Metal strip
3. Glass railing
4. Primary structure concrete
5. Luminaries
6. Wood panel
7. Acoustic panel
8. Air conditions

1. 预制玻璃纤维加固混凝土
2. 金属条
3. 玻璃围栏
4. 基本结构，混凝土
5. 照明灯具
6. 木板
7. 隔音板
8. 空调系统

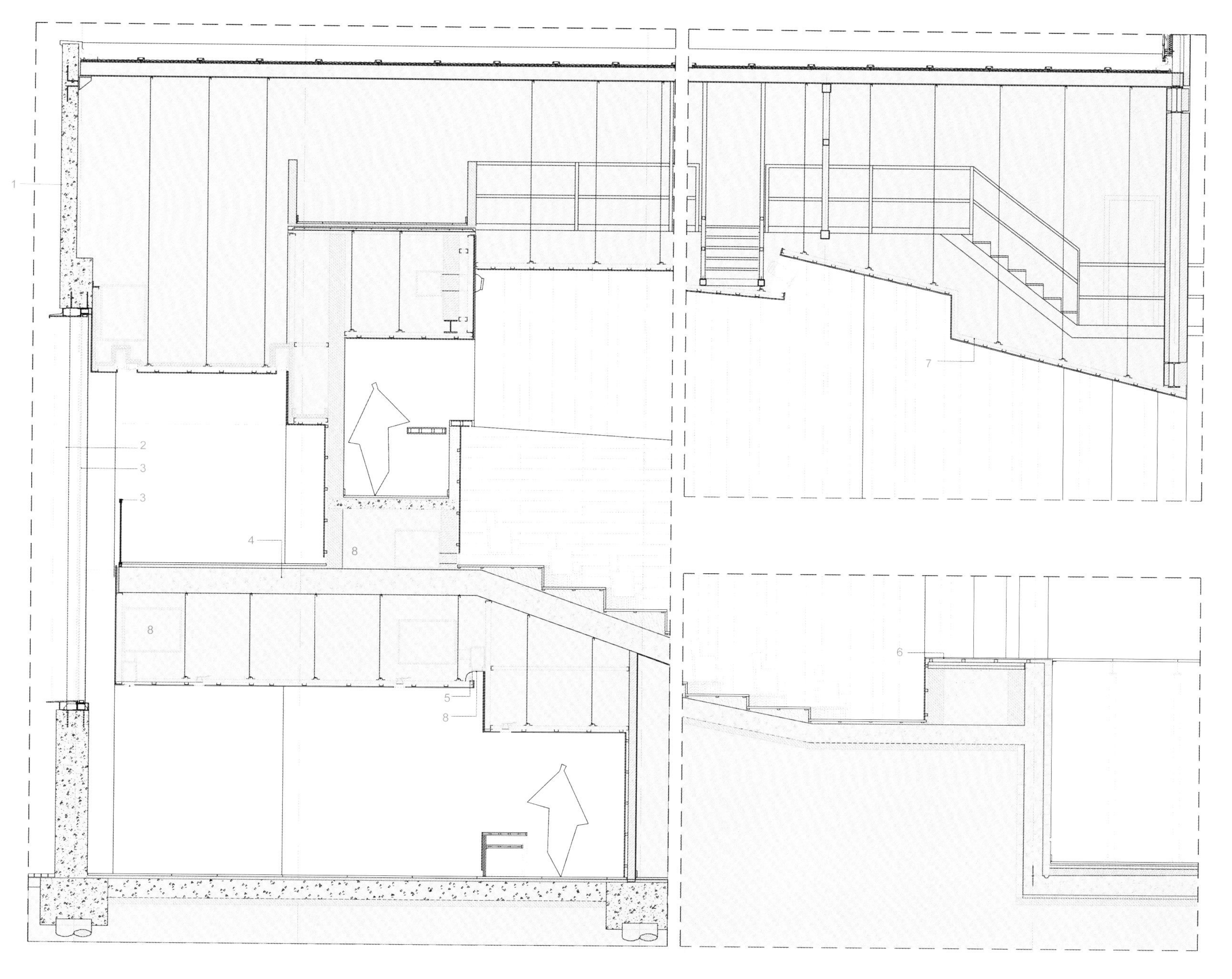

Eurostars Book Hotel

欧洲之星图书酒店

Location/地点: Munich, Germany/德国，慕尼黑
Architect/建筑师: Capella García Arquitectura and Schmid Architekten
Photos/摄影: ©Hotusa, ©FHolzherr
Built area/建筑面积: 10,670m^2
Completion date/竣工时间: 2012
Key materials: Façade – GRC concrete panels
主要材料：立面——玻璃纤维加固混凝土板

Overview

The Book Hotel, as its name suggests, is inspired by the pages of a book. The façade is made up of white leaves that shelter the windows of the bedrooms. Altogether it suggests a disordered but harmonious fluttering of blank pages, like riffling through the pages of a book with your thumb. A classic but organic composition, suited to the rhythm of the street where it stands, and yet standing out because of its chromatic purity and the kinetic effect conferred by the curved lintels of the openings. The rays of the sun throw shadows that progressively mark out the always changing cut-out volume of the building. As in a book, the hotel's formal calligraphy is in "black on white", plus several shades of grey, that reinforce the leitmotiv of writing. Both the exterior and the interior play mainly on this austere bitonality.

Access to the interior of the Hotel is on the corner of the building, through a wide double-height space that houses the lobby, reception

and bar. Other facilities such as the restaurant and the meeting rooms are on the ground floor, distributed around the rear courtyard.

The welcoming interior design of the common areas is based on the contrast between black and white, with a few touches of lilac. The interior design pays special attention to circulation within the Hotel, based on the vertical core with stairs and lifts in the angle of the building. Sequences of bi-tonal stripes, resembling linear barcodes, make up the flooring. The low-key passageways, flanked by white linear flashes along the dark walls, accompany the guest to the illuminated "open books", made up of a pair of adjacent entrances to bedrooms. Also, the walls of the passageways are decorated with texts of seven different literary genres, an identifying characteristic for each of the seven floors of the hotel.

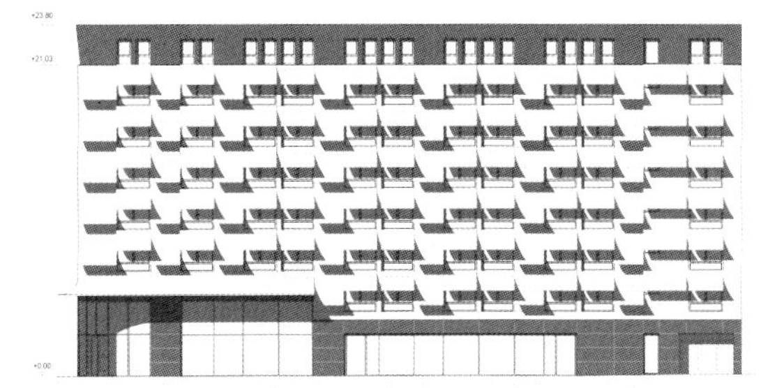

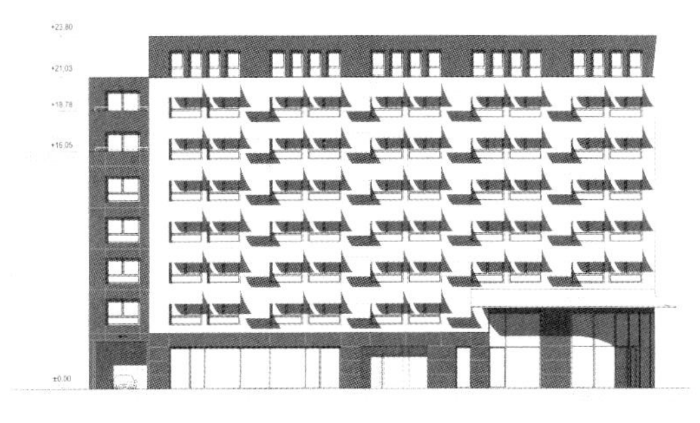

Detail and Materials

On the exterior, the elegant volumetrics of this L-shaped building are organised according to classical architectural practice, being divided into a plinth, central body and cornice. The plinth is formed by the ground floor and the cornice by the top floor of the building, which are both black in colour, and are thus differentiated tonally from the white central body, constructed using the revolutionary material FiberC, concrete reinforced with glass fibre, which had never before been used with such curvatures. The central body, which contains the bedrooms, is distinguished by striking projecting features by way of awnings, that resemble the pages of a book. The non-aligned arrangement of the awnings gives the façade considerable dynamism.

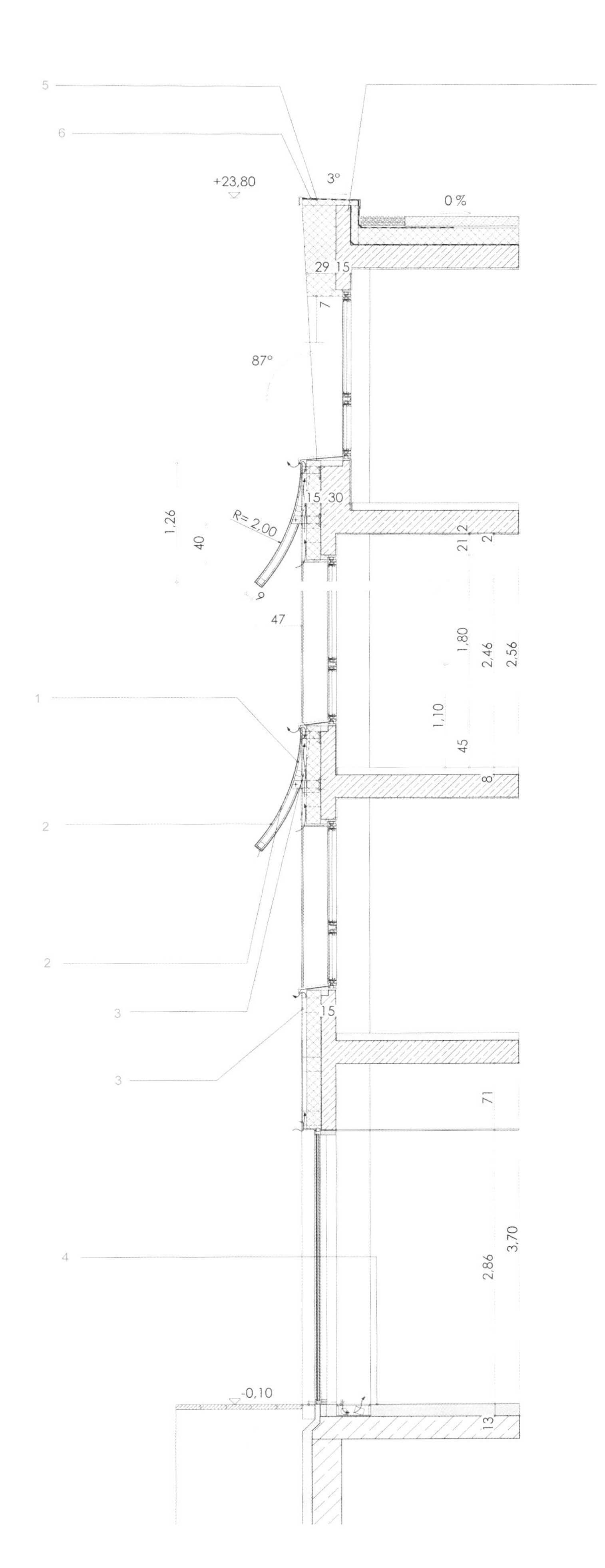

项目概况

正如名字所暗示的一样，图书酒店设计灵感来自图书的书页。建筑立面由卷起的白色叶片构成，它们为卧室的窗口提供了遮阳，仿佛无数空白的书页迎风飘扬，无序却又和谐，就像指间翻动的书页一样。经典而有机的建筑结构与其所在街道的节奏相吻合，同时又以纯粹的色彩和活泼的曲线效果脱颖而出。阳光照在楼面上的阴影形成了不断变换的视觉效果。正如图书一样，酒店的主色调也采用黑白两色，并且添加了若干灰色阴影来烘托效果。建筑的室内外设计全部以这种素雅的基调为主。

酒店的大门设在转角处，宽敞的双高空间内设置着酒店大堂、前台和酒吧。餐厅、会议室等其他设施也设在一楼，围绕着后院分散开。

室内公共区域友好的设计以黑白两色为基础色调，其间点缀着少量淡紫色。室内设计特别注重酒店内部的交通流线，以位于角落的楼梯和电梯为基础。地面由像条码一样的双色条纹组成。低调的走廊两侧是黑色的墙面，上面配有线形灯光，让宾客宛如置身于打开的书本。此外，走廊的墙面上还装饰着来自七种不同文学体裁的文字，正好标志出七层不同的楼层。

细部与材料

这座优雅的L形建筑采用了经典的建筑设计风格，分割成底座、中央楼体和飞檐三个部分。底座是建筑的底层，飞檐则是建筑的顶层，二者均呈现为黑色，与中间部分的白色区分开。白色楼体采用了革命性的材料FiberC（一种由玻璃纤维加固的混凝土），这种材料还从未以此种弧度被使用在建筑上。客房卧室的弧形遮阳篷模仿了书页的效果，十分引人注目。遮阳篷的不规则布局让外立面更富动感活力。

1. Fibre panel C
2. Curved fibre panel C, white iron FL
3. Coating fibre panel C, white iron FL
4. 13cm pavement layers
5. Bituminous membrane
6. Gutter
7. Vertical plate

1. 纤维板 C组
2. 弯曲的纤维板 C组，白口铁FL
3. 涂层纤维板C组，白口铁FL
4. 13cm 铺装层
5. 沥青膜
6. 排水沟
7. 竖板

Leyweg Municipal Office

利维格市政楼

Location/地点: The Hague, The Netherlands/荷兰，海牙
Architect/建筑师: Rudy Uytenhaak
Photos/摄影: Rudy Uytenhaak
Built area/建筑面积: 32,600m²
Completion date/竣工时间: 2011
Key materials: Façade – precast concrete elements
主要材料：立面——预制混凝土构件

Overview

As a result of the urban development plan for The Hague South West, the Escamp district of the city will experience a wave of renewal in the coming years. The main aim is to give the area a more varied and dynamic atmosphere.

One element of this plan is the new municipal office, which as the social and administrative heart of the district deserves to become an icon for The Hague South West/Escamp. Inspired by the genius loci of the plan by Willem Marinus Dudok: an open city, in which primary volumes define the space of a spacious "park city", the buildings are placed like ships throughout the park city: objects standing free in space rather than walled-in streets.

The municipal office, with its robust prow, will acquire a strong role as an orientation landmark in the "mental map" of the area. The independence of the municipal office is emphasised by the orientation of the upper section of the building as an object on a rising plane.

The design of the building has an autonomous form, but is in keeping with the existing buildings in its colouring and situation, representing a new interpretation of the original light and "modern" architectural heritage. The service building on the rising plane has an information square at its centre, opening up onto the various public functions and the collective office building.

On the ground floor, in addition to facilities functions, are the public service desks of the municipal services, information facilities and a library. The wedding hall, the staff restaurant and the meeting centre are located on the first floor. On a higher level, above the atrium roof, is the municipal health service, while the multi-purpose building unfolds around the triangular atrium. The upper nine storeys are made up of homes.

The Hague Municipal Council aims to be a transparent organisation. In this new municipal office, contact both with the outside world and within the organisation is enhanced by the building's triangular form. Because the workstations are placed in an open-plan office design around the atrium, staff have sightlines to each other and to other departments. By breaking open the corners of the triangular form, every department's area is provided with special points with outside views. There is a corridor circuit on each storey, so encouraging informal contact between employees and different departments. There are stairways at regular intervals in these corridors, so vertical circuits are also created.

In an inventive manner, the Leyweg municipal achieves a result that is expressive in its form, spatially surprising, flexible and functional, and in which the logic of sustainable construction has been optimally applied.

In the structural concept of the Leyweg municipal office, spaciousness, construction and installation technology are optimally attuned to one another, with sustainability and economic intelligence as the primary criteria. So, for example, all ducts and installations are integrated in a climate-control floor with concrete core activation, which also has acoustic and fire prevention functionality. Hanging ceilings are therefore not necessary, enabling a spacious total storey height. Using the principle of load-bearing exterior walls, the office areas are fully flexible in their layout, and therefore flexible in use. The meticu-

lous fine-tuning of all these aspects with one another results in a businesslike and efficient office building that makes use of simple, "proven" construction techniques.

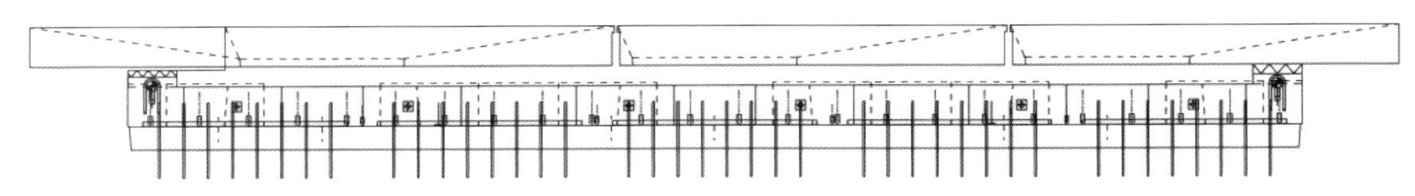

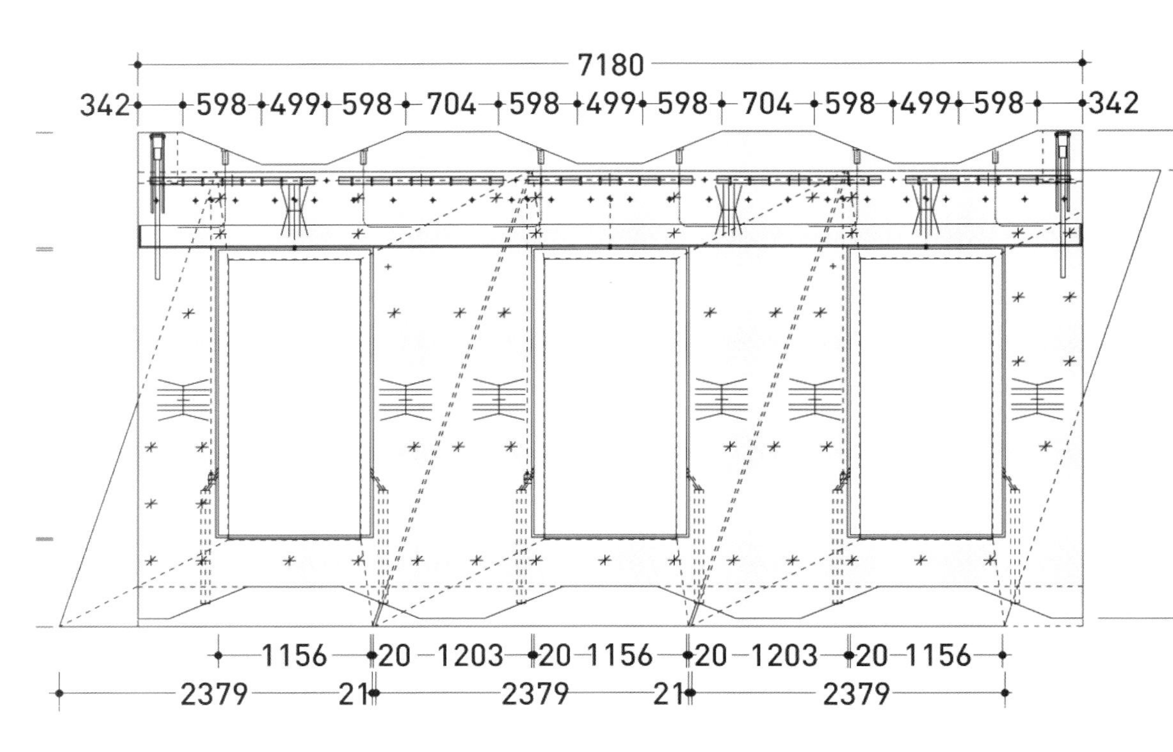

项目概况

作为海牙西南城区开发的一部分，未来数年，埃斯坎普区将经历一系列翻新改造，主要目标是为该地区带来更加丰富而活跃的城市氛围。

新建的市政办公楼就是这一规划中的主要项目之一。作为该地区的社会与行政中心，它将成为海牙西南城区的地标。项目从著名建筑师威廉•杜多克的“地方精神”概念中获得了灵感。在一个开放的城市中，即主要场所的特征将决定一个空间的特色和精神。所有建筑像船舶一样停泊于城市之中，它们是相互独立的，而不是被约束在街道围墙之内。

市政楼就像一个强健的船头，起到了强有力的指向地标效果。建筑的上半部分通过上扬的姿态进一步突出了市政楼的独立性。

建筑设计显得十分出挑，但是在色彩和位置上与周边的建筑仍然保持了一致，重新诠释了轻巧而现代的建筑传统。呈上扬姿态的服务楼在中央设有一个信息广场，可以举办多种公众活动，也是办公楼的集会场所。

除了通用服务设施之外，一楼还设有市政公共服务台、信息设施和一个图书馆。婚礼大厅、员工餐厅和会议中心都设在二楼。中庭上面的楼层是市政健康服务设施，而多功能楼则围绕着三角形中庭展开。最上方的九层楼全部是住宅。

海牙市议会致力成为一个透明的组织。在这座新市政楼内，三角造型不仅提升了建筑与外界的联系，还加强了内部的组织联系。由于办公台都放置在围绕着中庭的开放式办公空间内，员工能够清楚地看到彼此和其他部门。通过打开三角布局的角落，每个部门的办公区都拥有独特的外部视野。每层楼有一个环形走廊，鼓励员工和不同部门之间进行非正式的联系。走廊的固定间隔处设有楼梯，从而实现了垂直交通循环。

利维格市政楼通过创新的方式实现了引人注目的外观、令人惊喜的空间以及良好的灵活性和功能性，并且优化实行了可持续建造设计。

在利维格市政楼的结构设计中，以可持续性和经济成本为主要准则，空间感、结构性和安装技术实现了相互调和优化。例如，所有管道和安装设施都与拥有混凝土芯活化系统的气候控制楼板结合起来。楼板同时还具有隔音和防火功能。因此，建筑并不需要吊顶，从而保证了舒适的楼层高度。承重外墙的设计让办公区的布局更加灵活，便于实现各种配置。各方面精心细致的设计共同实现了这座类似于商务楼的节能政府办公楼。

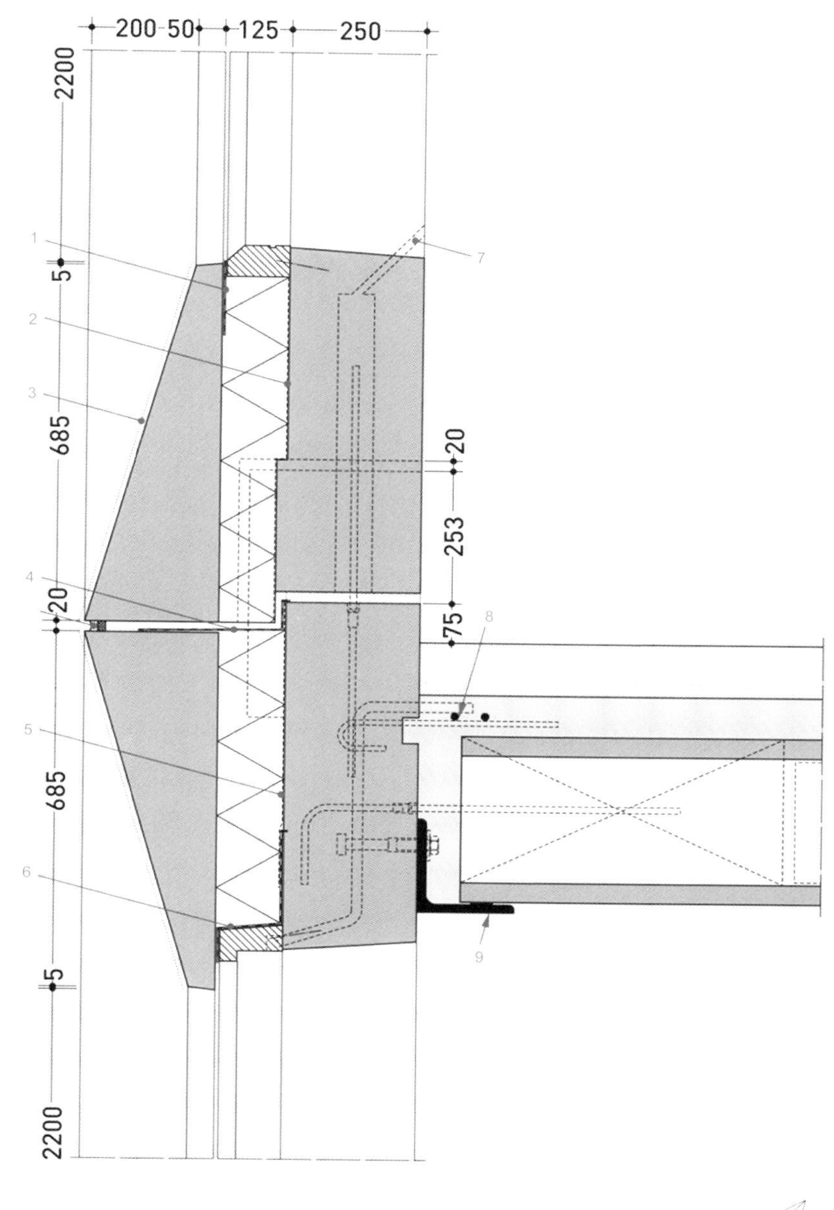

1.EPDM foil
2.PE foil
3.Radiate
4.EPDM foil
5.PE foil
6.EPDM foil
7.Pouring pole
8.Pull band
9.Hoeklijin 180x180x6

1. EPDM板
2. PE板
3. 遮阳板
4. EPDM板
5. PE板
6. EPDM板
7. 浇注孔
8. 拉力带
9. Hoeklijin型材 180x180x6

ROC Mondriaan Laak II

ROC蒙德里安学院二号楼

Location/地点: The Hague, The Netherlands/荷兰，海牙
Architect/建筑师: LIAG
Photos/摄影: Ben Vulkers
Built area/建筑面积: 11,000m²
Completion date/竣工时间: 2011
Key materials: Façade – precast concrete elements
主要材料：立面——预制混凝土构件

Overview

The new building is situated on a narrow plot, perched between the busy railway lines at The Hague HS Station and the Waldorpstraat. The position, close to the public transport node is an excellent strategic choice: the school campus that is formed here in conjunction with The Hague University of Applied Sciences, the student and entry level housing and the resulting social life, is directly connected to all forms of (public) transport and is situated in the middle of the city.

The building forms a powerful and colourful (a la Mondriaan) completion of the Leeghwater-plein. Seen from the train, it serves as a calling card for ROC Mondriaan. The building has a grid shaped concrete façade decorated in the ROC colours. The effect in combination with the windows results in a strong whole. An arcade follows the pedestrian route from the train station to the housing complexes and Megastores shopping mall. Large glass areas along the arcade provide a glimpse into the school, which serves to combine transparency and a display window for vocational training with increased public safety.

The building's is designed with a floor plan based on a framework, to facilitate future conversion into offices. By choosing spans without columns, from the façade to the cores, the layout of the school is adaptable for future schooling insights.

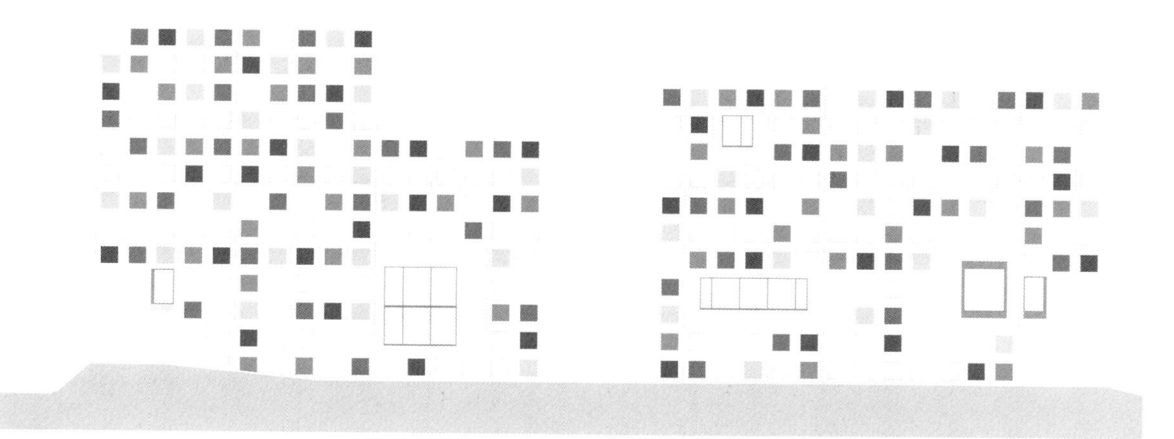

The building has two main volumes each with its own core and all support functions. This means that the complex could in the future be leased as two individual office building sections.

Detail and Materials

The entirely prefabricated building was manufactured in Germany. The squares on the grid shaped concrete façade feature the school's colours. Together with the "random" pattern of coloured panels and windows, the repetition of the exterior concrete elements creates a look without scale and a powerful whole.

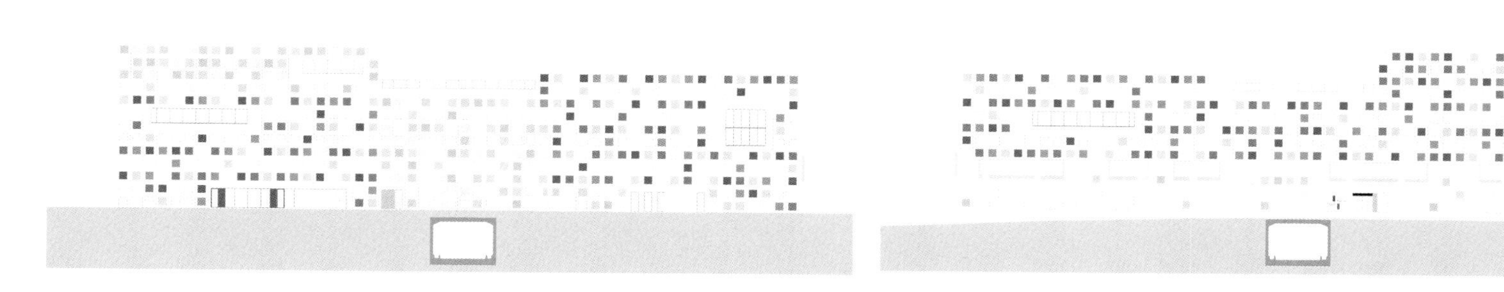

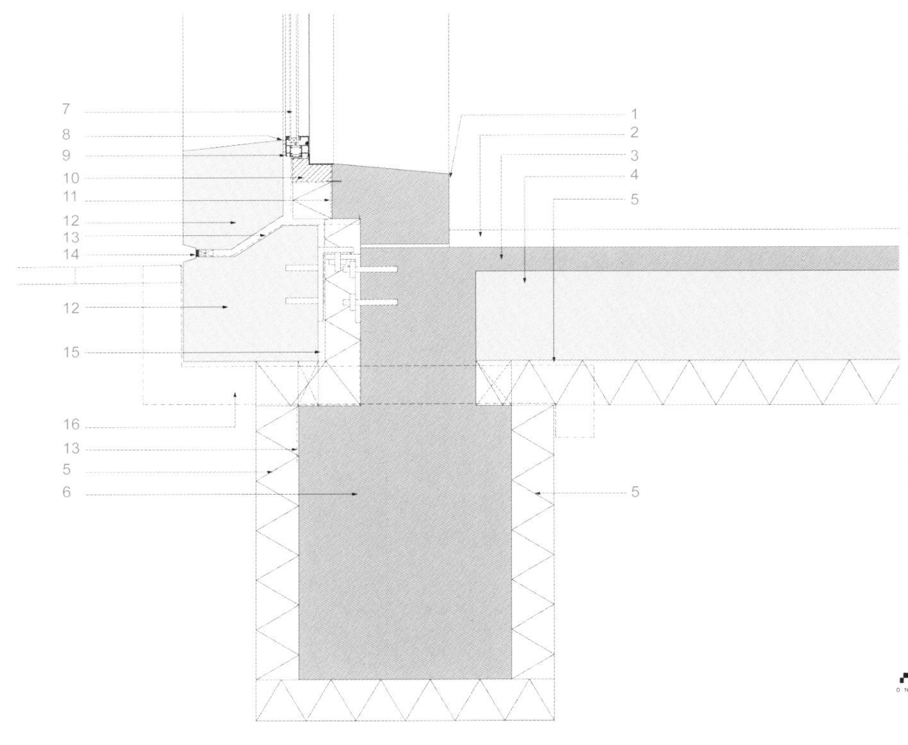

1. Plastering and painting
2. 50mm anhydrite screed
3. 70mm concrete topping
4. Concrete hollow core slab 260mm
5. Insulation
6. Reinforced concrete beam
7. Double glazed uv-resistant soundproof glass
8. Sealant on backer rod
9. Aluminium window frame
10. Wooden mounting frame with plastic in-lay
11. Façade membrane
12. Precast concrete facade element
13. Waterproof membrane
14. Elastic joint seal with plastic venttubes, 15x30mm
15. Ventilated cavity in vertical joints
16. Tube for ventilation of crawlspace

1. 石膏和涂料层
2. 50mm硬石膏砂浆
3. 70mm混凝土顶层
4. 混凝土空心板260mm
5. 隔热层
6. 钢筋混凝土梁
7. 双层防紫外线玻璃隔音玻璃
8. 泡沫密封条
9. 铝窗框
10. 木安装架，配有塑料插件
11. 立面膜
12. 预制混凝土
13. 防水膜
14. 弹性结合密封，配有塑料通风管，15x30mm
15. 垂直缝通风腔
16. 可爬行通风管

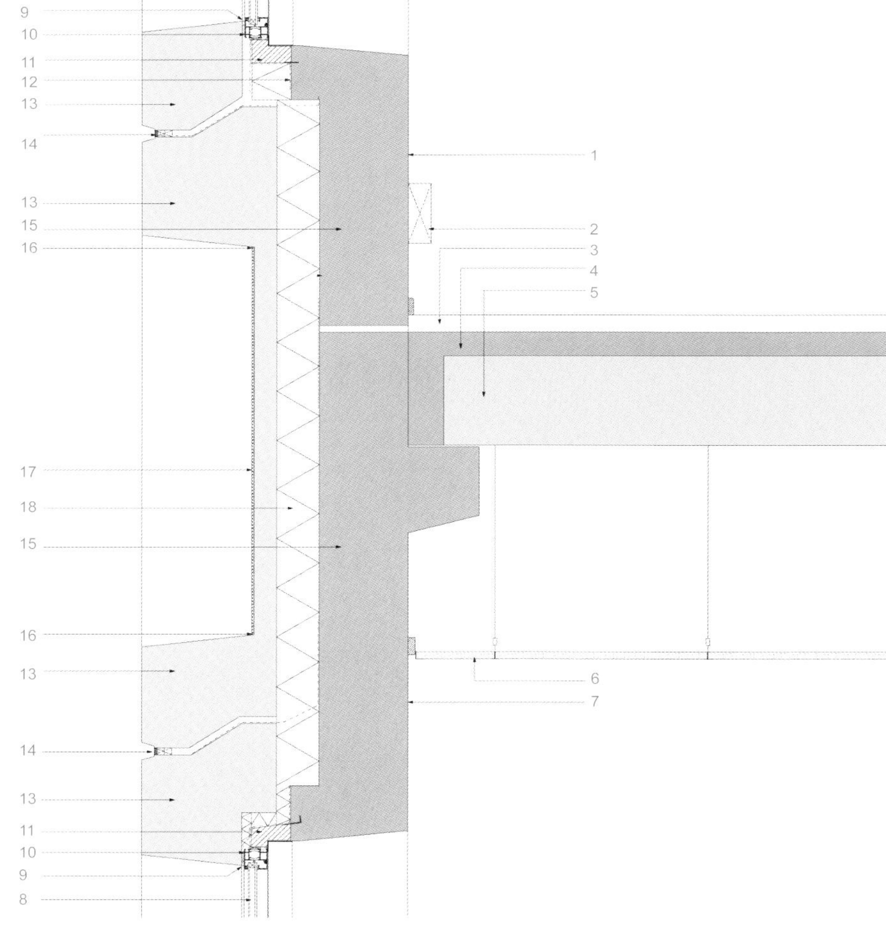

1. Plaster and painting
2. Wall cable tray
3. 50mm anhydrite screed
4. 70mm concrete topping
5. Concrete hollow core slab 260mm
6. Suspended ceiling
7. Plastering and painting
8. Double glazed uv-resistant soundproofglass
9. Sealant on backer rod
10. Aluminium window frame
11. Wooden mounting frame with plastic in-lay
12. Façade membrane
13. Precast concrete facade element
14. Elastic joint seal with plastic venttubes, 15x30mm
15. Bearing precast concrete cavity wall element
16. Sealant
17. betoglass
18. insulation

1. 石膏和涂料层
2. 墙上电缆槽
3. 50mm硬石膏砂浆
4. 70mm混凝土顶层
5. 混凝土空心板260mm
6. 吊顶
7. 石膏和涂料层
8. 双层防紫外线玻璃隔音玻璃
9. 泡沫密封条
10. 铝窗框
11. 木安装架，配有塑料插件
12. 立面膜
13. 预制混凝土
14. 弹性结合密封，配有塑料通风管，15x30mm
15. 承重预制混凝土空心墙构件
16. 密封
17. betoglass玻璃
18. 隔热层

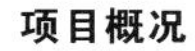

项目概况

新建筑坐落在一块狭窄的地块上，位于繁忙的海牙HS火车站和沃尔德普街之间。这一位置靠近公共交通节点，十分具有优势：校园与海牙应用科学大学、学生以及社交生活的主要区域——入口层联合起来，共同与各种形式的公共交通相互连接，并且占据了市中心的有利位置。

建筑在利尔沃特广场上形成了强烈而丰富的色彩。从火车上看，它就像是ROC蒙德里安学院的名片。建筑的格栅式混凝土立面以学院的代表色进行装饰。它们与窗口共同组成了强烈的视觉效果。一条拱廊沿着人行道将火车站与住宅群和大型购物中心连接起来。人们可以透过宽阔的玻璃立面看到学校内部，既保证了建筑的通透感，又可以作为职业培训教育的展示窗，增加公众的安全感。

1. Plastering and painting
2. Bearing precast concrete cavity wall element
3. Wall cable tray
4. 50mm anhydrite screed
5. 70mm concrete topping
6. Concrete hollow core slab 320mm
7. High density insulation
8. Betoglass
9. Precast concrete ceiling element
10. Double glazed uv-resistant soundproofglass
11. Sealant on backer rod
12. Aluminium window frame
13. Wooden mounting frame with plastic in-lay
14. Façade membrane
15. Precast concrete façade element
16. Waterproof membrane
17. Elastic joint seal with plastic venttubes, 15x30mm
18. Insulation
19. Ventilated cavity in vertical joints
20. Sealant
21. Precast concrete beam

1. 石膏和涂料层
2. 承重预制混凝土空心墙构件
3. 墙上电缆槽
4. 50mm硬石膏砂浆
5. 70mm混凝土顶层
6. 混凝土空心板260mm
7. 高密度隔热层
8. Betoglass玻璃
9. 预制混凝土天花板构件
10. 防紫外线隔音双层玻璃
11. 泡沫密封条
12. 铝窗框
13. 木安装架，配有塑料插件
14. 立面膜
15. 预制混凝土立面构件
16. 防水膜
17. 弹性结合密封，配有塑料通风管，15x30mm
18. 隔热层
19. 垂直缝通风腔
20. 密封
21. 预制混凝土梁

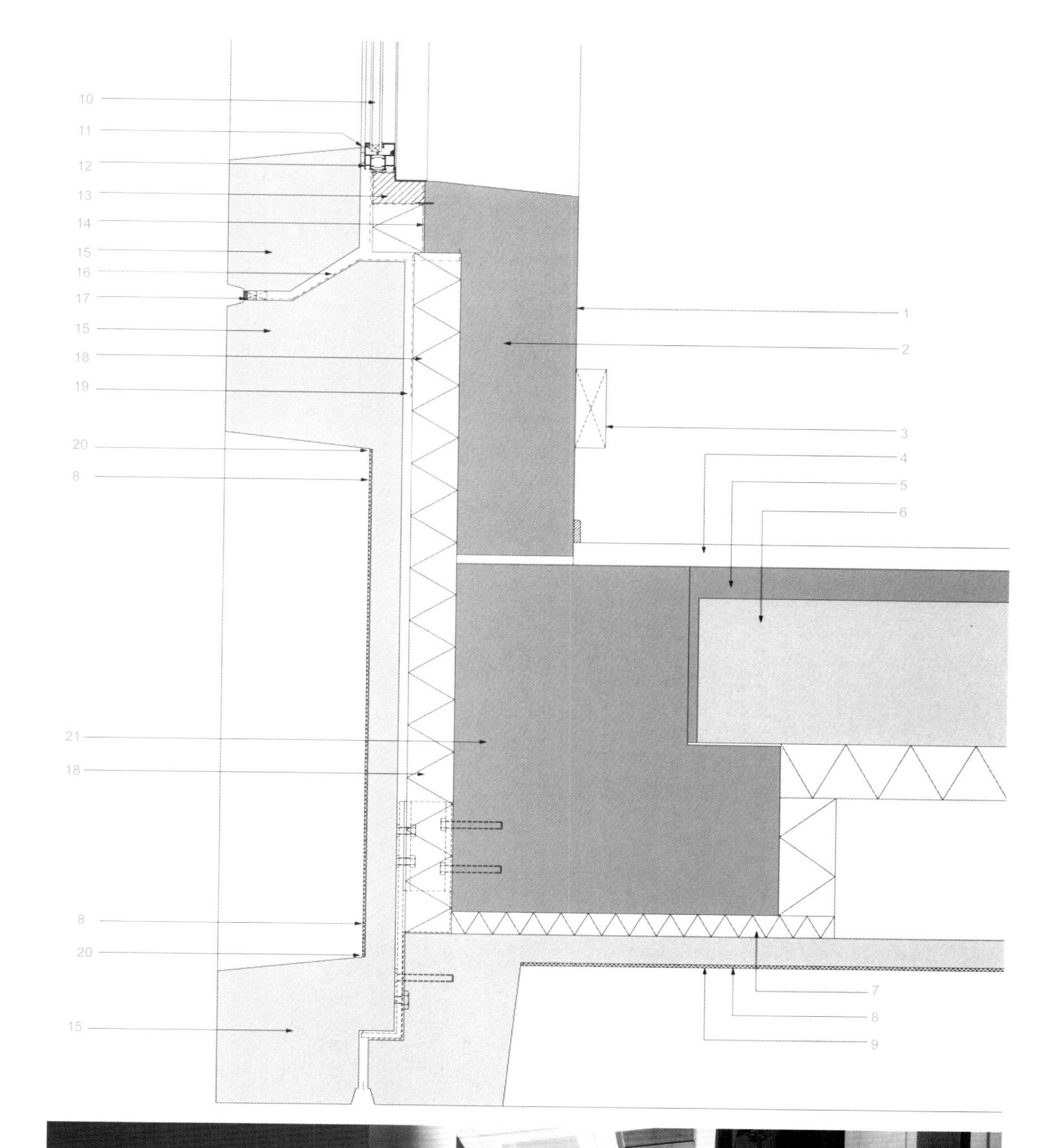

建筑的平面设计以建筑框架为基础，便于未来向办公建筑转变。无柱式空间设计让学校可以根据发展需要灵活地进行布局调整。

建筑包含两个主要空间，二者各有一个中央内核来支持所有功能。这种设计让建筑未来可以作为两座独立的办公楼出租出去。

细部与材料

整个建筑都采用预制安装的方式，在德国进行加工制造。混凝土方格中的色彩凸显了学校的特色。无规则的彩色面板和窗口与不断重复的外墙混凝土构件共同形成了强有力的整体效果。

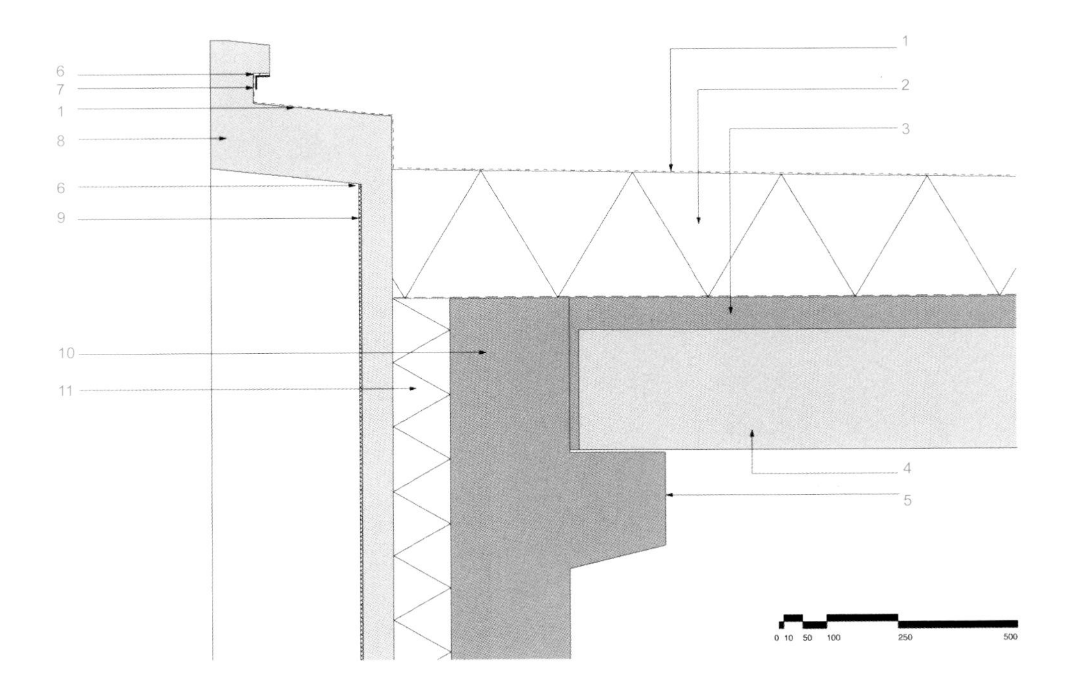

1. Bituminous roofing
2. Insulation with vapour barrier
3. 70mm concrete topping
4. Concrete hollow core slab 260mm
5. Plastering and painting
6. Sealant
7. Compression profile
8. Precast concrete façade element
9. Betoglass
10. Bearing precast concrete cavity wall element
11. Insulation

1. 沥青屋面
2. 隔热层，带有隔汽层
3. 70mm混凝土顶层
4. 混凝土空心板260mm
5. 石膏和调料层
6. 密封
7. 压缩型材
8. 预制混凝土立面构件
9. Betoglass玻璃
10. 承重预制混凝土空心墙构件
11. 隔热层

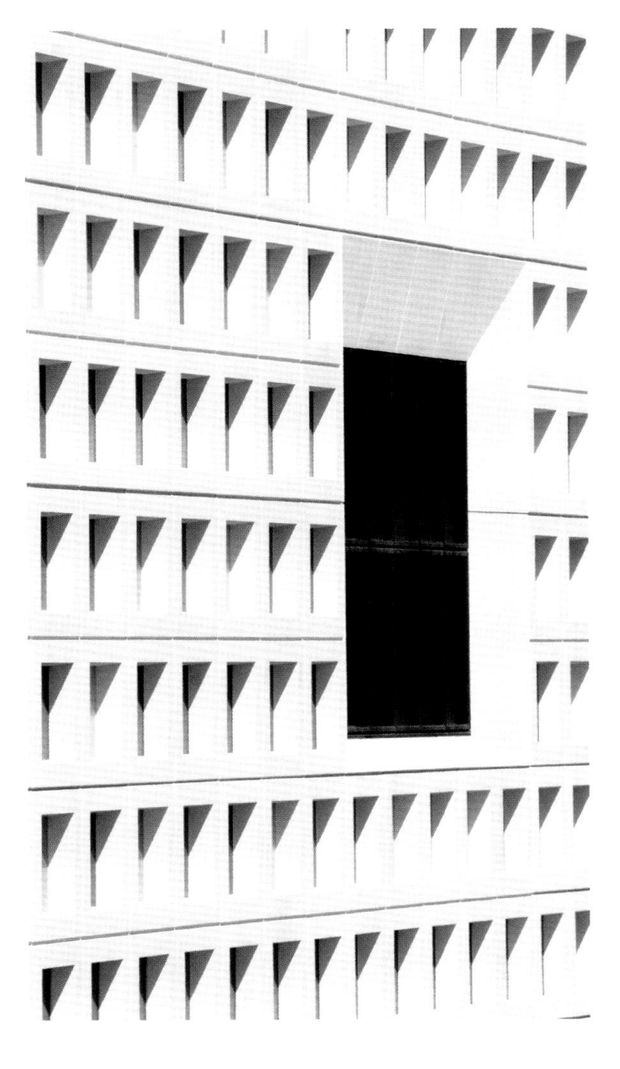

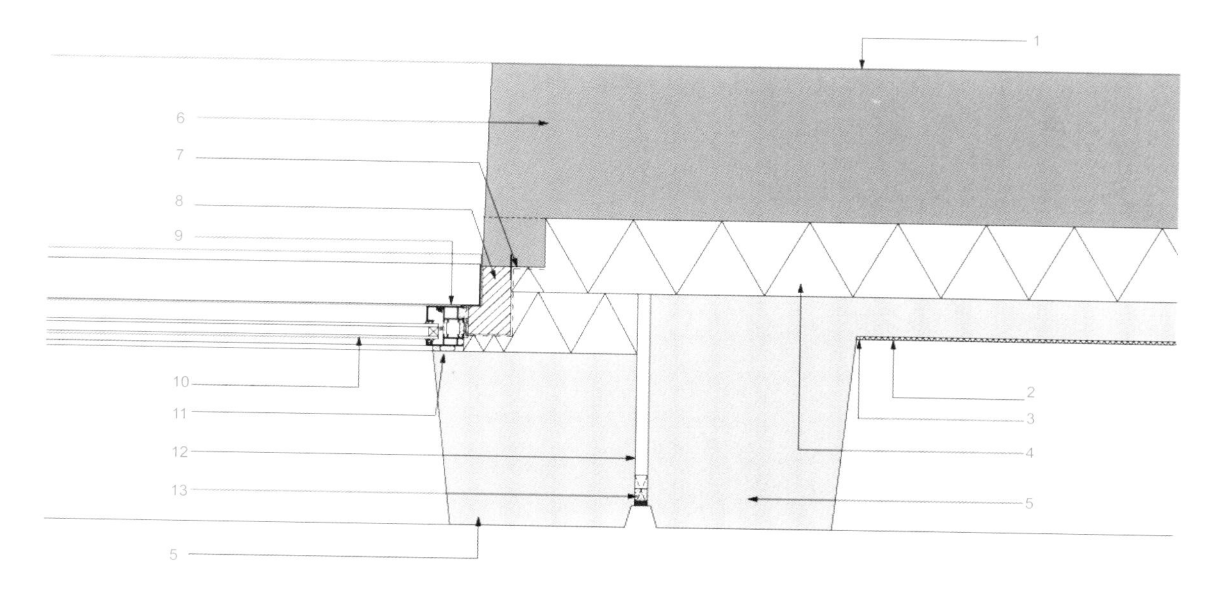

1. Plaster and painting
2. Betoglass
3. Sealant
4. Insulation
5. Precast concrete façade element
6. Bearing precast concrete cavity wall element
7. Façade membrane
8. Wooden mounting frame with plastic in-lay
9. Aluminium window frame
10. Double glazed UV-resistant soundproof glass
11. Sealant on backer rod
12. Ventilated cavity in vertical joints
13. Elastic joint seal with plastic venttubes, 15x30 mm

1. 石膏和涂料层
2. Betoglass玻璃
3. 密封
4. 隔热层
5. 预制混凝土立面构件
6. 承重预制混凝土空心墙构件
7. 立面膜
8. 木安装架，配有塑料插件
9. 铝窗框
10. 防紫外线隔音双层玻璃
11. 泡沫密封条
12. 垂直缝通风腔
13. 弹性接合密封，配有塑料通风管，15x30mm

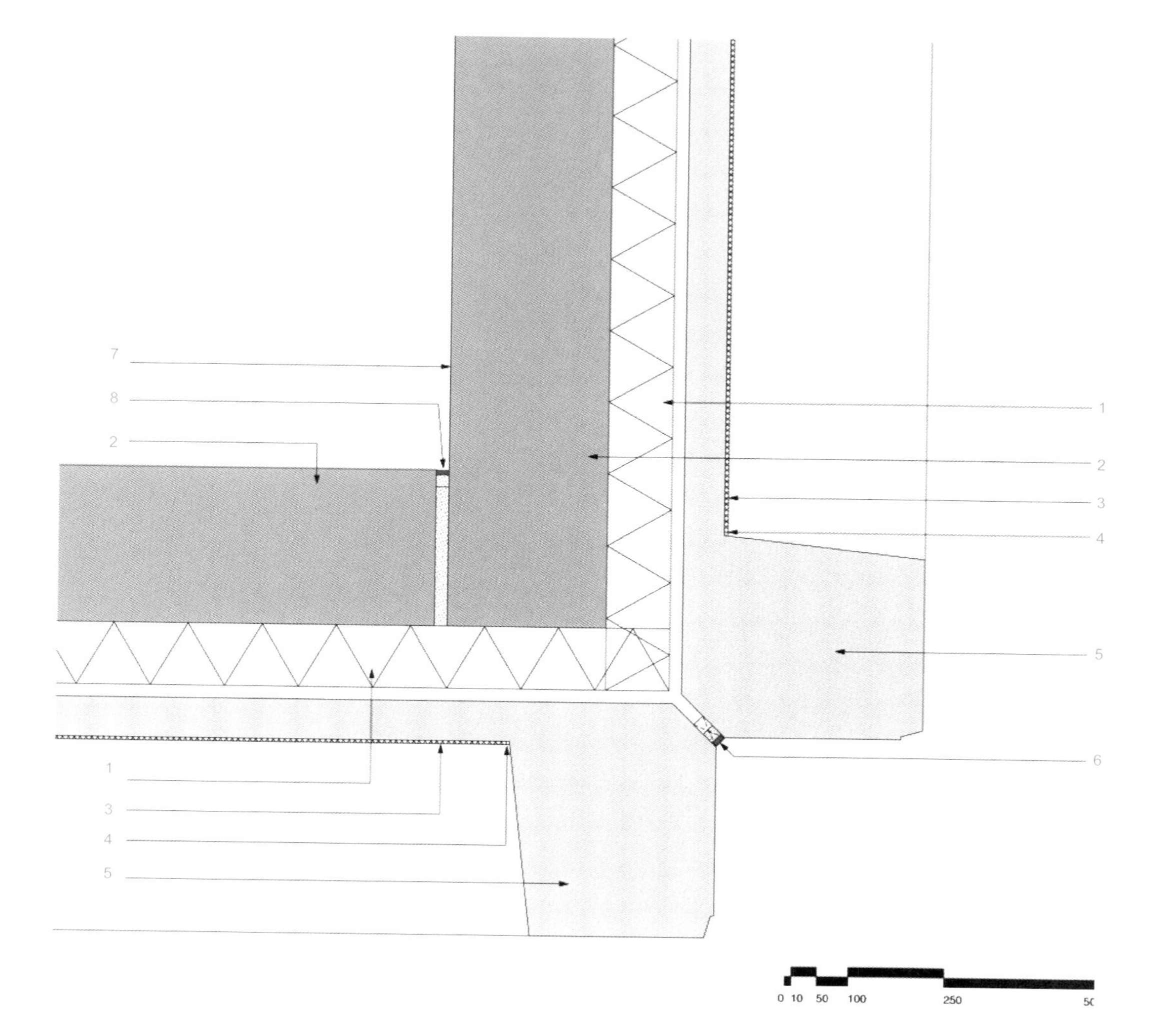

1. Insulation
2. Bearing precast concrete cavity wall element
3. Betoglass
4. Sealant
5. Precast concrete façade element
6. Elastic joint seal with plastic vent tubes, 15x30 mm
7. Plaster and painting
8. Sealant on backer rod

1. 隔热层
2. 承重预制混凝土空心墙构件
3. Betoglass玻璃
4. 密封
5. 预制混凝土立面构件
6. 弹性接合密封，配有塑料通风管，15x30mm
7. 石膏和涂料层
8. 泡沫密封条

ARGOS, Building for an Electrical Generator at a Cement Factory

阿格斯水泥厂发电站

Location/地点: Yumbo, Valle, Colombia/哥伦比亚，永布
Architect/建筑师: MGP Arquitecturay Urbanismo. Felipe González-Pacheco Mejía, Álvaro Bohórquez
Photos/摄影: Andrés Valbuena
Built area/建筑面积: 2,466m²
Key materials: Façade – prefabricated concrete pieces and steel plate
主要材料: 立面——预制混凝土构件、钢板

Overview

This project was the winner of an architectural contest looking for a design proposal of the "skin" for a technical building containing a self-generation electrical plant for a cement factory. With the resultant design, the factory wanted to generate and strengthen as well its corporative image. On the other hand, the main purpose became a mutual opportunity to generate an experimental laboratory of technical and artistic possibilities involving the material they produce, which is very low density concretes, used as the prominent factor in the design.

Detail and Materials

As a result, 2 great format concrete prefabricated pieces were designed, vertically

placed in a way to simulate the intertwined organic fiber textures used by the locals to fabricate handcraft objects, revealing the capacity that such a rigid material has to create flexible and mobile structures as well. The texture result generates a light and shadow effect that makes such a heavy and rigid material seem lighter.

The pieces consist of lightweight shells anchored to the structure through bolts and steel plates, leaving gaps between them act as openings and windows, limiting visibility into the interior.

The ground floor is cleared of solid façade, creating transparency and giving the building an acute sense of weightlessness which to pride of MGP has since been imitated by other projects in Colombia and later in a variety of materials such as wood.

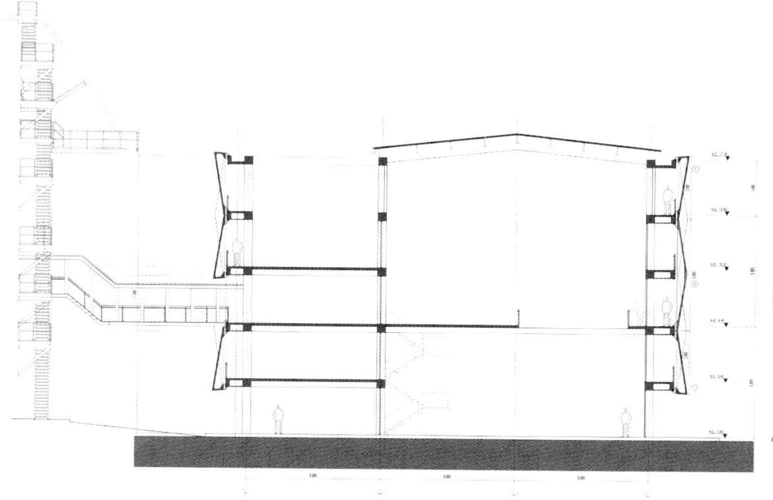

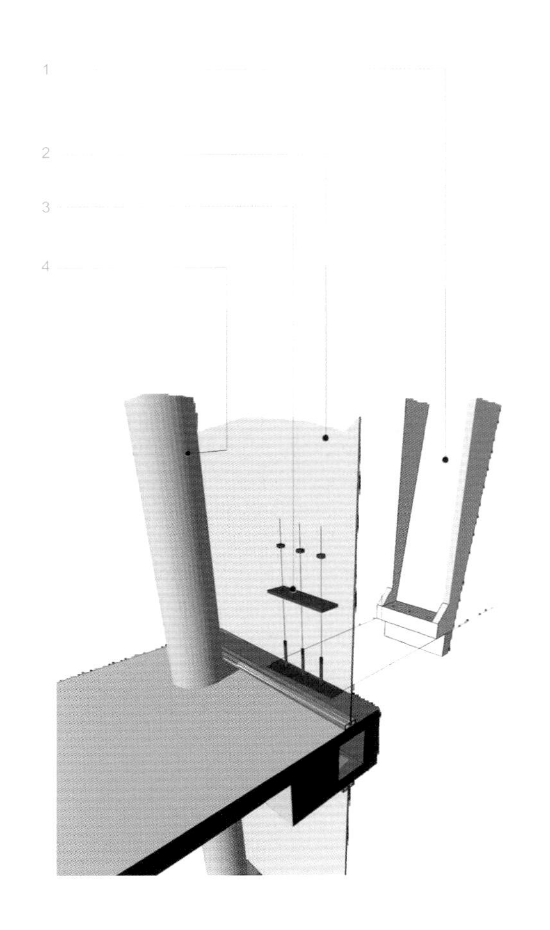

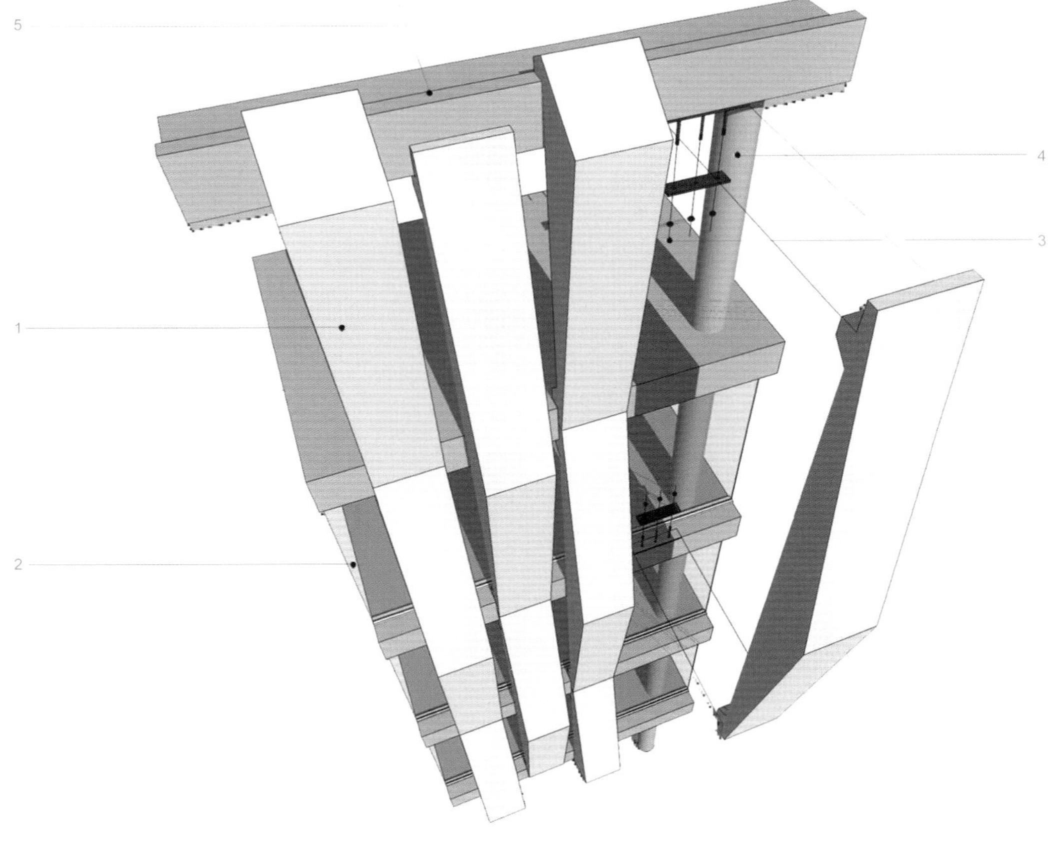

1. Prefabricated concrete pieces
2. Laminated glass
3. Metal anchor
4. Circular section concrete column
5. Concrete beam

1. 预制混凝土件
2. 夹层玻璃
3. 金属锚点
4. 圆截面混凝土柱
5. 混凝土梁

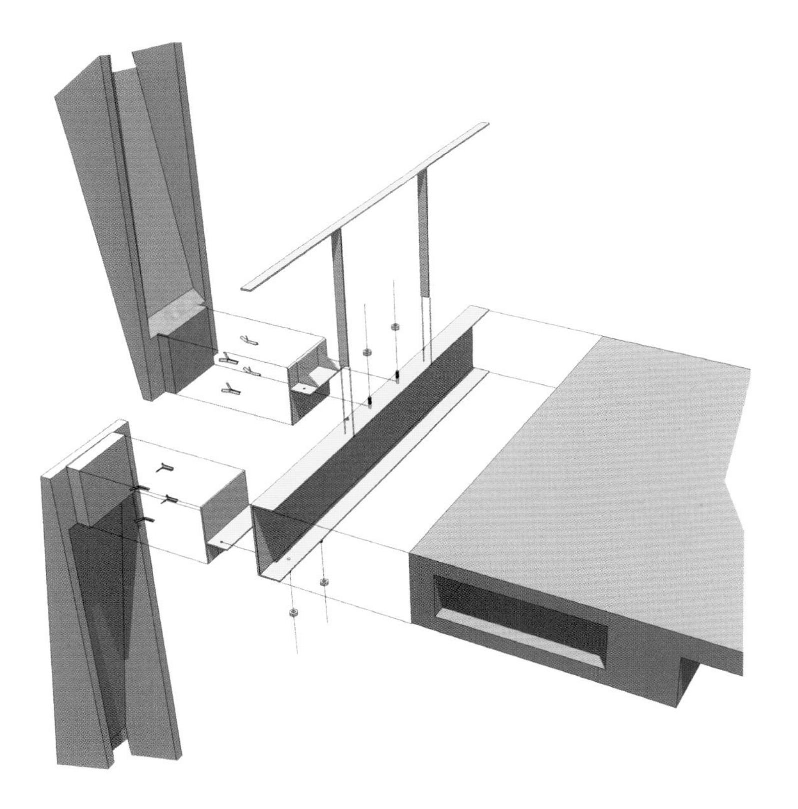

项目概况

本项目是建筑竞赛的优胜作品，为一家水泥厂的自助发电厂的技术楼进行了新的外墙设计。水泥厂希望通过新的设计来树立并强化它的企业形象。另一方面，项目的主要目标是通过使用低密度混凝土作为主要建筑材料，对其的技术性和艺术性进行全面的试验。

细部与材料

建筑师设计了两种尺寸的巨型预制混凝土构件，二者垂直摆放，形成了有机纤维纹理一样的交织形态，与当地的手工产品有异曲同工之妙，展现了混凝土材料灵活、细腻的一面。交织的纹理形成了独特的光影效果，让厚重而坚硬的混凝土看起来更加轻盈。

轻质外壳通过螺栓和钢板与建筑结构固定起来，二者之间的裂口被用作门窗开口，限制了室内的曝光性。

建筑底层并没有厚重的立面，而是营造出一种通透的效果，赋予了建筑一种奇妙的失重效果。随后，哥伦比亚的一些其他项目模仿了这种设计，并且用木材等其他材料实现了同种效果。

1. Roof in metallic tiling
2. Drain beam in concrete
3. Aluminum profile
4. Laminated glass
5. Prefabricated concrete
6. Concrete beam
7. Metallic anchor
8. Concrete column

1. 金属瓦面屋顶
2. 混凝土排水梁
3. 铝型材
4. 夹层玻璃
5. 预制混凝土
6. 混凝土梁
7. 金属锚点
8. 混凝土柱

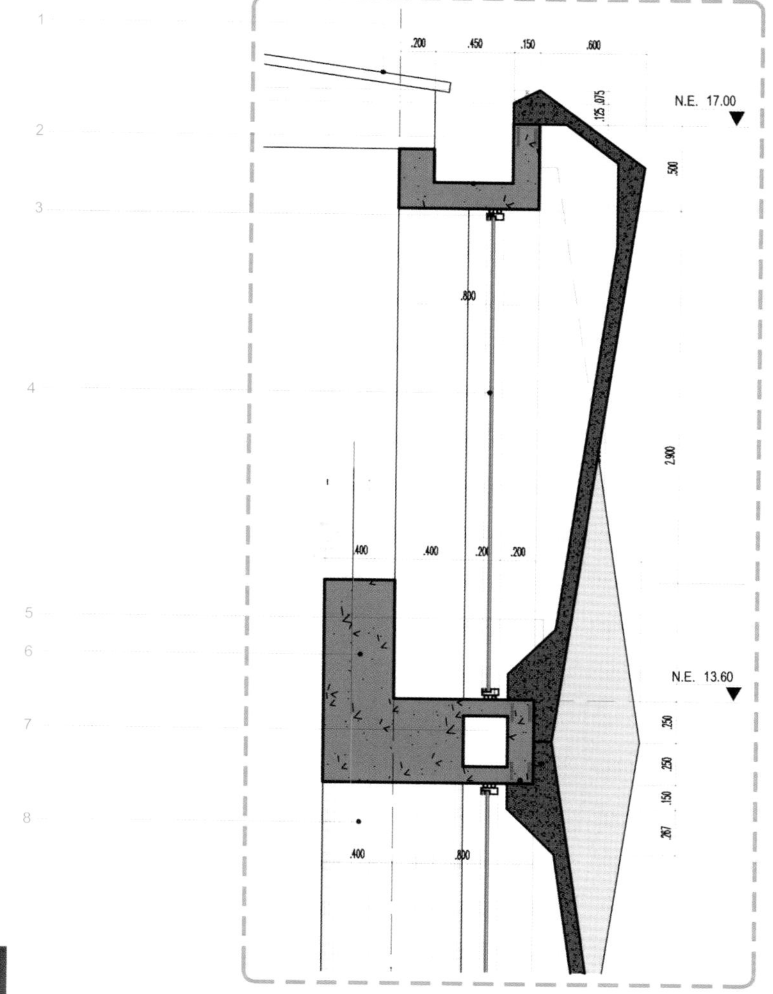

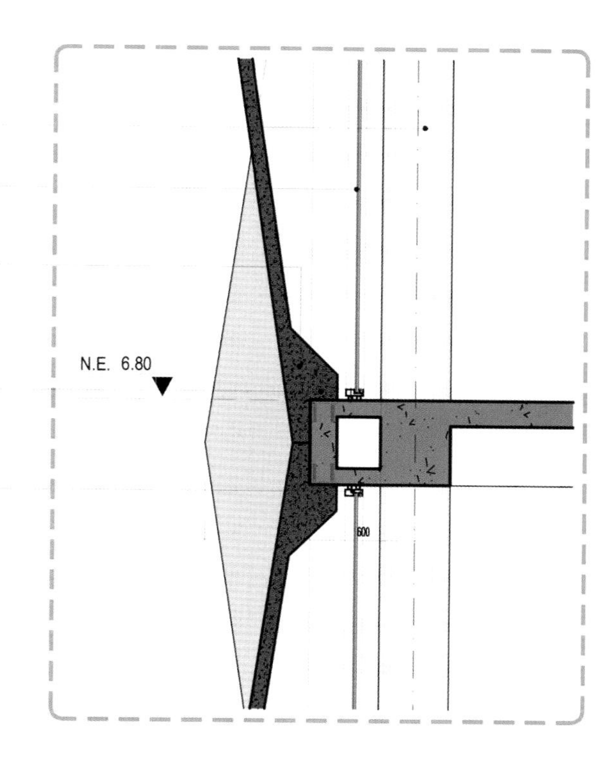

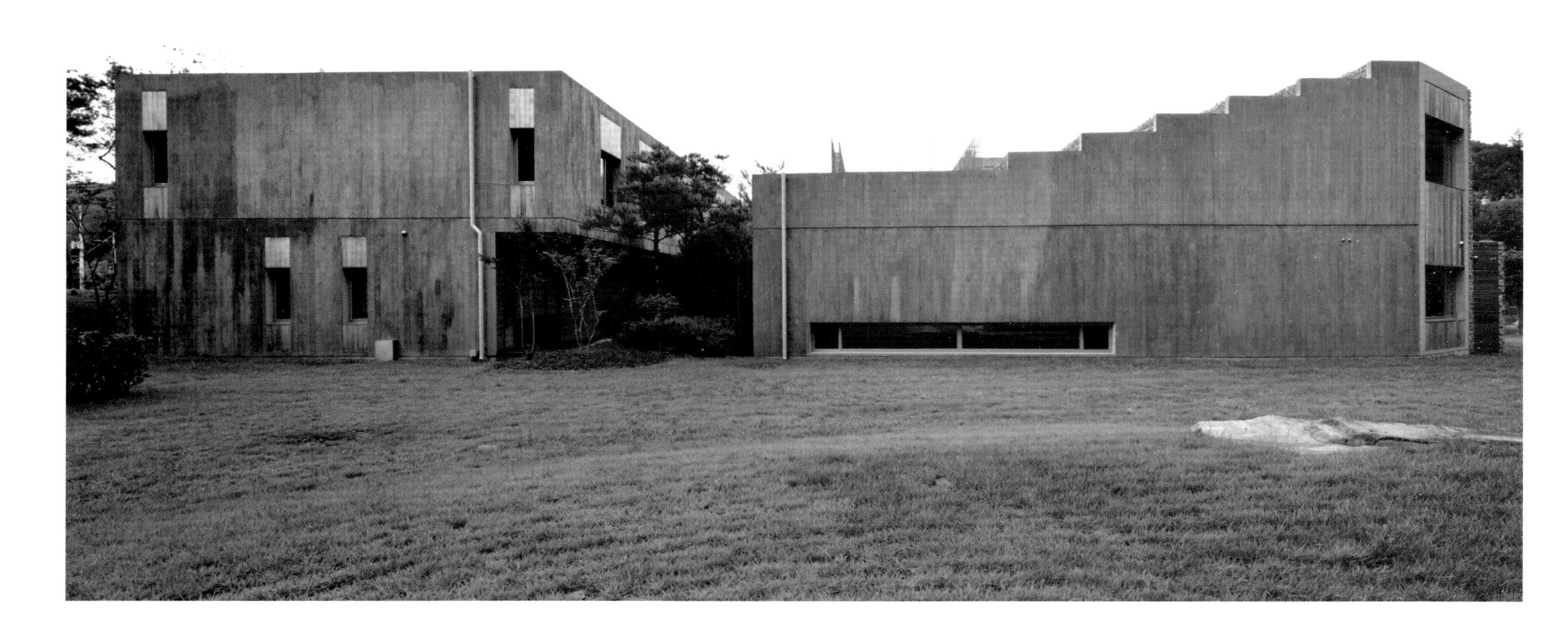

Hanil Visitors' Centre and Guest House

韩一访客中心与宾馆

Location/地点: Danyang-gun, Chungbuk, Korea/韩国，忠清北道
Architect/建筑师: BCHO Architects
Photos/摄影: Yong Gwan Kim
Site area/占地面积: 3,957 m²
Built area/建筑面积: 1,031.2 m²
Completion date/竣工时间: 2009
Key materials: Façade-fabric-formed concrete and recycled waste concrete
主要材料：立面——帆布式混凝土、回收再利用混凝土

Overview

The purpose of this project is to educate visitors about the potential for recycling concrete. In Korea concrete is one of the primary building materials so it is imperative to re-use, the otherwise wasted, concrete as buildings come down and are replaced. The Information Centre is an example of how to reuse this material in different types of construction, casting formwork types as well as re-casting techniques. Concrete has been broken and recast in various materials creating both translucent and opaque tiles. The displays will continue to evolve and change at the Information Centre as new techniques are developed. The gabion wall and fabric formed concrete which constitute the main façades of the building, was erected first, and the concrete left over from it was recycled in the gabion cages, on the rooftop for insulation from sun, and as a landscape material at the street and around the factory.

The site is located to the westernmost part of the factory, adjacent to Mt. Sobaek National Park. The existing land had been changed much to facilitate the movement of trucks to the cement factory. First of all, the architects tried to restore the damaged original mountains and forest. In order to revive the landscape, they brought in earth to fill the courtyard between the two buildings. The flow of the mountains from the west leads to the reception and cafeteria in the inner courtyard of the building. In the in-between spaces the architects allowed people to experience the mass of the building while watching the building shift around its central courtyard.

While following the linear placement and movement of land and earth, the architects came up with ideas for the new building façade. They applied canvas-like concrete walls to the east façade, evoking images of the adjacent forest.

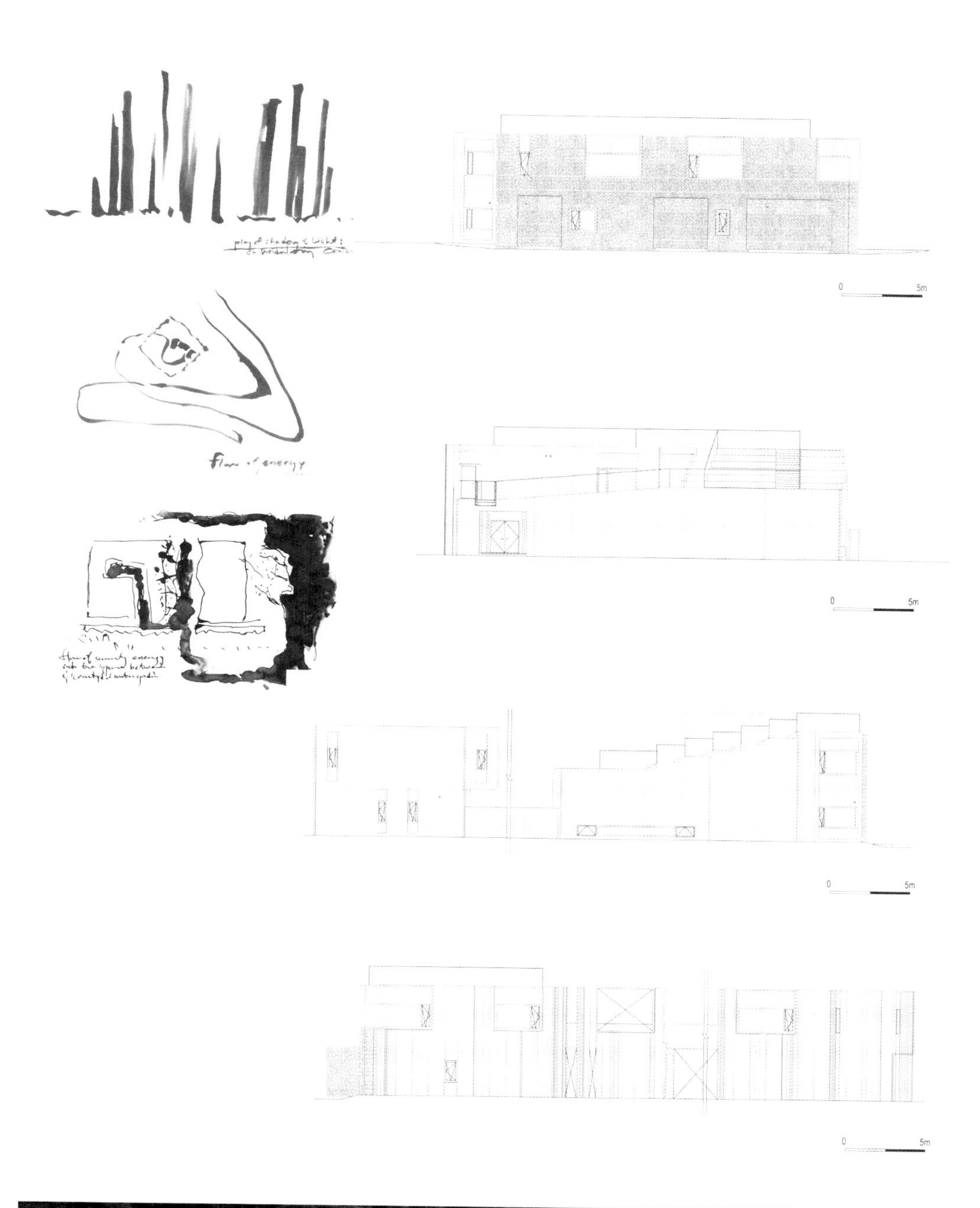

There are four openings in the eastern wall and long vertical windows have been created in their in-between spaces. Through the windows, one can see how concrete is produced at the factory. Behind the two larger openings, one can see the courtyard of the Visitors' Centre and the cafeteria next to the courtyard, which is encircled by a water garden.

Detail and Materials

Recycled Concrete Wall

By recycling waste concrete generated in erecting the eastern wall, a new concrete wall was erected on the opposite side. The footing used in erecting the fabric formed concrete wall was cut into pieces of 10-20cm and put into the gabion wire netting to be recycled as exterior finishing material for the southern façade. Recycling of waste concrete not only has eco-friendly and cost benefits, but also offers an antique feel, as dust and moss gather on the concrete with the passage of time.

Fabric Concrete Wall

In order to keep the physical properties of concrete intact and simultaneously express gentle curves, the fabric-formed concrete wall was developed in collaboration with C.A.S.T., based in Manitoba, Canada, after much research and consultations. Concrete moulds were created on the footing; curved forms were set using pipes; and high-strength fabric was placed on top like a mould. After embedding connecting fittings, concrete is poured, and pre-cast concrete is lifted so that it could be installed at the external wall in the east. While producing a non-bearing concrete wall with convex and concave curvatures, the architects conducted a variety of experiments, departing from stereotyped notions.

项目概况

项目希望通过自身向参观者介绍回收混凝土的潜能。在韩国，混凝土是主要的建筑材料，因此再利用混凝土是十分必要的，否则被拆除的建筑混凝土材料就会被废弃。信息中心的设计通过各种建造形式、铸造模架以及重铸技术实现了混凝土的再利用。混凝土被打碎，重铸成各种材料，形成半透明和不透明的砖块。随着新技术的开发，信息中心的展览会不断进化。建筑的外墙主要由石笼墙和纤维混凝土构成，在建成之后，留下的混凝土废料被回收起来，装在铁笼里，安装在屋顶遮阳或是作为街边和厂区内的景观元素。

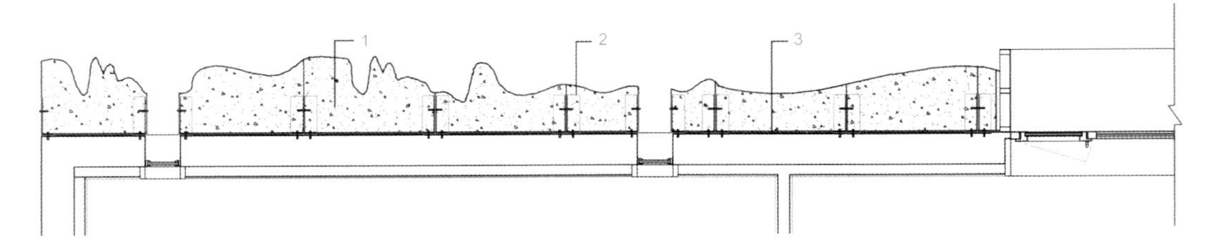

1. Precast fabric concrete wall
2. THK9 manufactured bracket
3. THK9 steel
4. THK300 recycled concrete Gabion wall

1. 预制帆布式混凝土墙面
2. THK9加工支架
3. THK9钢材
4. THK300回收再利用混凝土石笼墙

0 1m

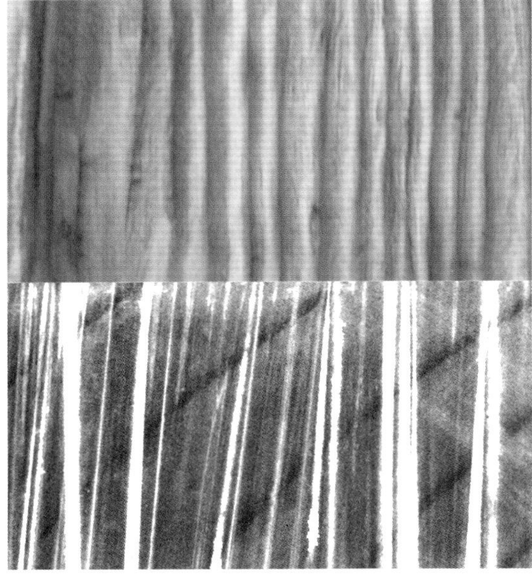

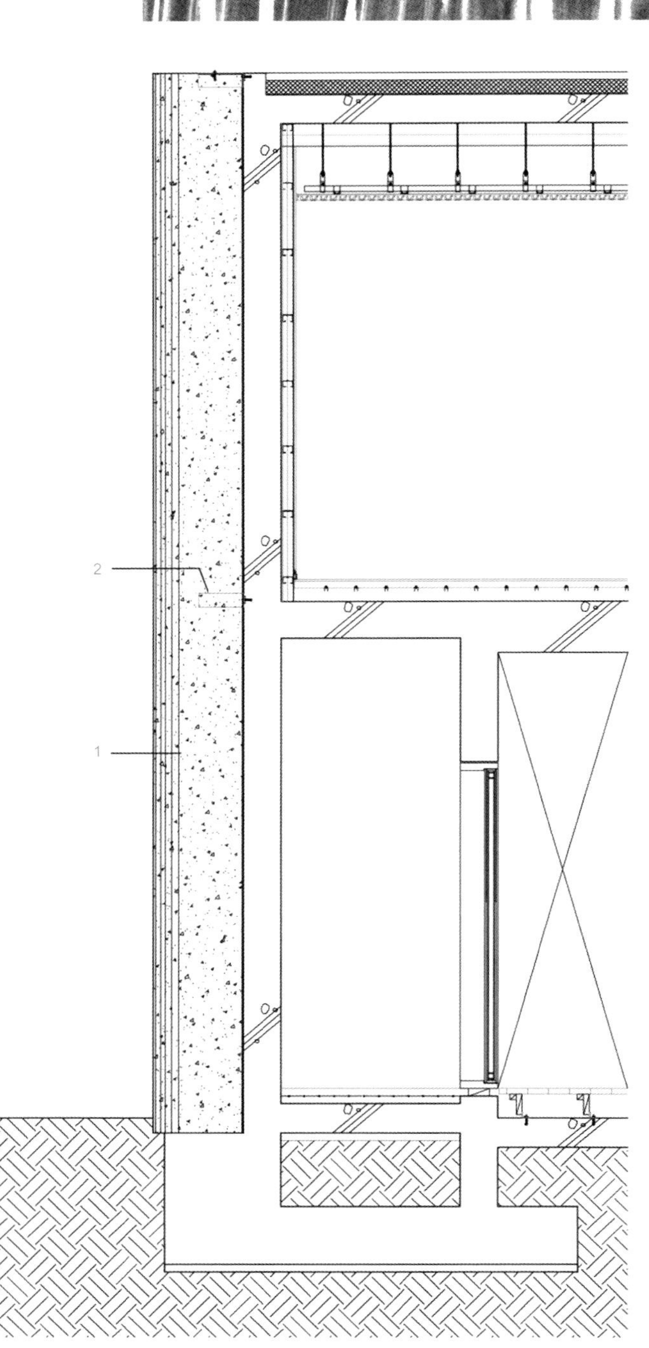

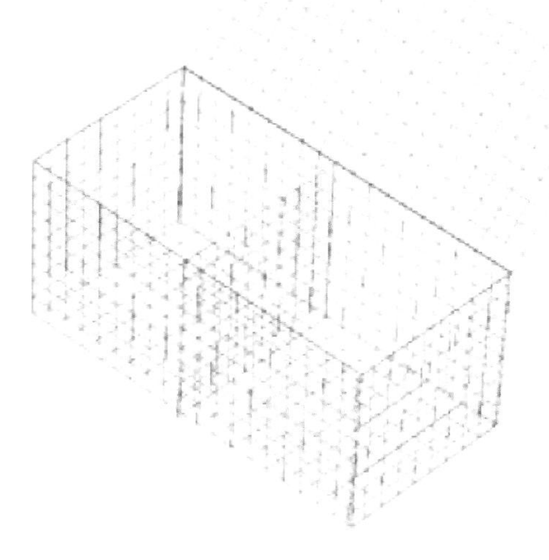

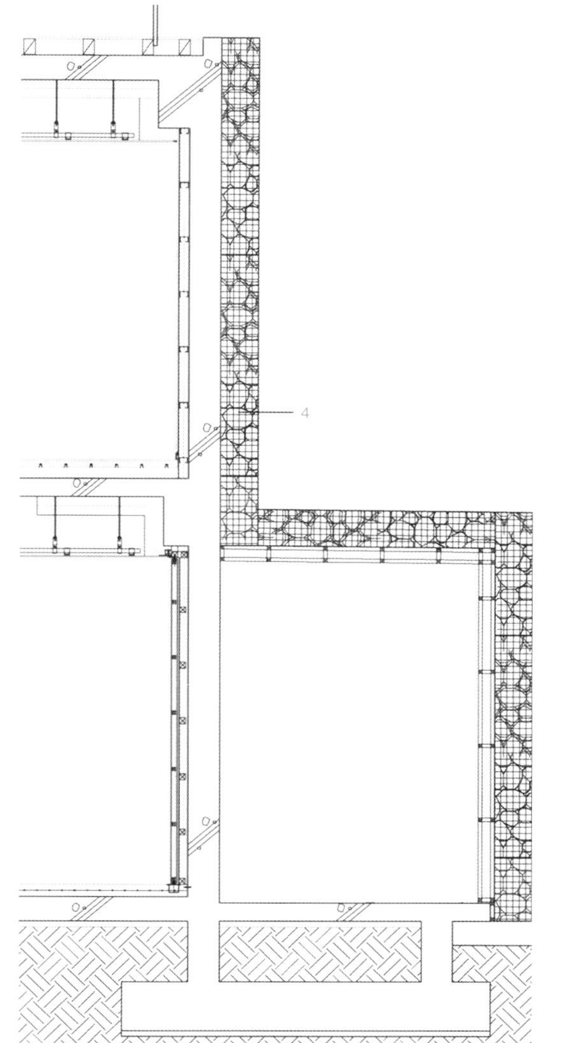

项目场地位于工厂的最西面，毗邻小白山国家公园。为了方便水泥工厂运输卡车的通行，原始地貌的变动很大。首先，建筑师试图修复被破坏的山丘和森林。为了复苏环境，他们运来泥土对两楼之间的庭院进行填充。西面山脉的走势一直延伸到内庭的接待前台和餐厅。在中间空间，人们可以感受建筑的体块和它围绕中庭发生的变化。

在遵循土地的线形配置和运动的前提下，建筑师设计了一种全新的建筑立面。他们在东立面应用了帆布样的混凝土墙，给人以森林的感觉。东侧墙面上共有四个开口，中间有细长的窗户。透过窗户，人们能看到混凝土在工厂中的制作过程。透过两个较大的开口，人们能看到访客中心的庭院和紧邻庭院的餐厅，后者被水景花园所环绕。

细部与材料

回收再利用混凝土墙

通过回收在建造东墙过程中所留下的混凝土废料，建筑师在对面建造了一座新的混凝土墙。帆布式混凝土墙的边角料被切割成10~12厘米的小块，回收到铁笼中用作南立面的外墙装饰。回收废料不仅环保节约，还能营造一种复古感，因为随着时间的流逝，灰尘和苔藓会聚集在混凝土表面。

帆布式混凝土墙

为了保持混凝土物理性能的完整性、自然地呈现曲线，帆布式混凝土墙的开发选择了与加拿大C.A.S.T.公司进行了合作。混凝土模架被建在基脚上，用管子设置出弧形，同时在上面铺上了高强度面料作为模子。在植入连接件之后，灌注混凝土，然后在东侧外墙上安装预制混凝土板。除了制作凹凸有致的非承重混凝土墙之外，建筑师还进行了一系列标新立异的实验。

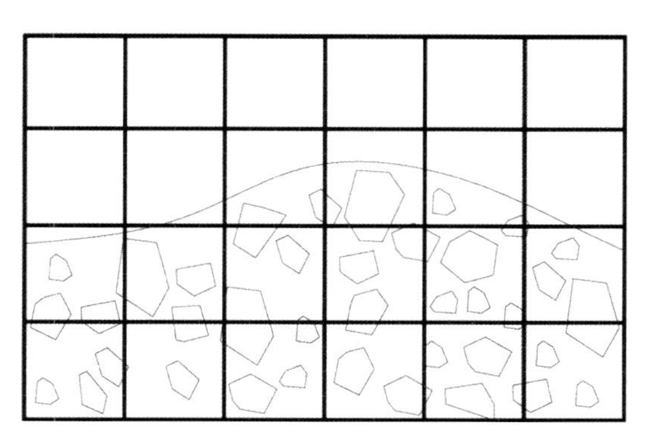

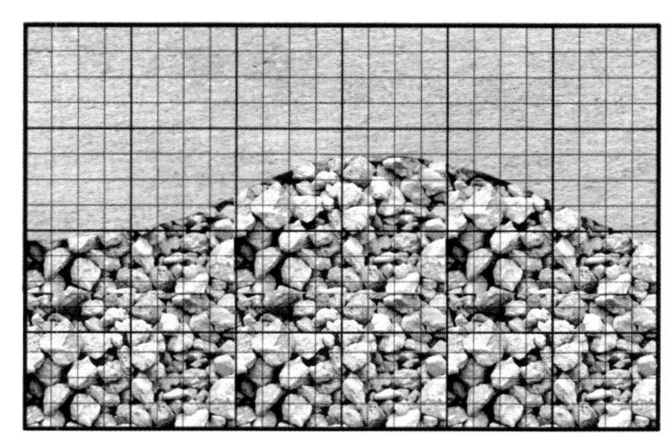

Cultural and Community Centre in Bergem

博杰姆文化社区中心

Location/地点: Bergem, Luxembourg/卢森堡，博杰姆
Architect/建筑师: Françoise Bruck, Thomas Weckerle
Photos/摄影: Lukas Roth
Site area/占地面积: 2,400m^2
Built area/建筑面积: 1,600m^2
Completion date/竣工时间: 2012
Key materials: Façade – precast concrete elements, aluminium, glass
主要材料： 立面——预制混凝土构件、铝、玻璃

Overview

A new cultural centre was to replace the dilapidated community centre in the heart of the village of Bergem. A condition of the commission was that clubs could continue their activities in the existing building until the new one was ready. Dominating the remaining free space on the plot, which borders a green zone, was a majestic walnut tree.

Bruck + Weckerle Architekten felt that this tree, sheltering its meadow for over 80 years, was such a feature of the village of Bergem that it deserved to be retained and displayed to its best advantages. They recognised the scope offered by this symbolic tree, when allied to subtlety of design, for creating a unique meeting point that would not only reconcile the natural and cultural worlds but also architecturally enrich the core of the village.

The design concept comprises a low, rectangular structure covering the entire remainder of the plot. Where the walnut tree stands, a semicircular recess is scooped out of that stark geometry – creating a transitional zone between indoors and outdoors. The building appears to welcome visitors with open arms and to embrace the walnut tree.

The welcoming stance of the semicircular inner courtyard, with a continuing façade built in structural glazing, contributes to the character of the building and draws the visitor to the heart of it. A curved wall, panelled in walnut, sweeps round to create a single uninterrupted space that surrounds the entire inner courtyard and leads to the main hall. The wall facing onto the tree is wholly glass, including glass doors that can be opened in fine weather to give access to the courtyard.

This spatial organisation offers many different ways of using the space. All the ancillary rooms are located behind the curved wall. The falling contour allows natural light in the basement,

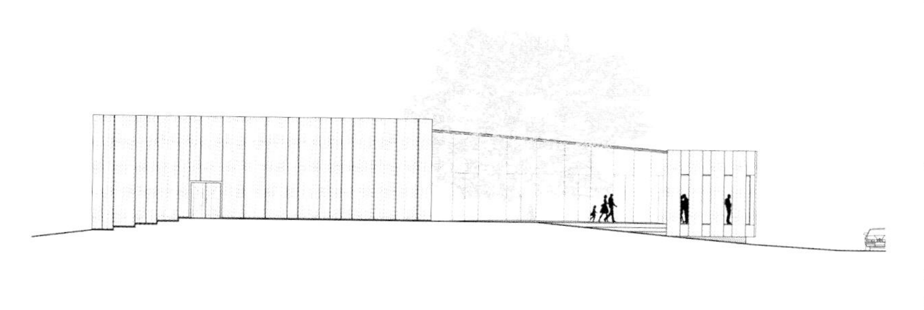

where the spaces for delivery are situated. They can be accessed any time from behind the building, independently from events and visitors. Also the clubrooms and the entire logistics are located in the basement. The building meets the modern requirements with respect to ecology, engineering and building technology (green roof, rain water usage, ground-coupled heat exchanger etc.).

Detail and Materials

As an external expression of the building's cultural role and its restrained festivity, a stylised red stage curtain serves as a metaphor for celebration, community and culture. Cast in concrete, it looks like a freeze-frame image, composed of nine individual molded precast concrete elements. It nevertheless seems to move back and forth as the light brightens or dims in time to the swaying branches. The windows are vertical bands of glass, splitting the curtain into strips as if it had simply been pushed to a side.

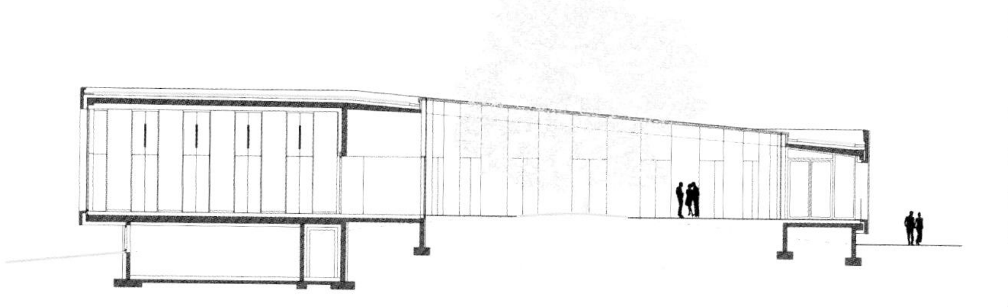

项目概况

新建的文化中心将取代之前位于博杰姆村中心破旧的社区中心。设计有一个先决条件：在新的文化中心完工之前，社团还将继续在旧社区中心内开展活动。在整个建筑场地中央，有一棵高大宏伟的胡桃树。

建筑师认为这棵树龄80多年的大树是博杰姆村的标志，必须被保留下来，好好利用。它们确认了大树所辐射的范围，然后将其巧妙地应用到设计之中，打造了一个独特的集会空间。这个空间不仅调和了自然世界与文化世界，还在建筑意义上让整个村庄变得丰富起来。

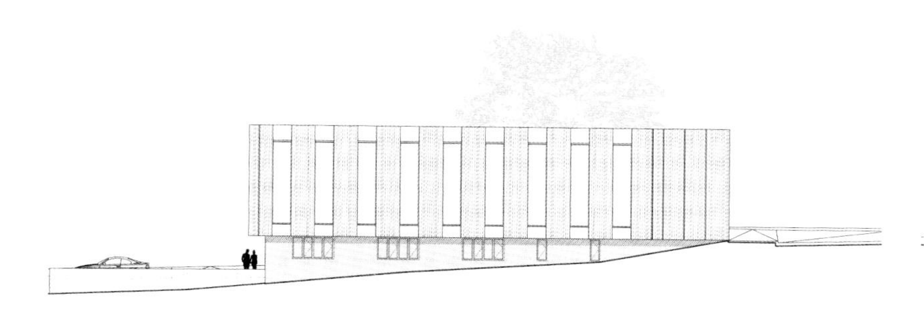

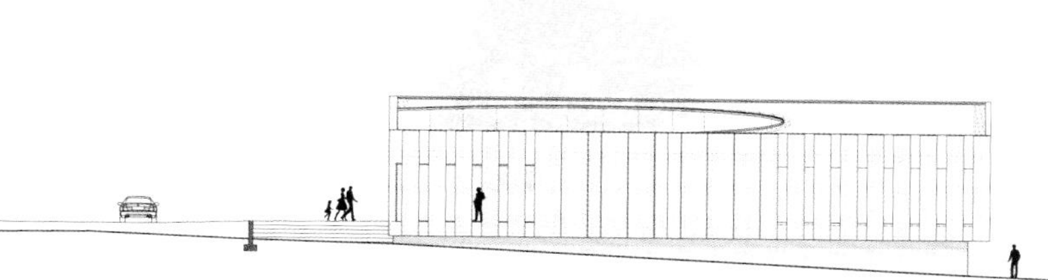

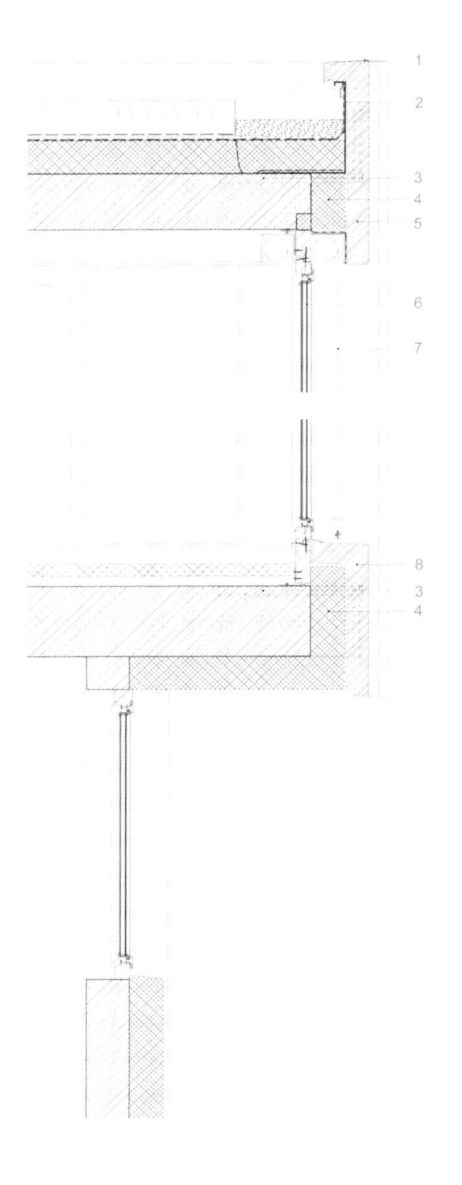

1. Aluminium profile
2. L-profile of sheet steel, 5mm
3. Recess in the concrete for HALFEN BRA-NJ
4. Insulation, 160mm
5. Precast concrete element C.3, 100mm
6. Sun protection glass, 6mm
7. Sun shading
8. Precast concrete element C.4, 100mm

1. 铝型材
2. L形钢板，5mm
3. 混凝土凹处，HALFEN BRA-NJ
4. 隔热层，160mm
5. 预制混凝土构件C.3，100mm
6. 遮阳玻璃，6mm
7. 遮阳板
8. 预制混凝土构件C.4，100mm

1. Precast concrete element B.0, 205mm
2. Insulation, 160mm
3. Cast-in-place concrete, 200mm
4. Air gap, 120mm
5. Acoustic walnut veneered MDF panel, 19mm
6. Precast concrete element C.4, 100mm
7. Sun protection glass, 6mm

1. 预制混凝土构件B.0，205mm
2. 隔热层，160mm
3. 现浇混凝土，200mm
4. 气隙，120mm
5. 隔音胡桃木中密度纤维板，19mm
6. 预制混凝土构件C.4，100mm
7. 遮阳玻璃，6mm

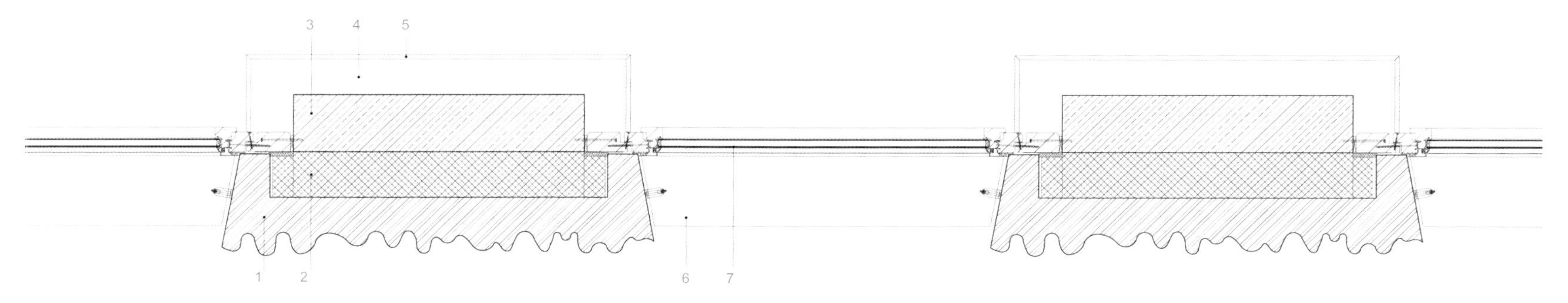

设计概念以低矮的直角结构将场地的其他部分覆盖起来。在胡桃树的所在地，建筑师开拓出一个半圆形空间，铺上了木地板，实现了室内外空间的过渡。建筑以开放友好的姿态迎接着人们的到来，同时也拥抱着胡桃树。

半圆形内庭的欢迎姿态与连续的结构玻璃墙面为建筑增添了特色，吸引着来访者进入建筑。胡桃木装饰的弧形墙面形成了统一而连续的空间，将整个内庭环绕起来，直通主大厅。朝向大树的墙面全部由玻璃制成。当天气晴好时，可以通过玻璃门直接进入庭院。

这种空间配置实现了空间应用的多样化。所有辅助房间都被设在弧面墙的后方。下坠的轮廓让自然光可以进入地下室的货运空间。任何时候，都可以由建筑后方进入地下室，不会影响中心举办的活动和来访者。同时，俱乐部聚会室和整个物流区都设在地下室。建筑符合现代环保建筑的要求，采用了绿色屋顶、雨水利用、地源热泵等技术措施。

细部与材料

为了体现建筑的文化内涵和低调的欢庆感，建筑外墙采用了红色混凝土幕墙，象征着欢庆、社区和文化。混凝土外墙看起来像是单个的图像，由9个独立的模塑预制混凝土构件组成。随着光线的明暗变化，它也似乎前后摆动。窗户采用垂直玻璃带，将幕墙切割成细条，就好像幕墙被推开了一样。

FDE Várzea Paulista School

FDE瓦尔泽亚保利斯塔学校

Location/地点: São Paulo, Brasil/巴西，圣保罗
Architect/建筑师: Forte, Gimenes & Marcondes Ferraz Arquitetos
Site area/占地面积: 6,344m²
Built area/建筑面积: 2,703m²
Key materials: Façade – precast concrete elements, perforated aluminum tiles
主要材料：立面——预制混凝土构件，穿孔铝板

Overview

As a result of a recent programme by FDE, Foundation for the Development of Teaching, the primary and secondary state schools built by the Government of the State of São Paulo have in common, as the choice of their constructive system, the industrialised components, the room programme and the leisure areas, the articulation between the spaces and the intention to create a comfortable place, with qualified architecture for the occupants of the schools and the teaching practice, combined with a extremely low budget.

Since the beginning, the intention was to create a large integration between the public and semi-public spaces, between the internal and the external sides. Thus, the complicated site was treated so as to provide a large access square to the school which, when their doors are open, is transformed into an agreeable space of gathering for the community. This space is complemented by the rest of the ground floor of the building, which has a multisport court, a covered multi functional space for several activities and a recreation space in the back part of the lot. During the weekends, all these articulated spaces serve as a leisure area for the local population that is so much in need of this kind of space.

The structure of the school is entirely composed of pre-molded concrete elements. This system, chosen for the control quality of execution, the speed in assembling and the

accessible cost, provides the character of the school. The structure is modular and corresponds to the dimensions of the main internal environments.

The building has a three storey block and another one just with the ground floor, where the multisport court is located, with a high ceiling. The other pavements are occupied by the classrooms, environment rooms, computer and storage rooms, besides the teachers' and the director's rooms. The covered multi functional space, on the ground floor, has double height and is totally open for the external leisure area.

Detail and Materials

The concrete structure of the building extrapolates its limits, also supporting the shadow elements (brise soleil). On the front part, open concrete elements with irregular openings are grouped so as to from a large mosaic which filters the light. This concrete mosaic creates interesting visual forms, both from the inside, from where it seems to frame the landscape, as from the outside, from where it looks like a giant panel. During the night, when the classrooms are lit, the mosaic doesn't look so strong and the school gains a lighter and more diaphanous aspect.

On the back part, the part with double height in the multi functional space, the shadow is created by perforated aluminum tiles. The multi functional space, which is the organising element of the entire school, is a place of multiple uses, used by the students for several activities, but also by the community on weekends and special events. It remains completely open for the external space and the shadow element is applied higher. The perforated tiles allow the light to enter during the day in such a way that they get practically transparent for those who are inside the galpão, enhancing the feeling that there is no clear limit between the internal and the external sides of the room. During the night, with the internal lighting, this sensation is inverted, and the observer who is on the external garden has the feeling that the school is totally open for the community.

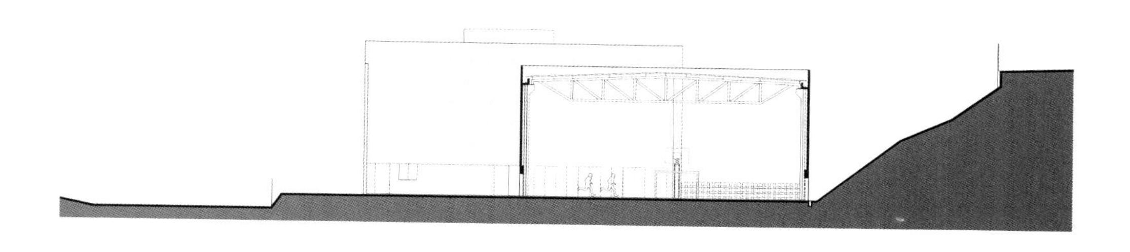

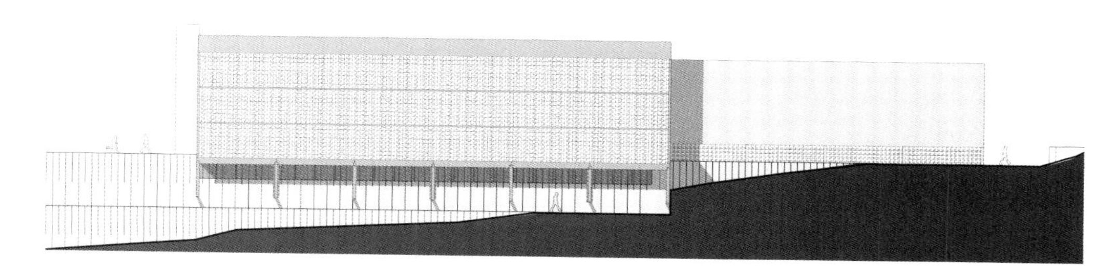

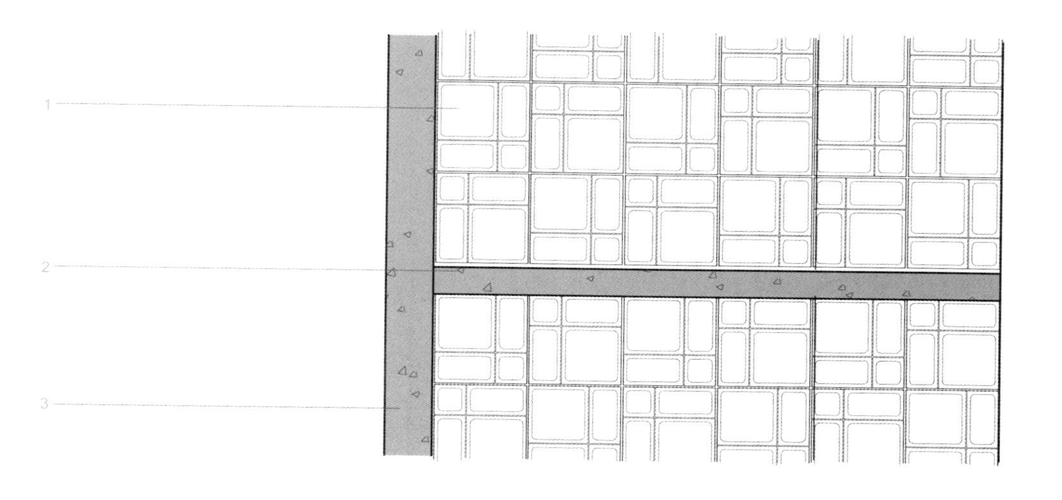

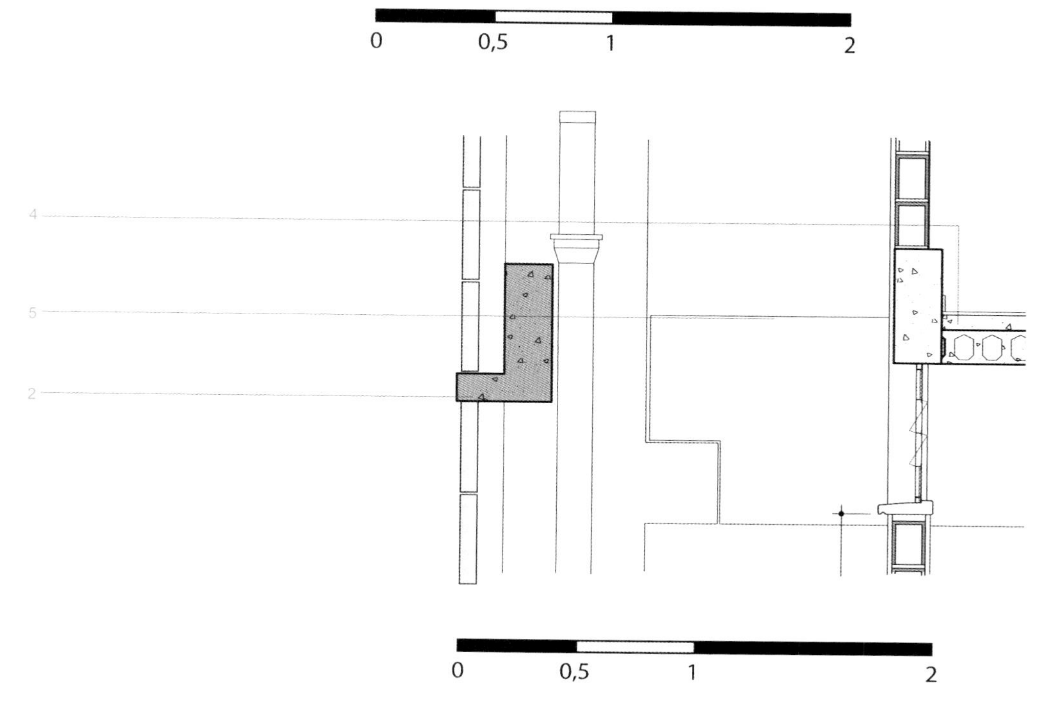

项目概况

由巴西教育发展基金会（FDE）发起、圣保罗州政府兴建的中小学有许多共同点：结构系统、工业化构件、空间规划和休闲区、空间的连接以及舒适的空间。学校不仅通过高品质的建筑为师生和教学活动提供了优质的环境，同时建造预算还很低廉。

项目的目标是将公共与半公共空间、室内与室外空间高度融合在一起。因此，建筑师在复杂的场地环境中设计了一个大型入口广场，当校门开放时，广场将变成一个宜人的社区集会空间。教学楼的一楼设有多种运动场馆，二者结合起来共同组成了多功能半室内活动区和休闲空间。在周末，这些相互连接的空间将为当地居民提供一个他们所急需的休闲区。

学校的建筑结构全部由预制混凝土构件组成。这一系统便于控制工程质量、安装快速、成本较低，形成了学校的主要特色。建筑采用模块化结构，并且与主要内部环境的维度相呼应。

建筑上方有三层楼体，底层还设有多功能运动场。其他空间由教室、环境控制室、电脑室、储藏室、教师办公室等组成。一楼的多功能空间有两层楼高，完全朝向户外休闲区开放。

细部与材料

建筑的混凝土结构向外延伸，同时还支持了遮阳元件。在建筑正面，开放的混凝土构件上有不规则的开口，形成了巨型马赛克图案，能过滤阳光。这层混凝土马赛克形成了有趣的视觉效果。从里面看，它将风景框了起来；从外面看，它看起来像一块巨大的屏风。晚上，教室的灯光点亮，马赛克幕墙会显得比较柔和，让学校看起来更加轻盈通透。

在建筑背面双高多功能活动区，穿孔铝材起到了遮阳作用。作为整个学校的组织元素，多功能空间具有多重用途：既可供学生活动，又能在周末和特殊活动时供社区使用。它完全朝向外部空间开放，因此遮阳元件更加轻质。白天，穿孔铝板让阳光进入门廊，使空间更加通透，模糊了室内外空间的界限。晚上，建筑内部的照明让这种感觉颠倒过来，花园里的人们会感到学校与社区完全融合起来。

1. Cast elements
2. Pre-fab concrete locking beam “L” shaped in support of cast elements
3. Concrete pre-fab pillars to the finishing of cast elements
4. Layer of solidarity among the slab
5. Pre-fab concrete beam

1. 浇筑元件
2. 预制混凝土梁，L形，用于支持浇筑元件
3. 预制混凝土柱，用于装饰浇筑元件
4. 面板之间的实体层
5. 预制混凝土梁

Agave Library

阿加弗图书馆

Location/地点: Phoenix, Arizona, USA/美国，凤凰城
Architect/建筑师: Willbruder + Partners
Photos/摄影: Bill Timmerman
Built area/建筑面积: 2,361m^2
Key materials: Façade – concrete masonry units, glass
主要材料: 立面——混凝土砌块，玻璃

Overview

The design of this 2,361-square-metre branch library for the City of Phoenix addresses issues of excellence and affordability in sustainable design. Impacted within a Planned Shopping Centre in north Phoenix behind a gas station, car wash, fast food restaurant, and supermarket, the Library's construction and material pallet quietly draws from, and (re)presents, the language of its retail neighbours.

Detail and Materials

Stacked bond concrete masonry units and glass enclose the simple rectangular volume of a hard-trowelled concrete floor with carpeted "area rugs", green sandblasted CMU walls, exposed gang-nail trusses, glu-lam beams, steel pipe columns, and sparingly used painted gyp-board interior partitions.

In the tradition of banks, post offices, courthouses, and city halls of fledgling western frontier towns, whose dignified, yet paper-thin street façades belie their utilitarian construction behind, the Library's "false front" mediates between its two realities: one of a limited budget, the other of the civic presence expected in a public institution. With its torquing false metal scrim curving along the site's eastern edge of 36th avenue, the Library's "cowboy front" gives scale, presence, and distinction commensurate with it position in the community. Constructed in the tradition of the old lathe houses of Phoenix's Desert Botanical Garden using off-the-shelf galvanised hat channels. The scale and form of the scrim also recalls the tradition of drive-in movie theatres so common across Post-War American suburbs.

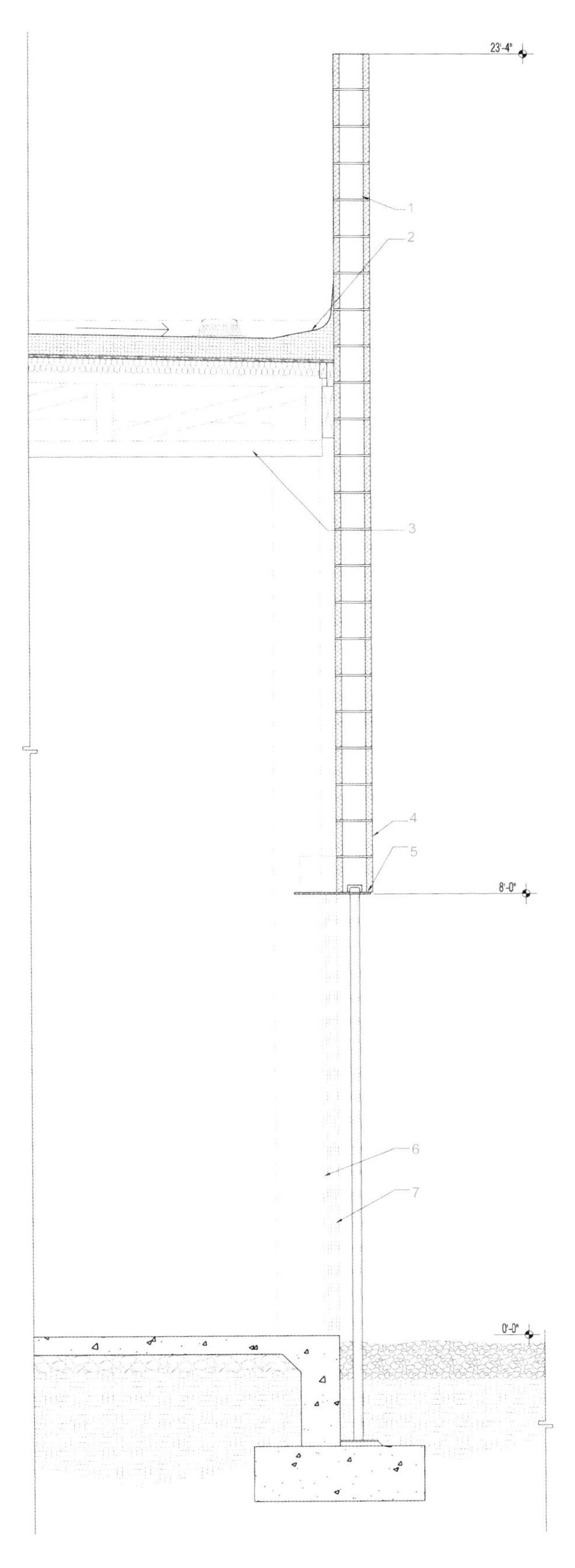

1. R-19 Integra block parapet
2. Foam roof
3. Wood truss structure
4. R-19 Integra block wall
5. 3/8" steel plate header per structural
6. TS column beyond per structural
7. 1" insulated glass beyond

1. R-19 Integra砌块护墙
2. 泡沫屋顶
3. 木桁架结构
4. R-19 Integra 砌块墙
5. 3/8"钢板盖
6. TS柱
7. 1"隔热玻璃

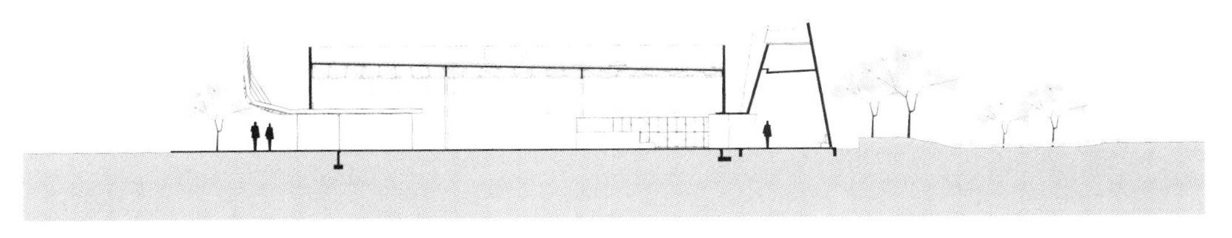

项目概述

这座位于凤凰城的图书馆总面积2,361平方米，在设计过程中充分应用了可持续设计。图书馆周边有一座规划中的购物中心，还被加油站、洗车场、快餐店、超市等零售设施所环绕，因此，它的施工和材料都充分借鉴了这些零售设施的设计风格。

细部与材料

对缝砌筑的混凝土构件和玻璃墙将简洁的长方体建筑结构包覆起来，地面采用硬质混凝土楼板。其他建筑材料包括绿色喷砂CMU墙面、外露组钉桁架、胶合板梁、钢管柱以及涂漆石膏板室内隔断。

在城西，施工中的银行、邮局、法院和市政厅等设施利用看似豪华的薄片围墙将后方的施工场景隐藏起来，而图书馆的“假墙面”则介于有限的预算和体面的公共形象之间。倾斜的假金属帘将建筑场地朝向36号街的东侧边缘遮挡起来，呈现出气势宏大的公共形象，充分体现了图书馆在社区中的地位。图书馆的建造采用了凤凰城沙漠植物园的传统设计，选用成品镀锌槽型钢。金属帘的尺寸和造型都令人想起在那种二战后美国城郊十分常见的汽车电影院。

Dolez

多莱斯公寓

Location/地点: Brussels, Belgium/比利时，布鲁塞尔
Architect/建筑师: B612associates, Li Mei Tsien & Olivier Mathieu
Photos/摄影: Serge Brison, B612associates
Site area/占地面积: 4,297m^2
Built area/建筑面积: 7,269m^2
Key materials: Façade – concrete blocks
主要材料：立面——混凝土砌块

Overview

The project is a new housing complex of 34 high-standard apartments of 2 and 3 bedrooms. It is located in the bottom of the valley between the so-called plateaus Avijl and Kauwberg, two high-value ecological sites in Uccle (a Brussels District). This is a very sensitive location as it is on the threshold between rural and urban spaces and a protected natural site.

The proposal was to reinforce the existing connections with the landscape. Its formal expression is based on a sound analysis of the site topography and the morphology of the built surrounding. The apartments are organised into two main buildings (with four floors each) around a communal garden. This disposition offers a high permeability from the public space to the central garden and the open natural space of the Kauwberg in the background. It also generates a large panel of different apartments that enjoy generous roof terraces, balconies and private gardens.

The two main volumes of the project slope down from the street in the direction of the Kauwberg and melt into the landscape as the green-roof gardens rejoin the soil. The project strives to promote the visual and physical relationship between the highly ecologically valued Avijl and Kauwberg plateaus and to reweave the urban constructions with the natural environment at the same time.

Creation of an ample, semi-natural communal garden at the heart of the site generates a relay between the existing natural sites and benefiting to all the building's occupants and neighbours. All the roofs are planted; they slope down to the bottom of the site to rejoin the undulating natural relief and to also express the interrelation of the building and the landscape. The impermeability of the site has been reduced from 82% to 45%, and if we consider the surfaces of the green roofs, it is even further reduced to 10%. In addition to their aesthetic qualities, the planted

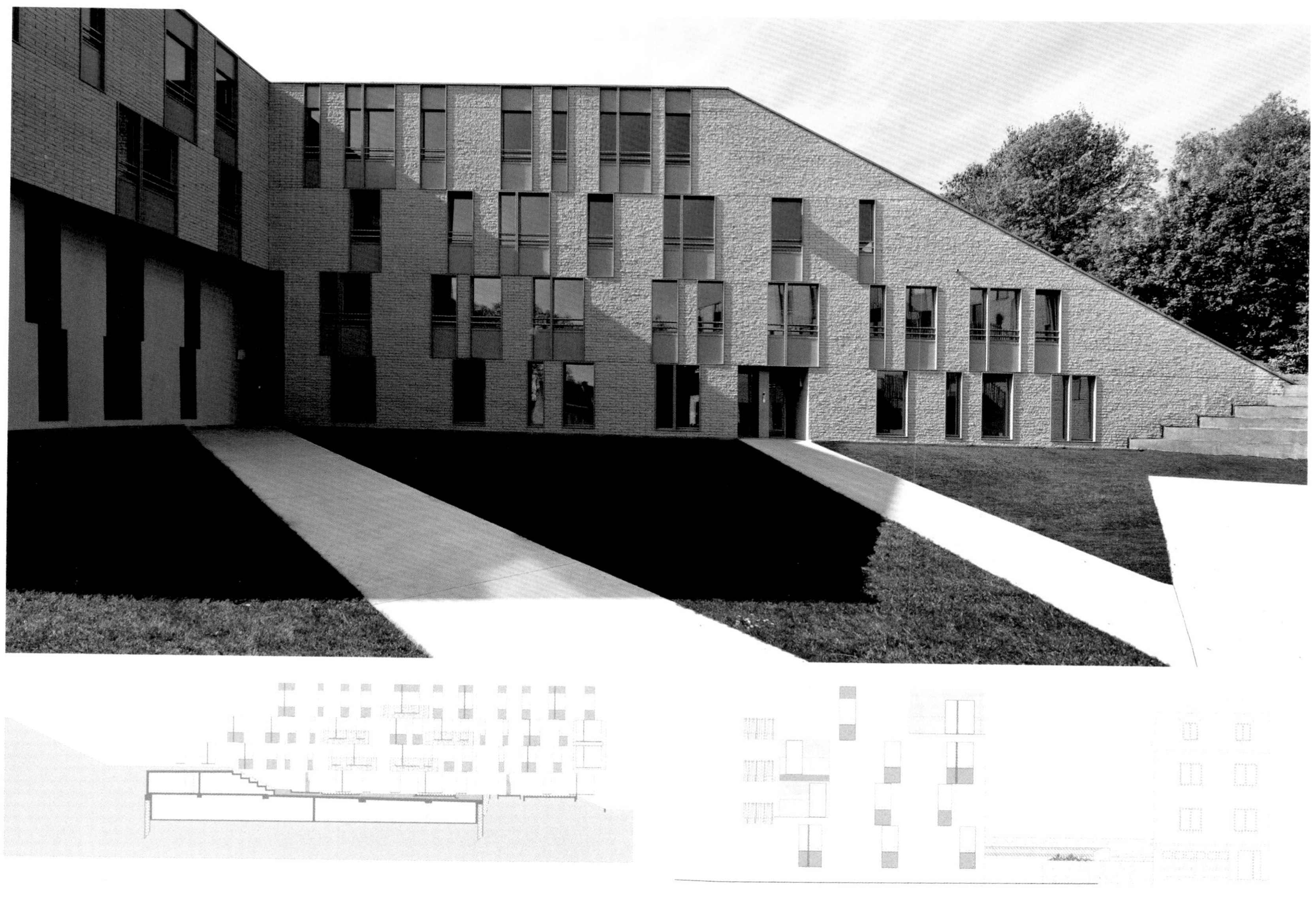

roofs improve greatly the thermal insulation. They contribute to reduce the surrounding CO2 and help to protect the roof.

Detail and Materials

The façades are composed as to echo the random, yet regular principles of natural patterns: the elegant windows alternate with light masonry piers in a non-defined but strict rhythm, while the aluminium balconies are inspired by the flow of the bamboo motive. All the chosen materials have a long lifespan and need low maintenance.

项目概况

这是一个高端公寓项目，拥有34套二室或三室公寓。项目位于阿维吉尔高原和考夫伯格高原（二者均为价值极高的生态保护区）之间的山谷中，其地理位置十分敏感，属于乡村、城市和自然保护区之间的连接地带。

设计方案试图强化场地与景观之间的已有联系。在对场地地形的有效分析和建筑周边环境的形态研究的基础上，建筑师进行了独特的造型设计。公寓分布在两座分别为四层高的主楼内，中间是一个公共花园。这种布局让公共空间能更好的享有中央花园和四周自然景观的美景，并且还让更多的公寓可以享受宽敞的屋顶露台、阳台和私人花园。

两座主楼随着地势而起伏，逐渐融入景观之中，通过绿色屋顶花园回归大地。项目力求促进与生态高原之间的视觉联系和物理联系，同时将城市建设植入自然环境之中。

场地中央茂密的半天然公共花园在已有的自然景观和建筑之间形成了过渡，使得所有居住者乃至整个区域都能从中获益。所有屋顶上都种植着绿色植物，它们顺势而下，与起伏的自然地形融为一体，同时也体现了建筑与自然景观之间的相互联系。场地的不渗透表面比例从82%减至45%。如果将绿色屋顶也纳入考虑的话，这一比例甚至小到10%。除了美观价值之外，绿色屋顶还大幅提高了建筑的隔热保温性能。绿色植物不但有助于减少周边的二氧化碳含量，还能保护屋顶。

细部与材料

建筑立面的构成与高低错落的自然地势相互呼应：优雅的窗口随着轻质混凝土砌块立柱的位置而变化，形成了不规则的韵律感；铝制阳台的设计灵感来自于竹子主题。建筑所选用的所有建材都具有经久耐用和便于维护的特征。

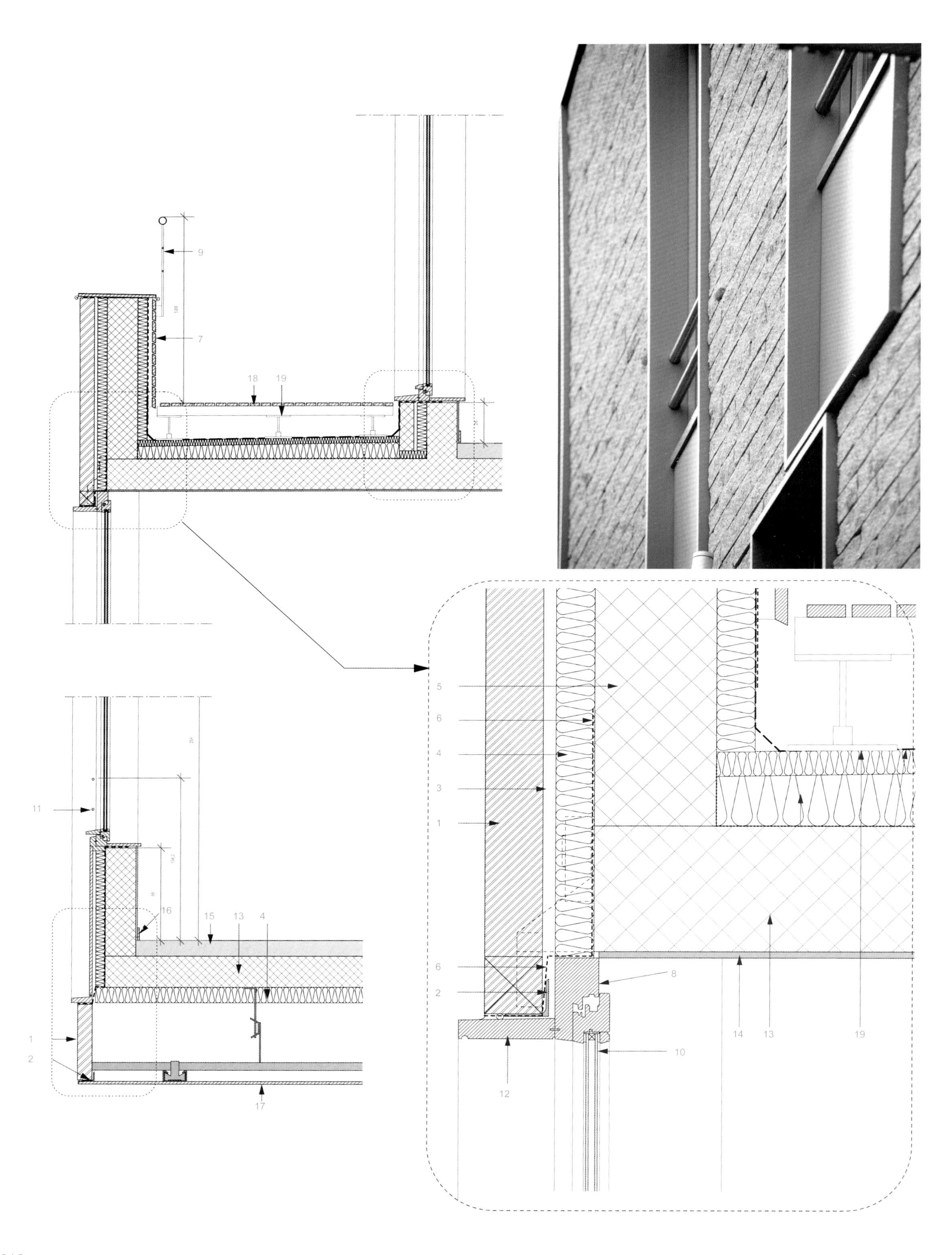
9
120
7
18
19
11
16
15
13
4
1
2
17
5
6
4
3
1
6
2
8
12
10
14
13
19

1. Masonry of concrete blocks
2. Steel profile
3. Void
4. Insulation 6cm
5. Structural wall
6. Waterproof membrane
7. Wooden façade cladding 6x2cm
8. Framework in Afzelia wood
9. Steel railing
10. Double glazing (k≤1.3)
11. Stainless steel railing
12. Framing in Afzelia wood
13. Structural floor
14. Plaster
15. Cover and coating
16. Plinths
17. Stainless steel panel
18. Itauba wooden terrace cladding
19. Synthetic structure of terrace

1. 混凝土砌体
2. 钢型材
3. 中空
4. 隔热层6cm
5. 结构墙
6. 防水膜
7. 木制立面包层6x2cm
8. 缅茄木框架
9. 钢栏杆
10. 双层玻璃（k≤1.3）
11. 不锈钢栏杆
12. 缅茄木框架
13. 结构楼板
14. 石膏
15. 涂层和包层
16. 底座
17. 不锈钢板
18. 热美樟木露台包层
19. 露台复合结构

Nilo 1700

尼洛1700购物中心

Location/地点: Porto Alegree, Brasil/巴西，波尔图阿里戈里
Architect/建筑师: OSPA Arquitetura e Urbanismo
Photos/摄影: Marcelo Donadussi
Built area/建筑面积: 10,775m^2
Completion date/竣工时间: 2012
Key materials: Façade – precast concrete and anodised aluminum
主要材料：立面——预制混凝土、阳极氧化铝

Overview

The project is a shopping centre with 12 shops facing one of the most valuable corners of Porto Alegree. The project was designed and built with industrial materials, and its macro-structure made of precast concrete optimising the construction runtime.

The gallery consists of the projected roof and the ground 1.55 m above the street level, so that it is not in visual conflict with the parking necessary, as the project has high traffic. This façade also provides interesting views and protection to the stores and users passing through the gallery.

The unique and commercial character necessary for this project was given with the use of an annodised aluminum cladding in different colours, which contrasts with the austerity of exposed concrete.

10

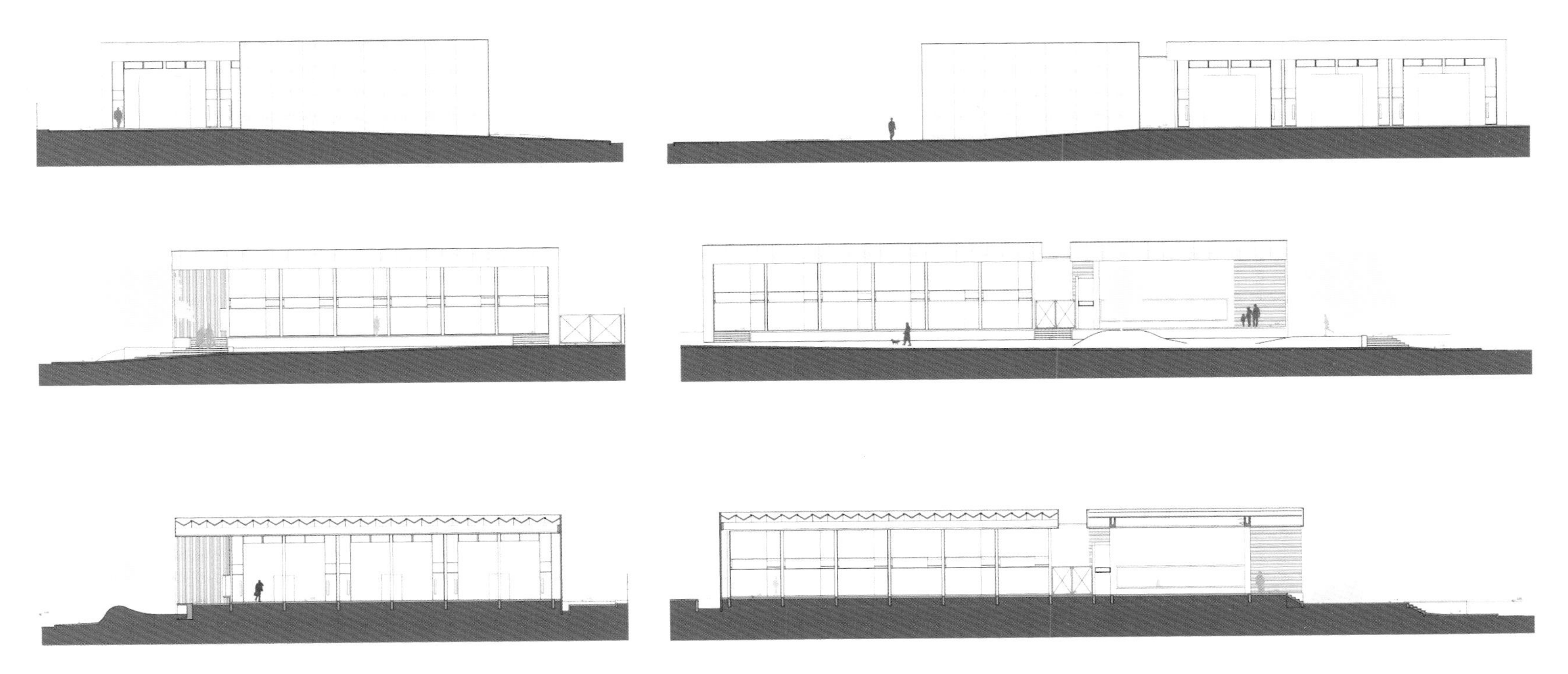

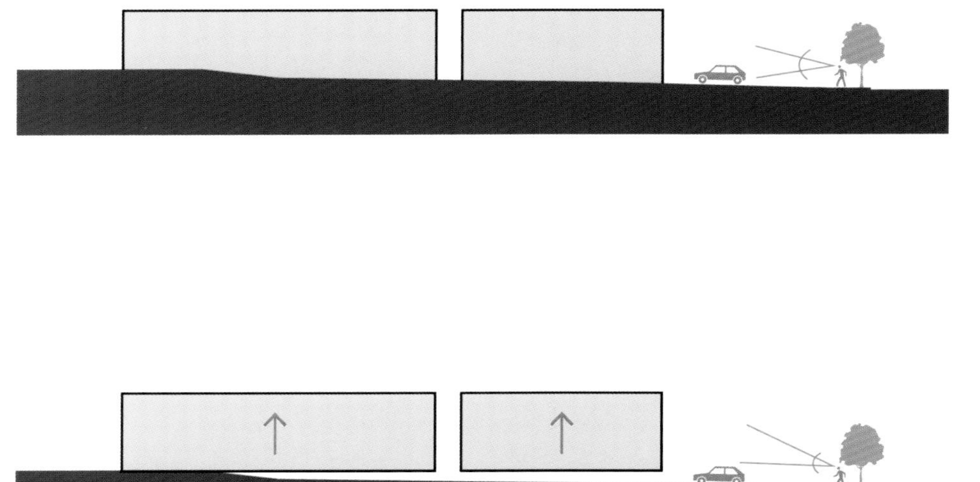

1. Basalt capstone
2. Waterproof flat sheets 150x120cm (according to the facades paging and Brasilit catalog)
3. Steel truss
4. V-Shaped precast concrete tile (V-43 Pré mold catalog)
5. Precast concrete beam support
6. U-shaped precast beam (Pré mold catalog)
7. Aluminium ceiling
8. Forecast for plaster ceiling
9. Metallic rolling door
10. Shop window, aluminium window frames 11. Precast structure - column 20x45cm
12. Tubular steel profiles with 40mm diameter for support of luminous panels (as detailed)
13. Luminous panel for store names (as detailed)
14. Level of finished internal floor
15. Basalt floor 41x41cm
16. Metallic floor grates
17. Precast element for side end and open floor circulation
18. Subfloor
19. Vegetation
20. Compacted soil

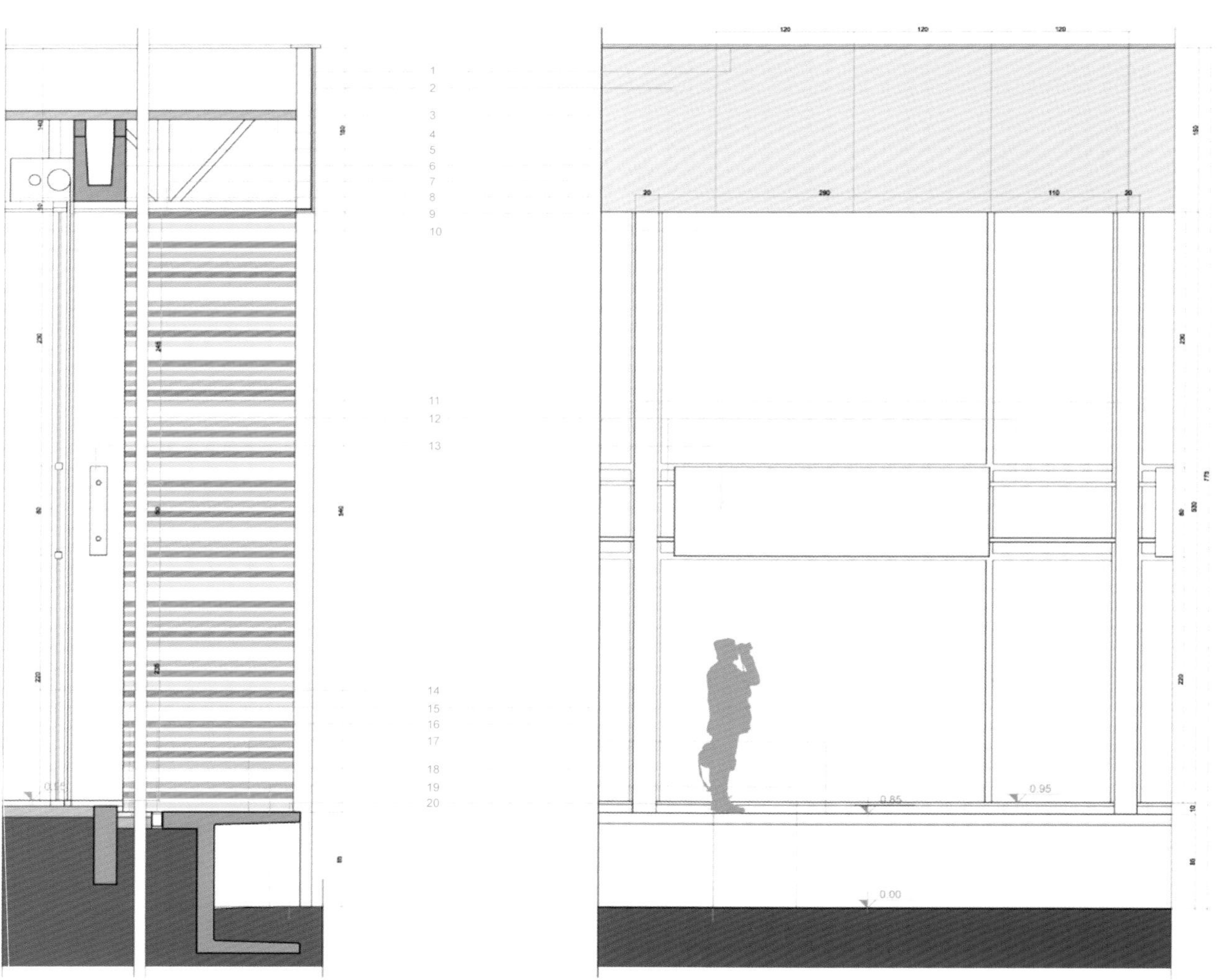

1. 玄武岩顶石
2. 防水板150x120cm
3. 钢桁架
4. V形预制混凝土块
5. 预制混凝土梁支架
6. U形预铸梁
7. 铝吊顶
8. 石膏吊顶
9. 金属卷帘门
10. 商店橱窗，铝窗框
11. 预制结构——柱20x45cm
12. 钢管型材，直径40mm，用于支撑照明板
13. 店铺招牌照明板
14. 室内装饰地面
15. 玄武岩铺装地面41x41cm
16. 金属地面格栅
17. 侧端和开放式楼面交通的预制构件
18. 底层地面
19. 绿植
20. 压实土

项目概况

项目是一座拥有12间店铺的购物中心，正对波尔图阿里戈里最有价值的几个街道转角处。项目的设计和建造以工业材料为主，其宏观结构由预制混凝土构成，有效优化了施工时间。

走廊由外伸的屋顶和高于街面1.55米高的地面构成，这避免了走廊与停车场的视觉冲突，因为购物中心的客流量很大。建筑立面的设计既提供了良好的视野，又能保护店铺和穿越走廊的行人。

几种不同色彩的阳极氧化铝板的运用让项目显得独特而充满商业气息，铝板与朴素的清水混凝土形成了有趣的对比。

Index 索引

Kim in-cheurl+ARCHIUM
www.archium.co.kr
Kimmel Eshkolot Architects
www.kimmel.co.il
LIAG architecten en bouwadviseurs
www.liag.nl
Lousinha Arquitectos
www.lousinhaarquitectos.pt
Love Architecture
www.love-home.com
Martid Mimarlık
www.martid.com
Massimo Mariani Architetto
www.massimomariani.net
MAYU architects
www.malonearch.com.tw
MDU Architetti
www.mduarchitetti.it
MGP Arquitecturay Urbanismo
www.mgp.com.co
MM_arquitectura
mm-arquitectura.com
modostudio
www.modostudio.eu
nps tchoban voss
www.nps-tchoban-voss.de
OSPA Arquitetura e Urbanismo
ospa.com.br
Otxotorena Arquitectos S.L.
www.otxotorenaarquitectos.com
Pencil Office
www.penciloffice.com
Pich-Aguilera Architects
www.picharchitects.com
Project Orange
www.projectorange.com
Rudy Uytenhaak Architectenbureau
www.uytenhaak.nl
Sanjay Puri Architects
www.sanjaypuriarchitects.com
Sol89
www.sol89.com
Superform
www.superform.si
Tabuenca & Leache, Arquitectos
www.tabuenca-leache.com
Taller Básico de Arquitectura
www.tallerbasico.es
Wildrich Hien Architekten
www.wildrich-hien.ch
will bruder+PARTNERS
willbruderarchitects.com
Zechner & Zechner
www.zechner.com

图书在版编目（CIP）数据

建筑材料与细部结构．混凝土 /（西）费尔南多编；常文心译．-- 沈阳：辽宁科学技术出版社，2016.3
ISBN 978-7-5381-9494-4

Ⅰ．①建…　Ⅱ．①费…②常…　Ⅲ．①混凝土　Ⅳ．①TU5

中国版本图书馆 CIP 数据核字 (2015) 第 272623 号

出版发行：辽宁科学技术出版社
（地址：沈阳市和平区十一纬路 29 号　邮编：110003）
印 刷 者：利丰雅高印刷（深圳）有限公司
经 销 者：各地新华书店
幅面尺寸：245 mmx290 mm
印　　张：20
插　　页：4
字　　数：200 千字
出版时间：2016 年 3 月第 1 版
印刷时间：2016 年 3 月第 1 次印刷
责任编辑：于峰飞
封面设计：蒋俊敏 吴杨
版式设计：蒋俊敏 吴杨
责任校对：周　文

书号：ISBN 978-7-5381-9494-4
定价：368.00 元

联系电话：024-23284360
邮购热线：024-23284502
http://www.lnkj.com.cn